GENE MANIPULATION IN PLANT IMPROVEMENT

GENE MANIPULATION IN PLANT IMPROVEMENT

16th Stadler Genetics Symposium

Edited by

J. P. Gustafson

USDA—ARS
University of Missouri
Columbia, Missouri

PLENUM PRESS • NEW YORK AND LONDON

Library of Congress Cataloging in Publication Data

Stadler Genetics Symposium (16th: 1984: University of Missouri–Columbia)
   Gene manipulation in plant improvement.

   Bibliography: p.
   Includes index.
   1. Plant-breeding—Congresses. 2. Plant genetic engineering—Congresses.
I. Gustafson, J. P. II. Title.
SB 123.S86   1984   631.5′23   84-18154
ISBN 0-306-41883-5

Printed in the United States of America

The editor would like to dedicate his effort in the
preparation of this publication to the memory of
Ronald Walsh McLean

ACKNOWLEDGMENT

The editor gratefully acknowledges the generous support of the following contributors: Anheuser Busch Company; College of Agriculture, University of Missouri; Division of Biological Sciences, University of Missouri; Donald M. Nelson Lecture Committee, University of Missouri; Graduate School, University of Missouri; Holden's Foundation Seeds, Inc.; Illinois Foundation Seeds, Inc.; International Maize and Wheat Improvement Center, Mexico; Louis M. Martini Winery; Monsanto; Northrup King Company; Pioneer Hi-Bred International, Inc.; School of Medicine, University of Missouri; United States Department of Agriculture - Agricultural Research Service, Cereal Genetics Research Unit, University of Missouri; and Sungene who made the 16th Stadler Genetics Symposium of over 660 participants from more than 28 countries a success.

The speakers, who spent a tremendous amount of time preparing their manuscripts and lectures are gratefully acknowledged. Without their expertise and dedication the Symposium could not have taken place.

I wish to thank R. W. Allard, K. J. Frey, V. A. Johnson, C. O. Qualset, J. R. Harlan, R. Busch, K. Kasha, E. T. Hubbard, R. Kaufman, E. A. Rupert, W. L. Brown and S. Borojevic who traveled great distances at their own expense to help in leading various sessions. The local chairpersons were tremendous in their effort to see that everyone in their respective sessions were well taken care of during the Symposium.

The behind-the-scene and on-site preparation was excellently handled by Hugh Keith and Joy Gasparovic from Conferences and Short Courses, University of Missouri, who tirelessly handled all of my peculiar requirements and made sure everything was extremely well organized.

Many thanks are due to Ms. Joyce Schooler and Ms. Regina Teson, University of Missouri, for their excellent secretarial help in handling all the correspondence and typing. Support from Kirk Jensen and Plenum Press was a tremendous help.

I would also like to acknowledge Robert Kaufman, Monsanto, for helping me in suggesting several speakers. Armon Yanders deserves a vote of thanks for making the poster session a success. A special thanks goes to A. J. Lukaszewski and K. Robertson.

                                                    J. P. Gustafson
May 7, 1984
Columbia, Missouri

CONTENTS

CONTENTS

PLANT BREEDING 1910-1984

Glenn W. Burton

USDA, ARS
Coastal Plain Experiment Station
Tifton, Ga.

The title of my assignment suggests that I should say something about the history of plant breeding from 1910 to 1984. Why no earlier than 1910? Perhaps it is because by 1910 we were beginning to apply Mendel's classical genetic papers, recognizing the gene and its role in plant breeding. Certainly plant breeding did not start in 1910.

We have good reason to date the beginning of plant breeding with the beginning of agriculture. It started when the women gathering food for their families recognized variation and "selected" seeds from the best plants for the next generation. Larger seeds or fruits and yield must have been top objectives. What a "break-through" was the discovery of a plant that did not shatter its seeds - that held its fruits to facilitate harvest.

The contribution of plant breeding to mankind since 1910 has been great. More than half of the phenominal increase in agricultural production can be credited to plant breeding. Those of us fortunate enough to be called geneticists or plant breeders can point with pride to the achievements of our profession. The papers we will hear at this symposium will help to explain our progress. The thousands of papers on gene manipulation, a few of which will be quoted, will impress us and particularly our administrators. They may help us to increase the efficiency of our craft (plant breeding). But if plant improvement is the ultimate goal of plant breeding, we must continually ask how can this new fragment of information help us to reach our goal?

Lest we become so enamoured with our sophisticated manipulation of the molecular components of the gene that we think all else is naught, let us compare the wild progenitors of our crop plants with the best we have today. Most of the change was made by the primitive plant breeders. They know no genetics and had none of the tools that most of us "must have" today. They had the same five senses that you and I possess and they used them. They knew their plants, probably better than most of us know ours. They realized, perhaps instinctively, that whatever controlled the characteristics of their crops was maleble, capable of change. Perhaps most important, they recognized the power of selection when applied to characters they could see in their plants. The change, the progress the primitive plant breeders made was at the molecular level. We now know that it had to be, thanks to research in the area of molecular biology. The selection, the screening, however, was made at the plant level. Every significant advance in yield, for example, must have involved many changes in the DNA sequences present in the plant. Interesting as this may be for you and me, the significant contribution of the early plant breeders for them and for us was that they changed a few weedy species into crops that efficiently feed mankind.

Instead of trying to trace the history of plant breeding and duplicating things that Dr. Duvick will no doubt cover in his address, permit me to share with you 50 years of plant improvement by a geneticist born in 1910. He grew up on a rented farm in Southwest Nebraska where he had learned the importance of yield and the toll that drought could take before he was 12. He had won the blue ribbon at the county fair with his collection of native grasses and had learned how to select seed corn while still in high school.

In the summer of 1931, while an agronomy major at the University of Nebraska, he served as "agronomist" at the small branch experiment station at Alliance, Nebraska. Taking notes and harvesting Coit Suneson's wheat, Triticum ssp., and H. M. Tysdal's alfalfa, Medicago spp., gave him his first taste of plant breeding.

It was F. D. Keim, head of Nebraska's Agronomy department, who sent him to Rutgers University in 1932 to earn his M.Sc. and Ph.D. degress while working half-time in Howard Sprague's Crops deparment. Under the inspiring leadership of Howard Sprague, he helped with the breeding of corn, Zea mays, small grains and alfalfa, the teaching, the state-wide testing, the seed certification program and the turf research carried on in the Crops department.

In 1934, for his Ph.D. thesis, he became an alfalfa breeder crossing the Hairy Peruvian variety with winterhardy types in an effort to combine their superior traits. The inheritance of flower color and a number of quantitative characters produced his thesis and a 350-page New Jersey bulletin, one of his first publications (Burton, 1937).

He also cut and weighed spaced plants of a wide based population of alfalfa in a modified plant to row breeding method that led to the development of the variety Atlantic.

His training at Rutgers included completing 5 courses in chemistry, 3 in math, 3 in botany, 2 in plant pathology, and enough additional courses for a major in crops and minors in plant physiology and soils. Equally important he says were the experience and inspiration provided by Howard Sprague for 4 1/2 years.

His farm experience and the training he had received in Nebraska and Rutgers stood him in good stead when he arrived at the Coastal Plain Experiment Station, Tifton, GA, April 30, 1936. The seventh professional person, a USDA agent later entitled "geneticist", his assignment was large--his job description was brief--he was to develop better grasses for the South. His first summer was spent with the help of a day laborer, fencing 5 acres of pasture land, installing an irrigation system and building a fieldhouse, in addition to acquainting himself with the species he was to breed.

With the exception of sorghum, Sorghum bicolor, nothing was known about the breeding behavior of the other species to be improved. The limited genetic improvement with cold season forage grasses had emphasized pasture type. The breeding method used called strain building consisted of pooling together low growing types (often selected in old pastures) that could persist under close grazing. The young geneticist recognized the importance of persistence and tolerance of close defoliation, but he believed that "improvement" should increase yield. His experience in New Jersey had convinced him that farmers interest in new varieties was proportional to the yield increase that might result from their use. He has continued to believe that forage crop improvement must increase yield of forage, or the meat and milk the forage will produce. With this conviction, much of his time has been spent measuring the yield of potential varieties that his program has developed.

His experience with hybrid corn, Zea mays, had convinced him that the best way to increase yield was to discover a means of putting the F_1 hybrid on the farm. His studies of bermudagrass, Cynodon dactylon, in 1936 convinced him that it was cross pollinated, highly heterozygous and that it could spread fast enough with stolons and rhizomes to permit vegetative propagation. The logical strategy for improvement seemed to be the development of an outstanding F_1 hybrid plant that could be propagated vegetatively on the farm. Therefore, in 1937, Tift bermuda (selected in a Tift County cotton patch) and two excellent 1936 introductions from South Africa were interplanted in isolation so they could intermate (Burton, 1954). The few seeds produced when these plants were selfed and the appearnce of their selfed progeny proved that most of the 5000 spaced plants from this intermating effort were F_1 hybrids. Many notes on rate of spread, disease resistance, seed head formation and vigor were used in the fall of 1937 to select 147 hybrids. Forage yields of a triplicated planting of these hybrids in 4-inch clay pots in the greenhouse in the 1937-38 winter failed to correlate with later field performance and proved a waste of time. The procedure, however, did supply excellent potted plants to be set one each, in the center of triplicated 4 x 24 foot plots in the spring of 1939. Five-foot alleys between the plots kept clean and weed controlled with a lot of hoeing by high school boys, enabled most of the clones to cover the 4 x 24 foot plots by the end of the season. Many notes, including rate of spread, forage yield, seed head protection, compatability with annual lespedeza, Lespedeza striata, and disease resistance taken on this planting pointed to five hybrids good enough to be included in large replicated plots with and without crimson clover, Trifolium incarnatum, and with rates of 4-8-4 fertilizer up to 1000 lb/A.

In June, 1942, a plot of hybrid number 35 growing beside common bermudagrass was obviously producing about twice as much forage but was producing no seed. A national pasture specialist, when he learned that vegetative propagation on the farm was planned, asked, "Who ever heard of planting a pasture grass vegetatively?" and went on to say, "I'd throw it away if it were mine". In 1943, in response to an urgent request from L. L. Patten, prominent Georgia farmer, hybrid 35 was named "Coastal" for the station where it was developed and released (Burton, 1954). In 1943, bermudagrass was still the South's worst weed and Mr. Patten's neighbors thought he had lost his mind. It took another 15 years of numerous well replicated experiments, with and without the help of other scientists to ascertain fertilizer and management requirements and develop planting methods, certification procedures, and proof of the excellence of Coastal bermudagrass. Tests and demonstrations by others across the South led to the planting of 10,000,000 acres of Coastal bermudagrass that is performing as well today as it did 40 years ago.

The excellent combination of genes carried by Coastal bermudagrass has made it a logical parent in future breeding efforts with bermudagrass. Out of several hundred F_1 hybrids with a winterhardy common bermudagrass from Indiana screened for yield, cold tolerance, etc., came one superior plant named Midland (Harlan and Burton, 1954). More winterhardy than Coastal, it has been planted on more than a million acres north of the Coastal Belt.

Out of 385 F_1 hybrids between Coastal and a highly digestible but cold susceptible plant of C. nlemfuensis screened first on sand dune soil but finally tested in grazing and feeding trials came Coastcross-1 (Burton, 1972). Sterile and with only above ground stolons, Coastcross-1 compared with Coastal is 12% more digestible and when grazed gives 30 to 40% better average daily gains (ADG) and liveweight gains per acre (LWG/A). Cuba is using a half million acres of Coastcross-1 to produce milk because it has produced more milk without supplement than other grasses tested. Its lack of winterhardiness has restricted it to Florida and the tropics.

Tifton 44, released in 1978, is the best of several thousand F_1s between Coastal and a winterhardy bermudagrass from Berlin, Germany (Burton and Monson, 1978). It combines the desirable traits of Coastal with extra quality and enough more winterhardiness to grow dependably 100 miles farther north. Its extra cold hardy genes makes Tifton 44 a better gene pool than Coastal for further breeding purposes.

Tifton 78-22 is an F_1 hybrid between Tifton 44 and Callie that combines most of the desirable traits of both parents with added hybrid vigor. The dominant genes for immunity to rust from Coastal are making Tifton 78-22 immune to this disease. Callie is very susceptible to rust.

Most golf courses and many football fields, athletic fields, and lawns across the South and in the tropics around the world are planted to Tifgreen, Tifway, and Tifdwarf bermudagrass, sterile C. dactylon x C. transvaalensis F_1 hybrids (Burton, 1982). Added to their excellent turf qualities is their failure to produce the pollen that brings pain to asthma and hay fever sufferers. They are also excellent for the sod production used to plant today's "instant turf" because their rhizomes left in the sod-field soil after stripping reestablishes the turf without planting.

Tifway II, released in 1981, and Tifgreen II, released in 1983, are M_1 radiation induced mutants of Tifway and Tifgreen created in 1971 by exposing dormant rhizome sections to gamma rays (Burton, 1982). Repeated tests have shown them superior to their normal parents in several important traits especially under low cost maintenance.

Breeding elephant grass, Pennisetum purpureum, during his first 10 years at Tifton, the young geneticist crossed a very leafy dwarf he had selected with a tall type to produce F_1 hybrids from which Merkeron, a tall type was selected. Merkeron, propagated in the tropics from stem cuttings and used as a green chop forage for milk cows, was also kept living in a grass nursery at Tifton. The interest in biomass production motivated research to prove that Merkeron could produce up to 40 metric tons per ha in one growing season at Tifton. The improved quality of the short Tifleaf 1 pearl millet suggested that short leafy elephant grass

should be evaluated with cattle. The dwarf elephant grasses were
gone but the dwarf genes preserved in Merkeron were not. Selfing
Merkeron has produced a number of dwarfs currently being tested
with livestock.

Breeding a disease resistant sudangrass, Sorghum bicolor,
was one of the challenges suggested for the young geneticist in
1936. Enroute from Bartley, Nebraska to Tifton in April, 1936,
the young geneticist stopped at the Hays Kansas Experiment Station
and obtained from D. A. Savage, seeds of Leoti sorghum, Sorghum
bicolor, his most disease resistant variety. In the summer of
1936, the disease resistant Leoti sorghum was crossed with susceptible sudangrass. The F_1 grown in the winter in a section of a
little horticultural greenhouse, produced enough seed to plant
35,000 F_2 plants in 1937. Six of these plants were as free of
disease as Leoti sorghum but they were also coarse-stemmed. The
resistant plants crossed again to sudangrass gave F_1s increased in
the winter and 30,000 F_2s planted in 1938. One of these F_2s with
fine stems was as resistant as Leoti. When selfed progeny of this
plant bred true for disease resistance and fine stems, it was
increased and became Tift sudan (Burton, 1950).

Tift sudan had a high content of the HCN glucoside. The development of low HCN lines of sudangrass in Wisconsin and the discovery of greater disease resistance in other sorghums motivated a
breeding effort directed toward combining low HCN, maximum disease
resistance and a uniform seed coat color in one variety. Eleven
years of crossing and testing were spent transferring the genes
responsible for these traits into one variety. No attempt was
made to measure yield. When the goal was achieved in Georgia 337
sudangrass and yield was measured, it yielded less than the commercial sorghum sudangrass hybrids and never was accepted by
farmers (Burton, 1964). Georgia 337 has been useful germplasm for
some sorghum breeders and may have been used as a parent in one
commercial sorghum-sudangrass hybrid.

Improving the seed set and ergot, Claviceps microcephala,
resistance of dallisgrass, Paspalum dilatatum, by breeding offered
some new problems. The striking uniformity of selections, introductions and their progeny indicated that it was an obligate
apomict. Crossing it on two other Paspalum species produced
vigorous ergot resistant F_1 hybrids but they were highly sterile
and the occasional seed produced gave rise to apomictic offspring
(Burton, 1943). Their bunchgrass growth habit and performance
under close clipping made their farm-use with vegetative planting
impracticable.

Irradiating seeds of dallisgrass to break apomixis produced
mutants in the M_1 generation (Burton and Jackson, 1962). None of
these were as good as the dallisgrass parent and when they gave

rise to uniform progeny, it was apparent that apomixis had not been broken.

A 40-chromosome introduction (common dallisgrass has 50 chromosomes) proved superior to common dallis in clipped plots and gave good results when grazed. It too was apomictic, was susceptible to ergot and set seed poorly. When it was attacked by chinch bugs, plans to release it were dropped and dallisgrass breeding was discontinued. Much was learned but attempts to improve dallisgrass by breeding had failed.

Breeding projects designed to improve several other species were dropped when their obligate apomixis and/or lack of adaptation made it seem advisable.

Pearl millet, Pennisetum americanum, a robust annual bunchgrass proved to be an ideal species for genetic and breeding method studies. Called cattail millet and used as a summer grazing crop in the South in 1936, it was drought tolerant and well adapted to infertile sandy soils. Its sexual reproduction and protogynous flowering habit facilitated crossing without emasculation and selfing by enclosing heads in paper bags. Its 7 pairs of large chromosomes and its many heritable traits made it well suited for cytogenetic studies.

Starr pearl millet, the first product of the breeding effort with this species was a synthetic variety produced by blending together several lines with similar traits (Hein, 1958). Its increased leafiness and its later maturity enabled it to yield up to 25% more than the cattail millet check it replaced.

The heterosis observed when certain inbred lines were crossed indicated that yields could be materially increased if F_1 hybrids could be put on the farm. The attrition of the weaker seedlings in plantings at the normal seeding rate suggested that 100% hybrid seed would not be required to obtain a 100% hybrid yield. It was reasoned that a mixture of 4 compatable inbred lines flowering at the same time would produce seed 75% of which would be a mixture of the 6 possible single crosses and 25% would be selfs and sibs. A six-year clipping yield trial planted to various inbred hybrid seed mixtures proved that a 50% hybrid seed mixture planted at the normal rate would yield as well as 100% hybrid seed (Burton, 1948).

For this chance hybrid breeding procedure to succeed, four inbred lines that would produce high yielding hybrids in all possible combinations were needed. Numerous clipping trials of diallels involving different inbred lines were required to isolate 4 inbreds that met these requirements. Seed harvested from a field planted to a mixture of equal numbers of pure live seeds of these 4 inbreds - Tift 13, 18, 23, and 26 contained 75% of hybrids and

25% of selfs and sibs. When compared in repeated clipping trials, this chance hybrid seed yielded as well as the double cross involving the same inbreds and outyielded the cattail check by 50%. Named Gahi 1 (short for Georgia hybrid) it was released in 1962 (Burton, 1962). The USDA National Foundation seed program increased the 4 inbreds in isolation and prepared the foundation seed mixture. Only seed grown from foundation seed under state certification could be called Gahi pearl millet, a restriction required due to loss of vigor in the next generation. For many years, Gahi 1 was the leading pearl millet variety grown in the South.

To facilitate seed production, Gahi 2 was produced by substituting dwarf lines for tall lines. These dwarfs, selected from selfed progenies of exotic material, different from each other so that hybrids between any two of them would be tall. This helped the tall hybrids to eliminate the dwarf selfs and sibs in the plantings grown from Gahi 2. The discovery of cytoplasmic male sterility (cms) in pearl millet and the failure of Gahi 2 to yield as well as Gahi 1 prevented its release to the public as an improved variety.

The chance hybrid plant breeding method used to increase pearl millet forage yields was later shown to be an effective method for increasing yields of sudangrass (Burton et al., 1954).

The search for cms in pearl millet was rewarded in 1955 when an F_2 plant from a cross between inbred 556 and Tift 23 was sterile when selfed but fertile when crossed with Tift 23, our best inbred (Burton, 1958). Repeated generations of mating the sterile plants with Tift 23 soon produced Tift 23A, the first cms pearl millet inbred. The excellence of Tift 23, including seed yield and the heterotic ability it had demonstrated in the production of Gahi 1 suggested that a single cross based on Tift 23A could be superior to Gahi 1.

In 1961, K. O. Rachie, Rockefeller Foundation millet coordinator in India, reported that 27 million acres of pearl millet were grown in the most arid portion of India. He stated that it was grown primarily for its grain for human food because it could out-yield other crops under those conditions. Beginning in 1962, Rachie led a coordinated millet improvement program that involved making and testing F_1 hybrids between Tift 23A and lines developed by Indian breeders. In February, 1965, Rachie reported that the Indian millet breeders and agreed to release HB1 hybrid millet, a cross between Tift 23A and D. S. Athwal's Bil 3B, because it had yielded 88% more grain than the best open pollinated checks from 11 to 31 degrees north (Burton, 1983). In 1965, India produced 3.5 million tons of pearl millet grain. In 1970, with the help of HB1, it produced 8 million tons.

The study of the inheritance of five dwarfs isolated from selfed progenies of exotic germplasm, revealed that one labeled d_2 was superior to the others in several respects (Burton et al., 1969). Controlled by a single recessive gene, d_2 reduced the internode length of tall lines enough to cut their height in half without reducing peduncle length, head exsertion, leaf number, and leaf or head size (Burton et al., 1969). Introduced into Tift 23A and B with frequent backcrosses, the d_2 gene produced Tift 23DA and Tift 23DB (Burton, 1969). When F_1 hybrids involving Tift 23A and Tift 23DA yielded the same, the use of Tift 23DA to facilitate hybrid seed production was indicated.

In 1972, Gahi 3, the first forage single cross was released (Burton, 1977). Using Tift 23A or Tift 23DA as its female parent, Gahi 3 had as its male parent, Tift 186, an excellent inbred selected from a South African introduction (Burton, 1977). Compared with Gahi 1, Gahi 3 is leafier, more disease resistant, later maturing, and more productive. The absence of the dominant fertility-restoring R gene in Tift 186 makes Gahi 3 male sterile, a desirable trait in a forage grass. Most Gahi 3 seed is produced on Tift 23DA, but because Tift 186 is tall, Gahi 3 is also tall.

Tiflate pearl millet, a synthetic produced by blending together 54 short day introductions from West Africa, requires a short-day (12 hours or less) to initiate seed head primordia (Burton, 1972). As a consequence, it fails to mature seed before frost in most of the continental U.S. Tiflate compared with Gahi 3 produces more forage in the fall or when cut for silage and gives a better seasonal distribution of its forage when grazed.

Tifleaf 1, released in 1975, is an F_1 hybrid between Tift 23DA and Tift 383 (Burton, 1980). Because both parents are dwarfed with the d_2 gene, Tifleaf 1 grows only half as tall as Gahi 3 and yields only about 80% as much dry matter. When grazed, however, it produces more ADG and LWG/A than Gahi 3, due to its much greater leafiness.

Bahiagrass, Paspalum notatum, a slow spreading perennial from South America is well adapted to the sandy soils of the deep South. Introductions available in the later '30s and early '40s all gave very uniform spaced progeny suggesting apomixis as their mode of reproduction. When a white stigma male-sterile mutant pollinated with pollen from a red stigma plant gave rise to many male-sterile white stigma plants with several red stigma hybrids carrying 60 chromosomes instead of the usual 40 its obligate apomixis was confirmed (Burton, 1948).

In 1941, county agent E. H. Finlayson described a bahiagrass growing wild near Pensacola, Florida, where the old Perdido docks had been located before they were destroyed by a hurricane

(Finlayson, 1941). A search for the origin of Pensacola bahiagrass led to the conclusion that it probably originated in northern Argentina and was brought to Pensacola, Florida in the digestive tracts of cattle shipped from Santa Fe, Argentina to Pensacola before the Perdido docks were destroyed (Burton, 1967). A space planting of this Pensacola bahiagrass grown at Tifton, Georgia showed it to be highly variable and to have smaller seeds, longer stolons, and greater frost tolerance than the South American introductions. It also set seed better and produced more seed than the tetraploid apomicts. Cytological studies found Pensacola bahiagrass to be a diploid (2n = 20) with regular meiosis and sexual reproduction (Burton, 1955).

When the sexual tetraploid necessary to break apomixis could not be found in the various introductions, Ian Forbes, Jr. was asked to try to create one by doubling the chromosome number in the sexual diploid, Pensacola bahiagrass. Among many plants from colchicine treated seed, Forbes found several plants or plant sectors with highly sterile heads (detected by chewing a few "seeds") that proved to be tetraploids (Forbes and Burton, 1957). Isolated from the diploids, they set seed reasonably well and when used as females were easily hybridized with the obligate apomicts. Extensive inheritance studies that followed indicated that in the autotetraploid bahiagrass, apomixis in most of the material was controlled by a recessive gene $\underline{a}$ that must be homozygous (Burton and Forbes, 1960). This hypothesis could not explain the inheritance of apomixis in all of the material studied. The occurrence of only 1 apomictic plant in 36 F_2s required large populations to produce the two good apomictic hybrids Tifton 54 and Tifton 91 that have not been released to date.

The extreme variability in F_2 populations from sexual x apomictic introductions indicated that the natural apomicts were very heterozygous. Most of the F_2s were inferior to their apomictic parent. Attempts to develop a superior sexual tetraploid population by intermating the best plants, has been hampered by the occurrence through cycle 5 of a high percentage of inferior plants.

Most plants of Pensacola bahiagrass are largely self-sterile and cross-fertile (Burton, 1955). This discovery suggested that commercial F_1 hybrid seed could be produced by harvesting year after year all seed produced in a field planted to alternate strips of two self-sterile cross-fertile clones. Two such clones, whose F_1 hybrids yielded 17% more dry matter than the Pensacola check in clipping tests were sought and found by screening several diallels. A pilot seed production field planted vegetatively produced seed to plant in a pasture that gave 17% more LWG/A than Pensacola bahiagrass when grazed for 4 years. Two of these hybrids, Tifhi 1 (Hein, 1958) and Tifhi 2, were released but the hand labor required to establish the seed fields and the cold injury to one parent clone in a severe winter (discovered after their release) kept them from becoming important on the farm.

In 1961, a population improvement program was begun with Pensacola bahiagrass starting with a mixture of seed from 39 farms as a wide gene pool (WGP) and seed from an isolated planting of 75 F_1 plants of Tifhi 2, a 2-clone hybrid, as the narrow gene pool (NGP). The objectives were to increase the forage yield and improve the efficiency of mass selection. The beginning breeding procedure consisted of cutting and weighing 1000 spaced plants of each population and intermating the top 20% for the next cycle (Burton, 1974). At cycle 6 when the NGP and WGP populations yielded the same and were equally variable, they were combined (Burton, 1982). By cycle 8, the first-year-yield of the spaced plant population was still increasing at an annual rate of 16.4% and it still possessed enough variability to indicate continued progress. The breeding method at cycle 8 (improved each year) contained 8 restrictions that made it 4 times more efficient than ordinary mass selection. Called recurrent restricted phenotypic selection (RRPS), its success resulted from the improved screening and intermating systems imposed by the 8 restrictions. In replicated seeded plots clipped several times per year for 2 years, cycle 8 has yielded as much dry matter as the best 2 clone hybrid between plants selected from cycle 4 and has yielded 40% more than the Pensacola bahiagrass check.

Finally, permit me to record a few of my convictions about my profession, plant breeding.

Plant germplasm as expressed in higher plants is maleable and capable of significant response to selection pressure. Plant breeding is the profession charged with the responsibility of modifying the germplasm of a few species to better serve mankind. If plant breeding is to be held responsible for more than half of the increased production required to satisfy the food, fiber, wood, etc. needs of twice as many people in 40 years, the efficiency of its methodology must be improved. The time required per unit of advance must be reduced. I believe it deserves as much support and attention as genetic engineering which at best can only be a plant breeding tool.

The efficiency of backcrossing, an old reliable tool for gene transfer, was greatly increased when we learned how to grow four generations of pearl millet per year instead of one or two (Burton, 1983). Five generations per year, our present goal, will soon be realized.

The effective screening systems that Homer Wells has developed for resistance to rust and Piricularia leafspot in pearl millet are

enabling us to rapidly transfer resistance from a wild <u>Pennisetum</u>
<u>monodii</u> plant to our best pearl millet inbreds.

Our recent discovery that pearl millet pollen will remain
viable for 6 months or more when collected in glassine bags, dried
to 6% moisture content at 40°C in a forced air oven, and kept dry
at -18°C in a standard food freezer will greatly increase the efficiency of our backcrossing programs (Hanna et al., 1983). Half of
our greenhouse space previously used to grow the recurrent parents
can be used for other projects.

One of our greatest needs is for better screens to isolate
the top plants in a population. We need improved screens for
resistance to all the important pests that attack our crops. An
evaluation of the losses due to some pests might justify dropping
resistance to them as a breeding objective (Burton and Wells, 1980).
At ICRISAT in India, in September, 1979, I saw lines of pearl millet
immune to downy mildew that had been selected from a highly susceptible landrace from Chad. Four cycles of recurrent selection
with an <u>effective screen</u> had produced lines that looked like their
parent except that they were immune to India's worst pearl millet
disease. R. J. Williams, the plant pathologist involved, expressed
the opinion that resistance to most diseases of most species could
be produced with an effective screen and recurrent selection.
Increasing evidence confirms his opinion.

If increased yield is an important objective in most breeding
projects, why not screen for yield rather than some component. The
poor correlations between components such as photosynthetic activity
and yield have been disappointing but should have been expected.
Why should a character as complex as yield be closely correlated
with any single yield component. More research on increasing the
efficiency of the yield screen is badly needed. Our research on
improving the yield screen for pearl millet forage hybrids allows
us to obtain 3 yields per minute with CVs for error of less than
10%. One important component of the current method of measuring
yield is the use of Yates 9 x 9 balanced lattice square design
that compares 81 entries with an efficiency twice that of the
randomized block (Burton and Fortson, 1965). Locating yield trials
on land uniformly cropped to a legume the previous year is another.

Much of the success of our RRPS breeding method for Pensacola
bahiagrass is due to the effectiveness of our screen for yield.
Equally important, I believe, is the intermating system restricted
to the selected plants. This procedure doubles the efficiency
obtained from ordinary mass selection. In addition, our current
RRPS method provides for a complete balanced intermating of all
selections and permits a screening of many more selection-
combinations than chain crosses or even field polycrosses. We have
been able to easily intermate selections in the annual pearl millet

by mixing together pollen collected from one head of each selection
and using the mixture to pollinate another head on each selection.
Equal quantities of seed from each cross mixed together provides
the plants for the next cycle of improvement. Most population
improvement programs could benefit from research designed to improve
their system of intermating selected plants.

Important as it is, I doubt that research on plant breeding
methods will ever receive the support presently enjoyed by genetic
engineering. Most research leading to more efficient plant breeding
has been conducted by plant breeders as a part of their breeding
programs. This may continue to be true and can be effective if
plant breeders are given enough support to be able to include such
research in their programs. In addition to the extra support they
must continually search for more efficient methods motivated by the
conviction that the best methods are yet to be found. I believe
that nothing is so good that it cannot be made better. I hope you
agree.

I wish to acknowledge the assistance of many high school
students and several excellent technicians, high school graduates
trained in the program. I must also acknowledge the assistance of
118 scientists, many in other disciplines and other research stations
whose names appear as coauthors of 40% of the 488 papers that bear
my name.

REFERENCES

Burton, Glenn W., 1937, The inheritance of various morphological
          characters in alfalfa and their relation to plant yields
          in New Jersey, New Jersey Agr. Exp. Sta. Bul. 628, 35 pages.
Burton, Glenn W., 1943, Interspecific hybrids in the genus Paspalum,
          Jour. of Heredity, 39:14-23.
Burton, Glenn W., 1948, The performance of various mixtures of
          hybrid and parent inbred pearl millet, Pennisetum glaucum
          (L.) R. Br., Jour. Amer. Soc. Agron., 40:908-915.
Burton, Glenn W., 1948, The method of reproduction in common bahia-
          grass, Paspalum notatum, Jour. Amer. Soc. Agron.,
          40:443-452.
Burton, Glenn W., 1950, Tift sudan, Ga. Crop Impr. News, 2:1-2.
Burton, Glenn W., 1954, Coastal bermudagrass, Ga. Coastal Plain
          Exp. Sta. Bull. N.S. 2.
Burton, Glenn W., 1955, Breeding Pensacola bahiagrass, Paspalum
          notatum, Agron. J., 47:311-314.
Burton, Glenn W., 1958, Cytoplasmic male-sterile in pearl millet,
          (Pennisetum glaucum) (L.) P. Br., Agron. J., 40:230.
Burton, Glenn W., 1962, Registration of varieties of other grasses -
          Gahi 1 pearl millet (Reg. No. 6), Crop Sci., 2:355-356.

Burton, Glenn W., 1964, Registration of Georgia 337 sudangrass (Reg.
    No. 111), Crop Sci., 4:666.
Burton, Glenn W., 1967, A search for the origin of Pensacola bahia-
    grass, Econ. Bot., 21:379-382.
Burton, Glenn W., 1969, Registration of pearl millet inbreds Tift
    23B$_1$, Tift 23A$_1$, Tift 23DB$_1$, and Tift 23DA$_1$, Crop Sci.,
    9:397.
Burton, Glenn W., 1972, Registration of Coastcross-1 bermudagrass,
    Crop Sci., 12:125.
Burton, Glenn W., 1972, Registration of Tiflate pearl millet, Crop
    Sci., 12:128.
Burton, Glenn W., 1974, Recurrent restricted phenotypic selection
    increases forage yields of Pensacola bahiagrass, Crop Sci.,
    14:831-835.
Burton, Glenn W., 1977, Registration of Gahi 3 pearl millet (Reg.
    No. 40), Crop Sci., 17:345-346.
Burton, Glenn W., 1977, Registration of pearl millet inbred Tift
    186 (Reg. PL7), Crop Sci., 17:488.
Burton, Glenn W., 1980, Registration of pearl millet inbred Tift 383
    and Tifleaf 1 pearl millet (Reg. PL8 and Reg. No. 60),
    Crop Sci., 20:292.
Burton, Glenn W., 1982, The Tif-turf bermudas, Parks & Recreation
    Resources, 1(9 & 10):9-12.
Burton, Glenn W., 1982, Improved recurrent restricted phenotypic
    selection increases bahiagrass forage yields, Crop Sci.,
    22:1058-1061.
Burton, Glenn W., 1983, Breeding pearl millet, in:"Plant Breeding
    Reviews," Vol. 1, Jules Janick, ed., The Avi Publishing Co.,
    Inc., Westport, CT, pp. 162-182.
Burton, Glenn W., and DeVane, Earl H., 1951, Starr millet, Southern
    Seedsman, pp. 17 and 68.
Burton, Glenn W., DeVane, E. H., and Trimble, J. P., 1954, Polycross
    performance in sudangrass and its possible significance,
    Agron. J., 46:223-226.
Burton, Glenn W., and Forbes, Jr., Ian., 1960, The genetics and mani-
    pulation of obligate apomixis in common bahiagrass,
    Paspalum notatum, 8th Intern'l Grassl. Cong., Reading,
    England, pp. 66-71.
Burton, Glenn W., and Fortson, James C., 1965, Lattice-square designs
    increase precision of pearl millet forage yield trials,
    Crop Sci., 5:595.
Burton, G. W., Hanna, W. W., Johnson, Jr., J. C., Leuck, D. B.,
    Monson, W. G., Powell, J. B., Wells, H. D., and Widstrom, N. W.
    1977, Pleiotropic effects of the tr trichomeless gene in
    pearl millet on transpiration, forage quality, and pest
    resistance, Crop Sci., 17:613-616.
Burton, Glenn W., and Jackson, J. E., 1962, Radiation breeding of
    apomictic prostrate dallisgrass, Paspalum dilatatum var.
    pauciciliatum, Crop Sci., 2:495-497.

Burton, Glenn W., and Monson, Warren G., 1978, Registration of
        Tifton 44 bermudagrass (Reg. No. 10), Crop Sci., 18:911.
Burton, Glenn W., Monson, W. G., Johnson, Jr., J. C., Lowrey, R. S.,
        Chapman, Hollis D., and Marchant, W. H., 1969, Effect of the
        d_2 dwarf gene on the forage yield and quality of pearl millet,
        Agron. J., 61:607-612.
Burton, Glenn W., and Wells, H. D., 1980, Use of near-isogenic host
        populations to estimate the effect of three foliage diseases
        on pearl millet forage yield, Phytopathology, 71:331-333.
Finlayson, E. H., 1941, Pensacola -- a new, fine-leaved bahia,
        Sou. Seedsman, Dec. issue.
Forbes, Ian, and Burton, Glenn W., 1957, The induction and some
        effects of autotetraploidy in Pensacola bahiagrass,
        Paspalum notatum var. saurae Parodi, Agron. Abstracts,
        ASA meetings, p. 53.
Hanna, W. W., Wells, H. D., Burton, G. W., and Monson, W. G., 1983,
        Long-term pollen storage of pearl millet, Crop Sci.,
        23:174-175.
Harlan, Jack R., and Burton, Glenn W., 1954, Midland bermudagrass -
        a new variety for Oklahoma pastures, Bul. No. B-416,
        Okla. Agr. Exp. Sta.
Hein, M. A., 1953, Registration of varieties and strains of pearl
        millet (Pennisetum glaucum (L.) R. Br.), Agron. J., 45:
        573-574.
Hein, M. A., 1958, Registration of varieties of strains of grasses,
        Agron. J., 50:399-401.

PROGRESS IN CONVENTIONAL PLANT BREEDING

Donald N. Duvick

Pioneer Hi-Bred International, Inc.

P. O. Box 85, Johnston, Iowa 50131

SUMMARY

Grain yields of wheat, maize, sorghum and soybeans have increased continually since 1930. Variety improvement has been responsible for 50% or more of the yield increase for each crop. Genetic improvements in yield potential have increased linearly for the past 25 years and show no sign of leveling off, for any crop. Improved varieties of the cereal grains: wheat, maize and sorghum, have in part been superior because they were able to efficiently utilize increased amounts of commercial nitrogen fertilizer. Increased yield potential of new varieties of all four crops generally is due to improvements in defensive traits such as standability, heat and drought tolerance, and pest resistance. Inheritance of the improved traits is usually genetically complex and is not well understood either in genetic or physiological terms. A list of most needed genetic improvements in new varieties will include such things as improvements in yield potential, stress tolerance, and pest resistance. Until the genetics and physiology of these traits are better understood it will be difficult for biotechnology to make major direct contributions to plant breeding. However biotechnology can make many useful small contributions to plant breeding at the present time and eventually it also will provide major assistance in explaining the genetics and physiology of important traits.

INTRODUCTION

Grain yields of major U.S. crops have increased continually since about 1930, in contrast to the 30 year period before 1930 when yields per unit area were essentially static (Tables 1-2, Figs. 1-4). The period 1930-1982 can be divided into two approximately equal periods, before and after 1955. For each of the cereal grains: wheat (<u>Triticum</u> spp.), maize (<u>Zea</u> <u>mays</u> L.) and sorghum (<u>Sorghum</u> <u>bicolor</u> (L.) Moench.), the average rate of gain is much higher after 1955 than before 1955. But for the grain legume, soybeans (<u>Glycine</u> <u>max</u> (L.) Merrill), pre- and post-1955 rates of increase in yield essentially do not differ. Thus for the cereal grains three contrasting periods of gain in productivity can be described: (1) 1900-1930. no increase in yields; (2) 1930-1955. clear-cut annual increases in yield per unit area; (3) 1955-1982. even steeper rates of increase in yield. For soybeans, these three periods cannot be demonstrated. (As noted above there is no marked increase in rate of gain in soybean yields, post-1955. There is essentially no record of soybean yields before 1930.)

TABLE 1. AVERAGE GRAIN YIELDS OF FOUR MAJOR U.S. CROPS IN THREE TIME PERIODS [a]

Crop	1900-1930 kg/ha	1930-1955 kg/ha	1955-1982 kg/ha
Wheat	887	987	1770
Maize	1666	1958	4841
Sorghum	1037 [b]	1004	2979
Soybeans	- [c]	1138	1636

[a]Data obtained from various volumes of USDA's "Agricultural Statistics", U.S. Government Printing Office, Washington, D.C.
[b]For years 1919-1930.
[c]No data

TABLE 2. AVERAGE ANNUAL RATES OF GAIN IN GRAIN YIELD OF FOUR MAJOR U.S. CROPS, IN THREE TIME PERIODS [a]

Crop	1900-1930 kg/ha/yr	1930-1955 kg/ha/yr	1955-1982 kg/ha/yr
Wheat	0	15	30
Maize	-3	57	144
Sorghum	-36 [b]	22	70
Soybeans	- [c]	17	20

[a]Data obtained from various volumes of USDA's "Agricultural Statistics", U.S. Government Printing Office, Washington, D.C.
[b]For years 1919-1930.
[c]No data.

Fig. 1. Annual average grain yield of U.S. wheat, 1930-1982. Source: USDA's "Agricultural Statistics". Straight lines indicate linear regressions of yield on years, calculated for 1930-1955 ($r^2 = 0.55$) and 1955-1982 ($r^2 = 0.80$).

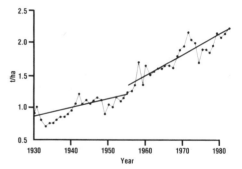

Fig. 2. Annual average grain yield of U.S. maize, 1930-1982. Source: USDA's "Agricultural Statistics". Straight lines indicate linear regressions of yield on years, calculated for 1930-1955 ($r^2 = 0.79$) and 1955-1982 ($r^2 = 0.89$).

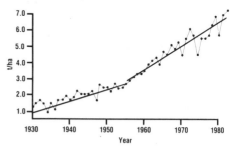

Fig. 3. Annual average grain yield of U.S. grain sorghum, 1930-1982. Source: USDA's "Agricultural Statistics". Solid straight lines indicate linear regressions of yield on year. calculated for 1930-1955 ($r^2 = 0.53$) and 1955-1982 ($r^2 = 0.66$). Dashed straight lines indicate linear regressions for 1955-1968 ($r^2 = 0.87$) and 1968-1982 ($r^2 = 0.07$).

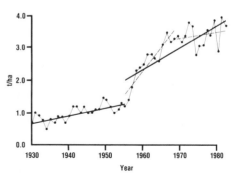

Fig. 4. Annual average grain yield of U.S. soybeans, 1930-1982. Source: USDA's "Agricultural Statistics". Straight lines indicate linear regressions of yield on year. calculated for 1930-1955 ($r^2 = 0.57$) and 1955-1982 ($r^2 = 0.75$).

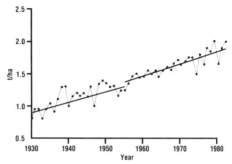

Causes Of Variation In Yield Trends

Causes of the contrasting rates of yield gain within each of
the cereal grains and of the difference between the cereals and
soybeans are no doubt manifold, but certain causes seem to be
obvious and important. Starting in about 1930 new and improved
varieties of wheat, soybeans and grain sorghum began to appear in
increasing numbers. Hybrid maize was introduced in the mid-1930's
and within about 10 years nearly all maize fields were planted to
hybrids. (Hybrid sorghum was introduced in the 1950's.) The
first hybrids generally yielded about 15-20% more than the open
pollinated varieties they replaced. New higher yielding hybrids
or varieties of all four crops were introduced continually through
succeeding years. Thus it is likely that advances in breeding
were responsible for at least part of the gains in productivity in
the period 1930-1955. A second major change during this period
was the replacement of the horse with the tractor. This change
resulted in some improvements in soil cultural conditions and in
timeliness of planting, cultivating and harvesting, thereby giving
rise to increased yields. It is probable however that variety
improvement was the most important cause of increased yields in
the 1930-1955 period (see references in next section).

Fig. 5. Annual U.S. consumption
       of nitrogen in commercial
       fertilizers, 1930-1980.
       Source: USDA's "Agri-
       cultural Statistics".
       Straight lines indicate
       linear regressions of
       million tons on year,
       calculated for 1930-1955
       (r^2 = 0.82) and 1955-1980
       (r^2 = 0.98).

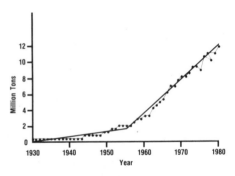

A major change in agronomic inputs for the cereal grains
started in the 1950's -- increasing amounts of nitrogen fertilizer
were applied to these non-leguminous crops each year. Calculations of rates of increase in nitrogen fertilizer usage in the
periods 1930-1955 and 1955-1980 show pre- and post-1955 differences much like those for yield gains in the cereal grains (Fig.
5). (Because total acreages of the cereal grains have varied
little compared to changes in nitrogen consumption since 1955, per
acre applications of nitrogen fertilizer probably have increased
at nearly the same rate as increases in total nitrogen fertilizer
consumption. The cereal grains are major recipients of commercial
nitrogen fertilizer.) It seems likely that the use of nitrogenous
fertilizers in increasing amounts and on increasing numbers of
acres has been an important cause of the sharp upturn in yields of
the cereal grains in the years since about 1955 (Ruttan and

Sundquist. 1982). This supposition is reinforced by the absence of any significant increase in the rate of gain for soybean yield in the post-1955 period, as noted above.

Although increasing use of nitrogen fertilizers certainly has been one important cause of recent yield increases, other factors also have had important effects. Improvements in farm machinery, herbicides and insecticides have provided important inputs. And improved varieties and hybrids of all crops have continued to appear as well. To calculate the proportionate importance of each of these factors is a formidable task. It has been done for at least one crop -- maize -- in one state -- Minnesota (Cardwell. 1982). That study indicated that both positive and negative inputs have affected Minnesota corn yields since 1930. Yields in Minnesota have increased from 2.0 t/ha in 1930 to 6.3 t/ha in 1980. Genetic improvements (the introduction of hybrids and continual improvements in hybrids) were credited with 58% of the 4.3 t/ha gain in yield. Net effects of increased applications of synthetic nitrogen fertilizer added 19% of the total gain; better weed control added 23%. However other changes such as increased soil erosion and damage by rootworm and corn borer subtracted from the potential yield gains, so the above cited positive percentage figures are somewhat smaller in their net effect, although their relative importance stays the same.

Estimates Of Genetic Improvements In Yielding Ability

Plant breeders have made several estimates of the proportion of total yield gains that is due to genetic improvements, i.e., to improved cultivars of the major field crops (Salmon, et al., 1953; Maunder, 1969; Reitz and Salmon, 1959; Wellman and Hassler. 1969; Hallauer, 1973; Russell. 1974; Hueg, 1977; Leudders, 1977; Duvick, 1977, 1984; Jensen, 1978; Wilcox, et al., 1979; Sim and Araji, 1981; Miller and Kebede, 1984; Schmidt, 1984; Specht and Williams, 1984). In general, they have estimated that genetic improvements are responsible for 50% or more of total gains (since about 1930) in yield per unit land area. Estimates for wheat, maize, grain sorghum and soybeans are in fairly good agreement in this regard although for each crop there is a relatively wide range in size of estimates.

For each crop it is pointed out that improvements in breeding are maximized as they interact with changes in crop culture. Thus, even though new semi-dwarf wheat varieties yield more than the old ones under both low and high nitrogen regimes, the improved straw strength and shorter stature of the semi-dwarf wheats give the new varieties much more yield advantage in well-fertilized fields than in fields deficient in nitrogen (Austin, et al., 1980). Likewise, modern maize hybrids are superior to those of 50 years ago when planted at the relatively low densities of

1930, but if hybrids are planted at today's high planting rates the advantage of new over old is greatly increased; indeed the yields and standability of old hybrids usually are reduced as densities are increased (Russell. 1974; Duvick, 1977 and Figs. 6 and 7).

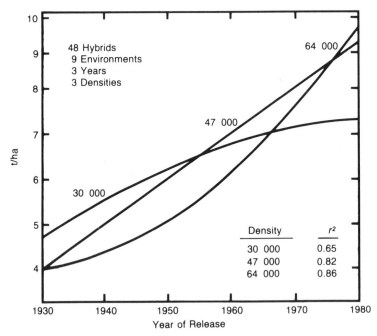

Fig. 6. Regressions of hybrid grain yield on year of hybrid release, for 47 maize hybrids and one open pollinated variety tested at three plant populations. The hybrids were released at intervals from 1934 to 1978. From Duvick, 1984.

Two Questions:

Two important questions need to be asked about total yield gains and about genetic improvements in yielding ability. First. what are the causes of the yield gains? Second, will the gains continue?

Will Yield Gains Continue?

Total yield gains. The second question can be answered more simply than the first, at least for the short term. Average U.S. yields of wheat. maize and soybeans are still increasing in linear fashion, taking into account year to year variation due to

weather.   There is is no indication of reduction in the rate of
increase (Figs. 1, 2, 4).   Grain sorghum yields on the other hand
have fluctuated a great deal in recent years, and since 1968 they
have shown no predictable trend (Fig. 3).   It is thought that a
combination of new disease and insect pests (starting in 1968) and
a shift in sorghum acres from irrigated to dryland conditions
(starting in 1973) have been the chief causes of the increased
year to year variation and the slackened rate of gain of sorghum
yields (Miller and Kebede, 1984).   Breeders have been able to
develop varieties with resistance to most of the pest problems but
there has consequentially been some reduction in their efforts to
develop varieties with higher yields.   But probably the shift from
irrigated to dryland culture has had the greatest negative effect
on sorghum yield trends, since side by side comparisons show that
newer hybrids yield more than the older ones, under either
irrigated or dryland conditions (Miller and Kebede, 1984).

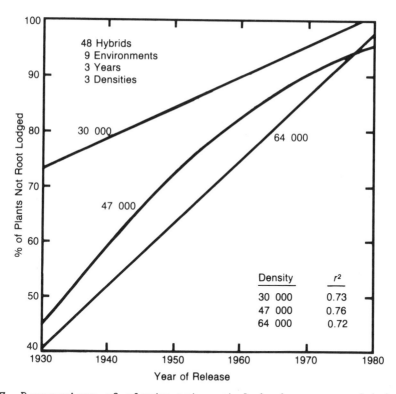

Fig. 7. Regressions of plants not root lodged on year of hybrid
        release for 47 maize hybrids and one open pollinated
        variety tested at three plant populations. The hybrids
        were released at intervals from 1934 to 1978.   From
        Duvick, 1984.

Yield gains from breeding. Progress in breeding for higher
yields -- the contribution from genetic inputs -- shows no sign of
leveling off for any of the four major grain crops: wheat, maize,
sorghum and soybeans. Studies of genetic progress generally
indicate constant or even increasing rates of increase in genetic
yield capacity, that is, in variety improvement. (See citations
in the previous section, and also Fig. 8 for an example in maize.)
Plant breeders intuitively agree with the conclusion that genetic
yield gains can continue to be made with present-day breeding
methods. If pressed to explain why, they usually will give the
following reasons: 1) The perfect variety has not yet been
developed in any crop. 2) For each trait needing improvement,
germplasm exists carrying the needed trait. 3) Present-day
breeding methods are quite capable of transferring the needed
traits into elite germplasm from which improved varieties can be
selected. (One should point out that in any crop, breeding gains
are saltatory. Periods of slow gain are interspersed with periods
of rapid gain, following identification of new, highly productive
cultivars and breeding stocks.)

Plant breeders also know however, that much time, work, and
ingenuity will be needed to effect these desired changes.
Increases in the yielding ability and general performance of new
varieties and hybrids come at the expense of ever-increasing
amounts of scientist years and research inputs. For example, I
have estimated that during the years 1930-1980, maize yield
potential (the genetic component) has increased at an average rate
of 1.4%/yr, while numbers of maize breeders have increased at an
average rate of 4%/yr (Duvick, 1984).

Yield gains from other inputs. There may be little opportunity for further help from two other important past causes of
higher yields -- increased use of nitrogen fertilizer and better
weed control. They probably are now being used at maximum
effectiveness, or very nearly so (Ruttan and Sundquist, 1982). Of
course, improved herbicides will continue to appear and nitrogen
use in the cereals may go up even more as newer varieties, able to
efficiently use more nitrogen, are developed. But the major gains
in yield due to these two additions seem to have been achieved.
Therefore it is likely that the proportionate importance of
breeding will increase during the next decade or two. It follows
that unless breeding gains increase at a faster rate than in
previous years, the rates of annual increase in achieved (total)
yield may begin to slow down, since plant breeding gains will be
responsible for an increasingly higher percentage of the total
gains. The time is ripe for introduction of new, more powerful
plant breeding tools, and for better understanding and use of
present day breeding techniques and genetic resources.

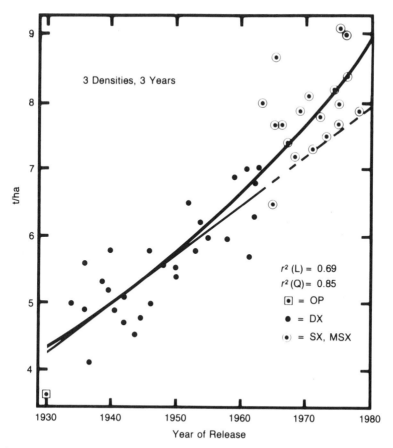

Fig. 8. Mean grain yields of 47 maize hybrids released during the
period 1934-1978, and of one open pollinated variety.
Square with enclosed dot indicates open pollinate (O.P.);
solid dots indicate double cross hybrids; circles with
enclosed dot indicate single cross hybrids. Quadratic
regression (Q) is based on all 48 entries (1930 to 1978);
linear regression (L) is based on the 29 non-single cross
entries (1930 to 1963). Dashed line projects linear
estimate to 1980, indicating that gains might have been
slower if recent hybrids were double crosses. From
Duvick, 1984.

What Factors Have Contributed To Yield Gains?

Causes of past yield gains can be divided into two kinds --
those due to breeding and those due to other inputs. As noted

above, non-breeding inputs such as nitrogen fertilizer, timeliness
of planting, better insect and pest control, and improved harvesting machinery have been important causes of increased yield during
the past 50 years, but variety improvement, interacting with
changes in agronomic practices, has probably been the most
important cause of higher yields.

Physiological reasons for high yields -- potential or actual
-- of the improved new varieties are still largely unknown in
fundamental detail (Boyer. 1982). Published experiments and
breeder experience indicate that for each crop rather different
constellations of changes seem to be important, but that none of
the changes seem really to be changes in potential to transform
sunlight and nutrients into dry matter when environmental factors
are not limiting. Rather, dry matter deposition has been shifted
to different organs, or physiologically limiting processes have
been changed or bypassed. Improvements are in secondary rather
than primary processes.

Thus, improved cultivars have for example (1) increased the
ratio of grain to straw without increasing total dry matter
production (wheat; Austin et al, 1980), (2) delayed the time of
leaf senescence thus allowing more time for grain fill and better
odds that stalk lodging will not occur before harvest (grain
sorghum; Miller and Kebede, 1984), (3) strengthened physiological
resistance to floral abortion (barrenness) caused by heat and
drought thus giving greater ability to consistently make high
yields under high planting rates (maize; Russell. 1974 and Figs. 6
and 9), or (4) they have utilized genes for determinate flowering
and upright plant habit to control plant height and erectness thus
producing varieties that can produce both high yields and a
harvestable crop (soybeans; Cooper. 1981). And for all crops,
improvements or adjustments in disease, insect and nematode
resistance have been essential to allow potential yield to be
expressed. Essentially, breeding improvements have improved
dependability of yield and thus have raised average yields
country-wide and year in, year out.

Most of the varietal improvements have been genetically
complex and are poorly understood, genetically. Some of the
disease and insect resistance improvements have been governed by
one or a few genes, but these simply inherited kinds of
resistance, unfortunately. have usually not lasted very long
(National Research Council, 1972). Variants of the pest organisms
soon multiplied and over-ran the "resistant" varieties. Examples
are stem rust (Puccinia graminis Pers. f. sp. tritici Eriks. and
Henn.) in wheat, cyst nematode (Heterodera glycines Ichinohe.) in
soybeans and sorghum downy mildew (Sclerospora sorghi (Kulk)
Weston and Uppal) in grain sorghum. The basic reasons for durable
vs. short-lived pest resistance are not known, but experience does

indicate that simply inherited resistance tends to be relatively short-lived, especially if it is heavily exploited.

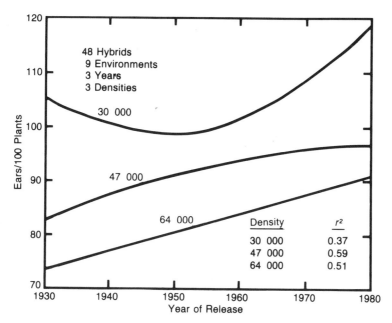

Fig. 9. Regressions of ears per 100 plants on year of hybrid release, for 47 maize hybrids and one open pollinated variety tested at three plant populations. The hybrids were released at intervals from 1934 to 1978. Note increase in prolificacy of newer hybrids at low population, and decrease in barrenness of newer hybrids at higher populations. From Duvick, 1984.

Traits Needing Genetic Improvement

Plant breeders, when asked to list traits in most need of genetic improvement, usually put high yielding ability in first place. In second place they usually will name, as a collective group, stress tolerance and disease, insect and nematode resistance. (The relative rankings within this group depend on the crop and the geographical location of the breeder.) In third place they usually list quality considerations such as milling and baking characteristics of wheat, grain color of sorghum, grain hardness of maize or resistance to seed coat cracking of soybeans.

Needed Improvements In Breeding Technology

Plant breeders, when asked to list the most needed improve-

ments in breeding technology, are likely to hesitate or to have no
response at all, largely because they find it hard to call on help
from technologies that are unfamiliar or unknown to them. But in
general they will complain about the lack of basic knowledge of
the physiology and genetics of durable pest resistance; about the
large number of years and locations needed to evaluate and
identify dependable yield and stress tolerance; about the long
time, in generations, needed to break up undesirable genetic
linkages and conversely to create and identify desired assemblages
of useful traits. Breeders of hybrid crops will complain about
their inability to accurately predict amounts of heterosis among
crosses of inbred lines. They also complain about the complete
failure of genetics and physiology to explain the basis for
heterosis -- hybrid vigor -- except in very general, essentially
descriptive terms.

I should like to point out a few ways in which I think
molecular biology and accessory sciences can help plant breeding.
First and foremost it can help give the breeder a solid foundation
of knowledge about gene action and the genetics of important
traits. I have already noted that it appears that most of the
agronomically important traits are controlled by many genes.
Molecular genetics and genetic engineering unfortunately are at
present usually constrained to working with single genes, in
higher organisms. But I think that if molecular genetic studies
can be combined with Mendelian and then with quantitative
genetics, increasingly difficult basic genetic questions will be
answered to the advantage, eventually, of practical plant
breeding. Some of the questions that need answering are, as noted
earlier: what makes durable -- long-lasting -- disease
resistance, what is the physiology, biochemistry and genetics of
stress tolerance, how can one identify useful gene combinations,
and how can one predict the highest yielding hybrid combinations
of inbred lines. An example of how molecular genetics could help:
understanding the interrelations between gene structure and gene
action of genes giving resistance to wheat stem rusts might help
us to design better, more durable types of single factor disease
resistance.

Useful shortcuts, utilizing new techniques from biotechnology, likely will be available to plant breeders long before the
above fundamental questions are answered. Variations of tissue
and cell culture already are providing disease free propagules of
certain vegetatively produced crops; propagation of selected
hybrid individuals is underway in forest crops and likely will be
used in some horticultural species. The first uses of molecular
biology for disease identification already are being overtaken by
a second generation of molecularly based specific identification
techniques. In the near future, plant variety protection and
identification -- proof of ownership -- will be aided by identifi-

cation tools provided by the new biotechnology. Any list made today will prove to have been too short and devoid of imagination in the next few years. The only limitation will be the ability of plant breeders to comprehend the uses they can make of the new biotechnology, and the ability of biotechnologists to understand what plant breeders really need. It is time now to get on with the work.

Concluding Remarks

I make one final suggestion. I suggest that from now on we avoid use of the term "conventional plant breeding". This term suggests, subtly, that present-day plant breeding is a static, unprogressive technology. Successful plant breeders always have been unconventional; they always have incorporated truly useful new techniques from the new sciences of the day; they will continue to do so. But they also have not wasted their time in trying to use new but impractical technologies once it became obvious that no practical advantage to plant breeding was to be found in them.

The only problem with this description of successful plant breeders wisely using the best of the new technologies is that one cannot be sure what the right new technologies are, except in hindsight. That, of course, is our dilemma -- and opportunity -- today.

LITERATURE CITED

Austin, R. B., Bingham, J., Blackwell, R. D., Evans, L. T., Ford, M. A., Morgan, C. L., and Taylor, M., 1980, Genetic improvements in winter wheat yields since 1900 and associated physiological changes, Jour. Agric. Sci., 94: 675-689.
Boyer, J. S., 1982, Plant productivity and environment, Science, 218: 443-448.
Cooper, R. L., 1981, Development of short statured soybean cultivars, Crop Sci., 21: 127-131.
Cardwell, V. B., 1982, Fifty years of Minnesota corn production: Sources of yield increase, Agron. J., 74: 984-990.
Duvick, D. N., 1977, Genetic rates of gain in hybrid maize yields during the past 40 years, Maydica, XXII: 187-196.
Duvick, D. N., 1984, Genetic contributions to yield gains of U.S. hybrid maize, 1930 to 1980, in "Genetic Contributions to Yield Gains of Five Major Crop Plants," W. R. Fehr, ed., Crop Science Society of America Special Publication No. 7, Madison, Wisconsin.
Hallauer, A. R., 1973, Hybrid development and population improvement in maize by reciprocal full-sib selection,

Egypt. Jour. Gen. and Cytol., 2: 84-101.

Hueg. W. F. Jr., 1977. Focus on the future with an eye to the past, in "Agronomists and Food: Contribution and Challenges," Amer. Soc. Agron. Special Publication No. 30. Madison, Wisconsin.

Jensen, N. F., 1978. Limits to growth in world food production, Science. 201: 317-320.

Leudders, V. C., 1977. Genetic improvement in yield of soybeans, Crop Sci., 17: 971-972.

Maunder. A. B., 1969. Meeting the challenge of sorghum improvement, in "Proceedings of the 24th Annual Corn and Sorghum Research Conference," John Sutherland and Robert Falasca. eds., American Seed Trade Association, Washington. D.C.

Miller. F. R. and Kebede. Y., 1984. Genetic contributions to yield gains in sorghum. 1950 to 1980. in "Genetic Contributions to Yield Gains of Five Major Crop Plants," W. R. Fehr. ed., Crop Science Society of America Special Publication No. 7. Madison, Wisconsin.

National Research Council. 1972. "Genetic Vulnerability of Major Crops," National Academy of Sciences, Washington. D.C.

Reitz. L. P., and Salmon, S. C., 1959. Hard red winter wheat improvement in the Plains: A 20-year summary. USDA Tech. Bull., 1192: 117.

Russell. W. A., 1974. Comparative performance of maize hybrids representing different eras of maize breeding. in "Proceedings of the 29th Annual Corn and Sorghum Research Conference," Delores Wilkinson, ed., American Seed Trade Association, Washington, D.C.

Ruttan, V. and Sundquist, W. B., 1982, Agricultural research as an investment; past experience and future opportunities, in "Report of the 1982 Plant Breeding Research Forum," Pioneer Hi-Bred International, Inc., Des Moines, Iowa.

Salmon, S. C., Mathews, O. R. and Leukel, R. W., 1953, A half century of wheat improvement in the United States, Adv. Agron., 5: 1-151.

Schmidt, John W., 1984, Genetic contributions to yield gains in wheat, in "Genetic Contributions to Yield Gains of Five Major Crop Plants," W. R. Fehr, ed., Crop Science Society of America Special Publication No. 7, Madison, Wisconsin.

Sim, R. J. R. and Araji, A. A., 1981, The economic impact of public investment in wheat research in the Western Region, Idaho Agric. Expt. Stn. Res. Bull., 116: 27.

Specht, J. E. and Williams, J. H., 1984, Contribution of genetic technology to soybean productivity - retrospect and prospect, in "Genetic Contributions to Yield Gains of Five Major Crop Plants," W. R. Fehr, ed., Crop Science Society of America Special Publication No. 7, Madison, Wisconsin.

Wellman, A. C. and J. B. Hassler, 1969, How technological change affects Nebraska crops, Quarterly Serving Farm. Ranch and Home, (Univ. of NE, College of Agric. and Home

Econ.) 16: 23-25.

Wilcox, J. R., Schapaugh, W. T. Jr., Bernard, R. L., Cooper, R. L., Fehr, W. R., and Neihaus, M. H., 1979, Genetic improvement of soybeans in the midwest, Crop Sci. 19: 803-805.

PHILISOPHY AND METHODOLOGY OF AN INTERNATIONAL

WHEAT BREEDING PROGRAM

S. Rajaram, B. Skovmand and B.C. Curtis*

CIMMYT, Apartado Postal 6-641
Delegación Cuauhtémoc, 06600, Mexico, D.F.

*Authors are: bread wheat breeder, triticale breeder and
Directors, respectively, of CIMMYT's Wheat Improvement Prog.

Introduction

The philosophical approaches adopted by plant breeders, as
well as the breeding methodologies used by them within a specific
philosophical context, constitute one of the most complex and
controversial issues in plant breeding today. While there is some
agreement upon classical procedures for the improvement of
self-pollinated vs. cross-pollinated crops, there is quite clearly
no general consensus among plant breeders as to which
philosophical approach is best, nor as to which breeding methods
are most appropriate in a given situation for a given crop.

Plant breeding philosophies tend to fall into one of two
categories: those that reflect a desire to enhance plant
performance (and thus crop production) on a site-specific basis,
and those that reflect an interest in improving crop production
under a wide range of growing conditions by developing cultivars
that will perform well across diverse environments. Of course,
within the latter philosophical context there is ample room for
site-specific breeding activities; the orientation of these
special breeding efforts, however, is always toward overcoming one
or more limiting environmental factors by incorporating into
widely adapted germplasm the specific genetic traits needed to
improve performance in a certain location.

The methodology of any plant breeding program for a
self-pollinated crop [such as wheat (Triticum aestivum L. em
Thell.)] must reflect the environmental realities of the target
area, such as temperature regimes, moisture availability, soil

33

fertility and pH conditions, disease and insect spectrums, as well
as cropping systems.

Numerous examples of very different, yet very successful,
plant breeding programs involving self-pollinated crops could be
discussed here to support the adoption of one or the other general
breeding philosophies. Most of these programs have a site-specific
breeding orientation, and many of them have had dramatic impacts
on per hectare yield potential within their geographic areas of
concern (Johnson and Schmidt, 1979; Bingham, 1979 and 1980;
Borojevic and Potocanac, 1966; Rao, 1980; Kronstad, 1983). In
addition, other breeding systems have been discussed elsewhere
(Fasoulas, 1973 and 1976; Joshi, 1979; and Qualset and Vogt,
1980). The purpose of this paper, however, is to describe the
breeding philosophy and methodologies considered by the authors to
be most appropriate in an international setting, i.e., the
breeding of improved germplasm for a wide range of production
environments.

The CIMMYT Wheat Improvement Program has had dramatic impacts
on global wheat production. Today, over 35 million hectares in the
developing world are planted to high-yielding wheat varieties with
CIMMYT germplasm in their pedigrees (Dalrymple, 1980). These
impacts have been documented by Borlaug (1968) and many others,
and will not serve as the focus of this presentation. Of course,
CIMMYT's is not the only successful international agricultural
research effort; Dr. G.S. Khush will be reporting to this
Symposium on the work of the International Rice Research Institute
(IRRI), which has had a tremendous impact on global rice
production. Yet CIMMYT's wheat breeding philosophy and methods
have been distinctive from the outset; the authors feel strongly
that the approaches developed and used so effectively by CIMMYT's
Wheat Improvement Program provide an excellent model for the
successful breeding of self-pollinated crops in an international
setting. Hence, our presentation to this Symposium constitutes an
illustration of this model. We thank the Stadler Genetic Symposium
Organizing Committee for inviting this discussion.

Philosophy of CIMMYT's Wheat Breeding Program

CIMMYT is engaged in the improvement of bread wheat (Triticum
aestivum), durum wheat (T. durum), triticale (X Triticosecale) and
barley (Hordeum vulgare). The breeding approach used by all four
crop programs is similar, with only slight modifications. To
simplify this presentation, only the bread wheat breeding program
will be discussed.

The current philosophy and breeding methodology of the bread
wheat program is to strive, in close collaboration with national
wheat research institutions, for the development of widely

adapted, high-yielding varieties characterized by semidwarf plant stature, photoperiod insensitivity, acceptable industrial quality, and resistance to such environmental stresses as drought, aluminum toxicity, heat and prevalent diseases [such as stem rust (Puccinia graminis f. sp. tritici), leaf rust (P. recondita), stripe rust (P. striiformis), septoria tritici blotch (Mycosphaerella graminicola), septoria nodorum blotch (Leptospharia nodorum), Helminthosporium spp. and barley yellow dwarf virus (BYDV)].

Our philosophy emphasizes the production of advanced lines suitable for various agroclimatic areas of the world where wheat is currently the major food crop, or where its consumption is rapidly rising. These agroclimatic areas include (Table 1):

- irrigated areas,
- rainfall areas with annual rainfall higher than 500 mm,
- drought stress, semiarid areas with annual rainfall less than 300 mm,
- acid-soil areas where wheat production is affected by aluminum toxicity, and
- warmer areas where sudden increases in temperature are common.

Disease complexes, soil types, and climatic conditions vary considerably from one area to the next. Thus, CIMMYT's breeding approach is to select widely adapted, high-yielding lines that also possess other necessary characteristics. CIMMYT's bread wheat breeding program is fully integrated, in that attention is given to combining various agronomic, pathologic and yield characteristics. However, to clarify our methodologies, the following major breeding objectives will be discussed separately:

- breeding for wide adaptation and high yield potential,
- breeding for disease resistance,
- breeding for aluminum toxicity resistance,
- breeding for drought resistance, and
- breeding for heat tolerance.

Toward these objectives, more than 8,000 simple and top crosses are made in Mexico each year to combine different traits and develop lines suitable for those areas of the world included under CIMMYT's operational mandate (Table 1). More than 1,000 winter and spring wheat parents are studied carefully for their agronomic, pathologic and quality characteristics before crosses are made. The selection of this parental material is not based on genetic analyses for disease resistance and yield components. Rather the mere presence of desirable traits provides sufficient reason for inclusion in the crossing program, and crosses are often made

Table 1. Major spring wheat producing areas of the developing
world with regions or countries and important diseases in
relation to bread wheat breeding at CIMMYT.

Areas	Regions and Countries	Important Diseases
Irrigated areas	Gangetic Plains of India, Indus Valley of Pakistan, Bangladesh, Lybia, Tarai of Nepal, Egypt, Mexico, Zimbabwe	Stem rust, leaf rust, stripe rust
Rainfed areas with rainfall more than 500 mm	Mediterranean North Africa, Middle East, China, Southern Cone countries, Andean and East African highlands, Central America	Stem rust, leaf rust, stripe rust, Septoria tritici, Septoria nodorum, BYDV, Fusarium spp., Helminthosporium spp., Xanthomonas translucens
Semiarid areas with rainfall less than 300 mm	Central India, certain areas of the Middle East, North Africa and Southern Cone countries	Bunt, loose smut
Acid areas affected by aluminum toxicity	Brazil	Stem rust, leaf rust, Septoria tritici, Septoria nodorum, Fusarium spp., Helminthosporium spp., Xanthomonas translucens
Warmer areas in the tropical zone	South and South East Asia, West Africa, Cerrados of Brazil, Central America and Caribbean	Leaf rust, stem rust, Helminthosporium spp, Xanthomonas translucens

without full knowledge of the inheritance patterns. Similarly, no combining ability studies are made for the crossing program. In our opinion, these excercises, while necessary for scientific publication and the understanding of genetic interactions, are not central to CIMMYT's primary mission: the improvement of germplasm.

Breeding for Wide Adaptation and High Yield Potential

The long term objective of CIMMYT's Wheat Improvement Program continues to be the development of germplasm that possesses:
- high yield potential,
- stable yield,
- wide adaptation,
- semidwarf characteristcs,
- photoperiod insensitivity, and
- disease resistance.

Combining these traits has so far resulted in an unprecedented rate of acceptance throughout the world of what have become known as high yielding varieties (HYVs), the majority of which originated from CIMMYT germplasm. Recently, Worrall, et al. (1980) analyzed 14 years of data from the International Spring Wheat Yield Nursery (ISWYN), a replicated yield trial of 50 high-yielding varieties and advanced lines distributed around the world each year by CIMMYT. Their analysis shows that CIMMYT wheats and their derivatives perform well in nearly every wheat growing environment, including semiarid zones. This analysis has been verified by Pfeiffer (1983), who concluded (again on the basis of ISWYN data) that significant progress has been made, and continues to be made, in breeding wheats for stable performance over a wide range of environments.

Wide adaptation, as well as high and stable yields are hallmark characteristics of CIMMYT bread wheat germplasm, and have been derived through application of the following methods:

1. Shuttle breeding using alternate sites--The CIMMYT breeding program in Mexico utilizes two crop cycles per year:

    a) A summer cycle in Toluca, at 17.4°N latitude and an elevation of 2,640 meters, where genotypes are screened for resistance to the rusts, as well as septoria and fusarium diseases. Agronomic characters and yields are also evaluated. Plantings are made in Toluca during May, when the day-length is increasing.

    b) A winter nursery in Ciudad Obregon in northwest Mexico, at 28°N latitude and an elevation of 40

meters, where breeding materials are screened for
resistance to stem rust and leaf rust, and
evaluated for their yield potential. Plantings are
made at this location in November, when day-length
is decreasing.

In addition to testing at these sites, advanced materials are
tested in Los Mochis, Rio Bravo and El Batan for leaf rust
resistance, in Patzcuaro for Septoria tritici resistance and in
Poza Rica and Santiago Ixcuintla for Helminthosporium resistance
(Figure 1).

Testing at Cd. Obregon helps in identifying genotypes with
high yield potential, while testing at Toluca, a typical highland
location, helps produce advanced high-yielding lines for highland
locations. El Batan, being a semiarid site, is good for the
screening and identification of lines suitable for environments
with limited moisture. A yield evaluation of all advanced lines to
verify their broad adaptation is conducted at each of these three
locations. The shuttle breeding approach conceived by Dr. N.E.
Borlaug in Mexico in the late 1940s was primarily intended to
reduce the time required for breeding wheat cultivars. However,
the system resulted in an even more important contribution: the
incorporation of wide adaptation and stability.

2.   Multilocation testing and cycling of widely adapted
     germplasm and pyramiding of genes carrying resistance to
     various pathogens into the crossing program--A schematic
     of how germplasm moves through CIMMYT's wheat breeding
     program is shown in Figure 2. National crop improvement
     programs, which grow and evaluate CIMMYT nurseries and
     thereby provide vital information for use in CIMMYT's
     crossing programs, are grouped into the following
     regions:
     - the rainfed areas of Mediterranean North Africa and
       the Iberian Peninsula;
     - the semiarid regions of the Middle East;
     - the highlands of East Africa;
     - Southern Africa;
     - the irrigated valleys of the Indian Subcontinent;
     - an East Asian region with optimum rainfall;
     - the rainfed areas of the Southern Cone countries of
       South America with acid and nonacid soils;
     - the Andean highlands of South America with high
       rainfall and a high incidence of disease;
     - the highlands of Central America including the High
       Plateau of Central Mexico;
     - the irrigated areas of northern Mexico and southern
       United States;
     - the rainfed areas of the northern United States and Canada;

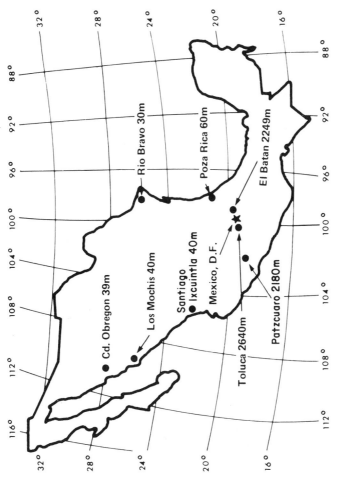

Figure 1. Location and elevations of experiment stations in Mexico at which CIMMYT conducts research.

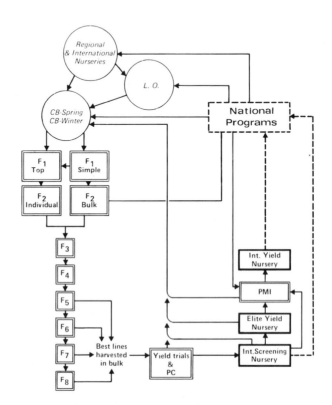

- Nurseries for use in CIMMYT Crossing program
= CIMMYT breeding nurseries
- CIMMYT international nurseries
--- National, CIMMYT regional, and other international programs

CB = Crossing block
L.O. = Observation lines
PMI = International multiplication plot
PC = Small plot multiplication

Fig. 2.   Movement of germplasm in CMMYT's Wheat Program.

- Large areas of the European Continent;
- Oceania; and
- areas in the tropical belt.

Evaluations of international screening nurseries, international yield trials and F_2 nurseries are made in many countries of these regions. Data from these nurseries help guide future crosses; the best performing commercial varieties and advanced lines of both spring and winter wheats are assembled in crossing blocks after initial screening for disease resistance and agronomic type in Mexico.

Successes related to widely adapted varieties:

Through application of the above techniques, a great deal of very useful germplasm has been produced and distributed through our international nursery system. The best examples are the varieties Pitic 62, Penjamo 62, Lerma Rojo 64, Siete Cerros 66 and Anza (released in the 1960s); Marcos Juarez INTA, Jupateco 73, Pavon 76, Nacozari 76 and CIANO 79 (released in the 1970s); and Glennson 81, Genaro 81, Ures 81 and Seri 82 (released to date in the 1980s). These materials were bred cooperatively by the Instituto Nacional de Investigaciones Agrícolas (INIA), Mexico, and CIMMYT. The variety Marcos Juarez INTA was developed cooperatively by Argentina and CIMMYT. The variety Anza is a Californian release after undergoing reselection there. The impacts of CIMMYT/Mexican-derived germplasm on per hectare wheat production has been remarkable in such countries as India, Pakistan, Turkey, Tunisia, Spain, Argentina, Mexico, Zimbabwe, Egypt and Bangladesh (Table 2). Eighteen years of international multilocation testing data clearly desmonstrate the wide adaptability, stability and high yield potential of CIMMYT-developed materials in the 14 regions of the world mentioned earlier. Data from these trials have been published annually by CIMMYT as "Results of the International Spring Wheat Yield Nursery (ISWYN)."

Breeding for Disease Resistance

One of the central objectives of the CIMMYT bread wheat program continues to be the development of disease resistant germplasm for worldwide distribution. Initially, our work on disease resistance emphasized stem rust in the 1950s and 1960s; we expanded our efforts in the 1970s to include breeding for resistance to leaf rust, stripe rust and septoria tritici blotch; our focus will be further expanded in the 1980s to include breeding for resistance to Helminthosporium spp., fusarium head scab (Fusarium spp.), barley yellow dwarf virus (BYDV) and Xanthomonas translucens.

Table 2.   Impacts of CIMMYT/Mexican semidwarf wheat germplasm on the per hectare wheat yield of 10 countries when compared to 1961-1965 pre-semidwarf base.[a]

Countries	1961-65 (tons/ha)	1970-72 (tons/ha)	1980-82 (tons/ha)	Growth Rates 1961-1965 to 1980-1982 (%)
India	0.84	1.30	1.59	3.5
Pakistan	0.83	1.15	1.62	3.7
Turkey	1.08	1.37	1.87	3.0
Tunisia	0.85	0.79	1.16	1.7
Spain	1.05	1.29	1.74	2.8
Argentina	1.54	1.43	1.68	0.5
Mexico	2.09	2.89	4.00	3.6
Zimbabwe	2.00	3.57	4.50	4.5
Egypt	2.62	2.96	3.34	3.0
Bangladesh	0.62	0.90	1.85	6.1

[a] Calculated from FAO Tape of Production Statistics, 1983.

The strategy CIMMYT employs for breeding disease resistant
materials has been discussed in detail by Borlaug (1953), Rajaram
and Dubin (1977), and Dubin and Rajaram (1981). In 1981, Rajaram
and Torres presented a paper to the American Society of Agronomy
stressing: 1) utilization of broad-based resistance with effective
combinations of resistance genes, 2) identification and use of of
dilatory resistance (such as slow rusting), and 3) the production
and deployment of multilineal cultivars for disease control.

Multilocation testing--CIMMYT's most
important strategy for identification of resistance:

International multilocation testing does not provide an
indication of the number and type of resgenes present in a
cultivar, but it does permit the classification of cultivars for
their spectrum and magnitude of resistance (Torres and Rajaram,
1983). Genetic studies have suggested that wheat genotypes
resistant to rust diseases in many dissimilar localities
(indicated by low coefficients of infection) often contain
multiple genes for resistance (Rajaram and Luig, 1972). However,
it is not suggested that this polygenic resistance is race
nonspecific.

The international wheat nursery system provides multisite
disease data and allows the calculation of average coefficients of
infection (ACI) for stem rust, leaf rust, stripe rust, septoria
tritici blotch, septoria nodorum blotch, BYDV, Helminthosporium
spp. and Fusarium spp. The lowest ACIs are used in simple and
three-way crosses to pyramid the "resgenes." The segregating
progenies of these crosses are then exposed under field conditions
to all available races of the three rusts, as well as to septoria
leaf blotch. It is our strong belief that individual plants and
lines resistant to the rusts and septoria should be selected in
the field at the adult stage when the multiplicity of pathogenic
races prevails.

Identification and utilization of
slow rusting or dilatory resistance:

Dilatory resistance to the rusts (Browning, et al., 1977),
commonly known as slow rusting (Caldwell, 1968), refers to a
phenomenon in which the rate of disease development is slower
compared to that of a susceptible cultivar; the final disease
intensity is also less. This type of host-parasite interaction is
characterized by a susceptible host reaction.

Cultivars Torim 73 and Pavon 76 (released in 1973 and 1976,
respectively) have maintained a stable leaf rust reaction combined
with high yields. There are leaf rust races capable of attacking
these varieties in the seedling stage, but without posing a threat

to their productivity. Leaf rust intensities of 5 to 30 percent
have been recorded in the late growth stages of these varieties.
The experience with Torim 73 and Pavon 76 resulted in an intensive
search for slow-rusting wheat cultivars and a vigorous breeding
program to transfer this trait.

Slow-rusting varieties in Mexico:

Advanced lines and cultivars are rated for leaf rust
reactions three times throughout the epidemic period in Mexico.
The varieties that exhibit a disease intensity lower than the
susceptible check are entered into CIMMYT's crossing program.
Experimental sites like El Batan and Cd. Obregon, where there is
good leaf rust development, are utilized for classifying the
slow-rusting character.

Figure 3 shows the responses of the slow-rusting cultivars
Veery "S", Pavon "S", Harrier "S" and Tanager "S", compared to
that of the susceptible check cultivar (INIA 66) in El Batan,
1980. These cultivars have shown slow-rusting resistance to leaf
rust during the past eight seasons.

Advances made in breeding resistance to rusts:

The CIMMYT bread wheat program originally focused on breeding
cultivars resistant to stem rust, and the resistant cultivar
Newthatch, derived from Hope/Thatcher//2x Thatcher, played a basic
role in stabilizing resistance to this disease. The
Newthatch-derived varieties Yaqui 50, Chapingo 52 and Chapingo 53
(Borlaug, 1968) that were released 30 years ago have remained
resistant to stem rust under field conditions. The modern
semidwarf varieties, such as Pavon 76, CIANO 79 and Genaro 81, are
the descendants of the original tall, stem rust-resistant
cultivars, such as Yaqui 50 and Chapingo 52. The important point
is that there has been no threat of a stem rust epidemic on
semidwarf wheats in Mexico for the last 20 years. This is a clear
indication of the effectiveness of the Yaqui 50 gene complex in
(1) imparting resistance to stem rust, and (2) in providing a
stabilizing effect on the stem rust fungus.

In contrast to the stem rust situation, leaf rust pathogens
have caused significant losses in Mexico, such as occurred in the
great epidemic of 1977 in northwest Mexico. However, such
slow-rusting cultivars as Torim 73 and Pavon 76 have been
providing effective resistance for more than eight years. There
has been no threat of a leaf rust epidemic in northwest Mexico
since 1979, primarily because of the introduction of the variety
CIANO 79, and later releases like Glennson 81, Sonoita 81, Genaro
81 and Ures 81. Preliminary results indicate that Sonoita 81,
Genaro 81 and Ures 81 are also slow-rusting types.

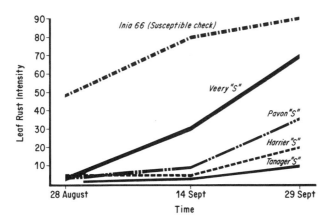

Fig. 3. Slow rusting lines of bread wheat showing leaf rust intensity over time compared to INIA 66 (susceptible check) at El Batan.

More than 1000 advanced lines are produced annually in the
CIMMYT bread wheat program. These lines result from crossing
cultivars that are selected on the basis of multilocation testing
for yield, adaptation and disease resistance, as well as
slow-rusting characteristics (Figure 3). Most of these advanced
lines are returned to national programs for retesting. In this
way, many semidwarf varieties are developed cooperatively by
national programs and CIMMYT.

Dr. J.M. Prescott of CIMMYT has compared the performance of
these semidwarf varieties to improved tall types, which are
developed by national programs alone, and local land race types.
The data for stem rust, leaf rust and stripe rust are presented in
Tables 3, 4 and 5, respectively. These data indicate that the
levels of resistance to stem rust and leaf rust are satisfactory
in the semidwarfs as compared to local land races and the improved
tall types. However, the level of resistance to stripe rust as
compared to the improved tall types is not yet satisfactory.

Advances made in breeding
resistance to septoria tritici blotch:

In 1970 the advanced CIMMYT bread wheats were highly
susceptible to septoria tritici blotch under field conditions in
various areas of the world where the disease is endemic. Since
then, concerted efforts have been made by CIMMYT to identify and
incorporate germplasm with acceptable resistance to this disease.
The varieties from the USSR, USA, Brazil and Argentina used in
crossing for Septoria tritici blotch resistance are given in Table
6. Some of the advanced lines with good septoria tritici blotch
resistance are presented in Table 7. It should be emphasized that
the yield potentials of the lines listed in Table 7 have moved
closer to such widely adapted high-yielding varieties as Ures 81,
Glennson 81 and CIANO 79.

Our efforts to improve resistance to BYDV, Helminthosporium
spp., Fusarium spp. and Xanthomonas translucens began in the early
1980s and no progress report can be provided at this stage.

Breeding for Drought Resistance

Thirty-seven percent of the total area in all developing
countries consists of semiarid environments in which available
moisture constitutes the primary constraint on wheat production.
The Middle East/North Africa region constitutes about 59% of these
semiarid Third World environments, with the largest area located
in Turkey. In South Asia and East Asia, the countries having the
largest semiarid areas are India and China, respectively.
Argentina ranks first in area among the Latin American semiarid
regions.

Table 3. Bread wheat Regional Disease Trap Nursery (RDTN) data
(A.C.I.) from approximately 50 locations in 30
countries. Source: Prescott, J.M., unpublished.

	Stem rust (Puccina graminis tritici)				
	1978	1979	1980	1981	Avg.
Local	21.8	21.8	21.1	19.6	21
Improved Tall	9.2	8.6	9.9	9.3	9
Semidwarf	7.8	4.7	5.1	6.2	6

Table 4. Bread wheat Regional Disease Trap Nursery (RDTN) data
(A.C.I.) from approximately 50 locations in 30
countries. Source: Prescott, J.M., unpublished.

	Leaf rust (Puccina recondita)				
	1978	1979	1980	1981	Avg.
Local	39.9	31.1	41.6	28.1	35
Improved Tall	19.1	11.4	14.1	7.8	13
Semidwarf	12.1	7.1	8.7	6.0	8

Table 5. Bread wheat Regional Disease Trap Nursery (RDTN) data
(A.C.I) from approximately 50 locations in 30 countries.
Source: Prescott, J.M., unpublished.

	Stripe rust (Puccina striiformis)				
	1978	1979	1980	1981	Avg.
Local	11.9	18.3	22.4	18.6	18
Improved Tall	7.9	6.5	9.2	6.9	8
Semidwarf	9.2	8.6	10.0	9.9	9

Table 6. Sources of resistance to _Septoria tritici_ used in
          CIMMYT's breeding program.

Varieties from the USSR	Varieties from the USA	Varieties from Brazil	Varieties from Argentina
Kavkaz	Chris	IAS 55	TZPP
Aurora	Era	IAS 58	
Bejostaja 1	FKN	IAS 63	
		IAS 62	
		CNT 7	
		CNT 8	
		PF 70254	
		Maringa	
		Carazinho	
		Lagoa Vermelha	
		Gaboto	

Table 7. _Septoria tritici_-resistant semidwarf bread wheat
          cultivars with good yield and good agronomic type.

	Toluca, 1982		Cd. Obregon, 1982–1983		
	Septoria tritici[a]	Ht. (cm)	Yield (kg/ha)	Yield Ck. Var.	Check Variety
KVZ-HD2009	22	36	95	6198	6389	CIANO 79
KEA 'S'	26	33	90	6194	6517	URES 81
MAYA'S'-MON'S'	25	32	90	5720	6554	URES 81
BOW 'S'	12	32	85	6759	7117	URES 81
SNB 'S'	25	34	90	6509	6611	URES 81
GOV-AZxMUS'S'	24	35	90	5113	6389	CIANO 79
FINK 'S'	25	26	90	5793	6517	URES 81
CNO67-MFDxMON'S'	23	34	100	5491	6389	CIANO 79
LIRA 'S'	12	36	80	6650	6474	GLENNSON 81
TZPP (Resistant Check)	21	34	110			
CNO 79 (Susceptible Check)	44	95	80			

[a] Two readings were taken at 75 and 96 days after sowing.

To plant breeders, the term "drought resistance" is related
to a moisture stress environment and refers the ability of one
genotype to be more productive with a given amount of soil
moisture than other genotypes (Quizenberry, 1981). Hurd (1969) and
Schmidt (1980) presented excellent reviews of breeding systems for
rainfed wheat improvement in semiarid regions.

Smith (1982), in discussing drought resistance, suggested
that the ideal variety would have a highly stable yield combined
with high yield potential. His "ideal," which he thinks is
difficult but not impossible to achieve, is represented in Figure
4. In his correlation studies, the higher yield potential of
short-statured genotypes was achieved at the expense of yield
stability. Smith suggested a "middle of the road" genotype that
would give high yield potential in all environments, except those
characterized by either extremely high or extremely low production
conditions.

CIMMYT's philosophical approach to breeding for drought
resistance was expressed by Rajaram and Nelson (1982); they
believe that wide adaptation, high yield and stability are
necessary in any variety to be recommended for semiarid regions.
The variety must be able to give relatively high yields under low
input conditions, but must possess the genetic potential for
higher yields should environmental conditions such as moisture and
fertilizer availability, improve. Hence, the authors believe that
Smith's ideal type (Figure 4) is appropriate for any wheat
breeding program, and the methodology is available for obtaining
it.

Joint CIMMYT/ICARDA
drought-resistance breeding program:

Our drought-resitance breeding program is a cooperative
venture between CIMMYT (locations: Cd. Obregon and El Batan,
Mexico) and ICARDA (location: Aleppo, Syria). A schematic
representation of germplasm flow and the handling of segregating
generations is shown in Figure 5. Spring x winter crosses are made
in Mexico and F_2 populations are grown in Cd. Obregon and in
Aleppo. High fertility and optimum moisture are applied in the F_2,
F_5, F_6, F_7 and F_8 generations. The F_3 and F_4 generations are grown
under low fertility and reduced moisture conditions. Selected
materials are exchanged between Mexico and Syria for reselection.

The following characters are important drought-resistance
selection criteria, and are used in selecting materials grown
under limited moisture and low fertility:

- longer leaf duration,
- moderate tillering ability,

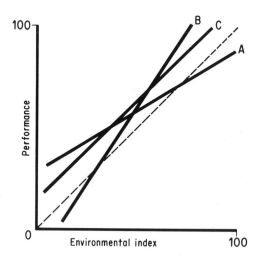

Regression line 'A' represents a genotype with above average stability and below average mean yield. Many of the older, standard height varieties fall into this category.

Regression line 'B' represents a genotype with below average stability and above average mean yield. Many of the newer, semi-dwarf varieties fall into this category.

Regression line 'C' represents and 'ideal' genotype with average stability and above average yield.

Fig. 4.   Generalized representation of yield stability for
          different categories of genotypes (Smith, 1982).

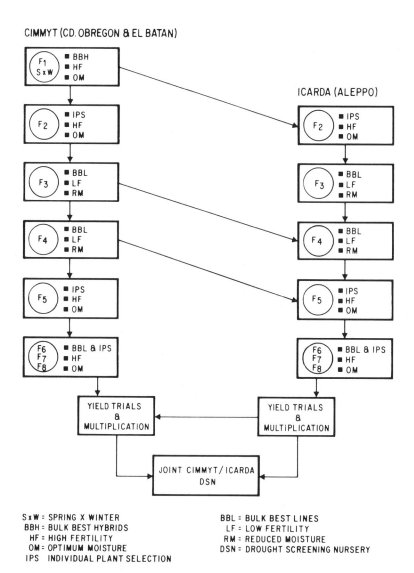

CIMMYT (CD. OBREGON & EL BATAN)

F1 SxW
- BBH
- HF
- OM

ICARDA (ALEPPO)

F2
- IPS
- HF
- OM

F2
- IPS
- HF
- OM

F3
- BBL
- LF
- RM

F3
- BBL
- LF
- RM

F4
- BBL
- LF
- RM

F4
- BBL
- LF
- RM

F5
- IPS
- HF
- OM

F5
- IPS
- HF
- OM

F6 F7 F8
- BBL & IPS
- HF
- OM

F6 F7 F8
- BBL & IPS
- HF
- OM

YIELD TRIALS & MULTIPLICATION

YIELD TRIALS & MULTIPLICATION

JOINT CIMMYT/ICARDA DSN

SxW = SPRING X WINTER
BBH = BULK BEST HYBRIDS
HF = HIGH FERTILITY
OM = OPTIMUM MOISTURE
IPS INDIVIDUAL PLANT SELECTION

BBL = BULK BEST LINES
LF = LOW FERTILITY
RM = REDUCED MOISTURE
DSN = DROUGHT SCREENING NURSERY

Fig. 5. Schematic representation of germplasm movement for
collaborative drought-resistance breeding to CIMMYT
and ICARDA (Aleppo).

    - grain plumpness,
    - high test weight, and
    - yield, per se.

Comparisons are made to a check variety optimally represented in
the nursery.

Selections during the irrigated and well-fertilized
generations, F_5 through F_8, are conducted normally; rust epidemics
are created and susceptible plants are discarded; high tillering
capacity, good head development, good leaf retention capability,
and grain plumpness are the agronomic characters evaluated in
individual plant and line selections.

Advanced lines normally produced under high fertility and
optimum moisture are subjected to low fertility and reduced
irrigation in Cd. Obregon and El Batan. The objective is to select
genotypes that are high yielding under optimum conditions and
reasonably high yielding under suboptimum conditions. Experiments
conducted in 1982 at Cd. Obregon suggest that lines can be
selected which would be high yielding in both situations
(Table 8).

The first International Drought Screening Nursery (DSN) was
assembled and sent in 1983 to cooperating national programs. At
this time, results of the CIMMYT/ICARDA drought-resistance program
cannot, of course, be presented; in future years, however, we hope
to present the results of this cooperative venture in a forum
similar to this one.

Breeding for Aluminum Toxicity Resistance

Among the soils adversely affected by mineral toxicities
and/or deficiencies are the highly leached acidic oxisols and
ultisols, which are characterized by toxic levels of soluble
aluminum (Al) and manganese. Aluminum is normally quite severe
below pH 5.0. In such situations, cell division is prevented and
root growth is thus inhibited.

Aluminum toxic soils comprise approximately one billion
hectares in the tropical and subtropical areas of Brazil,
southeast China, southeast Asia and Central Africa (Van Wambeke,
1976). Currently, such areas are either underdeveloped for
agriculture or, where cultivated, are of very low productivity.
The traditional approach to the mineral stress problems of acidic
soils has been to raise the pH level through liming; this is often
expensive and sometimes impractical, particularly where strongly
acidic subsoils exist. Fortunately, however, tolerance to Al
toxicity has been found to be genetically controlled in a
considerable number of crop species.

Table 8. Yield performance of 6 wheat lines under 2 irrigation regimes in Cd. Obregon, Sonora, Mexico, 1982.

Variety	Yield under 5 irrigations		Yield under 2 irrigations	
	kg/ha	Rank	kg/ha	Rank
Genaro 81	7461	1	4745	2
Lira'S'	6920	2	4757	1
Veery'S'	6819	3	4378	5
Neelkant'S'	6758	4	4585	4
Junco'S'	6346	5	4681	3
Tanager'S'	5202	6	3637	6

LSD(5%) 639
CV 6%

LSD(5%) 711
CV 10%

Brazilian wheat breeders pioneered the development of wheats
with tolerance to Al toxicity. This work was initiated in 1925
(Beckman, 1976) and, over the last five decades, has resulted in a
large number of resistant lines. Unfortunately, the Brazilian
wheats with Al toxicity resistance were of poor agronomic type and
had low maximum genetic yield potentials. In comparison, CIMMYT
materials possessed broad adaptation and high genetic yield
potential, but without exception were susceptible to toxic levels
of aluminum and were poorly adapted to acidic soils.

Cooperative CIMMYT/Brazilian
shuttle breeding program:

Over 10 years ago, the CIMMYT Wheat Improvement Program began
to breed high-yielding, broadly adapted, semidwarf wheats with
tolerance to high levels of aluminum. This effort quickly evolved
into a cooperative project between CIMMYT and three Brazilian
wheat breeding programs: 1) EMBRAPA, at the Passo Fundo National
Wheat Research Center, 2) FECOTRIGO, at Cruz Alta, and 3) OCEPAR,
at Cascavel (Figure 6). Crosses between Al tolerant Brazilian
wheats and CIMMYT/Mexican semidwarf germplasm are made in Mexico,
and selections from segregating generations are made at two
locations: 1) Cd. Obregon, and 2) Toluca, Mexico. In addition,
selected progenies undergo laboratory screening at El Batan.

'Screening for Al toxicity tolerance, per se, is accomplished
at the three Brazilian locations and the laboratory at El Batan;
the Cd. Obregon location is used to select for agronomic type and
rust resistance; the Toluca station is used primarily for
screening and selecting lines resistant to stripe rust, leaf rust,
Septoria tritici, Fusarium nivale and BYDV. The lines selected in
Cd. Obregon are divided into five sets: three are sent to Brazil,
one goes to the aluminum screening laboratory at El Batan, and one
is planted in Toluca. Results from Brazil regarding the resistance
or susceptibility of these lines are transmitted by telex to
CIMMYT. This information, as well as that coming from the
laboratory, is used to select Al resistant plants in Toluca under
field conditions. The selected progenies are then planted in Cd.
Obregon during the next cycle and a subset is sent to the
laboratory for Al screening. The results are used to again select
progenies in the field. This procedure continues until homogeneity
of progenies is achieved.

Research results:

Since the beginning of the program, many advanced lines have
been produced that combine the Al tolerance of Brazilian wheats
with the high yield potential of CIMMYT's semidwarfs. A
considerable number of these wheats also appear to possess a broad
spectrum of resistance to a complex of leaf spotting diseases,

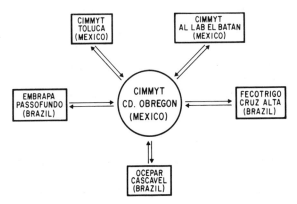

Fig. 6. Aluminum shuttle breeding scheme.

including <u>Fusarium</u> <u>nivale</u>, <u>Septoria</u> <u>tritici</u> and <u>Helminthosporium</u> <u>tritici</u> <u>repentis</u>. Two crosses can be discussed that appear to represent a significant yield improvement combined with an acceptable level of aluminum tolerance.

<u>Alondra "S"</u>--This is a high-yielding semidwarf moderately tolerant to Al toxicity under field conditions. It is derived from the cross D6301 - Nainari 60 x Weique - Red Mace/CIANO 67^Z - Chris (CM 11683). Alondra's tolerance to aluminum is believed to be inherited from Weique-Red Mace, since the other parents are susceptible. Alondra possesses good agronomic type, high yield potential, and has been extensively used in crossing with Al tolerant Brazilian wheats. Many advanced lines involving crosses with Alondra have been produced with acceptable yield potential in Brazil. This variety has been released in Brazil as Alondra 4546.

<u>Thornbird "S"</u>--This advanced line is derived from a cross made by EMBRAPA, IAS 63 - Alondra x Gaboto - Lagoa Vermelha (F 11915), and cooperatively selected by CIMMYT, EMBRAPA and FECOTRIGO through the shuttle breeding program. In the advanced yield trials of EMBRAPA, it became obvious that this line possesses a higher yield potential than traditional varieties under Brazilian conditions (Ottoni Soza de Rosa, personal communication). Thornbird "S" has yielded up to 4.5 tons/ha in Passo Fundo, compared to 2.5 tons for Maringa, a variety widely grown in Southern Brazil.

Breeding for Heat Tolerance

It is becoming clear that heat tolerance is an essential characteristic for wheats grown in certain areas of the world. Smith (1983) discusses the importance of heat tolerance in conjunction with drought resistance for the Great Plains area of the U.S.A. In collaboration with the national crop improvement programs of certain tropical to subtropical countries, CIMMYT has recently embarked on producing wheats with the necessary heat tolerance for warmer wheat production environments.

Varietal information on heat tolerance is scarce and CIMMYT has initiated a testing program to identify varieties and advanced lines having tolerance to high temperatures. In screening for heat tolerance, we are using one of our basic methodologies, that of multilocation international testing. The hub of this screening program is Cd. Obregon, where planting of heat tolerance nurseries is delayed until January so that the early growth stages occur during a period of rapidly rising temperatures (February, March and April). Traditionally, wheat is planted in Cd. Obregon during November and harvested in late April or early May.

In our screening for heat tolerance, the following characteristics are used as selection criteria:

- longer leaf retention,
- high tillering capacity,
- acceptable spike fertility,
- 1,000 grain weight,
- hectoliter test weight, and
- yield, per se.

Varietal differences exist for these characteristics and, if they can be combined successfully, then a favorable grain yield should result which could then be used as a heat tolerance index.

Our program is in the initial stage of heat tolerance screening and no crosses have been made to date for this purpose. We do plan to initiate breeding for heat tolerance within the next two years.

Conclusion

It 'is our strong belief that a balance among wide adaptation, yield stability and high yield potential is central to an international plant breeding program. These characteristics can be achieved through multilocation international testing and/or shuttle breeding, and a wise use of information from those areas for which the final product is intended. Once these core characteristics are developed and the central breeding scheme established, additional traits, such as resistance or tolerance to environmental stresses, quality characteristics and various agronomic attributes can be combined and recombined as necessary.

Durability of disease resistance is highly sought in widely adapted materials, and we believe that the incorporation of dilatory resistance (such as slow rusting) would prove rewarding. Our concept of drought resistance breeding results from the synthesis of work done in various other locations, and is designed to produce advanced lines and varieties that will yield well under both droughty and nondroughty conditions.

Acknowledgments

We thank Dr. Reynaldo Villareal and Mr. Tiff Harris for the help they provided in preparation of tables, graphs and the manuscript.

References

Beckman, I., 1976, Cultivation and breeding of wheat in south Brazil, in: "Proceedings of a Workshop on Plant Adaptation to Mineral Stress in Problem Soils," M.J. Wright, ed., Beltsville, Maryland.

Bingham, J., 1979, Wheat breeding objectives and prospects, Agricultural Progress, 1-17.

Bingham, J., Trends in wheat breeding, in: "Proceedings of the Australian Plant Breeding Conference," Adelaide, February 14-18, 1983, 1-15.

Borlaug, N.E., 1953, New approach to the breeding of wheat varieties resistant to Puccinia graminis tritici, Phytopathology, 43:467.

Borlaug, N.E., 1968, Wheat breeding and its impact on world food supply, in: "Proceedings of the Third International Wheat Genetics Symposium," K.W. Finley and K.W. Shephard, eds., 1-36, Australian Acadamy of Science, Canberra.

Borojevic, S., and Potocanac, J., 1966, The development of the Yugoslav programme for creating high yielding wheat varieties, Contemporary Agriculture, 11-12:7-36.

Browning, J.A., Simons, M.D., and Torres, E., 1977, Managing host genes: epidemiologic and genetic concepts, in: "Plant Disease: An Advanced Treatise, Vol. 1," J.G. Horsfall and E.B. Cowling, eds., 191-212, Academic Press, New York.

Caldwell, R.M., 1968, Breeding for general and/or specific plant disease resistance, in: "Proceedings of the Third International Wheat Genetics Symposium," K.W. Finley and K.W. Shephard, eds., 263-272, Austral. Acad. Sci., Canberra.

Dalrymple, D.G., 1980, "Development and spread of semi-dwarf varieties of wheat and rice in the United States: an international perspective," Agricultural Economic Report Number 455, USDA, Washington, D.C.

Dubin, H.J. and Rajaram, S., 1981, The strategy of the International Maize and Wheat Improvement Center (CIMMYT) for breeding disease resistant wheat: an international approach, in: "Strategies for the Control of Cereal Diseases," J.F. Jenkyn and R.T. Plumb, eds., 28-35. Blackwell Scientific Publication.

Fasoulas, A., 1973, "A new approach to breeding superior yielding varieties," Pub. 3, Aristotelian University of Thessaloniki, Greece.

Fasoulas, A., 1976, "Principles and methods of plant breeding," Pub. 6, Aristotelian University of Thessaloniki, Greece.

Hurd, E.A., 1969, A method of breeding for yield of wheat in
    semiarid climates, Euphytica, 18:217-226.

Johnson, V.A., and Schmidt, J.W., 1979, Role of classical breeding
    procedures in improvement of self-pollinated crops, Presented
    to: "The World Soybean Research Conference, II," Raleigh,
    North Carolina.

Joshi, A.B., 1979, Breeding methodology for autogamous crops, The
    Indian Journal of Genetics and Plant Breeding, 39:567-578.

Kronstad, W.E., 1983, Wheat breeding: a never-ending battle, in
    "Proceedings: Oregon State University Agricultural Conference
    Days," 46-54.

Pfeiffer, W., 1983, "Ertragsleistung und ertragsstabilitaet von
    commerweizen auf regionaler und globaler ebene - analyse
    einer serie von internationalen sortenversuchen ueber 15
    jahre und 973 umwelten," Ph.D. thesis, Stuttgart-Hobenheim.

Qualset, C.O., and Vogt, H.E., 1980, Efficient method of population
    management and utilization in wheat breeding, in: "Proceedings
    of the International Wheat Conference, Madrid, Spain,"166-188.

Quizenberry, J.E., 1981, Breeding for drought resistance and plant
    water use efficiency, in: "Breeding Plants for Less Favorable
    Environments," M.N. Christiansen and C.F. Lewis, eds.,
    198-212, Wily-Inerscience, New York.

Rajaram, S., and Dubin, H.J., 1977, Avoiding genetic vulnerability
    in semidwarf wheats, Annals of the New York Academy of
    Science, 287:243-254.

Rajaram, S., and Luig, N.H., 1972, The genetic basis for low
    coefficients of infection to stem rust in common wheat,
    Euphytica, 21:363-376.

Rajaram, S., and Nelson, W.L., 1982, Wheat production systems in
    arid and semiarid regions, Presented to: "Conference of
    International School of Agriculture," Royal Dublin Society,
    Ireland.

Rajaram, S., and Torres, E., 1981, Breeding for disease resistance
    in wheat, Presented to: "Symposium, American Society of
    Agronomy," Atlanta, Georgia, USA.

Rao, M.V., 1980, A coordinated approach to crop improvement:
    India's experience with wheat, in: "Proceedings of the
    International Wheat Conference, Madrid, Spain," 68-82.

Schmidt, J.W., 1980, Breeding systems for rainfed wheat
    improvement in semiarid regions, in: "Proceedings of the
    International Wheat Conference, Madrid, Spain," 30-37.

Smith, E.L., 1982, Heat and drought tolerant wheats of the
    future, in: "Proceedings of the National Wheat Research
    Conference, Beltsville, Maryland," 141-147.

Torres, E., and Rajaram, S., 1983. International testing as a
    means to implement durable resistance, in: "Durable
    Resistance in Crops," L. Lamberti, J.M. Waller and N.A.
    Vander Graaff, eds., 363-367, Plenum Publishing Coorporation.

Van Wambeke, A., 1976, Formation, distribution and conseqences of
    acid soils in agricultural development, in: "Proceedings of a
    Workshop on Plant Adaptation to Mineral Stress in Problem
    Soils," M.J. Wright, ed. 15-24. Beltsville, Maryland.

Worrall, W.D., Scott, N.H., Klatt, A.R., and Rajaram, S., 1980,
    Performance of CIMMYT wheat germplasm in optimum and
    suboptimum production environments, in: "Proceedings of the
    International Wheat Conference, Madrid, Spain," 5-29.

IRRI BREEDING PROGRAM AND ITS WORLDWIDE IMPACT ON INCREASING

RICE PRODUCTION

Gurdev S. Khush

Plant Breeder and Head, Plant Breeding Department

International Rice Research Institute, Los Baños
Laguna, Philippines

INTRODUCTION

Rice, Oryza ssp., is the world's most important food crop. It was grown on 145 million hectares which yielded 435 million tons of grain in 1982. It is the principal food of more than half of the world's population. Although rice is grown in about 90 countries spread over 6 continents, more than 92% of rice is produced and consumed in Asia. Yields of rice are disappointingly lower in the less developed tropical regions, where most of the rice is grown, as compared with yields from temperate areas. In tropical Asia, for example, the average yield is 2.0 t/ha as compared with average yields of more than 5 t/ha obtained in such temperate zone developed countries as Japan, Korea, USA, Australia, Spain, Italy and others.

Rice is grown under a wide range of agroclimatic conditions and a broad array of natural enemies and adverse growing conditions that combine to reduce the yield of rice.

o It is grown across a gradient of water regimes ranging from upland hill slopes to a maximum water depth of about 5 meters in river deltas of Bangladesh, India, Thailand and Vietnam.

o It is produced on a wide array of soils and under a wide range of solar energy regimes.

o Numerous diseases and pests severely reduce rice yields in the tropics.

61

o   Suboptimum temperatures -- low temperatures during
    cooler months or high elevations and high temperatures
    in the drier climates -- reduce rice yields.

0   Grain characteristics which affect palatability, milling
    recovery, and appearance influence rice yields.

The rice breeding programs in the developing countries of
Asia, Africa and Latin America lack trained manpower, germplasm
resources, literature and access to modern crop improvement
technology.

The rice breeding program of the International Rice Research
Institute (IRRI) is one of the biggest crop improvement programs
in the world which endeavors to develop and help develop improved
germplasm for the diverse rice growing conditions. It is an
interdisciplinary team effort where plant breeders work with
scientists in other disciplines such as plant pathologists,
entomologists and agronomists in improving the rice plant in many
important attributes. It has five interrelated components as
follows:

(1) Germplasm collection, conservation and evaluation.
(2) Development of improved germplasm.
(3) Methodology development for crop improvement.
(4) Training and staff development.
(5) International cooperation in varietal improvement.

GERMPLASM COLLECTION, CONSERVATION AND EVALUATION

1. Collection

Germplasm collection work was started in 1961. Since then,
cultivated and primitive varieties of Oryza sativa as well as
African cultivated rice, O. glaberrima and wild species of Oryza
have been continuously collected. We have also assembled a
collection of genetic testers and mutants of rice. There are now
more than 70,000 entries in our germplasm bank. The largest
collection belongs to cultivated varieties of Oryza sativa
(>65,000). We estimate there are about 120,000 rice varieties
either grown or preserved in the various germplasm collection of
the national rice improvement programs. We are in the process of
collecting the seeds of remaining varieties through five year
collaborative collection arrangements with the national rice
improvement programs (IRRI, 1983). The IRRI collection includes
about 2600 accessions of O. glaberrima, 1100 wild taxa and more
than 690 genetic testers and mutants (Chang, 1983). Included in
the collection are seeds of improved varieties of post IR8 era as
well as many improved breeding lines with specific traits.

2. Preservation

Seeds of rice collections are preserved under three storage conditions:

(a) airconditioned room temperature and relative humidity of 45%. Seeds remain viable for up to five years.
(b) 4°C and relative humidity of 35%. Seeds are expected to remain viable for 40-50 years.
(c) -10°C and relative humidity of 35%. Seeds should stay viable for up to 100 years.

3. Evaluation

IRRI accessions are evaluated for 50 morphological traits during seed multiplication, for grain quality, for disease and insect resistance and for tolerance to unfavorable environments such as drought, floods, problem soils and adverse temperatures. In all, data on 92 traits is being recorded for each accession. Each year approximately 30,000-35,000 seed samples are supplied to IRRI scientists for evaluation purposes who identify donor parents for the traits to be incorporated into the breeding materials.

Because of the expanding scope of the germplasm collection and increasing volume of work, IRRI's germplasm bank was designated as a separate entity under the name of International Rice Germplasm Center (IRGC) in 1983.

DEVELOPMENT OF IMPROVED GERMPLASM

This is the biggest component of our program and major resources are allocated to develop germplasm with improved characteristics. Improvement is sought for various problem areas as well as for various cultural types.

1. Rice Improvement for Problem Areas

We develop improved germplasm for the following problem areas.

1. Agronomic characteristics
2. Grain quality
3. Disease resistance
4. Insect resistance
5. Nutritional content
6. Drought tolerance
7. Adverse soil tolerance
8. Deepwater and flood tolerance
9. Temperature tolerance

   Agronomic characteristics. Improvement of yield potential - a
most important agronomic trait - received major attention in the
earlier years. The traditional varieties of rice are tall, leafy
with weak stems, and have harvest index of about 0.3. When nitrogen
fertilizer is applied they grow even taller and lodge severely;
yield decreases instead of increasing.

   The improved varieties, on the other hand, are highly responsive
to the nitrogenous fertilizers as they are lodging resistant. They
are characterized by short stature (about 100 cm), have sturdy stems,
high tillering ability, and have dark green and erect leaves which
utilize solar energy efficiently. They produce greater amounts of
dry matter and have harvest index of 0.5. Therefore their yield
potential is much higher and they can produce yields of 8-9 tons/ha
under favorable growing conditions. The nitrogen response curve
for a tall tradutional variety (Peta) and an improved variety (IR8)
is shown in Figure 1. The development of improved plant type was a
major breakthrough. It doubled the yield potential of tropical
rice in one step. This was brought about by incorporation of a

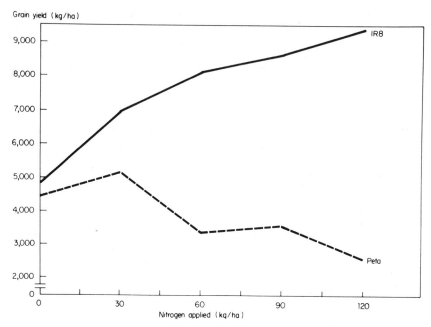

Fig. 1. Nitrogen response of Peta and
         IR8 during 1966 dry season at
         IRRI

single recessive gene, sd 1, from variety Dee-geo-woo-gen which
shortens the plant stature. It appears that very closely linked
to this dwarfing gene is a gene or a block of genes which confer
high tillering ability and erect leaves as these traits segregate
together with the short stature. The merit of this improved plant
type convinced most of the national rice improvement programs to
immediately initiate crossing programs to develop varieties with
short stature.

Since this initial jump, the improvement of yield potential
has been modest, as our resources were utilized in solving other
pressing problems of rice improvement to be discussed shortly.
However, we believe that further increases in yield potential are
possible and we are now devoting some resources to raise the yield
potential further. However, it is obvious that the yield potential
of available modern vareities far exceeds the average yields in
most of the developing countries.

Development of improved germplasm with shorter growth duration
has received major attention in our program. Most of the traditional
varieties grown in tropical and subtropical Asia matured in 160-170
days and many were photoperiod sensitive. IR8 and subsequent
improved varieties such as IR20, IR24 and IR26 are photoperiod insensitive and mature in 130-135 days. The growth duration was
reduced still further in IR28, IR30 and IR36 (Table 1) which
mature in 110 days. Because of shorter growth duration, high yield
and high yield stability, IR36 has been accepted very widely in
rice growing countries of Asia. Its widespread acceptance has led
to increased cropping intensity on Asian farms. It is now possible
to grow two crops of rice where only one was possible before or
farmers who plant IR36 can grow an additional upland crop after
rice. Efforts are continuing to develop materials with even
shorter growth duration. Recently released IR58 matures in 100
days only. Under optimum conditions the maximum yield potential of
the short duration varieties is comparable or only slightly lower
than the varieties with medium growth duration (130-135 days), but
as shown in table 2, their per day productivity is much higher.

Grain Quality. Grain quality determines, to a large extent,
the market price and consumer acceptability. Preferences for grain
shape, grain size and eating quality frequently vary. The physical
properties (size, shape, and translucency of grain) influence the
milling recovery. The cooking quality is determined by the physico-
chemical properties of the starch. Starch fraction, which constitutes 90% of the grain, is a polymer of glucose. The proportions
of its linear fraction (amylose) and the branched fraction (amylo-
pectin) and its gelatinization temperature determine in part, the
eating quality of rice.

Table 1. Main characteristics of IR varieties.

Variety	Recommended for	Growth duration (days)	Height (cm)	Amylose content	Gelatinization temperature	Grain size and appearance
IR5	Rainfed	140	130	High	Intermediate	Medium long bold
IR8	Irrigated	130	100	High	Low	Long bold
IR20	Irrigated	125	110	High	Intermediate	Medium long slender
IR22	Irrigated	125	90	High	Low	Long slender
IR24	Irrigated	120	90	Low	Low	Long slender
IR26	Irrigated	130	100	High	Low	Medium long slender
IR28	Irrigated	105	100	High	Low	Long slender
IR29	Irrigated	115	100	Glutinous	Low	Long slender
IR30	Irrigated	110	100	High	Intermediate	Medium long slender
IR32	Irrigated	140	105	High	Intermediate	Long slender
IR34	Irrigated	130	125	High	Low	Long slender
IR36	Irrigated	110	85	High	Intermediate	Long slender
IR38	Irrigated	125	100	High	Intermediate	Long slender
IR40	Irrigated	120	100	High	Intermediate	Medium long slender
IR42	Irrigated	135	110	High	Low	Medium long slender
IR43	Upland	125	110	Low	Low	Long slender
IR44	Irrigated	130	110	High	Low	Long slender
IR45	Upland	125	100	High	Intermediate	Long slender
IR46	Rainfed	130	110	High	Intermediate	Long slender
IR48	Irrigated	140	120	Inter-mediate	Low	Long slender
IR50	Irrigated	105	90	High	Intermediate	Long slender
IR52	Rainfed	115	95	High	Low	Long slender
IR54	Irrigated	120	95	High	Low	Long slender
IR56	Irrigated	110	90	High	Low	Long slender
IR58	Irrigated	100	80	High	Low	Medium long slender
IR60	Irrigated	108	95	High	Low	Long slender

Table 2. Yield of promising early maturing lines evaluated at
          IRRI during 1981 Dry and Wet seasons.

Selection	Growth duration (days)	1981 D. S.		1981 W. S.	
		Total yield (t/ha)	Yield per day (kg)	Total yield (t/ha)	Yield per day (kg)
IR8455-78-1-3-3	100	6.2	79.5	4.6	59.9
IR9729-67-3	100	7.2	92.3	5.1	65.4
IR9752-71-3-2	98	7.5	98.7	4.7	61.8
IR15429-268-1-2-1	97	6.8	90.7	5.1	68.0
IR19729-5-1-1-3-2	97	6.1	81.3	4.3	57.3
IR19735-5-2-3-2-1	100	6.5	83.3	4.9	62.8
IR19743-25-2-2-3-1	96	6.4	86.5	4.6	62.1
IR19743-40-3-3-2-3	97	5.8	77.3	4.6	61.3
IR19746-28-2-2-3	97	6.0	80.0	4.5	60.0
IR36 (check)	108	6.9	80.2	4.7	54.6
IR42 (check)	135	5.9	52.2	4.7	41.6

The majority of rice varieties grown in tropical Asia have
high amylose content (>25%) and cook dry and fluffy. However, the
preferred rice varieties have an intermediate level of amylose
(20-23%); they cook moist and remain soft when cool. Varieties
with an intermediate amylose content should have universal
acceptance in areas where indica rices are grown. All the japonica
varieties have a low amylose content and the consumer in temperate
areas where japonicas are grown almost exclusively prefer low
amylose varieties which are soft cooking.

A great majority of IR varieties have high amylose (Table 1).
Only IR24 and IR43 have low amylose, IR29 is glutinous and IR48
is the only IR variety which has intermediate amylose. However, it
is late maturing and has not been adopted widely by the Asian
farmers. We have emphasized the development of improved germplasm
with intermediate amylose and in our replicated yield trials
conducted during the dry and wet seasons of 1983, more than 50% of
the entries had intermediate amylose content. It is hoped that in
the near future many improved varieties with intermediate amylose
content will become available to the rice farmers of the world.

Disease and insect resistance. Rice plant is subject to the
attack of numerous diseases and insects. The most common diseases
are blast, sheath blight, bacterial blight, tungro, grassy stunt,
and ragged stunt and the most important insects are brown planthopper, green leafhopper, stem borers, and gall midge. Varietal
resistance is essential for yield stability. Therefore, major

resources in our program have been allocated for incorporating genes
for disease and insect resistance. We have developed improved germplasm with multiple resistance to as many as five diseases and four
insects. For example, IR36 is resistant to blast, bacterial blight,
tungro, brown planthopper, green leafhopper, stemborers, and gall
midge. Several other IR varieties (Table 3) and numerous breeding
lines with multiple resistance to major diseases and insects have
been developed and we continue to emphasize the incorporation of
diverse genes for resistance into our improved germplasm. We have
not found good sources of resistance to sheath blight but the search
continues. The importance of multiple resistance to diseases and
insects in imparting yield stability is indicated in figure 2. The
yields of IR8, a susceptible variety, fluctuate greatly due to
pressure of diseases and insects, whereas the yields of multiple
resistant IR36 and IR42 show little variation from year to year.

Nutritional content. The protein content of brown rice varies
from 7.5 to 8.5%. We screened over 26,000 entries of the germplasm
bank and the brown rice protein varied from 4.3 to 18.2%. A modest

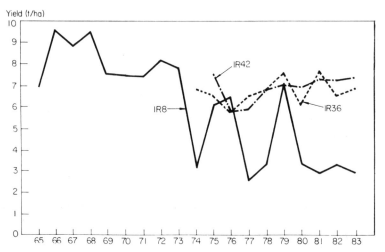

Fig. 2.   Yields of IR8, IR36, IR42 in dry season
replicated yield trials at IRRI.

Table 3. Disease and insect reactions[1] of IR varieties in the Philippines.

Variety	Blast	Bacterial blight	Grassy stunt	Tungro	BPH[2] Biotypes 1	2	3	Green leaf-hopper	Stem borer	Gall midge
IR5	MR	S	S	S	S	S	S	R	MS	S
IR8	S	S	S	S	S	S	S	MR	S	S
IR20	MR	R	S	MR	S	S	S	R	MR	S
IR22	S	R	S	S	S	S	S	S	S	S
IR24	S	S	S	S	S	S	S	R	S	S
IR26	MR	R	S	MR	R	S	S	R	MR	S
IR28	R	R	R	R	R	S	R	R	MR	S
IR29	R	R	R	R	R	S	R	R	MR	S
IR30	MS	R	R	R	R	S	R	R	MR	S
IR32	MR	R	R	R	R	S	R	R	MR	S
IR34	R	R	R	R	R	S	R	R	MR	S
IR36	R	R	R	R	R	R	S	R	MR	R
IR38	R	R	R	R	R	R	S	R	MR	R
IR40	R	R	R	R	R	R	S	R	MR	R
IR42	R	R	R	R	R	R	S	R	MR	R
IR44	MR	R	S	R	R	R	S	R	MR	S
IR46	R	R	S	R	R	S	R	R	MR	S
IR48	MR	R	R	R	R	R	S	R	MR	S
IR50	MS	R	R	R	R	R	S	R	MR	S
IR52	MR	R	R	R	R	R	S	R	MR	–
IR54	R	R	R	R	R	R	S	R	MR	–
IR56	R	R	R	R	R	R	R	R	MR	–
IR58	R	R	R	R	R	R	S	R	MR	–
IR60	R	R	R	R	R	R	R	R	MR	–

1/ R = Resistant, MR = Moderately resistant, S = Susceptible.   2/ BPH = Brown planthopper

breeding program to raise the protein content of improved rice
varieties was started in 1969 but due to extremely low heritability
of this trait we could not make any progress. Because of the
uncertainty in payoff in breeding for protein against the urgency
of breeding for yield, pest resistance and shorter growth duration,
breeding for protein content was relegated to low priority. Now
we just analyze our replicated yield trial entries for protein
content and those with low scores are excluded from further evaluation as varietal possibilities. However, we have found that our
short duration varieties are more or less equal in yield to those
of longer duration, but have consistently higher protein content.
For example, protein content of early maturing IR36, now grown on
an estimated 11 million hectares worldwide and IR58 which is even
earlier than IR36, is 1% higher than that of IR8 and IR42, which
take 25-30 days longer to mature (Table 4).

Drought tolerance. Drought tolerance is essential for stable
yields in nearly all rice growing environments which do not have
dependable irrigation. Major differences have been noted among
rice varieties for drought tolerance. This trait is being incorporated into the germplasm being developed for upland and drought
prone rainfed lowland situations. Even the varieties adapted to
deepwater areas grow under moisture stress conditions during early
stages of growth and need drought tolerance. IR36 and IR42, which
have been recommended for irrigated situations, have moderate level
of drought tolerance.

Table 4. Grain yield and brown rice protein content of four rices
varying in growth duration, from replicated yield trials,
1978-82 dry seasons.

Variety or line	Growth duration (days)	1978	1979	1980	1982	Mean
		Brown rice proteint (%)				
IR42	131	8.2	8.3	8.8	7.8	8.3
IR8	124	7.1	7.3	7.7	7.7	7.4
IR36	110	8.6	8.5	10.2	8.8	9.0
IR58	104	8.8	8.6	10.2	9.2	9.2
		Grain yield (t/ha)				
IR42	131	6.5	6.6	6.4	6.7	6.5
IR8	124	3.4	7.2	3.4	5.4	4.9
IR36	110	6.0	6.9	5.5	6.3	6.2
IR58	104	5.8	6.9	5.7	6.3	6.2

Adverse soil tolerance. Millions of hectares of land which
are otherwise suitable for growing rice remain uncultivated because
of soil toxicities caused by salt, alkali, acid, or excess organic
matter. Additionally, vast areas exist where deficiencies of zinc,
phosphorus, and iron and excesses of iron, aluminum, and manganese
limit rice yields. Entries from the germplasm collection have been
screened and sources of tolerance to various soil stress conditions
have been identified. These donors have been utilized in the
hybridization program and improved materials with multiple stress
tolerance have been developed. In addition, all the elite breeding
lines are evaluated for tolerance to soil stresses and several of
them, even though not specifically bred for tolerance, turn out to
have tolerance to several problem soils. IR36 and IR42 are two
varieties which have good level of tolerance to several problem
soils (Table 5) although no deliberate attempt was made to incorporate these traits into these varieties.

Deepwater and flood tolerance. In the low lying areas of large
river deltas in Asia, water stagnates for periods of 2-4 months
and water depth may vary from 50 cm to over 3 or 4 meters. The only
crop which can be grown under such conditions is rice. Rice varieties adapted to these conditions have the ability to elongate with
the rising water level and stay above water. We are developing
improved germplasm with elongation ability and other desirable
traits such as photoperiod sensitivity for the deepwater areas.
For the areas where flash floods occur to submerge the crop for a
few days and then recede, we have identified some primitive varieties
of rice such as FR13A which can be submerged for 10-12 days and
still survive. We are incorporating this trait into improved
germplasm being developed for the submergence prone areas.

Temperature tolerance. Rices grown in the temperate areas or
those grown at higher elevations must have cold tolerance. Rice
grown during off season such as boro rices of India and Bangladesh
are exposed to low temperatures. The growth stage at which rice
varieties are subjected to low temperature may be different depending upon the locality. Boro rices, for example, are exposed to low
temperature at early growth stages, but in the temperate areas, low
temperatures prevail both at the early vegetative stage and the
reproductive stage. We have identified numerous varieties with
good levels of cold tolerance at different stages and improved
materials with high yield potential and cold tolerance have been
developed. Such materials have been used extensively in Korea,
where rice production has risen sharply.

2. Rice Improvement for Different Cultural Types

Rice is grown under diverse ecological conditions where
cultural and production practices vary considerably and adapted
varieties have distinct characteristics. Rice growing environments

Table 5. Reaction of some IR varieties to adverse soil conditions on the scale 0-9[1].

| Variety | Wetland Soils | | | | | | | Dryland Soils | |
| | Salt | Alkali | Peat | Iron | Boron | Deficiencies | | Deficiency | Toxicities |
						Phos-phorous	Zinc	Iron	Aluminum and manganese
IR5	4	7	0	6	4	5	5	4	5
IR8	3	6	5	7	4	4	4	4	4
IR20	5	7	4	2	4	1	3	4	4
IR28	7	5	6	4	4	3	5	6	5
IR36	3	3	3	3	3	7	2	2	4
IR42	3	4	5	3	2	3	4	6	5
IR48	4	7	5	6	0	5	5	4	3

[1]/ 0: no information; 1: almost normal plant; 9: almost dead or dead plant.

have been classified into five major categories; irrigated, rainfed lowland, deep water, upland, and coastal wetlands. These major categories have been divided into several subcategories as shown in table 6. The table also shows the area under each major category of rice in the world. The world wide distribution of area according to different cultural types is shown in Figure 3. By far the largest area (53%) is under irrigated conditions. These are the most favored environments and productivity is high. More than 75% of the worlds rice production comes from these favored environments. In the initial years of the institute, varietal improvement work was primarily focused on development of improved germplasm for irrigated conditions and that is where the major impact has been made.

Table 6. Variability in rice growing environments. Area of each major category is indicated in parentheses.

1. Irrigated (77 m. ha.)
   a. Irrigated with favorable temperature
   b. Irrigated, low temperature, tropical zone
   c. Irrigated, low temperature, temperate zone

2. Rainfed lowland (33 m. ha.)
   a. Rainfed shallow favorable
   b. Rainfed shallow drought prone
   c. Rainfed shallow drought and submergence prone
   d. Rainfed shallow submergence prone
   e. Rainfed medium-deep waterlogged

3. Deepwater (12 m. ha.)
   a. Deepwater (50 cm to 100 cm water depth)
   b. Very deepwater (100 cm water depth)

4. Upland (19 m. ha.)
   a. Upland with long growing season and favorable soil factors (LF)
   b. Upland with long growing season and unfavorable soil factors (LV)
   c. Upland with short growing season and favorable soil factors (SF)
   d. Upland with short growing season and unfavorable soil factors (SV)

5. Tidal wetlands (5 m. ha.)
   a. Tidal wetlands with perennially fresh water
   b. Tidal wetlands with seasonally or perennially saline water
   c. Tidal wetlands with acid sulfate soils
   d. Tidal wetlands with peat soils

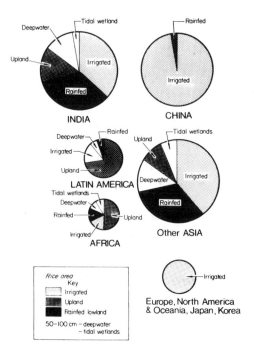

Fig. 3.   Area under different types of rice culture
          in different regions of the world.

Fig. 4.   Stepwise improvement in rice varieties. Right to
          left. Leb Mue Nahng, a tall traditional variety,
          is photoperiod sensitive, susceptible to diseases and
          insects and has harvest index of 0.3. Peta, a semi-
          improved variety is tall, photoperiod insensitive,
          matures in 155 days, is susceptible to diseases and
          insects and has harvest index of 0.35. IR8, an
          improved short statured variety matures in 135 days,
          is susceptible to diseases and insects and has harvest
          index of 0.5. IR36, an improved, short statured
          variety matures in 110 days, is resistant to diseases
          and insects and has harvest index of 0.5.

In the 1980's we face twofold challenges. For the favored
environments (irrigated and rainfed shallow favorable), we must
develop materials with higher yield potential, greater yield
stability, shorter growth duration, and superior grain quality.
We must also be prepared to face the threats from the new disease
or insect problems or the new biotypes and races of the existing
pests if and when they arise. For the unfavorable environments
(rainfed lowland, deepwater, upland and tidal wetlands), we must
develop rice varieties with higher yield potential while retaining
those traits responsible for their adaptability to these specific
ecosystems.

Further improvement of germplasm for favored environments.
During the last twenty years, the productivity advances in the
tropical rices have been brought about in several steps.

1. Development of photoperiod insensitive, nitrogen responsive,
    short statured varieties such as IR8 with sturdy stems, high
    tillering, and dark green leaves. These varieties have
    higher dry matter production and higher harvest index (0.5
    vs. 0.3 to 0.35 for the old varieties) and are capable of
    yielding 8-9 tons/ha under ideal conditions.

2. Incorporation of genes for multiple resistance to diseases
    and insects into improved plant type materials such as IR36
    and IR42 which have greater yield stability.

3. Development of shorter growth duration varieties with short
    stature and multiple resistance to disease and insects. In
    addition, these shorter duration varieties have very rapid
    growth rates and improved harvest index so that they are
    capable of producing high yields. This stepwise improvement
    is shown in Fig. 4.

As pointed out by Frey (1971) there is no evidence that a yield
plateau has been reached for any crop species. Yield in rice is a
function of total dry matter and harvest index. Therefore rice
yields can be increased beyond the present levels by increasing
total dry matter or harvest index or both. The studies on sink-
source relationship in rice have led rice physiologists to believe
that physical environment is not limiting to increasing rice yield
beyond the current levels. In order to raise the yield ceiling the
sink size of the crop should be increased by genetic manipulation
(developing genotypes with greater number of spikelets and or
larger grains) or by cultural practices (manipulating plant populations to obtain higher number of panicles per unit area). It
should also be feasible to increase the harvest index to 0.6

Thus, in the rice ideotypes believed to increase the yield
potential further, we must retain the desirable features of the

modern varieties but must also incorporate the following varietal traits:

(1)   Increased biomass production

For this purpose, we should select for:

(a)   Fast leaf area development
(b)   Low maintenance respiration
(c)   Adequate growth duration (about 120 days)

(2)   Increased sink size

(a)   Large spikelet number per shoot
(b)   Large grain size

(3)   Increased harvest index

It is obvious that the yield improvements in rice so far have been brought about through conventional breeding, primarily through improvement of plant type, and further improvement in their yield potential may be possible through increasing physiological efficiency of the improved plant type varieties. Physiological characteristics having a bearing on yield are related to source and sink size and harvest index. The inheritance of most of these is not understood clearly and genes controlling these traits and other desirable plant characteristics are likely to have desirable and undesirable linkage relationships. Breeding approaches that will help retain the desirable genetic linkages, but break the undesirable (repulsion phase) linkages and improve the selection efficiency in discriminating between high and low yielding genotypes would lead to development of higher yielding rice varieties. Therefore, we are adopting some non-conventional breeding methods such as heterosis breeding, population breeding, biparental and disruptive mating, anther culture, and single seed descent methods, aided by modified selection and evaluation procedures that should help meet these goals.

Improvement of germplasm for unfavorable environments. We started breeding programs for developing improved germplasm for upland conditions in 1970, for deep water situations in 1974, for rainfed lowland environments in 1976, and for tidal wetland in 1982. Thus we now have distinct breeding programs for each of the rice cultural types. It must be admitted however, that up to now we have not been able to develop improved germplasm with higher yield potential for most of the unfavorable environments. Some materials developed for irrigated culture such as IR42 have been found suitable for growing under rainfed lowland as well as tidal wetlands.

Therefore, while defining and developing breeding strategies for unfavorable environments, we are keeping the following points in mind.

o   Ideal high yielding plant types for the unfavorable
    environments have not been clearly defined.
o   Adaptability of the genotypes developed for the specific
    rainfed ecosystem is most important, hence it is essential
    that we retain as much as possible, the adaptability traits
    of locally adapted varieties.
o   Adaptability and productivity traits possess negative
    correlations, hence the problems of repulsion phase genetic
    linkages must be overcome.
o   Inheritance of varietal traits, such as drought tolerance,
    submergence tolerance, elongation ability, and tolerance to
    mineral deficiencies and toxicities needed in the varieties
    for unfavorable environments is complex and not fully known.
o   Selection for local adaptability can be done only by
    screening and evaluating the breeding materials in target
    environments, most of which do not occur in the Philippines.
o   Available screening procedures to select for tolerance to
    various stresses do not always give consistent results.

To expedite our efforts for developing improved germplasm for the unfavorable environments, we have taken the following steps.

o   We have assigned one plant breeder to develop improved
    germplasm for each of the five cultural types.
o   We are in the process of further defining and characterizing
    the unfavorable environments.
o   We have identified the representative sites for selection
    and evaluation of breeding materials.
o   We are using recurrent selection and population improvement
    methods utilizing genetic make sterility.
o   We have adopted shuttle breeding approach for evaluating
    the breeding materials under target environments.
o   We have developed facilities for handling selected crosses
    through Rapid Generation Advance (RGA) procedures.

We hope the strategies and approaches outlined above will help expedite the development of superior germplasm with higher yield potential for the unfavorable rice growing environments. This will help the millions of disadvantaged farmers located in these areas.

METHODOLOGY DEVELOPMENT FOR CROP IMPROVEMENT

IRRI scientists have developed or modified many procedures and methods for the rapid development and evaluation of improved germplasm. IRRI's trainees and visitors are exposed to these procedures

and new findings are also publicised through appropriate channels
such as research highlights, annual reports, International Rice
Research Newsletter (IRRN), IRRI Research Paper Series (IRPS), or
journal articles.

Some of the salient contributions are listed below.

(1)   Development of screening techniques for evaluating large
      number of germplasm entries to identify the donor parents
      for disease and insect resistance and for tolerance to
      various environmental stresses such as drought, floods, low
      temperature, and problem soils and adaptation of these
      techniques for evaluating breeding materials (Khush, 1977).
(2)   Development of procedures for incorporating multiple disease
      and insect resistance into improved germplasm (Khush, 1978).
(3)   Studies on the inheritance of resistance to diseases and
      insects to identify diverse genes for resistance (Khush, 1980)
(4)   Design and modification of a vacuum emasculator for facili-
      tating hybridization work (Herrera and Coffman, 1974).
(5)   Development of procedures for handling large breeding
      nurseries (Jennings et al., 1979).
(6)   Research on hybrid rice to exploit the phenomenon of heterosis
      for raising the yield potential. Our results indicate that
      significant heterosis exists in rice and we are developing the
      hybrid rice technology to exploit the heterosis (Virmani et
      al., 1981; Virmani et al., 1982).
(7)   Use of recurrent selection and population improvement using
      genetic male sterility to improve the quantitatively inheri-
      ted traits. Recurrent selection is used primarily to
      promote recombination and to increase the frequencies of
      favorable genes for quantitatively inherited traits. Although
      this breeding approach has been extensively used in cross
      pollinated crops, its principles are equally valid for auto-
      gamous crops (Hallauer, 1981). We have induced monogenic
      recessive male sterile mutants of rice (Singh and Ikehashi,
      1981) which are being utilized to overcome the problems of
      intermating in the recurrent selection programs (Chaudhary
      et al., 1981).
(8)   Use of biparental and disruptive mating schemes to improve
      specific traits. Biparental mating involves intermating of
      selected plants in F_2 so as to accumulate favorable genes and
      to break linkages, thereby releasing a greater reservoir of
      genetic variability to enable the breeder to exercise
      selection (Joshi, 1979). Such intermating has been found to
      push the population means in F_3 and break the genetic
      linkages, especially in crosses showing heterosis in the F_1
      and least reduction in F_2. The latter indicates the existence
      of fixable gene effects and additive gene action.

Disruptive mating involves the intermating of unalikes in segregating generations. It leads to greater opportunity for crossing
over, which releases latent variation by breaking up the predominantly repulsion phase linkages (Thoday, 1960). Frey (1982)
proposed use of disruptive mating under a situation when a breeder
uses germplasm from exotic and especially wild or weedy sources and
faces the problem of linkage drag, whereby an undesirable allele
is brought into the breeding population because it is linked to a
favorable gene.

We have successfuly utilized biparental mating to develop
breeding lines with long stigmas, for our hybrid rice program. The
trait was introduced from a wild species of Oryza.

(9) Investigations on the feasiblity of utilizing anther culture
     to practice haploid breeding. We have not been able to utilize
     this technique because of problems of regeneration from the
     haploid calli, but the research continues (Zapata et al., 1983).

(10) Selection of useful mutants through somaclonal variation. We
     have selected very useful improved plant type variants from
     the somatic cell cultures of tall traditional salt tolerant
     rice varieties (Yoshida, unpublished).

(11) Single seed descent method of breeding has been successfully
     used at IRRI to breed photoperiod sensitive rices. For this
     purpose we have developed Rapid Generation Advance (RGA)
     facilities and the procedure for handling the breeding
     materials have been outlined by Ikehashi and HilleRisLambers
     (1977).

(12) For exploiting the wild germplasm resources, we have evaluated
     the collections of various wild species of Oryza and located
     useful genes. Resistance to grassy stunt from O. nivara was
     transferred to O. sativa by backcrossing and several grassy
     stunt resistant varieties were developed (Khush et al., 1977).
     We are now in the process of exploring the possibility of
     transferring genes for brown planthopper resistance from
     O. australiensis to cultivated.

(13) For developing the improved materials for the unfavorable
     environments which do not occur in the Philippines, we have
     adopted shuttle breeding approach. This involves making the
     crosses between appropriate parents and growing of F_1s at IRRI,
     planting of F2 populations under target environments at
     selected locations in collaborating countries and making the
     selections at maturity, growing of F3 progenies at IRRI and
     evaluating them for selected traits such as grain quality and
     disease or insect resistance, sending selected F4 lines back
     to the collaborators for evaluation under target environments.
     This approach is being followed for developing improved germ-
     plasm for rainfed lowland and deep water areas in Thailand,
     Bangladesh, and India (Khush, 1983).

TRAINING AND STAFF DEVELOPMENT

The training program for familiarizing rice breeders with the
latest methodologies and techniques of crop improvement was initiated in 1962. Since then 518 individuals from 38 countries have
participated in one of the several training activities. At present
we have four types of training programs to cater to the varied
training needs of rice growing countries.

1. Non-Degree Training Program

Rice breeders come to IRRI for periods varying from 3 to 18
months and work with rice breeders in the ongoing breeding programs.
They may participate in all the breeding operations or work in a
specialized program such as germplasm conservation or grain quality
evaluation. To date 112 individuals have participated in the non-
degree training programs.

2. GEU Training Program

This formalized training program was initiated in 1975. A
group of 25-30 young rice scientists participate in this four and
half month training course. It is generally held from February to
the middle of June to coincide with a rice cropping season.
Trainees are given formal lectures on techniques of crop improvement,
they participate in all the field operations and laboratory work
related to Genetic Evaluation and Utilization (GEU) of rice germ
plasm and make crosses for their own programs. During some years
we conducted two GEU courses. To date 340 trainees from 27 countries
have participated in this training program.

3. Degree Training Programs

Degree training programs leading to M.Sc and Ph.D programs
were initiated in cooperation with the nearby Los Baños campus of
the University of the Philippines. Now we have entered into
agreements with several other universities in the developing and
developed countries of the world for cooperative training. The
students take their course work at the University of the Philippines
or one of the other collaborating universities and conduct their
thesis research at IRRI under the supervision of one of the IRRI
plant breeders. To date 28 students have studied for master's and
17 for Ph.D. degrees at IRRI and 18 are now enrolled for the degree
programs.

4. Post Doctoral and Senior Research Fellowships Program

Young Ph.D.'s as well as experienced rice breeders working with
the national rice improvement programs spend one or two years at IRRI
working on specific projects related to rice improvement and partici-

pate in the ongoing rice breeding operations. To-date 23 rice
breeders have worked as Post-Doctoral or Senior Research Fellows
in rice breeding.

We feel it is extremely important for us to have well trained
cadres of national program scientists with whom we can cooperate on
well focused rice improvement activities. We do our best to accomo-
date the training needs of the rice growing countries. In spite of
this alumni network, developed over the last 22 years, we are
receiving an increasingly large number of requests for manpower
development.

INTERNATIONAL COOPERATION IN VARIETAL IMPROVEMENT

The main objective of the IRRI breeding program is to help the
national program scientists in developing improved varieties of
rice for varied growing conditions. Towards this end, rice
germplasm at various stages of development is supplied at the request
of rice scientists all over the world. Since 1962 the germplasm
bank has provided more than 91,000 seed packets of rice accessions
to researchers in more than 100 countries of the world.

More than 179,342 seed packets of IRRI breeding lines have been
supplied upon the request from 87 countries (Table 7) since 1963.
These include seeds of early generation segregating lines, advanced
generation fixed lines, and named varieties. Breeders in developing
countries frequently ask us to make crosses with specific objectives
for them. We make the crosses and send either the F_1 or F_2 seeds
to them. Many breeders ask for F_2 seeds of IRRI crosses which are
supplied freely. Many of our visitors and trainees request the
seeds of promising lines or early generation materials and take
the samples with them.

International nurseries provide an excellent mechanism for
exchange and distribution of germplasm internationally. First
international nursery, the International Rice Blast Nursery (IRBN)
was started in 1965. International Rice Yield Nursery (RYN) and
International Rice Observation Nursery (IRON) were initiated in
1972 and 1973, respectively. This program was expanded and organized into International Rice Testing Program (IRTP) in 1975. It
now composes, distributes and reports on 23 nurseries (Table 8)
which are grown in over 60 countries. In 1983, 1,195 sets of these
nurseries were sent to more than 700 collaborating scientists in
61 countries. More than 60,000 seed packets of IRRI breeding
materials are distributed each year through these nurseries. The
international nurseries also facilitate the exchange of germplasm
from one national program to another; for example, a good line
developed in India can be nominated to a specific nursery and
distributed to other countries. In this way Indonesian breeders
have the opportunity to receive improved breeding materials from

Table 7.   Number of seed samples sent from IRRI to other countries.

Year	Germplasm bank	Wild taxa	IRRI Breeding Lines	
			Directly	Through IRTP
1963	2296	155	–	–
1964	2355	221	2296	–
1965	1608	227	3033	–
1966	1052	233	6000	–
1967	1764	789	12400	–
1968	5000	–	12000	–
1969	5800	287	15998	–
1970	5660	91	N.A.	–
1971	2300	325	N.A.	–
1972	2500	610	6865	–
1973	9777	456	7618	–
1974	2603	167	11492	–
1975	3347	696	4729	50000
1976	5226	250	10365	42867
1977	4126	–	9220	36583
1978	7316	331	14948	69087
1979	3260	1086	16817	75677
1980	4142	N.A.	8240	68769
1981	9337	539	6771	68157
1982	7724	417	8124	53315
1983	9616	873	22426	65778

India and vice versa.

Seeds of breeding materials are also exchanged through collaborative research projects. For example, a collaborative project
aimed at determining the biotypes of brown planhopper involves the
exchange of differential varieties and selected resistant entries
between IRRI and several national rice improvement programs.

In summary, we share our germplasm at all stages of development freely with other rice scientists anywhere in the world. To
date, we have not denied even a single seed request. In the
earlier years, IRRI used to name varieties but because of our close
partnership role in varietal development with national rice improvement programs, we discontinued the policy of naming varieties in
1975. Evaluation of elite IRRI germplasm and naming of varieties
is the responsibility of the national programs who can release any
IRRI breeding line under any name by acknowledging its source. The
Government of the Philippines continues to release IRRI bred lines
under IR designation. Thus, eleven IR varieties were named by IRRI

Table 8.   International rice testing program nurseries for 1983.

		Trials
Nurseries for Target Environments		
Irrigated		
Yield	- IRYN-VE	International Rice Yield Nursery-Very Early
	IRYN-E	International Rice Yield Nursery-Early
	IRYN-M	International Rice Yield Nursery-Medium
Observational	- IRON	International Rice Observational Nursery
Rainfed		
Upland		
Yield	- IURYN-E	International Upland Rice Yield Nursery-Early
	IURYN-M	International Upalnd Rice Yield Nursery-Medium
Observational	- IURON	International Upland Rice Observational Nursery
Rainfed lowland		
Yield	- IRRSWYN	International Rainfed Rice Shallow Water Yield Nursery (50 cm water depth)
Observational	- IRRSWON	International Rainfed Rice Shallow Water Observational Nursery
	- IRDWON	International Rice Deep Water Observational Nursery (50-100 cm water depth)
	- IFRON	International Floating Rice Observational Nursery (100 cm water depth)
	- ITPRON	International Tide-Prone Rice Observational Nursery
Nurseries for Specific Stresses		
Temperature	- IRCTN	International Rice Cold Tolerance Nursery
Soil	IRSATON	International Rice Salinity and Alkalinity Tolerance Observational nursery
	Acid Upland	Acid Upland Screening Set
	Acid Lowland	Acid Lowland Screening Set
Diseases	- IRBN	International Rice Blast Nursery
	IRBBN	International Rice Bacterial Blight Nursery
	IRTN	International Rice Tungro Nursery
Insects	- IRBPHN	International Rice Brown Planthopper Nursery
	IRWBPHN	International Rice Whitebacked Planthopper Nursery
	IRSBN	International Rice Stemborer Nursery
	Rice Thrips	Rice Thrips Screening Set

between 1966 and 1975. Since then, 15 IR varieties (IR36 to IR60)
have been released by the Philippine Government. These IR varieties
have been recommended either under IR designation or under local
name by many rice growing countries.

In addition, numerous IRRI bred lines which perform well in
IRTP nurseries or local trials are promoted to national trials of
the collaborating countries. Many best performing entries are
recommended as varieties. To date, more than 100 IRRI bred lines
have been released as varieties in countries of Africa (Tables 9
and 10), Latin America (Table 11) and Asia (Table 12).

IRRI also provides short term consultancies in varietal
development area and IRRI breeders keep close contacts with their
national program counterparts. IRRI is requested to station expat-
riate rice breeders in some countries for a few years to organize
the local breeding programs.

Table 9. IR varieties recommended in Africa.

Variety	Countries where recommended
IR8	Benin, Senegal, Tanzania, Cameroon, Nigeria, Togo, Kenya, Zaire, Niger
IR5	Ivory Coast, Guinea Bissau, Upper Volta, Liberia, Nigeria, Uganda, Cameroon, Tanzania
IR20	Gambia, Guinea Bissau, Upper Volta, Nigeria
IR22	Gambia, Niger
IR24	Cameroon
IR28	Cameroon, Gambia
IR30	Nigeria
IR36	Central African Republic
IR42	Nigeria, Mali, Niger, Senegal, Ghana
IR46	Cameroon

Table 10. IRRI breeding lines recommended in Africa.

Breeding Lines	Country where recommended
IR269-26-3-3-3	Mali, Niger, Nigeria (Faro 26)
IR442-2-58	Benin, Guinea Bissau, Mali Senegal, Togo, Ghana
IR578-95-1-3	Ghana, Nigeria
IR579-48-1-2	Egypt, Kenya
IR627-1-31-4-3-7 (Faro 22)	Nigeria
IR655-79-2 (Faro 27)	Nigeria
IR790-35-5-3 (Faro 7)	Nigeria
IR1529-680-3	Benin, Ivory Coast, Upper Volta, Mali, Niger, Senegal, Togo, Gambia, Ghana
IR1561-228-3	Mauritania, Kenya, Egypt
IR1820-208-3	Ghana
IR2035-120-3	Nigeria
IR2061-288-3-9	Nigeria
IR2823-339-5-6	Senegal
IR4422-98-3-6-1	Liberia, Sierra Leone, Nigeria
IR2373-P 339-2	Ghana, Nigeria, Niger, Sierra Leone
IR2042-178-1	Togo, Benin
IR2061-522-6-9	Cameroon
IR7167-33-3-5	Cameroon
IR9129-K4	Cameroon

Table 11. IRRI varieties and breeding lines recommended in Latin
           America.

Variety or Breeding Line	Countries where recommended
IR8	Brasil, Colombia, Dominican Republic, Ecuador, Panama, Peru, Venezuela
IR8 (Milagro Filipino)	Mexico
IR22	Brazil, Colombia, Costa Rica, Nicaragua, Paraguay, Peru, Venezuela
IR22 (INIAP-2)	Ecuador
IR22 (Navolato A 71)	Mexico
IR930-31 (CICA 4)	Belize, Bolivia, Brazil, Colombia, Costa Rica, El Salvador, Guatemala, Honduras, Jamaica, Mexico, Nicaragua, Paraguay, Venezuela
IR930-31 (INIAP-6)	Ecuador
IR930-31 (Avance 72)	Dominican Republic
IR930-31-10 (Chancay)	Peru
IR930-2-6 (Naylamp)	Cuba, Peru
IR930-31-10 (IRGA 408)	Brazil
IR160-27-4 (Sinaloa A 68)	Mexico
IR442-2-50 (Huallanga)	Peru
IR442-2-58 (BR 2)	Brazil
IR579-48-1 (IR 100)	El Salvador, Nicaragua, Honduras
IR665-4-5-5 (IR665)	Brazil, Nicaragua
IR822-81-2 (CR 1113)	Costa Rica, El Salvador, Honduras, Nicaragua, Paraguay, Panama, Belize
IR837-20-3-6 (Bamoa A 75)	Mexico
IR837-46-2 (Piedras Negras A74)	Mexico
IR841-63-5 (IR841)	Argentina, Brazil
IR1052	Guyana
IR1055 (N)	Guyana
IR1529-430-3 (IR1529)	Bolivia, Cuba, Mexico
IR2053-206-1	Bolivia
IR2053-205-2-6-3	Mexico
IR1545-339-2-2	Ecuador
IR11248-52-2-3-3	Brazil
IR8208-146-1 (Pesagro 101)	Brazil
IR2058-78-1-3-2-3 (Pesagro 102)	Brazil

Table 12.   IRRI breeding lines and varieties recommended in Asia.

Country	Breeding Lines/Varieties
Philippines	IR5 to IR60 (26 varieties in all)
Indonesia	IR5, IR8, IR20, IR26, IR28, IR30, IR34, IR36, IR38, IR42, IR50, IR52, IR54, IR56, IR2071-621-2, IR2307-247-2-2-3
Vietnam	IR5, IR8, IR20, IR22, IR26, IR28, IR30, IR36, IR38, IR42, IR48, IR1561-228-3-3, IR1529-680-3-2, IR2070-734-3, IR2071-179-3, IR2307-247-2-3, IR2823-399-5-6, IR2797-115-3, IR9129-192-2-3-5, IR2070-199-3-6-6
India	IR8, IR20, IR22, IR24, IR28, IR30, IR34, IR36, IR50, IR5-114-3, IR579-48-1, IR579-97-2-2-1, IR442-2-50, IR442-2-58, IR442-2-24, IR665-79-2-4, IR1561-216-6
China	IR8, IR24, IR26, IR28, IR36, IR837-36, IR1561-228-3-3, IR2061-464-2-4-5
Bangladesh	IR5, IR8, IR20, IR28, IR532-1-176, IR272-4-1, IR2053-87-3-1, IR2071-199-3-6, IR2793-80-1, IR758-15-2
Burma	IR5, IR8, IR20, IR22, IR24, IR28, IR34, IR42, IR751-592, IR1529-680-3
Nepal	IR5, IR8, IR400-29-9, IR2061-628-1-6-4-3, IR2071-124-6-4, IR3941-4-Plp-2B, IR2298Plp-B3-2
Malaysia	IR5, IR8, IR5278, IR5-250
Kampuchea	IR36, IR42
Pakistan	IR8, IR6-156-2, IR841-36-2
Laos	IR36, IR38, IR253-4
Thailand	IR253-4
Sri Lanka	IR262-43-8
Korea	IR667-98
Iraq	IR8, IR26

BREEDING OPERATIONS

Of the five rice breeders, one handles the breeding program for irrigated rice, one works on upland rice, one is assigned to rainfed lowland rice, one is in charge of the breeding work for deepwater and tidal wetlands. The fifth plant breeder works on hybrid rice and coordinates the work on other innovative breeding methods. One geneticist is in charge of operations related to germplasm collection, conservation, and distribution. Two scientists handle the international nurseries.

Each season's hybridization block is composed of materials nominated by different breeders. All the crosses are made at one place and all the F1 hybrids are grown in one common F1 nursery. However, from F2 onward nurseries are grown under appropriate pressure. For example, breeding nurseries for irrigated conditions, particularly the segregating materials are managed to expose them to maximum disease and insect pressures. Nurseries for upland and rainfed lowland drought prone areas are grown under moisture stress conditions. The yield trials are also conducted under respective environmental conditions. However, seed increase and breeder seed production of named varieties and elite breeding lines is carried out together.

In addition to field evaluation for various stresses, breeding materials are screened in the greenhouses by plant pathologists and entomologists for disease and insect resistance. Agronomists and soil scientists evaluate the materials for tolerance to drought and problem soils and plant physiologists handle the screening work for cold tolerance. Thus we maintain close liaison with scientists in other disciplines.

We make 2000-2500 crosses per season or about 4000-5000 crosses a year (Fig. 5). Many of these crosses are made at the request of our collaborating scientists and we send them the F2 seed. Thus, one breeder may make 700-800 crosses a year for his program. We make a large number of topcrosses using the single cross F1's as female parents. F2 populations from many of the single cross F1's are never grown. Rather, F2 populations from the multiple crosses in which the single cross F1 was the parent are grown instead. The F2 populations from multiple crosses have greater diversity than the F2 progenies of single crosses. Moreover, it is advisable to resort to multiple crosses if traits from more than 2 parents are to be combined together.

We grow F2 populations from 400 to 600 crosses a year (Fig. 5) and make about forty to fifty thousand plant selections. The number of pedigree nurseries rows grown each year has varied between 60,000 and 120,000 a year (Fig. 5). Approximately 800 breeding lines a year

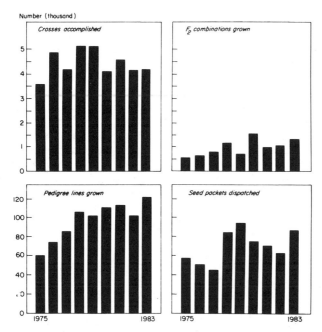

Fig. 5.   Volume of IRRI breeding program.

are evaluated for yield potential in replicated yield trials.
Promising lines are evaluated in nationwide coordinated trials and
are evaluated in international nurseries. In 1983, approximately
1,600 breeding lines and IR varieties were included in IRTP
nurseries. More than 22,426 seed samples of IRRI breeding lines
were sent abroad on the basis of requests from national program
scientists. We send between 50,000 and 80,000 seed packets of IR
varieties and IRRI breeding lines abroad through various exchange
mechanisms discussed in an earlier section (Fig. 5 and Table 7).

WORLDWIDE IMPACT OF IRRI BREEDING PROGRAM

     It will not be an exaggeration to say that IRRI breeding
program has been a pace-setter for other rice breeding programs,
particularly those in the developing parts of the world. As soon
as the first improved plant type variety for the tropics was
released by IRRI in 1966, most of the national programs started
developing the materials with similar plant type. When IRRI
released disease and insect resistant varieties, national programs
organized the host resistance programs. There is hardly any rice
breeding station in the developing world where one would not find
an IRRI trained rice breeder. These IRRI trained breeders use IRRI
developed methodology and use IRRI developed materials for testing

or as parents in their hybridization programs.

Twenty-six IR varieties are now planted to vast acreages in
countries of Asia, Africa and Latin America. In addition, national
programs have released more than 100 varieties from the IRRI bred
lines (Tables 9-12). The IR varieties and the national varieties
developed from IRRI bred materials are now planted to over 30
million hectares of rice land in the world (Herdt and Capule, 1983).
These varieties generate an additional income of 3-4 billion dollars
a year for the rice farmers. IR36 alone is planted to about 11
million hectares of rice land every year. No other variety of any
other crop has been planted that widely before. A quinquennial
review team of 12 outside experts, which evaluated the research
and training programs of IRRI in 1982, pointed out, "The impact of
IR36 alone would more than justify the investment in IRRI since its
establishment 21 years ago."

A recent survey of the 370 improved varieties released by
national programs in 36 countries revealed that 74% were semidwarfs
and a third of these were IRRI bred lines. In the ancestry of
almost 70% of 183 locally bred semidwarfs, an IRRI bred line was
the parent (Hargrove and Cabanilla, 1980). This shows how heavily
IRRI bred materials are utilized either as direct introductions or
in the breeding programs as parents all over the world.

In recent years, Chinese scientists have developed hybrid rice
varieties which are now planted to about 7 million hectares in that
country. The restorer parents of almost 90% of these hybrids are
IRRI bred varieties or breeding lines (Lin and Yuan, 1980). IR26
is the most important restorer used in China.

As a result of widespread adoption of improved varieties and
technology, most of the countries of Asia have become self sufficient in rice. Our host country, the Philippines, used to import
rice until 1977. It became self sufficient in 1978 and started
exporting rice from 1979. Eighty-four percent of the Philippine
rice area is now planted to improved varieties and almost all of
that to IR varieties. The progress in increasing the rice production in the Philippines is shown in Figure 6.

Similary, Indonesia used to import upwards of 2 million tons
of rice a year up to 1981. Its rice production started going up
rapidly (Fig. 7) after the wide scale adoption of improved varieties,
particularly IR36 in 1977. Fully 70% of the rice area in Indonesia
is now planted to improved varieties and most of these are IRRI
bred materials. Indonesia became self sufficient in rice in 1983.

The impact of improved varieties and technology have been
equally dramatic in other countries as well. Large areas are now
planted to improved varieties and in almost all the major rice

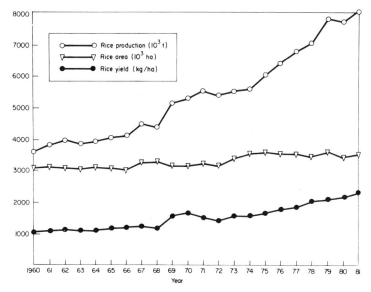

Fig. 6.   Paddy production, area and yield,
          Philippines. 1960-1981.

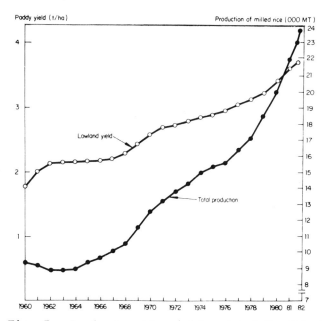

Fig. 7.   Lowland paddy yield and total milled
          rice production, 1960-1982.
          Indonesia.

growing countries, and in average yields have gone up greatly.
In some countries (Philippines, Indonesia, Burma, for example) the
yields have increased by as much as 70% since the introduction of
improved varieties.

SUMMARY

    The rice breeding program of IRRI is one of the biggest crop
improvement programs in the world. Its main components are:
germplasm collection, conservation, and evaluation; development of
improved germplasm; development of methods for crop improvement;
training and staff development; and international cooperation in
varietal improvement. We have collected seeds of more than 70,000
rice varieties, which are being preserved in our germplasm bank.
We are evaluating these entries for 50 morphoagronomic traits,
resistance to diseases and insects, and tolerance of physical and
biological stresses.

    We have developed improved germplasm capable of yielding two
to three times higher than the traditional varieties. Adaptability
traits such as multiple resistance to diseases and insects, and
tolerance of drought, problem soils, and floods have been incorporated into the improved germplasm. These improved materials have
been shared with national rice improvement programs who have
selected and released more than 100 varieties from these IRRI-
developed materials. These varieties are now planted on more than
30 million hectares of rice land in Asia, Africa, and Latin America.
Major yield increases have occurred in many countries as a result
of the introduction of these improved varieties.

    We have developed screening methods for evaluating germplasm
for resistance to various diseases and insects and for tolerance
of various stress conditions. Procedures for developing germplasm
for multiple resistance and stress tolerance have been developed.
We have studied the feasibility of application of innovative
breeding methods such as recurrent selection and population improvement, haploid breeding, single seed descent, hybrid breeding, wide
crosses, and shuttle breeding in rice improvement. We have trained
518 rice breeders from 38 rice growing countries in the techniques
of crop improvement.

REFERENCES

Chang, T. T., 1983, Global role of the International Rice Germplasm
    Center at the International Rice Research Institute, in: "1983
    Rice Germplasm Conservation Workspho." International Rice
    Research Institute, Los Baños, Philippines, pp. 57-65.
Chaudhary, R. C., Heinrich, E. A. and Khush, G. S., 1981, Increasing
    the level of stem borer resistance using male sterile
    facilitated recurrent selection in rice, Int. Rice Res. Newsl.,
    6(5):7-8.
Frey, K. J., 1971, Improving crop yields through plant breeding,
    in: "Moving off the yield plateau," Eastin J. D. and Munson
    R. D., eds., pp. 15-58, Am. Soc. Agron., Madison, Wis., U.S.A.
Frey, K. K., 1982, Breeding approaches for increasing crop yields,
    Paper presented at International Rice Research Conference,
    April 19-23, 1982, Int. Rice Res. Inst., Los Baños, Philippines.
Hallauer, A. R., 1981, Selections and breeding methods, in: "Plant
    Breeding II," Frey, K. J., ed., The Iowa State Univ. Press,
    Ames, Iowa, U.S.A.
Hargrove, T. R., and Cabanilla, V., 1980, Sources of semidwarfism
    in locally developed varieties, Int. Rice Res. Newsl.,
    5(1):3-4.
Herdt, R. W., and Capule, C., 1983, Adoption, spread and production
    impact of modern rice varieties in Asia, International Rice
    Research Institute, Los Baños, Philippines.
Herrera, R. M., and Coffman, W. R., 1975, Emasculation of rice by
    vacuum extraction, Proc. Crop Sci., Soc. Philippines, 5:12-14.
Ikehashi, H., and HilleRisLambers, D., 1977, Single seed descent
    with the use of rapid generation advance, Int. Rice Res.
    Conf., April 24-28, 1977, International Rice Research
    Institute, Los Baños, Philippines.
International Rice Research Institute, 1983, Rice Germplasm
    Conservation Workshop, Los Baños, Philippines.
Jennings, P. R., Coffman, W. F., and Kauffman, H. E., 1979, Rice
    Improvement, International Rice Research Institute, Los
    Baños, Philippines.
Joshi, A. B., 1979, Breeding methodology for autogamous crops,
    Ind. J. Genet. Plant Breed., 39:567-578.
Khush, G. S., 1977, Disease and insect resistance in rice, Adv.
    Agron., 29:265-341.
Khush, G. S., 1978, Breeding methods and procedures employed at
    IRRI for developing rice germplasm with multiple resistance
    to diseases and insects, Trop. Agri. Res. Series, 11:69-76.
Khush, G. S., 1980, Breeding rice for multiple disease and insect
    resistance, in: "Rice Improvement in China and other Asian
    Countries," pp. 220-237, International Rice Research Institute,
    Los Baños, Philippines.
Khush, G. S., 1983, International collaboration for improving
    rainfed lowland rice, in: "Proc. Workshop on Rainfed Lowland
    Rice," Rajendra Agri. Univ. Patna, India (In press).

Khush, G. S., Ling, K. C., Aquino, R. C. and Aguiero, M. V., 1977,
     Breeding for resistance to grassy stunt in rice, in: "Proc.
     3rd Int. Cong. SABRAO," Canberra, Australia, Plant Breeding
     Papers, 1:4(b):3-9.
Lin, S. C., and Yuan, L. P., 1980, Hybrid rice breeding in China,
     in: "Innovative Approaches to Rice Breeding," pp. 35-51,
     International Rice Research Institute, Los Baños, Philippines.
Singh, R. J. and Ikehashi, H., 1981, Monogenic male sterility in
     rice: induction, identification and inheritance, Crop Sci.,
     21:286-289.
Thoday, J. M., 1960, Effects of disruptive selection. III.
     Coupling and repulsion, Heredity, 14:35-39.
Virmani, S. S., Chaudhary, R. C. and Khush, G. S., 1981, Current
     outlook on hybrid rice, Oryza, 18:67-84.
Virmani, S. S., Aquino, R. C. and Khush, G. S., 1982, Heterosis
     breeding in rice (Oryza sativa L.), Theor. Appl. Genet.,
     63:373-380.
Zapata, F. J., Khush, G. S., Crill, J. P., Heu, M. H., Romero,
     R. O., Torrizo, L. B. and Alejar, M., 1983, Rice anther
     culture at IRRI, in: "Cell and Tissue Culture Techniques
     for Cereal Crop Improvement," pp. 27-46, Science Press,
     Beijing, China and International Rice Research Institute,
     Manila, Philippines.

IDEOTYPE RESEARCH AND PLANT BREEDING

Donald C. Rasmusson

Department of Agronomy and Plant Genetics
University of Minnesota
St. Paul, MN 55108

Crop yield is closely related to the genetic yield potential
of the cultivar. Evans (1983) defined genetic yield potential as
the yield of a cultivar grown in environments to which it is adapted,
with nutrients and water non-limiting, and with pests, diseases,
weeds and other stresses effectively controlled. In this context
the plant breeder may opt to enhance yield potential or breed to
reduce the impact of yield-limiting factors. It is not trivial to
ask whether breeding to enhance genetic yield potential affords an
opportunity equal to breeding to reduce yield-limiting factors.
Pertinent to the answer is the fact that yield gains have been made
in the major crops where breeding has been done over a period of
time using both approaches; and it is encouraging that yield gains
have been sizeable and sustained (Duvick, 1977; Austin et al.,
1980; Riggs et al., 1981; Wych and Rasmusson, 1983; Wych and Stuthman,
1983). We should take satisfaction in and applaud conventional
plant breeding.

Conventional breeding to enhance yield potential has, for the
most part, entailed yield testing to select winners after crossing
parental genotypes which were themselves yield trial winners. In
researching individual characters to enhance yield potential, which
we call ideotype research, we are searching for a way to manipulate
genes and the plant characters they control to enhance yield which
will effectively complement conventional yield breeding.

I do not wish to imply that ideotype breeding to enhance yield
potential is new or that it affords unique opportunities and should
become dominant to conventional breeding, but I do suggest a small
shift in emphasis in both the amount of attention devoted to
enhancing genetic yield potential and in the way it is done.

95

My presentation will deal with research I have done in cooperation with colleagues and graduate students who have worked on
the barley (Hordeum vulgare L.) breeding project at the University
of Minnesota. This report is not intended to provide an in depth
treatment of plant ideotype research It is intended to describe
our research which has had the dual objective of developing barleys
with increased genetic yield potential and obtaining information of
value to plant breeders and other plant researchers. Our experience in barley ideotype research should provide clues to what other
geneticists and breeders might expect in similar research.

IDEOTYPE CONCEPT

From time to time a special idea or research strategy emerges
that leads to a burst of enthusiasm and a sizeable research effort.
The concept of an ideal plant or an ideotype, as Donald (1968)
called it, was not really new; rather it was an effective elabora-
tion on an older idea accompanied by a new terminology. He suggested that there was a better way to breed for yield potential and
that greater progress would be achieved if a breeder had a goal
towards which he would direct his breeding effort. The suggested
goal was an ideotype or ideal plant described in terms of individual plant characters which would be yield promoting; such characters will be referred to as ideotype characters.

The ideotype concept is valuable for several important reasons.
It encourages plant breeders to think about yield and how it is
achieved. In part, because of the emergence of the ideotype concept, breeders think about capturing more light energy by altering
leaf area, leaf area duration and leaf angle. Discussion of subjects like yield component compensation and allometry are becoming
routine among substantial numbers of plant breeders. Similarly,
plant physiologists working alone and with plant geneticists and
breeders tend to have a sharper focus on enhancing yield and characters that affect yield because of the ideotype concept.

In contemplating a strategy of breeding for yield potential
via the ideotype approach, one major problem emerges. It is the
level of our understanding about plant growth and the role of
individual morphological and physiological characters in enhancing
yield, including the nature of character interrelationships.

The status of our current knowledge about morphological and
physiological characters and ideotype breeding depends on the
context within which the question arises. In one context, the
situation is encouraging. An impressive array of studies (Wallace
et al., 1972; Evans and Wardlaw, 1976; Wilson, 1981; Evans, 1983;
Rasmusson and Gengenbach, 1983) provide evidence on the behavior of
morphological and physiological characters and, for many of these
characters, their specific role in enhancing yield has been de-

scribed. Some plant physiologists and plant breeder-geneticists
can describe an ideal plant type and give good reasons for choosing
a specific combination of characters.

In another context, the situation is not so encouraging.
Most plant breeders are skeptical and believe there is insufficient information about which characters influence yield and the
optimum phenotype for these characters. Accordingly, conventional
plant breeders, myself included, are reluctant to emphasize individual characters or the several characters that might constitute
an ideotype, especially if it means a cutback in conventional
breeding. The lack of knowledge, as viewed by most plant breeders,
is emphasized in a forceful way if one attempts to assist a colleague in genetic engineering who is interested in yield and in
how to identify, locate, and then relocate yield genes. In this
context, we are in fact woefully short of needed information, not
only about which traits to improve, but about the genetics of
these characters.

If we accept the view that a lack of knowledge is the principal factor limiting ideotype breeding or breeding for individual
characters, it is reasonable to suggest that plant breeders and
geneticists commit time and resources to learn about the role of
individual characters in enhancing yield and doing breeding with
them. With the emergence of molecular biology and its implications for plant breeding, it is hard to over-emphasize the need
for additional genetic information.

With the premise that yield is the product of many individual
characters, two questions come to mind: (1) What characters afford
the greatest opportunity for enhancing yield; and (2) what character
interrelationships--resulting from compensation among yield components, allometric or character size relationships, and pleio-
tropy--exist that may enhance or hinder ideotype breeding?

Selecting Ideotype Characters

Research on whether or not a character enhances yield and
affords good opportunity for selection is seldom definitive.
Different results with various research methods, genetic backgrounds, and environments contribute to a continuing dilemma.
Nonetheless, past and on-going research make it possible to identify characters that are worthy of study and a plant breeding effort.
Breeders who require unequivocal experimental evidence before using
individual characters in an attempt to enhance yield potential may
not become involved, but if they will accept reasonable scientific
arguments, several characters emerge as candidates for a breeding
effort. It is well for breeders to keep in mind that methods of
conventional plant breeding have been tried, and have proven
themselves over time. We rarely have unequivocal experimental
evidence of their merit.

Table 1. Characters with potential for increasing grain yield
          by ideotype breeding in small grains.

Leaf Characters	Infloresence Characters
Leaf size	Number of spikes
Leaf angle	Awn length
Number of leaves	Kernel number
Leaf area duration	Kernel weight

Culm Characters	Root Characters
Number of culms	Density of penetration
Survival of culms	Rate of penetration
Diameter of culms	Volume
Vascular bundles	
Height	

Adaptation	Other
Photoperiod response	Biological yield
Duration of grain-filling	Harvest index
Maturity	

Table 2. Listing of correlated responses associated with
          selection for four ideotype characters in barley.

Character selected	Breeding cycles completed	Changes associated with selection for ideotype characters
High head number	8	Few and small kernels, small stems and leaves, lodging susceptibility
Erect leaf	5	Lodging resistance, late maturity, loss of flex in stem
Semidwarf (sdw gene)	10	Late maturity, small kernels, leaf rust susceptibility, poorer malting quality
Multiple awn	6	Few and small kernels

Promising ideotype characters have been identified and are being researched in various crops. Several characters, that are being utilized in ideotype breeding in small grains, are given in Table 1. A rationale can be presented for breeding for an optimum expression of these characters based on their effect on such factors as photosynthesis, transpiration, transport, storage capacity and efficiency of production. Obviously a definitive answer about the worth of each character is something for which we strive but rarely achieve.

Character Interrelationships

A challenging aspect of ideotype research or breeding is the complex interrelationships that exist among plant characters. Plant breeding literature is over-loaded with correlations testifying to the occurrence of associations among characters. Our experience in barley only serves to reinforce the observation that character associations are the rule, not the exception. Table 2 summarizes our experience with four ideotype characters. In each program, involving 5 to 10 cycles of breeding, ultimate use of the ideotype character has been hindered by correlated responses of several characters. These correlated responses can hinder or even rule out progress all together. It is becoming clear that understanding character interrelations, which we describe with terms like component compensation, proportionality of size, and pleiotropy, is critical if ideotype breeding is to be done with a degree of sophistication.

In our barley research, we have encountered or purposely studied component compensation, proportionality of size among plant parts (allometry), and pleiotropy. In barley and probably in all crop plants these factors, which are the basis for character associations, are not sufficiently understood or appreciated.

Component compensation is probably most familiar in the context of yield component compensation in the small grains. The data in Table 3 illustrate what is common knowledge, i.e., an increase in one yield component tends to be accompanied by reductions in other components. These data are from investigations of high and low head number lines in three barley populations which resulted from five cycles of breeding. Increases in head number were of little consequence in enhancing yield potential because reductions in both kernel number and kernel weight occurred and off-set the increases in head number. It is likely that such interrelationships are common and that compensating mechanisms have an impact on all aspects of growth and development.

An implicit assumption of the ideotype approach is that yield

Table 3. Yield component compensation among components of grain
          yield in barley (adapted from Benbelkacem et al.,
          1984).

Head number class	N	Head number (m^{-2})	Kernel weight (mg)	Kernel number $(head^{-1})$	Grain yield $(q\ ha^{-1})$
		----------- POPULATION 1 -------------			
High lines	6	305	33.7	47.1	37.2
Low lines	6	272	35.3	49.6	37.9
		----------- POPULATION 2 -------------			
High lines	6	300	31.1	41.2	32.3
Low lines	6	263	34.1	45.0	32.5
		----------- POPULATION 3 -------------			
High lines	6	348	32.7	41.3	36.8
Low lines	6	303	34.0	46.5	36.7

promoting characters can be manipulated and ultimately assembled
in a single genotype. An example of this is Donald's (1968) wheat
(Triticum aestivum L.) ideotype which was a uniculm plant with a
short, stout stem; small, erect leaves; and a large erect spike.
The incorporation of all of these attributes into a single wheat
genotype has yet to be demonstrated. In fact, there is a good
possibility that assembling Donald's wheat ideotype falls outside
the realm of morphogenic possibility.

     Living organisms tend to have a high degree of proportion-
ality of size among different organs. This balance or symmetry
definitely occurs in the small grains. In barley, Fowler and
Rasmusson (1969) found that size of adjacent leaves on the same
culm were highly correlated, and Yap and Harvey (1972) and Hamid
and Grafius (1978) found positive correlations between leaf area,
culm diameter and kernels per spike. The latter researchers also
reported that size of morphological structures was negatively
correlated with tiller number. High tillering genotypes tended to
have small morphological parts.

     To illustrate and reinforce this important concept of propor-
tionality of size of organs or symmetry of size among plant parts,
three data sets have been included in Tables 4 and 5, each set involving seven plant characters. Two of the three sets of data
were from two barley populations obtained by crossing parental
lines which differed for several morphological characters; the
third data set is an analysis of 73 cultivars from the same investigation. The correlations between culm diameter and the six charac-

Table 4. Phenotypic correlation coefficients between culm diameter
          and six other characters in three data sets — population 1
          (N = 82), population 2 (N = 82), and cultivars (N = 73)
          (Doley, 1983).

	Flag leaf area	Penultimate leaf area	Height	Head length	Kernels per head	Culms per meter of row
Culm diameter						
Pop. 1	.45**	.33**	.23*	.67**	.64**	-.38**
Pop. 2	.61**	.77**	.33**	.33**	.35**	-.70**
Cultivars	.47**	.71**	.48**	.56**	.71**	-.53**

*,** Significantly different from zero at the 0.05 and 0.01 levels,
    respectively.

Table 5. Size relationships between two barley characters. The num-
          bers in the table are ranks of genotypes for kernels per
          head in extreme high and low culm diameter genotypes (Doley,
          1983).

	Rank for kernels per head		
	Cultivars N = 73	Population 1 N = 82	Population 2 N = 82
Five genotypes with largest culm diameter			
1	1[†]	2	12
2	11	21	36
3	6	9	1
4	8	6	36
5	12	4	2
Average	8	8	17
Five genotypes with smallest culm diameter			
1	71	76	68
2	73	75	50
3	70	52	75
4	54	69	16
5	58	36	58
Average	65	62	53

[†] Rank is from highest kernel number to lowest.

ters in Table 4 are similar for the three data sets. The correlations between all size characters were positive, whereas the correlations between culm diameter and culm number were negative. We pursued the association between culm diameter and kernels per head by examining genotype rankings for kernels per head within the extreme culm diameter classes in the three data sets. The rankings (Table 5) show that genotypes with large culms tended to have large numbers of kernels per head and vice versa. There are exceptions to be sure, but the message is nonetheless clear, i.e., high numbers of kernels per head occur on plants with large culms. It is pertinent to point out that population 1 and 2 were phenotypically diverse for the seven characters as were the 73 cultivars.

Two important principals emerge from this and other research on the same subject:

1. Size of organs on a plant are proportional, i.e., there is a developmental cohesiveness or symmetry such that plants have large, intermediate, or small parts, but not a mix of the three. Some character combinations are mandated while others are precluded.

2. The size of a plant and its several organs appears to be inversely related to number of plant parts. In cereals these developmental relationships preclude having large organs – culms, leaves, heads - on plants with a large number of culms.

In our research, the finding of essentially the same character associations in the cultivar experiment as in the segregating populations appears to preclude linkage as a basis for the associations. The explanation likely resides in developmental symmetry and presumably in pleiotropic genes which control development. Hamid and Grafius (1978), working with barley, invoked "Sinnott's law", which is that organ size is proportional to apex size, to explain the proportional size relationship. They reasoned that apex size is relatively constant and hence it gives rise to structures that are relatively similar in size.

Because of proportionality of size among parts or allometric relationships, whichever terminology you prefer, a plant breeder is precluded from breeding the ideotype of his choice, i.e., some combinations of characters are not morphogenetically compatible. Of course, as geneticists we acknowledge genetic variation and think of tendencies and probabilities rather than absolutes.

BREEDING EXPERIENCE WITH IDEOTYPE CHARACTERS

The inheritance of ideotype characters that we have worked with in our research program has ranged from single-gene to polygenic,

with varying effects of environment. In all of these programs we
have used cyclic crossing accompanied by selection to incorporate
genes for the desired characters into the genetic background of
six-rowed midwestern barley. Each cycle has consisted of a cross
or backcross followed by selection for the character, and in some
cases, for agronomic type. The number of generations of selfing
and selection before initiating another breeding cycle depended on
the character, but frequently involved five generations or until F_5
or F_6. Each cycle has required about 3 years.

The results with the five characters, which will be presented,
have certain similarities, but there are differences which should
be illustrative of what can be expected in ideotype research and
breeding.

Leaf Angle

In the mid-sixties there were reports (Tanner et al., 1966;
Pearce, 1967) of a yield advantage in barley due to erect leaf. In
1969, we began leaf angle research with three erect leaf stocks.
Two were dropped because their erect leaf progenies had agronomically debilitating characters associated with erect leaf angle. We
have continued to research the erect leaf character, which we found
in CI 6146, and to date we have completed five cycles of breeding,
i.e. five cycles of crossing followed by selection. The erect leaf
character appears to be under control of a major gene, but it is
difficult to score and we do not observe discrete classes.

In the course of breeding for erect leaf angle we have obtained yield data at different stages in the program. CI 6146, the
source of erect leaf angle, is an inferior germplasm and it yielded
only 59% of the check cultivar, 'Dickson' (Table 6). Three erect
leaf lines obtained after three cycles of breeding produced grain
yields that were about 70% of the Dickson check. In the latest
evaluation, 1983, 22 erect leaf lines from three populations were
similar to 'Morex' in grain yield. Additional testing may show
that some of them are superior to Morex.

The 1983 leaf angle lines, which resulted from five cycles of
breeding, are much superior to the original source stock (CI 6146),
as well as the third cycle leaf angle lines, which was evaluated in
1977 (Table 6). However, it is impossible to know whether or not
the leaf angle character is influencing yield. With each breeding
cycle the germplasm has become increasingly similar to that of
midwestern barley and hence yields that are similar to Morex are
expected. It can be concluded that yield-limiting background genes
have been removed from the erect leaf germplasm.

The erect leaf lines now appear ready for use in research and
breeding. We can now have confidence that one can measure the

Table 6. Grain yield of erect leaf barleys compared to check cul-
          tivars (Barker, 1970; Chapko, 1984).

	Yield in percent of check		
	St. Paul	Crookston	Average
-------- erect leaf stock, 1969 --------			
Dickson			100
Erect line stock:			
CI 6146			59
----- 3 cycles of breeding, 1977 ------			
Dickson	100	100	100
Erect leaf lines:			
M74-175	77	60	67
M74-176	72	65	68
M74-177	72	71	72
----- 5 cycles of breeding, 1983 -------			
Morex	100	100	100
Erect leaf lines:			
Pop. 1 (N = 6)	105	102	103
Pop. 2 (N = 8)	100	108	104
Pop. 3 (N = 8)	103	95	98

effect of the leaf angle character relatively free of linkage with
deleterious genes. However, pleiotropy associated with the leaf
angle gene may be the deciding factor in dictating the future of
the erect leaf character. Erect leaf plants have erect spikes, as
well as leaves, and there is less flex in the stem. One can not
predict what effect this altered morphology, and likely anatomy,
will have on agronomic performance.

Grain-filling Duration

     A research program on grain-filling duration and its relation-
ship to grain yield was initiated in 1970. Given the interest in
the character, it is surprising how little research has been done,
although Aksel and Johnson (1961) had reported that a long vegetative period in barley was preferable to a long grain-filling
period.

     A search involving 400 genetically divergent barley genotypes

identified a few which were similar in maturity but with different
anthesis dates (Rasmusson et al., 1979). Utilizing this germplasm,
research was conducted to assess the consequence of a shift in the
amount of time devoted to vegetative growth versus grain-filling.
It seemed to us that this objective required genetic stocks with
similar maturity.

Data are presented in table 7 for barley genotypes in three
classes, i.e., short, intermediate and long grain-filling duration.
The typical midwestern cultivar represented by Dickson, 'Manker'
and 'Larker' are in the intermediate class. They devoted about 33%
of their growth and maturation period to grain filling. In contrast
the short duration genotypes devoted 25% and the long duration
genotypes about 39% of the time to grain filling. These genotypes
ranged from being very inferior agronomic types (e.g., very low
yield and extreme susceptibility to lodging) to commercial cultivars.

Two and three cycles of breeding were used to incorporate
variability for grain-filling duration into improved genetic
backgrounds. Ultimately, genotypes having long and short grain-
filling duration were selected in four populations and evaluated
for yield (Table 8). Morex, the check, reached maturity in 34 days
and 36% of its growth period was in grain filling. In contrast,
the long grain-filling duration lines devoted 40% and the short

Table 7. Duration of growth periods of nine barley cultivars
          (Rasmusson et al., 1979).

| Genotype | Duration of grain-filling period | Mean of five field environments | | |
		Duration of vegetative period	Duration of grain-filling period	Total time from planting to maturity
		---------------- days --------------------		
CI 5827	Short	62.4	21.3	83.6
CI 5809	Short	61.0	20.6	81.6
Dickson	Intermediate	55.6	25.1	80.7
Manker	Intermediate	53.6	26.1	79.8
Larker	Intermediate	54.3	26.5	80.8
Primus	Long	49.8	29.9	79.6
CI 6573	Long	49.1	30.8	80.1
CI 5926	Long	48.6	30.8	79.4
Vaughn	Long	48.1	32.2	80.4
L.S.D. (0.05)		0.7	1.0	0.9

Table 8. Average maturity, grain-filling duration, grain-filling
          index, and grain yield of four barley populations (Metzger
          et al., 1983).

	N	Maturity (days)	Grain-filling durations (days)	Grain-filling index	Grain yield (q ha^{-1})
Morex		84.3	30.5	.36	44.1
Pop. 1					
Long duration	5	83.9	33.9	.41	45.0
Short duration	5	85.9	29.3	.34	45.5
Pop. 2					
Long duration	5	82.5	32.4	.39	44.2
Short duration	5	82.2	27.4	.33	41.0
Pop. 3					
Long duration	4	86.8	36.6	.42	39.9
Short duration	4	87.6	31.5	.36	42.4
Pop. 4					
Long duration	4	87.3	31.1	.36	44.2
Short duration	4	89.3	28.1	.31	41.4
Average					
Long duration	18	85.1	33.5	.40	43.3
Short duration	18	86.2	29.1	.34	42.6

duration lines 34% of their life cycles to grain filling. On an
individual line basis, the long duration lines in the four populations spent an average of 3 to 7 days longer in the grain-filling
period than the short duration lines.

In comparing differences in grain-filling duration, it is apparent that all of the original variability (Table 7) was not
transferred to the progeny (Table 8). It is a quantitative character and is difficult to transfer through repeated breeding cycles.
However, the objective of developing genetically similar lines
having the same maturity date but differing in anthesis date was
achieved, at least in a modest way.

Grain yield was not associated with length of the grain-
filling period in any of the four populations (Table 8). Thus we
failed to demonstrate any value in substituting duration of one
growth period for another.

It may be that the grain-filling period and the vegetative
period are equally important to obtaining high yield, and that a
shift of days from one period to the other is neutral in effect.
However, the study had sufficient limitations to argue against
drawing firm conclusions. We completed only two and three cycles
of breeding with the four populations and we observed differences
among short and long duration lines for height, lodging and synchrony
of flowering. Clearly the long and short duration lines did not
share a common genetic background.

Multiple Awns

The influence of the awn on grain yield of barley has been
studied in some detail and, in general, awns have been advantageous.
Schaller et al. (1972) and Qualset et al. (1965) showed that full-
and half-awned lines of 'Atlas' barley produced more grain than
quarter-awned and awnless lines. Their computations showed that
the awns of a full-awned Atlas type contributed 19.3% of the total
grain yield.

In addition to awn length, variation exists in barley for awn
number. In a special stock, the e gene causes glumes to bear an
awn which results in three awns per spikelet, whereas normal
barley has one (Tsuchiya, 1974). Johnson et al. (1975) found that
the e-line with the large awn-like glume had greater net photosynthesis, dark respiration, and transpiration than its isogenic
normal-glume counterpart.

Barley lines differing in amount of awn tissue were developed
by backcrossing, using Dickson barley as the recurrent parent, and
an agronomically inferior donor which carried the e gene for awned
glumes. We refer to the large glumes with the attached awn as the
"multiple awn" character. The backcrossing procedure was easy
since the character is fully expressed in all environments.

The normal awn and the e-type awn were similar in appearance
(Rasmusson and Crookston, 1977). Both types of awns were plenti-
fully supplied with stomata on their abaxial surface. Light
microscope examinations indicated that the normal and e-type awns
were identical in internal cellular arrangements. Hence, we
expected the extra awns to function the same as normal.

Grain yields of Dickson and lines from five different backcross generations are shown in Table 9. In the 5 years of testing,
all of the multiple awn lines yielded less than Dickson. The
multiple awn lines from the first backcross were especially low in
yield ranging from 50 to 67% of Dickson, but no trend in yield
level was apparent from the second to the sixth backcross generation. The maximum yield of the multiple awn lines was only 88% of
Dickson. In marked contrast, yields of the normal awn lines ob-

Table 9. Yield of Dickson and backcross derived multiple awn and
normal awn lines (Rasmusson and Crookston, 1977).

Year	Dickson yield	Backcross generation	Yield in percentage of Dickson								
			Multiple awn lines					Normal awn lines			
			1	2	3	4	5	1	2	3	4
	q ha^{-1}										
1971	32.8	BC_1F_4	67	51	50	--	--	--	--	--	--
1972	49.5	BC_2F_4	82	82	83	--	--	--	--	--	--
1973	46.8	BC_3F_5	80	83	79	77	--	--	--	--	--
1974	26.9	BC_5F_5	82	80	70	88	82	126	122	112	98
1975	29.6	BC_6F_4	73	73	67	64	82	111	118	--	--

tained from the fifth and sixth backcross generations ranged from
98 to 126% of Dickson. This was a disappointing result. We hoped
that the extra awn tissue would increase yields.

In an effort to understand the basis for the low yields of
the multiple awn lines we found that they differed significantly
from their normal awn sister lines in kernels per head and kernel
weight (Table 10). The most striking difference was in kernels
per head, and this component appears to be the primary reason for
the lower yields. With the large reduction in kernel number we
expected heavier kernels in the e-lines, but this was not the
case.

Possible explanations implicate the character or the gene or
both. The large glumes could be a competing sink and competition
during the critical development time could result in weakened
florets which would ultimately fail to develop. An alternative is
that the e gene is pleiotropic and directly influences growth,
especially kernels per head and kernel weight. Plants having the
e gene were slightly smaller, on the average, so the gene may
hinder growth processes in general and specifically kernel development.

This is a case where laboratory research (Johnson et al.,
1975) suggested that a character, multiple awn, would contribute
to higher grain yield. Our research, which utilized breeding and
field trials, identified serious drawbacks and provided no reason
for optimism.

Table 10. Characteristics of BC_7F_4 multiple awn and normal lines (Rasmusson and Crookston, 1977).

	Number lines	Nodes/ head	Kernels/ head	Kernel weight
				--mg--
Normal awn	10	21.7 ± 0.8	50 ± 2.4	35 ± 1.8
Multiple awn	10	21.8 ± 0.9	34 ± 0.6	32 ± 2.1

Table 11. Tiller number, head number, and yield of barley genotypes. Data are averages from two locations and plant spacings corresponding to 34, 67 and 101 kg/ha (Pfund, 1974).

	Tiller number	Head number	Yield
	----- per plant ----		g ha^{-2}
Cultivars			
Conquest	3.7	2.5	45.0
Manker	3.5	2.4	48.5
Cree	4.9	3.1	42.8
Average	4.0	2.7	45.4
High Head Number Genotypes[‡]			
Minn. 95	6.4	3.9	36.0
Minn. 263	5.7	3.5	39.8
Minn. 269	6.6	3.7	38.3
Minn. 271	6.7	3.9	35.3
Average	6.4	3.8	37.5

[‡] These genotypes are from two cycles of breeding for high tiller and head number using sources XH-263 and Sel. 6194-63.

Head Number

Head number is a character that has received a great deal of attention and its potential for increasing grain yield in barley was recognized in 1923 by Engledow and Wadham. A change in head number alters the contribution of the primary components to final grain yield, and it alters the leaf and canopy display. Clearly, breeding for high head number has the potential for increasing yield. We began a program to increase head number in midwestern six-rowed barley in 1965.

Two higher-tillering genotypes, XH-263, a two-rowed barley, and Sel. 6194-63, a six-rowed barley which had a two-rowed parent, were the germplasm sources. Two-rowed barleys have more heads than six-rowed barleys and are thus convenient sources of high tiller and head number. These germplasm sources and their initial high tiller number progenies were very inferior yield-wise to commercially grown cultivars. This can be seen in Table 11, which provides a comparison of three cultivars and high tiller number lines obtained after two cycles of breeding. The high tiller number genotypes had 60% more tillers and 41% more heads than the cultivars. What was especially disappointing was the nearly 20% lower yield of the high tiller number genotypes.

In the ensuing years we completed four cycles of breeding with the Sel. 6194-63 germplasm source and seven cycles of breeding with the XH-263 source (Benbelkacem et al., 1984). The current status of the program is shown in Table 12. In populations 1 and 2, originating from Sel. 6194-63, the highest head number class had 20% more heads than the lowest class. In populations 3 and 4, originating from XH-263, the highest class had 14% more heads than the lowest class. These head number differences are between classes and hence are conservative in indicating individual genotype differences.

In the course of breeding for high head number we found that increases in tiller number were only 36% effective in terms of adding additional heads. This was because tiller mortality was high and increased linearly with increased tiller number. While this value is low, it is similar to that reported by Evans and Wardlaw (1976). They said that up to two-thirds of the tillers in cereals may be wasted, although they ensure considerable scope for compensation early in the life cycle.

Mean yield of the four tiller number classes within each population was similar (Table 12). In populations 1 and 2 the intermediate classes tended to be highest in yield, while in populations 3 and 4 yield was highest in the high tiller number classes. But in all cases the differences were small. Some individual lines, in all populations, exceeded the checks in yield, but non-significantly.

High tiller and head number were achieved at a cost, namely fewer kernels and reduced kernel weight. Because of this component compensation, increasing head number did not increase yield pro-

Table 12. Head number and yield of barleys which differ in head
           number (Benbelkacem et al., 1984).

Population number	Number of breeding cycles	Head number class	Number of lines	Head number m^{-2}	Grain yield $q\ ha^{-1}$
1	4	Highest	5	417	49.3
		2nd high	5	398	49.5
		3rd high	5	369	49.5
		Lowest	5	344	45.3
		LSD (.05)		28	2.2
2	4	Highest	5	454	45.8
		2nd high	5	412	47.9
		3rd high	5	393	46.9
		Lowest	5	376	45.2
		LSD (.05)		33	N.S.
3	7	Highest	5	377	51.2
		2nd high	5	356	50.6
		3rd high	5	342	48.8
		Lowest	5	332	49.3
		LSD (.05)		28	N.S.
4	7	Highest	5	382	45.7
		2nd high	5	351	45.0
		3rd high	5	343	44.8
		Lowest	5	335	43.2
		LSD (.05)		33	N.S.

portionately. Also, the high tillering barleys had a low percentage of plump kernels and greater susceptibility to lodging. The
increased lodging was apparently due to smaller diameter stems.

Even so, the weight of the evidence favors continued research
on the high tiller number character. The most persuasive evidence
was the improvement of the high tiller number lines during the
several cycles of breeding. This improvement assured a good genetic
background for the tiller-number genes, which was essential if the
character was to be evaluated adequately and if the genes are to
be used in breeding programs. Successful transfer of the high
tillering character was gratifying since heritability of tiller
number is low. Added inducement for continuing the research is
that relatively high head number is a feature of modern barley

cultivars. Wych and Rasmusson (1983) found higher head number in modern cultivars in the midwest of the U.S.A. In Europe, Gymer (1981) found that new cultivars have 10 to 15% more heads than older cultivars and Ekman (1981) estimated that modern barleys had 14% more heads.

Our breeding strategy in the future will focus on obtaining smaller increments of increased tillering than in the past. A tiller number increase of 15% and a head number increase of 5% compared to current cultivars, for example, may be more conducive to high yield and favorable expression of other characters than larger increases. Grafius et al. (1976) have suggested that small genetic changes which maintain the integrity of superior gene combinations are advantageous when adding new features to a germplasm. Also, there are pragmatic reasons for seeking modest increases in tiller number. Successful midwestern cultivars must have good lodging resistance and plump kernels. At the present time these traits are not attainable in barley lines with very high tiller and head numbers. In addition, it appears advantageous to seek lower tiller mortality as discussed by Simmons et al. (1982), although this will be a difficult task.

Semidwarfism

In 1957, a short-strawed barley mutant stock was obtained from Norway (Ali et al., 1978) and in the interviening years seven to nine cycles of crossing and selection have resulted in phenotypically attractive semidwarf barleys.

Dramatic improvements have been made over the years in the germplasm. The initial stock was wholly unsuited for the midwest of the U.S. In the beginning it was late maturing, the straw and peduncle were weak, the spike did not fully emerge from the boot, and it was highly susceptible to disease.

Evidence of progress in placing the gene in an improved genetic background was obtained during the course of the breeding program and in thesis studies. In statewide yield trials over a 13-year period dating from 1970 to 1983, the best semidwarf lines were initially inferior to the best normal height lines, but in recent years, the semidwarfs and normal height lines have been similar in yield (Figure 1). After five to seven cycles of breeding, semidwarf and normal height lines from five populations gave similar grain yields (Table 13). Lodging was only rarely a factor in these trials and had little influence on the yield results.

Two commercial semidwarf cultivars, 'Kombar' and 'Kombyne,' tracing to this germplasm, were developed by Dr. Robert Matchett of Northrup King and Co. (Matchett and Cantu, 1976). They have other semidwarf lines nearly ready for release. No semidwarf barleys

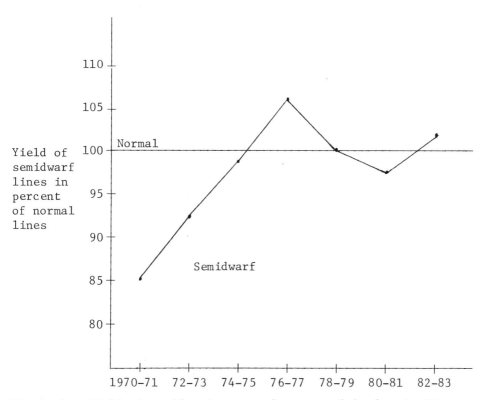

Figure 1.   Yield of semidwarf compared to normal barley in MN
            statewide trials, 1970 to 1983. The percentages are
            based on the three highest yielding normal height lines
            and the three highest yielding semidwarf lines.

Table 13. Yield and other characteristics of semidwarf and normal
           barley in five populations (Ali et al., 1978).

	Number lines	Plant height	Grain yield	Harvest index
		cm	q ha^{-1}	
Semidwarf	41	65	34.9	0.48
Normal	41	82	33.9	0.42
Difference		-17	1.0	0.06

have been released in the midwest of the U.S.

Linkages with deleterious genes were a major factor in the unusually poor performance of the initial semidwarfs that we observed
in the 1960's. However, linkage does not appear to provide a
sufficient explanation for the experience with the semidwarf gene
(*sdw*). After 25 years of breeding, the semidwarf barleys in the
Minnesota program are later maturing than tall counterparts, have
marginal malting and brewing quality, and have a tendency for
greater susceptibility to disease. We speculate that the *sdw* gene
is pleiotropic and affects characters other than height.

DISCUSSION AND CONCLUSIONS

Our ideotype research on barley has been both rewarding and
disappointing. We are disappointed that we have not succeeded in
obtaining new cultivars with enhanced genetic yield potential via
ideotype characters. Good progress in our conventional breeding
effort assures continued tough competition for our ideotype breeding. We are firmly convinced, however, that genetic enhancement
via research with individual characters is advantageous and should
be strengthened. Based on what we have learned and prospects for
future yield gains, we are satisfied that our investment of time
and resources has been appropriate.

Our current thinking about ideotype research or ideotype
breeding and the status of our program can be summarized as follows:

1)   Character interrelationships are highly integrated and more
     complex than generally thought.

     Proportional or size symmetry among plant parts, competi-
tion among parts leading to component compensation, and other

character associations due to control by a common gene (genic pleiotropy) contribute to a very complex system. The system is highly integrated and dynamic with the final phenotype dependent on earlier ontogenetic events as well as the environment encountered.

Based on our results, we can expect attempts to breed for one character at a time to be hindered or helped by correlated responses. Because size symmetry is mandated, certain character combinations that appear attractive may be precluded altogether. For example, having small narrow leaves to encourage light penetration into a canopy may preclude having a large spike with large kernels. On the other hand we observed possible beneficial pleiotropic effects of the gene for erect leaf angle. Erect leaf genotypes appeared to have less flex in the straw, and possible greater resistance to lodging.

Research on character interrelations should enter a new phase that recognizes association as a universal feature. It is not enough to report finding of associations. We need research that identifies and describes causal relationships among characters and research that provides information about genes and how they interact.

2)   An ideotype or several ideotypes should be postulated for each crop by plant breeder-geneticists and plant physiologists.

Ideotype thinking encourages bold projections and reaching out for genetic diversity. Crop ideotypes have been postulated for all of the major field crops and should be done for all crops, and there should be a greater emphasis on finding the optimum expression for individual characters. These models are useful because they stimulate interest and discussion of genetic yield enhancement, and they encourage research by plant breeders, plant geneticists and plant physiologists. Such models should ultimately lead to new cultivars with greater genetic yield potential.

3)   Ideotype research and breeding are good choices for plant breeder-geneticists.

This is especially so for those who desire to contribute knowledge about plant characters and genetics as well as to augment conventional yield breeding. Future gains in plant breeding depend on increased knowledge and finding ways to utilize previously untried genes. Public breeder-geneticists should give priority to such dual objectives and it appears quite reasonable to urge private breeders to assist with germplasm enhancement via ideotype breeding.

4)   Enhancement of genetic yield potential should include modest
      as well as dramatic changes in ideotype characters.

      When I think of the height of grain sorghums of just a
couple of decades ago in comparison to today's knee-high
cultivars, I marvel at what genetic diversity permits plant
breeders to accomplish. Considering the breadth of opportunities it is clear that current cultivars do not have optimum
ideotypes. We have the option of making modest changes in
morphology and physiology as well as dramatic changes. Perhaps we too often disregard potential gains from modest changes.
These would likely be easier to handle in the breeding program
and the gains might be very worthwhile. Dramatic changes may
be precluded in some crops, where selection has been practiced
for decades, but for other crops dramatic changes in phenotype
may lead to large gains in genetic yield potential. Adams'
(1982) restructuring of the dry bean, Phaseolus vulgaris L.,
plant is a current example of the opportunity afforded by
making dramatic changes via ideotype breeding.

5)   A gene bank or character bank for each crop ought to be estab-
      lished.

      A gene or character bank that would be available to all
researchers could lead to greater progress, and such a bank
should encourage cooperation and reduce duplication of effort.
The stocks in the gene or character bank would be freed of
deleterious linkages via breeding. It is envisioned, for
example, that genes for high kernel number would be tried in
many genetic backgrounds and in many environments. as a consequence of being available in a gene bank. This broad usage
is needed to realize potential gains.

      We have gene banks for disease resistant stocks and cyto-
genetic stocks such as monosomics and trisomics. Doesn't it
make equally good sense to have a bank for genes that will
enhance genetic yield potential?

6)   A sizeable breeding program may be necessary to bridge the gap
      between the unimproved gene pool, where the ideotype researcher
      will seek genetic diversity, and the gene pool that can lead
      to new cultivars.

      In reality plant breeders have two gene pools, at least
in crops that have been bred for several decades. One is the
working gene pool which provides parents commonly used in
crossing blocks. This pool has been improved year by year and
breeding cycle by breeding cycle. It is more diverse than
some think and will provide for genetic progress for many
years to come, but it is frequently too narrow in the context

of ideotype breeding. The second pool is largely neglected and the gap between the two pools is growing. Parts of the small grain "world collections" are examples of little used gene pools.

Undoubtedly many ideotype characters are labeled yield-negative or -neutral because of insufficient breeding. Deleterious linkages are likely when a character is sought in an unimproved gene pool. Without a substantial breeding effort, the desired genes are in effect unavailable. This challenge was demonstrated conclusively in our programs on leaf angle, semidwarf stature and head number. In each case the germplasm source and the progeny from the first cycles of breeding were decidely inferior and showed no promise.

After the genes for individual characters are incorporated into good genetic backgrounds, it may be necessary to resort to conventional breeding to obtain yield gains. This should mean many crosses and character combinations. We have evaluated ideotype characters in modest experiments with narrow germplasm bases without very much success.

7)   We have not been successful to date in developing higher yielding barleys using the ideotype approach.

Our semidwarf gene may not be compatible with good malting quality, nor has it given enhanced yield in the midwest compared to tall counterparts, although its success in western cultivars is encouraging. However, we are enthusiastic about future gains from our ideotype program as we now have genes for erect leaf angle, high head number, high kernel number and high kernel weight in good genetic backgrounds.

REFERENCES

Adams, M.W., 1982, Plant architecture and yield breeding, Iowa State J. of Res., 56:225-254.

Aksel, R., and Johnson, L.P.V., 1961, Genetic studies on sowing-to-heading periods in barley and their relation to yield and yield components, Can. J. Genet. Cytol., 3:242-259.

Ali, Mohamed A.M., Okiror, S.O., and Rasmusson, D.C., 1978, Performance of semidwarf barley, Crop Sci., 17:418-422.

Austin, R.B., Bingham, J., Blackwell, R.D., Evans, L.T., Ford, M.A., Morgan, C.L., and Taylor, M., 1980, Genetic improvements in winter wheat yields since 1900 and associated physiological changes, J. Agr. Sci., 94:675-689.

Barker, R.E., 1970, Leaf angle inheritance and relationships in barley, M.S. Thesis, Univ. of Minnesota, St. Paul.

Benbelkacem, A., Mekni, M., and Rasmusson, D.C., 1984, Tiller
    number and yield in barley, Crop Sci., In Press.
Chapko, L.B., 1984, Leaf angle and its effect on yield in barley,
    M.S. Thesis, Univ. of Minnesota, St. Paul.
Doley, W.P., 1983, Allometric relationships in spring barley,
    M.S. Thesis, Univ. of Minnesota, St. Paul, MN.
Donald, C.M., 1968, The Breeding of Crop Ideotypes, Euphytica,
    17:385-403.
Duvick, D.N., 1977, Genetic rates of gain in hybrid maize yields
    during the past 40 years, Maydica, 22:187-196.
Engledow, F.L., and Wadham, S.M., 1923, Investigations on yield in
    the cereals, J. Agric. Sci., Camb., 13:390-437.
Ekman, R., 1981, Biomass component studies in barley, their corre-
    lation to some yield characters and estimation of desirable
    effect from 50 years of barley breeding, in: "Proc. 4th Int.
    Barley Genetics Symp.," Edinburgh, pp. 104-111.
Evans, L.T., and Wardlaw, I.F., 1976, Aspects of the comparative
    physiology of grain yield in cereals, Advances in Agronomy,
    28:301-359, Academic Press, Inc., New York.
Evans, L.T., 1983, Raising the yield potential: by selection or
    design, in: "Genetic Engineering of Plants," p. 371, Kosuge, T.,
    Meredith, C.P., and Hollander, A., eds., Plenum Press, New York.
Fowler, C.W., and Rasmusson, D.C., 1969, Leaf area relationships
    and inheritance in barley, Crop Sci., 9:729-731.
Grafius, J.E., Thomas, R.L., and Barnard, J., 1976, Effect of
    parental component complementation on yield and components of
    yield in barley, Crop Sci., 5:673-677.
Gymer, P.T., 1981, The achievement of 100 years of barley breeding,
    in: "Proc. 4th Int. Barley Genetics Symp.," Edinburgh. pp.
    112-117.
Hamid, Z.A., and Grafius, J.E., 1978, Developmental allometry and
    its implication to grain yield in barley, Crop Sci., 18:83-86.
Johnson, R.R., Willmer, C.M., and Moss, D.N., 1975, Role of awns in
    photosynthesis, respiration, and transpiration of barley spikes,
    Crop Sci., 15:217-221.
Matchett, R.W., and Cantu, O.P., 1976, Barley breeding, Barley
    Newsletter, 20:14-16.
Metzger, D.D., Czaplewski, S.J., and Rasmusson, D.C., 1984, Grain-
    filling duration and yield in spring barley, Crop Sci. In press.
Pearce, R.B., Brown, R.H., and Blaser, R.E., 1967, Photosynthesis
    in plant communities as influenced by leaf angle, Crop Sci.
    7:321-324.
Pfund, J.H., 1974, Optimum culm number in barley, Hordeum vulgare
    L., Ph.D. Thesis, Univ. of Minnesota, St. Paul.
Qualset, C.O., Schaller, C.W., and Williams, J.C., 1965, Performance
    of isogenic lines of barley as influenced by awn length, linkage
    blocks, and environments, Crop Sci., 5:489-494.
Rasmusson, D.C., and Crookston, R.K., 1977, Role of multiple awns
    in determining barley yields, Crop Sci., 17:135-140.

Rasmusson, D.C., and Gengenbach, B.G., 1983, Breeding for physio-
    logical traits, p. 231-254, in: "Crop Breeding," D.R. Wood ed.,
    Amer. Soc. of Agron., Madison, WI.
Rasmusson, D.C., McLean, I., and Tew, T.L., 1979, Vegetative and
    grain filling periods of growth in barley, Crop Sci., 19:5-9.
Riggs, T.J., Hanson, P.R., Start, N.D., Miles, D.M., Morgan, C.L., and
    Ford, M.A., 1981, Comparison of spring barley varieties grown in
    England and Wales between 1880 and 1980, J. Agric. Sci., 97:599-
    610.
Schaller, C.W., Qualset, C.O., and Rutger, J.N., 1972, Isogenic
    analysis of the effects of the awn on productivity of barley,
    Crop Sci., 12:531-535.
Simmons, S.R., Rasmusson, D.C., and Wiersma, J.V., 1982, Tillering
    in barley: Genotype, row spacing, and seeding rate effects,
    Crop Sci., 22:801-805.
Tanner, J.W., Gardner, C.J., Stoskopf, N.C., and Reinbergs, E.A.,
    1966, Some observations on upright-leaf type small grains,
    Can. J. Sci., 46:690.
Tsuchiya, T., 1974, Preliminary results on genetic studies of
    outer glume characters in barley, Barley Genet. Newsl., 4:76-78.
Wallace, D.H., Ozbrun, J.L., and Munger, H.M., 1972, Physiological
    genetics of crop yield, Advances in Agronomy, 24:97-146,
    Academic Press, Inc., New York.
Wilson, D., 1981, Breeding for morphological and physiological
    traits, in: Plant Breeding II, p. 233, Frey, K.J. ed., Iowa
    State University Press, Ames.
Wych, R.D., and Stuthman, D.D., 1983, Genetic improvement in
    Minnesota - adapted oat cultivars relesed since 1923, Crop Sci.,
    23:879-881.
Wych, R.D., and Rasmusson, D.C., 1983, Genetic improvement in
    malting barley cultivars since 1920, Crop Sci., 23:1037-1040.
Yap, T.C., and Harvey, B.L., 1972, Inheritance of yield components
    and morpho-physiological traits in barley, Hordeum vulgare L.,
    Crop Sci., 12:283-286.

PHYSIOLOGICAL ASPECTS OF VARIETAL IMPROVEMENT

L. T. Evans

CSIRO, Division of Plant Industry
Canberra, A.C.T., Australia

INTRODUCTION

Increased crop production derives from both agronomic and
varietal improvement, and from their continuing interactions. For
example, cheaper nitrogen fertilizers created a need for shorter
cereals less prone to lodging but more dependent on herbicidal
control of weeds. In turn the denser crops possible with greater
fertilizer use open up new opportunities and criteria for selection,
such as tolerance of closer spacing and, possibly, smaller, more
upright leaves with faster photosynthesis per unit area. Varietal
improvement, therefore, both creates and responds to agronomic
change.

Plant physiological research contributes to agronomic change
as well as to varietal improvement. For example, work on plant
nutrition has led to more effective fertilizer practice; on water
relations to better irrigation management; on plant growth substances to selective herbicides and regulants. With varietal
improvement, however, empirical selection for better crop adaptation, yield potential or quality can proceed effectively without
reliance on selection criteria or screening techniques from crop
physiology, and the objective of this paper is to explore ways in
which such progress can be enhanced by physiological understanding.

I shall begin by considering the essentially comparative
nature of crop physiology and then, after a brief consideration
of other plant breeding objectives, concentrate on two major components of increased yield potential, namely photosynthetic rate
and changed partitioning associated with shorter stems. These
were chosen to illustrate both the opportunities and the problems

to be faced in using physiological analysis to enhance varietal
improvement.

THE COMPARATIVE ELEMENT IN CROP PHYSIOLOGY

Whereas plant physiology explores the full range of plant
behaviour, crop physiology concentrates on the ways in which
varieties and related genotypes differ and in which one may excel
another in particular environments or stress conditions.

Varietal differences

Physiological comparisons between older and newer, higher
yielding or better adapted varieties have often been used to
identify characteristics or processes which may have contributed
to crop improvement. But varieties usually differ in many features, with the result that rarely, if ever, have varietal comparisons established a particular physiological characteristic as
the cause of yield enhancement or better adaptation, rather than
being merely correlated with it. The greater the number of
varieties compared, the fewer the processes that can be analysed,
and the greater the likelihood of identifying non-causal characteristics. Comparing wheat (*Triticum aestivum*) varieties of differing
height, for example, Quarrie (1983) found that leaves of shorter
varieties had a greater capacity to respond to water stress by
producing more abscisic acid, but when near-isogenic lines were
used to check this conclusion, the association was found to be
fortuitous (King et al., 1983).

When fewer varieties are compared, their differences can be
examined in greater depth, but the problem of deciding which of
these are of greatest significance to yield or adaptation remains.
There have been many studies comparing tall and dwarf varieties of
wheat, for example, often with only one variety of each differing
not only in height but also in the presence or absence of awns,
adaptation to the environment and other characteristics. Not
surprisingly, conclusions as to the effects of dwarf stature on
yield potential have varied considerably. Comparing tall and dwarf
varieties both of which were adapted to local conditions, Wattal
and Asana (1976) came to quite different conclusions from Konar
and Asana (1975) who compared a local tall variety with a non-
adapted dwarf wheat. Gale (1983) considers that varietal comparisons are responsible for much of the present confusion over the
pleiotropic effects of the Rht dwarfing genes. Moreover, as
Quisenberry (1982) has pointed out, before a trait is used as a
selection criterion, its relationship with productivity should be
firmly established; "differences in productivity between genotypes
with extreme expressions of the trait should not be construed as
the establishment of a relationship".

Varieties and their wild progenitors

Varietal comparisons have recently been extended for several crop plants by the inclusion of closely related wild species, e.g. with wheat (Evans and Dunstone, 1970; Khan and Tsunoda, 1970b), rice (*Oryza spp.*) (Cook and Evans, 1983) and cowpeas (*Vigna unguiculata*) (Lush and Evans, 1981). The wider perspective resulting from the inclusion of wild progenitors in these comparisons has been helpful. The rise in harvest index has become more striking, particularly in comparison with the absence of any rise in photosynthetic rate.

However, while such comparisons provide a retrospective view of what changes have already occurred during domestication and improvement, they cannot establish whether the changes actually caused the improvement nor to what extent further advance in the same directions will be useful. For example, increased yield potential has come, to date, from increase in harvest index rather than in biomass production. Should we therefore suggest further progress along the same path because it has been effective, or conclude that further progress in that direction is likely to be limited. Austin et al. (1980a) estimated the likely limit to further rise in the harvest index of wheat, which should allow yield potential to be increased by a further 25%. As that limit is approached, further increase in yield potential will depend on detecting and exploiting genetic variation in biomass production.

Isogenic lines

Wheareas varieties often differ in many characteristics, the use of near-isogenic pairs of lines allows the physiological effects of major genes to be analysed. The fact that isogenic lines have not been widely used by crop physiologists so far may reflect the view of many plant breeders that the improvement of adaptation or yield mostly involves polygenic systems, and that relatively few major genes are known which would be suited to such an approach.

Nevertheless, even when the fine tuning of adaptation in a breeding program is accomplished via the polygenic system, the use of isogenic lines with major genes for response to daylength or vernalization may clarify the role of such responses, and may open up new opportunities for selection. The height of wheat varieties had, for many years, been reduced by polygenic selection, but when the availability of cheaper nitrogenous fertilizers made it necessary to reduce height more drastically, the major genes for height were used with great effect.

Major genes for other characteristics, such as leaf shape (e.g. okra leaf in cotton (*Gossypium hirsutum*), leaf inclination,

presence of awns, lysine content etc. are also being used in breeding programs, and isogenic lines will be of great value in allowing
their physiological effects to be analysed. An additional advantage
of isogenic lines is that they permit an analysis of the effects on
major gene expression both of genetic and cytoplasmic background
and of the physical environment.

For example, experiments with several pairs of isogenic lines
have shown that the effects of awns vary with the genetic background in both barley (*Hordeum vulgare*) (Scharen et al., 1983) and
wheat. Vogel et al. (1963) suggested that the presence of awns may
be more important in dwarf than in tall wheats, but this was not
confirmed by Olugbemi et al. (1976). Isogenic lines have also
allowed the interaction between awns and environmental conditions
to be examined. Awned plants significantly outyielded awnless
plants under dry but not under wet conditions (Evans et al., 1972),
and temperature also influences the effect of awns on grain development in wheat (Olugbemi and Cook, in preparation). Photosynthesis
by the awns is carried out in a favorable environment for irradiance and CO_2 and provides a source of assimilates close to the ear
which is particularly significant for the more distal spikelets
(Evans et al., 1972), and relatively protected from water stress.
Grain number may be less but grain size is often greater in the
presence of awns (Patterson et al., 1962, 1975). Awns may result
in cooler crop canopies (Ferguson, 1974) and may attract more
cytokinins to the grains (Michael and Seiler-Kelbitsch, 1972).
On the other hand, the developing awns may compete with developing
florets for assimilates (Rasmusson and Crookston, 1977), while
exposed awns may trap water and thereby accentuate disease problems
or sprouting in the ear under some conditions (King and Chadim,
1983). The net effect on yield of the gene for awns therefore
varies to a considerable extent with environmental conditions as
well as with genotype, and this is likely to be true of many other
major genes whether transferred by conventional or molecular plant
breeding.

Wider comparisons

Coming as they do from many taxonomic families, crop plants
span much of the range in plant behaviour and can fill many environmental niches. Physiological comparisons between different
crops have often been made, but these assume a different significance as recombinant DNA techniques develop because, in principle,
the genetic resources of all plants, indeed of the whole biosphere,
may become available for plant improvement.

Many crop physiologists have examined the possibility of
selecting in C_3 crop plants for at least some of the characteristics
of C_4 photosynthesis. Extensive screening for lines or mutants
with reduced photorespiration, as in C_4 plants, has not yielded

any success. In any case, photorespiration may well play an adaptive role in the temperate environments in which C_3 plants are dominant, because otherwise the selective advantage accruing to plants with faster photosynthesis due to reduced photorespiration would be very great.

It now seems likely that the topological requirements for the site of carboxylation inevitably confer oxygenase activity on rubisco (ribulose-1,5-bisphosphate carboxylase-oxygenase), and that the best way round this is the compartmentation achieved by the C_4 Kranz leaf anatomy. Rubisco from C_4 plants has higher K_m (CO_2) values than that from C_3 plants (Yeoh et al., 1981), but this would be disadvantageous in C_3 plants without the accompanying anatomical compartmentation. The absence of photorespiration in C_4 plants is associated with higher maximum rates of photosynthesis and crop growth and frequently with higher yields than in C_3 crops. But it does not follow, as is often assumed, that their higher yield and faster growth are due to faster photosynthesis. The relative growth rates of C_4 plants are lower than those of C_3 plants at cool temperatures, and exceed them only at high temperatures and irradiances. Moreover, their faster growth may be due as much to the more complete export of assimilates from the leaves for investment in new leaf growth as to the higher rates of photosynthesis per se. As for the higher yields, Gifford (1974) has shown that these are, to a considerable degree, associated with the longer seasons and higher irradiances in which C_4 crops are often grown and to which they are better adapted.

The real advantage of the C_4 pathway of photosynthesis is that it allows such environments to be more fully exploited, just as C_3 crops take greater advantage of lower irradiances and temperatures. However, many C_3 crops, such as rice, thrive in the tropics, while some C_4 plants have adapted to cool temperate conditions. There may be some scope, therefore, for the reciprocal transfer of specific characteristics between C_3 and C_4 crop plants by genetic engineering techniques, but it would seem a better strategy to maintain their complementary adaptive advantages and modify their other qualities. For example, rather than attempting to adapt wheat to lowland tropical conditions it might soon be more practicable to incorporate bread-making characteristics into one of the C_4 cereals.

Models and extrapolations

The preceding discussion of the C_4 syndrome was introduced to highlight the problem of selecting a single characteristic from a coordinated adaptive complex for incorporation in quite different plants, a problem to be faced by genetic engineers in the future more acutely than by plant breeders in the past.

One possible approach to the problem is the use of simulation
models to test in advance the impact and interactions of a novel
characteristic. An example of this approach is provided by
Landivar et al. (1983a) who used the reasonably well validated
GOSSYM model for cotton crops to assess the impact of the gene
for more dissected 'okra' leaves, without changing the morphogenesis logic in the model. In this case the predicted effects
could be compared with several known effects of the okra leaf gene,
and on the whole the agreement was reasonable. However, my
impression of this exercise is that it told us no more than we
would have anticipated from the physiological understanding built
into the logic of the model. The quantitative agreement with
actual field behaviour could be taken as indicating that the major
effects of the okra gene have been accounted for in the model, but
this could be a hazardous and stifling assumption, particularly
with genes of less predictable effect. The history of sciences is
replete with examples of confident calculations - e.g. on the
impossibility of continental movement, evolution, hydrogen bombs,
or mass flow in the phloem - which ultimately failed because they
overlooked major unrecognized factors, but in the meantime closed
off opportunities for research. The use of simulation models to
assess genetic engineering opportunities would be fraught with
comparable hazards. Modelling of the advantages of more inclined
leaves in crops, for example, has concentrated almost entirely on
light distribution within the canopy, whereas the entry of CO_2 and
its diffusivity within the canopy, the position and microclimate
of the inflorescences, and the patterns of assimilate distribution,
leaf senescence and mobilization of reserves may be just as important in determining the effect of leaf inclination on crop yield.
Nevertheless, forward projection from our present physiological
base is needed, regardless of complications, if the full potential
of genetic engineering is to be realized.

PHYSIOLOGICAL COMPONENTS OF VARIETAL IMPROVEMENT

Besides continuing selection for resistance to pests and
diseases, four other important components of selection are for:
1. phenological adaptation,
2. quality factors,
3. tolerance or resistance to environmental stress,
4. yield potential.

1. Phenological adaptation

Phenological adaptation is achieved by progressively adjust-
ing the life cycle of a crop to take maximum advantage of seasonal
climatic conditions and agronomic changes while minimizing the
adverse effects of environmental stresses. Both the overall
length of the life cycle and the timing of steps such as inflores-

cence initiation and flowering are crucial to phenological acaptation. The fact that many crop plants had spread to environments quite different from those of their origin before the physiology of flowering was understood indicates how effective empirical selection for phenological adaptation can be, and it seems likely that further fine tuning of such adaptation will best be done empirically.

Equally, however, physiological analysis of the control of flowering has undoubtedly helped by identifying the critical stages in each crop life cycle, the range of behaviour at each stage, and the most significant environmental factors. The recognition of the importance of daylength at flower initiation, for example, clarified the problems of adapting soybeans (*Glycine max*) and sorghum (*Sorghum vulgare*) to higher latitudes in the USA and of preventing flowering in crops such as tobacco (*Nicotiana tabacum*) and sugar cane (*Saccharum officinarum*). Major genes for daylength and vernalization responses are known in many crops, e.g. in wheat (Pugsley 1963, 1965) and as in the case of cereal dwarfing there may well be advantages in the use of major genes after a period of reliance on polygenic selection (Pugsley, 1983; Qualset, 1979).

2. Quality factors

By contrast with phenological adaptation, many of the changes in quality of the harvested products have been achieved by the use of major genes, enlightened by physiological analysis. Examples include lysine content in cereals, the elimination of deleterious components such as the Kunitz trypsin inhibitor, lectin, β-amylase and lipoxygenase from soybeans (Hymowitz, 1983), and the selection of high diastase barley in the presence of abscisic acid (Allison et al., 1976).

In the case of bread-making quality in wheat, varietal comparisions led to the identification of a specific sub-unit of the glutenins which is strongly associated with quality, and whose heritability is such that the subunit could readily be traced from the earliest high quality British wheats to the most recent breadmaking varieties (Payne et al., 1979). This subunit is not the only source of bread making quality in wheat, but it's identification illustrates not only how physiological analysis can aid plant breeders, but also how quality may be more amenable than environmental adaptation to genetic engineering.

3. Performance in unfavourable environments

Most crops are at some stage exposed to unfavourable environmental conditions which may limit their yield, often quite severely. Boyer (1982) has estimated that environmental stresses depress crop yields in the USA by almost 70%, far more than either diseases, pests or weeds. This estimate is a residual term in the difference

between record and average yields, however, and probably over-
estimates the effects of stress because the record yields are the
result of quite rare environmental conditions combined with levels
of inputs which would often be uneconomic. Nevertheless, genetic
improvement of tolerance or resistance to stresses such as extreme
heat or cold, chilling, drought, salinity and nutrient deficiency
or toxicity has played a major role in raising crop yields and in
extending the area on which they can be grown. Although there may
have been no increase in the extreme level of winter hardiness
over the last 50 years (Marshall, 1982), tolerance to less extreme
cold has been increased in many crops and has played an important
part in improving the early growth of crops such as maize (*Zea
mays*), soybean and cotton, and increasing their yield (e.g. Cardwell,
1982).

Plant breeders have long looked to physiologists for the
resolution of resistance to various stress conditions into major
components, and for the development of screening techniques enabling selection for these components. There has been no lack of
comparative physiological research identifying the many component
processes that appear to contribute to survival or productivity
under the different kinds of stress, as may be seen in the books
edited by Mussell and Staples (1979), Turner and Kramer (1980),
Paleg and Aspinall (1981), Christiansen and Lewis (1982) and Raper
and Kramer (1983).

Many screening tests have also been devised, yet there remains
a notable lack of examples of their effective use in plant breeding
programs for stress resistance (Marshall, 1982; Quisenberry, 1982).
Perhaps it is too early to expect this, especially when genetic
analysis has not yet complemented the physiological resolution. Or
perhaps plant breeders, and many crop physiologists too for that
matter, have yet to be convinced that selection for individual
components of stress resistance, especially those closer to the
molecular and cellular levels, is likely to result in better performance of the crop in the field (Passioura, 1981). The greater
need at present is not for more screening techniques but for evidence that selection for at least some of the characters of potential importance in resistance to stress actually improves performance under the relevant conditions in the field, and at least as
rapidly as empirical selection does. Morgan (1983) has shown that
selection for high osmoregulatory ability in wheat can result in
substantially greater yield under drought conditions than in lines
with low osmoregulation, but whether such selection will lead to
better performance than that by existing adapted varieties, or
faster improvement than that by empirical selection, remains to be
seen. The same holds for selection for greater vascular resistance,
resulting in the conservation of more water until the later stages
of crop growth (Richards and Passioura, 1981).

4. Yield potential

Yield is the ultimate outcome of the whole life cycle of the crop, and of the rates, durations and interlinkages of many processes at all stages of development, any one of which may be limiting to yield in a particular environment. No one process provides the key to greater yield potential, and different processes may limit the yield of one variety at different sites, or of different varieties at one site (e.g. Crosbie and Mock, 1981).

Where economic conditions are favourable for yield increase, e.g. for wheat in Europe, maize in the USA or rice in Korea, the already high yields continue to increase steadily. A substantial proportion of these increases has been due to greater genetic yield potential, defined as the yield of a cultivar grown in environments to which it is adapted, with nutrients and water nonlimiting, and with pests, diseases, weeds, lodging and other stresses effectively controlled. Only under such conditions can we assess progress in the assembling of productivity genes as distinct from genes for adaptation to environment, adaptedness to modern agronomy, and resistance to pests and diseases.

The improvement in yield potential of wheat varieties released in the USA, U.K. and other European countries over the last 50 years or more has been examined by van Dobben (1962), Silvey (1978), Jensen (1978), MacKey (1979), Austin et al. (1980a), Evans (1981) and others, while that for maize in the USA has been examined by Russell (1974) and Duvick (1977, 1983). Overall, the gain in yield potential has accounted for about half of the increased yield of wheat, and for even more than that (57-89%) in maize. There is no sign in any of this work that yield potential is approaching a limit in either crop. Indeed, it would appear from Silvey's (1978) analysis that the proportional contribution by greater yield potential to the rise in wheat yield is increasing, whereas during the early stages of yield advance better agronomy was probably more important.

Thus, current methods of raising yield potential by empirical selection appear to be effective. With maize, progress appears to have been steady, whereas with wheat it has been irregular, almost step-like (Mackey, 1979; Evans, 1981; Frey, 1981; Kulshrestha and Jain, 1982), a major recent step being associated with the reduction of stature using the dwarfing genes. However, there are many paths to high yield potential, as was apparent among the highest yielding modern wheat varieties examined by Austin et al. (1980a), in which the major component was high ear number (61% greater than the mean) in Benoist, grain number per ear (22% greater) in Hobbit, and kernel size (17% greater) in Huntsman. Yet the highest yielding variety of all was none of these, but one which avoided extremes, as was also the case in the comparisons by McNeal et al. (1974).

Moreover, as a comparison of rice varieties used in the Philippines over the last 70 years indicates, the constellation of favoured characters may change over the years, and even reverse, in response to changing agronomic conditions (Evans et al., 1984).

PHOTOSYNTHETIC RATE

Selection for greater photosynthetic rates in crop plants would seem, at first sight, to be a logical and direct route to greater yield potential. Most crop biomass derives from photosynthesis. Crop yields often bear a close relation to irradiance at certain stages during growth, especially when variations in temperature and daylength are small, e.g. for wheat (Evans, 1978), rice (Evans and De Datta, 1979) and maize (Jong et al., 1982). Similarly, exposure to a higher concentration of CO_2 at certain stages may also increase yield substantially in wheat (Krenzer and Moss, 1975; Fischer and Aguilar, 1976), rice (Cock and Yoshiad, 1973), soybeans (Hardman and Brun, 1971) and other crops.

Thus, there is strong evidence that crop yields may be limited by photosynthesis at certain stages and in some conditions. One might expect, therefore, that at least indirect selection for enhanced photosynthesis would have occurred during varietal improvement, but there is no evidence that it has yet done so. In wheat, for example, modern high yielding varieties produce no more biomass than older varieties of comparable duration, as shown by van Dobben (1962), Austin et al. (1980a, 1982) and Kulshrestha and Jain (1982).

Nor is there evidence of indirect selection during varietal improvement for increase in the maximum rate of photosynthesis. Varieties of higher yield potential have not been found to have higher light-saturated rates of CO_2 exchange per unit leaf area (CER) in wheat (Evans and Dunstone, 1970; Khan and Tsunoda, 1970a; Dunstone et al., 1973), maize (Duncon and Hesketh, 1968), sorghum (Downes, 1971), pearl millet (*Pennisetum americanum*) (Lavergne et al., 1979), sugar cane (Bull, 1971), cotton (El-Sharkawy et al., 1965), cowpea (Lush and Rawson, 1979) or rice (Cook and Evans, 1983; Evans et al., 1984), nor in pasture grasses such as tall fescue (*Festuca arundinacea*) (Cohen et al., 1982; Nelson et al., 1975). Indeed, for many of these species the wild relatives and primitive varieties displayed higher CER than the modern ones. For beans (*Phaseolus*) (Peet et al., 1977) and soybeans (Buttery et al., 1981) the evidence is equivocal. Bhagsari and Brown (1976) reported a higher CER in modern groundnut varieties than in wild *Arachis* species, but their rates and rankings correlated poorly between years and growing conditions, and the CER of the only closely related wild species, *A. monticola*, was close to the varietal means.

Moreover, although it has been possible to select directly for higher CER, this has not so far resulted in increased yield in maize (Moss and Musgrave, 1971; Crosbie and Pearce, 1982), soybean (Ford et al., 1983), alfalfa (*Medicago sativa*) (Hart et al., 1978) or peas (*Pisum sativum*) (Mahon, 1982). In view of the evidence that photosynthesis may often limit crop yield yet selection for higher CER has neither occurred indirectly nor succeeded directly so far, a closer physiological analysis is required. Two aspects will be considered.

Duration of photosynthetic activity

The discussion above has focussed on the photosynthetic capacity of leaves, i.e. on the maximum CER in young leaves on actively growing plants under favourable conditions. These rates usually fall as the leaf ages, and differences between cultivars in the duration of active photosynthesis may be more significant than those in maximum CER. Among the wheats, for example, the faster rate in the flag leaves of the wild diploids soon falls, whereas the lower rate in the modern varieties is maintained (Evans and Dunstone, 1970; Austin et al., 1982), possibly because of continuing demand from more prolonged grain growth. As a result, the rankings for CER are eventually reversed and they then relate more closely to yield, as also in cowpeas (Lush and Rawson, 1979), soybeans (Buttery et al., 1982) and rice (Cook and Evans, 1983). Maize improvement has likewise been associated with increased "stay green" (Duvick, 1983), while the crop simulation studies by Landivar et al. (1983b) suggest that an increase in the duration of photosynthetic activity by cotton leaves has promise for increasing lint yield.

There are, of course, complications. One is that the greater duration of photosynthetic activity must be matched by the potential for greater duration of grain or fibre growth. Grain growth in wheat may cease even though conditions are favourable and ample assimilate is still available (Sofield et al., 1977). Another complication is that greater yield may require greater withdrawal of nitrogen from the leaves, leading to earlier rather than later senescence.

Leaf area - photosynthetic rate compensation

Another reason why selection for higher CER has so far not resulted in higher yielding varieties, and why it has not been apparent during crop improvement to date, could be that there is often a negative relation between maximum CER and leaf area, with the latter being more closely related to yield. In wheat, for example, a pronounced negative relation between flag leaf area and CER has been found by Evans and Dunstone (1970), Gale et al. (1974), Planchon and Fesquet (1982), and Austin et al. (1982). Negative

correlations between leaf area and CER have also been found in soybeans (Kaplan and Koller, 1977; Hesketh et al., 1982) rice (Oritani
et al., 1979), peas (Mahon, 1982) and tall fescue (Cohen et al.,
1982).

Although this negative association is not always apparent, it
is easy to see how it can arise. The proporation of biomass allocated to leaf growth does not vary greatly between varieties in
many crops, and is partitioned between greater area on the one hand
and greater dry weight per unit leaf area (specific leaf weight or
SLW) on the other. SLW and the amount of nitrogen dm^{-2} of leaf are
often positively related to CER, hence the frequently negative
relation between the area and CER of a leaf.

It should however be possible to break this relation by selecting for high CER when this is not associated with high SLW, and is
therefore less likely to lead to reduced leaf area. This will depend on weakening the relation between SLW and N dm^{-2} or the very
close relation often found between N dm^{-2} and CER, e.g. in wheat
(Evans, 1983) and rice (Cook and Evans, 1983). The most potent
way of achieving this would be by selection for rubisco with a
higher specific activity, because this enzyme consitutes such a
major fraction of leaf protein. Several earlier reports of substantial differences in the specific activity of rubisco have not
been substantiated, but a recent analysis (Evans and Seemann, 1983)
offers some hope in wheat. However, it should not be forgotten
that photosynthetic efficiency has already been subjected to prolonged natural selection, especially under low levels of nitrogen,
whereas some other processes conferring high yield potential have
not.

CHANGED PARTITIONING OF CROP BIOMASS

Crop growth rate, photosynthesis and biomass accumulation have
so far resisted genetic improvement, although they have been greatly increased by agronomic measures such as fertilizer application.
Genetic improvement in yield potenital has come from greater partitioning of the biomass into the harvested organs, resulting in a
rise in the harvest index. Several quite different factors have
contributed to this rise, such as reduced investment in other
organs, changed balance between the phases of the life cycle,
greater remobilization of reserves, faster storage and enhanced
competitiveness of the harvested organs. The following discussion
focuses on only one of these, namely reduced investment in the
stems, taking wheat as an example.

DWARFING GENES IN WHEAT

Of all the genes deployed in plant breeding programs in recent years, those for dwarf stature in cereals have probably had
the greatest impact on world food production. Dwarfing greatly
reduces the likelihood that crops will lodge when they have been
heavily fertilized. Beyond that, however, it is widely believed
that reduced stem height has been associated with increase in
yield potential, although it is not at all clear just how this has
been achieved. The various ways in which the dwarf wheats could
use the assimilates saved by the reduced height (Rht) genes will
be considered first, and then some of the complications due to the
gibberellic acid insensitivity (Gai) associated with the major
dwarfing genes in wheat.

Potential benefits of shorter stems

The stems of shorter varieties may sometimes be as heavy as
those of taller ones, but across a wide range of varieties Hunt
(1979) found reduction in stem height to be accompanied by a proportional reduction in final stem weight. Gale and Flintham (1983)
found the Rht 1, 2 and 3 genes to reduce stem height by 17, 15 and
46% respectively, and we (Cook and Evans, unpublished) have found
them to reduce stem weight more or less in proportion to height in
isogenic lines. Thus, substantial savings of assimilate should be
available in the dwarfs for reinvestment in other organs.

That these savings are eventually used for grain growth in
high yielding varieties is suggested by the results of Austin et
al. (1980a) with a range of tall old and dwarf new varieties of
winter wheat, in which the increase in grain yield matched the
reduction in straw weight under both high and low fertility conditions. Just how this was achieved is not altogether clear, and
may vary with both variety and growing conditions. For example,
the savings from stem growth might be used immediately to increase
growth of the differentiating ear within the same shoot, or to
support greater tillering, tiller survival or root growth in the
rest of the plant, or they might be stored and remobilized later
to support greater grain growth.

Partitioning within shoots

Most rapid stem growth, the timing of which is important in
relation to these alternative uses of assimilates, occurs after
formation of the terminal spikelet and before anthesis. Although
effects of differences in assimilate use before the period of most
rapid stem growth are not ruled out, it seems likely that this
latter is the period when assimilate allocation is crucially changed.

The number of spikelets is fully determined by then, and many

comparisons between tall and dwarf varieties, isogenic lines and
segregating progeny have revealed no differences in spikelet number
associated with the Rht genes, except in earlier varietal comparisons where differences in daylength and vernalization responses
were involved, and in the experiments of Flintham and Gale (1983)
with the Rht3 gene in random F_4 lines. Therefore one of the earliest possible consequences of the Rht genes is faster growth of the
inflorescences, provided they have the capacity to respond in this
way. Lupton et al. (1974) found ear length and dry weight to be
greater from an early stage in dwarf plants than in tall ones among
random F_3 and F_4 progeny of two crosses. Comparing Cappelle-Desprez,
Hobbit and several F_4 lines, Brooking and Kirby (1981) found that
over the 2-3 week period before anthesis the reduction in stem
growth in the dwarfs was matched by the absolute increase in ear
growth, so that the total growth of stem plus ear was the same in
talls and dwarfs, the 12% reduction in stem growth rate being
equivalent to a 26% increase in ear growth rate. In this case,
apparently all of the savings from reduced stem growth were immediately invested in greater ear growth. Carbon-14 experiments by
Rawson and Evans (1971) and Makunga et al. (1978) have shown that
of the ^{14}C assimilated in the period just before anthesis, the
proportion found in the ears at anthesis, as well as at maturity,
is substantially higher in dwarf than in tall varieties. As a
result of such changes in partitioning, the ears of dwarf wheats
at anthesis can be heavier than those of comparable tall lines
(Thorne, 1982), but this depends considerably on the environmental
conditions.

Clear evidence that more florets per spikelet reach anthesis
in dwarfs than in talls is still lacking. Brooking and Kirby
(1981) and Bremner and Davidson (1978) found no advantage of dwarf
stature for the number of floret primordia initiated. Despite
this, however, the most consistently reported association with
reduced stem height in wheat is an increase in the number of grains
set per ear. This has often been found in varietal comparisons
(e.g. Makunga et al., 1978), but the most convincing evidence has
been obtained with segregating progeny. Gale and Flintham (1983)
found the Rht 1, 2 and 3 genes to be associated with reductions
in plant height of 17, 15 and 46% respectively, and with increases
in grain number per ear of 16, 21 and 39% respectively. However,
these increases were partly offset by decreases in kernel weight
of 4, 5 and 14% respectively. In fact, kernel weight is often
somewhat smaller in the presence of dwarfing genes, especially
when grain number per ear is increased (e.g. Makunga et al., 1978).
This is to be expected if much of the assimilate saved by reducing
stem growth is invested immediately in greater ear development
rather than reserved to support subsequent grain growth.

Only in the latter case would we expect to find greater accumulation of reserves in the dwarfs before anthesis, and their greater

remobilization after it. To date there is no clear evidence that
this occurs. Radley (1970) and Wojcieska and Slusarczyk (1975)
found more sugars in the flag leaves or stems of dwarf compared
with tall lines, but this was only after anthesis, while the data
of Dougherty et al. (1975) are confounded by differences in sampling times and conditions. In our experiments with near-isogenic
lines, the total extractable carbohydrates in the stems at anthesis
fall rather than rise as height is reduced, remaining approximately
constant as % of dry weight regardless of height. Nor has the loss
of stem weight between anthesis and maturity been found to be consistently greater in dwarf varieties (Hunt, 1979; Austin et al.,
1980a), even when respiratory losses are allowed for (Rawson and
Evans, 1971; Austin et al., 1980b). However, Thorne (1982) found
increased movement of ^{14}C from stem to grain after anthesis in
dwarf compared with tall wheats.

Thus, as far as changed partitioning within shoots is concerned, most of the evidence suggests that the assimilate saved by
reducing stem height is immediately invested in greater ear development, resulting most commonly in increased grain setting.

Partitioning to other organs

Dwarf wheats are often described as high tillering, and may
be, but to what extent this is due to the presence of the Rht genes
rather than extra fertilizer is not altogether clear. Near-isogenic
lines with Rht 3 on an April Bearded background had more tillers
(Gale and Flintham, 1983), but King et al. (1983) found no effect
of Rht 1, 2 and 3 on tillering in the early stages of growth.

In fact, far more tillers are initiated than survive in many
wheat crops, and the timing of stem growth is such that the assimilate savings would be expected to increase the survival more than
the initiation of tillers. This was found by Lupton et al. (1974b),
but not by Gale and Marshall (1973) or Pearman et al. (1978), nor
by Gale (1983) with isogenic lines in drilled plots.

The effect of dwarf stature on root growth is equally unclear,
but cannot be reviewed here. Varietal differences in rooting
pattern have been ascribed to dwarf stature, but work with isogenic
lines has generally not supported such conclusions.

Gibberellic acid insensitivity (Gai)

We have seen that the ways in which yield potential may be
enhanced by dwarf stature in wheat can be largely explained in
terms of changed partitioning of assimilates, mostly within each
shoot. However, the major dwarfing genes in wheat have other
effects which may have contributed substantially to the problems
of using these genes effectively in breeding programs.

Whereas the rice and maize dwarfs respond to applied gibberellins by stem and leaf elongation, the most widely used Rht genes
in wheat are manifested by, or tightly coupled with, gibberellic
acid insensitivity for stem and leaf elongation. This was first
shown by Allan et al. (1959), not only for dwarfs with Rht 1 and 2
genes, but also for those with Rht 3 derived from the variety Tom
Thumb. However, several other Rht genes in wheat, from mutagenesis
programs, are responsive to applied gibberellins (Singh et al.,
1978), and Hu and Konzak have claimed that recombination between
Rht and Gai genes can be found (Konzak, 1982). Such recombination
was not found by Gale and Marshall (1979), who consider gibberellic
acid insensitivity and reduced height to be pleiotropic effects of
the same genes, hence their notation as Gai/Rht. This insensitivity
to gibberellic acid has proved to be extremely useful in the
identification of Rht genes.

The inability of Rht 1-3 dwarfs to respond to gibberellins is
associated not only with shorter stems but also with a higher content of endogenous gibberellins, as shown by Radley (1970) and
Singh et al. (1978). Applied gibberellins are known to influence
tillering (Gale and Marshall, 1973), raise photosynthetic rate
(Gale et al., 1974) and reduce grain setting (Radley, 1980), and
the higher level of endogenous gibberellins in the dwarfs could
well have similar effects. Reduction of endogenous gibberellin
levels by the application of growth retardants such as CCC has often
been found to increase grain set in wheat (Hofner and Kuhn, 1982).
Thus, grain set in dwarf wheats could be adversely affected by
Gai/Rht genes through endogenous gibberellin levels while at the
same time being potentially favoured by the freeing of assimilates.

In fact, use of the Rht genes in many wheat breeding programs
was plagued in the early stages by a high degree of floret infer-
tility, and subsequent selection to overcome this may well have
acted by reducing either the high levels of endogenous gibberellins
or the sensitivity of grain setting to them. This was probably
necessary not only to restore floret fertility to the level found
in the tall wheats, but also to extend it beyond that in order to
take advantage of the additional assimilate freed by reducing stem
growth.

While this must remain conjectural, since empirical selection
for other changes may also have been taking place, it is supported
by the fact that the increased grain setting associated with the
Gai/Rht genes may not be apparent in some isogenic lines with
genetic backgrounds adapted to tall genes (e.g. Joppa, 1973) whereas it may be striking in both adapted dwarf varieties and random
segregating lines (Flintham and Gale, 1983; Gale and Flintham,
1983; Gale, 1983). This suggests that the effect of the Gai/Rht
genes on grain number and yield potential must depend to a considerable degree on the genotype into which they are introduced, as

has indeed been the case. Their effect also depends on the external as well as on the genetic environment. In our experiments with
near-isogenic lines, for example, increased partitioning to the ear
in dwarfs may not be apparent when irradiance is high, or temperatures are cool, or competing tillers are absent, all of these being
conditions where the internal competition for assimilates was
reduced.

The Gai/Rht genes also have several other effects, on photosynthesis, respiration, grain protein, α-amylase activity etc.
These effects cannot be discussed here, but those which already
have been suffice to illustrate some of the problems faced by the
crop physiologist in trying to help the plant breeder. In particular, there is the need to understand the full range of effects of
the Rht genes and the ways in which these interact to influence
yield, adaptation and quality. But what material will permit this?

If adapted varieties are compared, it is likely that the
genetic backgrounds of the Gai/Rht genes will have been modified
not only to minimize their undesirable effects and enhance their
useful ones, but also in other ways. On the other hand, if isogenic lines are used, the Gai/Rht genes will be on a genetic background which was selected for tall genes and may therefore restrict
the expression of the dwarfing genes, e.g. on grain setting. More
than one pair of isogenic lines is needed therefore, and preferably
reciprocal groups of lines with both the Rht and rht genes on both
tall and dwarf-adapted backgrounds.

CONCLUSIONS

1. Even with only a brief look at photosynthetic rate and stem
height, two characteristics with major impact on yield potential,
the interrelations of the processes known to be involved are so
complex, and the likelihood of still unrecognized effects is so
great, that empirical selection for greater yield potential seems
likely to be a surer path to better crops than selection for component processes. With awns, likewise, there are so many potential
benefits and disadvantages, whose relative effects depend on both
the genetic and physical environments, that empirical selection
seems the surest way forward. Delight in complexity is an occu-
pational hazard for crop physiologists, but to ignore complexity
is as hazardous as to rejoice in it.

2. Thus, besides identifying the component processes and developing techniques to screen them, the crop physiologist must analyse

their inter-relations with one another, especially the counter-
productive (and often counter-intuitive) associations. These may
negate the positive effects of selection, as in the case of photosynthetic rate through its negative associations with leaf area
and duration, or of dwarfing through its adverse effect on grain
set, until ways of breaching such associations can be found.

3. Physiological analysis is likely to help plant breeders only
when it attempts to integrate all the effects of the proposed
selection criteria at the higher levels of biological organization.
Central to this is the recognition that, as Donald (1981) has argued,
characteristics such as height or competitiveness that command
attention within a segregating population may not be conducive to
productivity within a crop community. The use of simulation models
may help to predict the integrated effects on crop growth and yield
of particular characteristics, but their output is no surer than
the physiological understanding on which the model is based. For
the same reason models cannot tell us how much headroom remains
for further progress by plant breeding for yield potential.

4. The resolution of adaptation to unfavourable environmental
conditions into many component processes, and selection for these
individually, has yet to be proved superior to empirical selection
in the rate and extent of improvement achieved. Many selection
criteria and screening techniques have been proposed, especially
at the cellular level, but effective adaptation to stress requires
the coordinated action of many processes at the levels of the
whole plant and crop community.

5. Consequently, the plant breeder's approach of gradually modifying already adapted and productive genotypes seems likely to continue to be the surest way forward. Nevertheless, the impact of
the dwarfing genes on cereal production suggests that use of major
genes for other physiological characteristics, particularly quality
factors, may have much to offer provided the genetic background
can be modified to allow their productive expression.

6. Plant breeding programs cannot wait upon physiological analysis
of their progress and, moreover, must advance simultaneously on
many fronts. Varietal comparisons are therefore not only retrospective but also confounded by many differing traits, and even
for the Gai/Rht genes it is almost impossible for the physiologist
to reconstruct how their impact on yield potential has been expressed and enhanced. If crop physiology is to play its part in helping the techniques of genetic engineering to be applied effectively
in plant breeding, it will need to make more creative use of near-
isogenic lines. These techniques may soon allow the introduction
into crop plants of major genes for physiological characters from
quite foreign sources, greatly enhancing not only the opportunities
but also the need for physiological analysis of varietal improvement.

REFERENCES

Allan, R. E., Vogel, O. A., and Craddock, J. C., 1959, Comparative
    response to gibberellic acid of dwarf, semi-dwarf and
    standard short and tall winter varieties, Agron. J., 51:737.
Allison, M. J., Cowe, I., and McHale, R., 1976, A rapid test for
    the prediction of malting quality in barley, J. Inst. Brewing,
    82:166.
Austin, R. B., Bingham, J., Blackwell, R. D., Evans, L. T., Ford,
    M. A., Morgan, C. L., and Taylor, M., 1980a, Genetic
    improvements in winter wheat yields since 1900 and associated
    physiological changes, J. Agric. Sci., 94:675.
Austin, R. B., Morgan, C. L., Ford, M. A., and Blackwell, R. D.,
    1980b, Contributions to grain yield from pre-anthesis
    assimilation in tall and dwarf barley phenotypes in two
    contrasting seasons, Ann. Bot., 45:309.
Austin, R. B., Morgan, C. L., Ford, M. A., and Bhagwat, S. G., 1982,
    Flag leaf photosynthesis of Triticum aestivum and related
    diploid and tetraploid species, Ann. Bot., 49:177.
Bhagsari, A. S., and Brown, R. H., 1976, Photosynthesis in peanut
    Arachis genotypes, Peanut Science, 3:1.
Boyer, J. S., 1982, Plant productivity and environment, Science,
    218:443.
Bremner, P. M., and Davidson, J. L., 1978, A study of grain number
    in two contrasting wheat cultivars, Aust. J. Agric. Res.,
    29:431.
Brooking, I. R., and Kirby, E. J. M., 1981, Interrelationships
    between stem and ear development in winter wheat: the
    effects of a Norin 10 dwarfing gene, Gai/Rht$_2$, J. Agric. Sci.
    Cambr., 97:373.
Bull, T. A., 1971, The C$_4$ pathway related to growth rates in sugar
    cane, in: "Photosynthesis and Photorespiration," M. D. Hatch,
    C. B. Osmond and R. O. Slatyer, eds., Wiley, New York.
    pp. 68-75.
Buttery, B. R., Buzzell, R. I., and Findlay, W. I., 1981,
    Relationships between photosynthetic rate, bean yield and
    other characters in field grown cultivars of soybean, Can. J.
    Plant Sci., 61:191.
Cardwell, V. B., 1982, Fifty years of Minnesota corn production:
    sources of yield increases, Agron. J., 74:984.
Christiansen, M. N., and Lewis, C. F., eds, 1982, "Breeding Plants
    for Less Favorable Environments," Wiley, New York.
Cock, J. H., and Yoshida, S., 1973, Changing sink and source
    relations in rice Oryza sativa L. using carbon dioxide
    enrichment in the field, Soil Sci. Plant Nutr., 19:229.
Cohen, C. J., Chilcote, D. O., and Frakes, R. V., 1982, Gas
    exchange and leaf area characteristics of four tall fescue
    selections differing in forage yield, Crop Sci., 22:709.
Cook, M. G., and Evans, L. T., 1983, Some physiological aspects of
    the domestication and improvement of rice Oryza spp, Field

Crops Res., 6:219.

Crosbie, T. M., and Mock, J. J., 1981, Changes in physiological
     traits associated with grain yield improvement in three
     maize breeding programs, Crop Sci., 21:255.

Crosbie, T. M., and Pearce, R. B., 1982, Effects of recurrent
     phenotypic selection for high and low photosynthesis on
     agronomic traits in two maize populations, Crop Sci.,
     22:809.

Dobben, W. H. van, 1962, Influence of temperature and light
     conditions on dry matter distribution, development rate,
     and yield in arable crops, Neth. J. Agric. Sci., 10:377.

Donald, C. M., 1981, Competitive plants, communal plants, and
     yield in wheat crops, in: "Wheat Science-Today and Tomorrow,"
     Evans, L. T., Peacock, W. J., eds., Cambridge Univ. Press,
     Cambridge. pp. 223-247.

Dougherty, C. T., Rooney, K. R., Scott, W. R., and Langer, R. H. M.,
     1975, Levels of water soluble carbohydrate in the preanthesis
     ear of wheat, and grain set per spikelet, N. Z. J. Agric. Res.,
     18:351.

Downes, R. W., 1971, Relationship between evolutionary adaptation
     and gas exchange characteristics of diverse Sorghum taxa,
     Aust. J. Biol. Sci., 24:843.

Duncan, W. G., and Hesketh, J. D., 1968, Net photosynthetic rates,
     relative leaf growth rates, and leaf numbers of 22 races of
     maize grown at eight temperatures, Crop Sci., 8:670.

Dunstone, R. L., Gifford, R. M., and Evans, L. T., 1973,
     Photosynthetic characteristics of modern and primitive wheat
     species in relation to ontogeny and adaptation to light, Aust.
     J. Biol. Sci., 26:295.

Duvick, D. N., 1977, Genetic rates of gain in hybrid maize yields
     during the past 40 years, Maydica, 22:187.

Duvick, D. N., 1983, Genetic contributions to yield gains of U.S.
     hybrid maize, 1930-1980, in: "Genetic Contributions to Yield
     Gains of Five Major Crop Plants," Fehr. W. R. ed. CSSA
     Special Publication No. 7, Madison, Wis. in press.

El-Sharkawy, M., Hesketh, J. D., and Muramoto, H., 1965, Leaf
     photosynthetic rates and other growth characteristics among
     26 species of Gossypium, Crop Sci., 5:173.

Evans, J. R., 1983, Nitrogen and photosynthesis in the flag feaf
     of wheat Triticum aestivum L., Plant Physiol., 72:297.

Evans, J. R., and Seemann, J. R., 1984, Differences between wheat
     genotypes in specific activity of RUBP carboxylase and the
     relationship to photosynthesis, Plant Physiol., in press.

Evans, L. T., 1978, The influence of irradiance before and after
     anthesis on grain yield and its components in microcrops of
     wheat grown in a constant daylength and temperature regime,
     Field Crops Res., 1:5.

Evans, L. T., 1981, Yield improvement in Wheat: empirical or
     analytical?, in: "Wheat Science - Today and Tomorrow," Evans,
     L. T. and Peacock, W. J. eds., Cambridge Univ. Press,

Cambridge. pp. 203-222.
Evans, L. T., Bingham, J., Jackson, P., and Sutherland, J., 1972,
     Effect of awns and drought on the supply of photosynthate and
     its distribution within wheat ears, Ann. Appl. Biol., 70:67.
Evans, L. T., and De Datta, S. K., 1979, The relation between
     irradiance and grain yield of irrigated rice in the tropics,
     as influenced by cultivar, nitrogen fertilizer application
     and month of planting, Field Crops Res., 2:1.
Evans, L. T., and Dunstone, R. L., 1970, Some physiological aspects
     of evolution in wheat, Aust. J. Biol. Sci., 23:725.
Evans, L. T., Visperas, R. M., and Vergara, B. S., 1984,
     Morphological and physiological changes among rice varieties
     used in the Philippines over the last seventy years, Field
     Crops Res., 8:105-124.
Ferguson, H., 1974, Use of variety isogenes in plant water use-
     efficiency studies, Agric Meteorol., 14:25.
Fischer, R. A., and Aguilar, I., 1976, Yield potential in a dwarf
     spring wheat and the effect of carbon dioxide fertilization,
     Agron. J., 68:749.
Flintham, J. E., and Gale, M. D., 1983, The 'Tom Thumb' dwarfing
     gene, Rht 3, in wheat. 2. Effects on height, yield and
     grain quality, Theo. and Appl. Genet., 65: in press.
Ford, D. M., Shibles, R., and Green, D. E., 1983, Growth and yield
     of soybean lines selected for divergent photosynthetic ability,
     Crop Sci., 23:517.
Frey, K. J., 1981, Capabilities and limitations of conventional
     plant breeding, in: "Genetic Engineering for Crop Improvement,"
     Rachie, K. O., and Lyman, L. M., eds., Rockefeller Foundation,
     New York. pp. 15-62.
Gale, M. D., 1983, The role and potential of dwarfing genes in
     wheat, in: "New Genetical Approaches to Crop Improvement,"
     Siddiqui, D. A., ed., in press.
Gale, M. D., Edrich, J., and Lupton, F. G. H., 1974, Photosynthetic
     rates and the effects of applied gibberellin in some dwarf,
     semi-dwarf and tall wheat varieties Triticum aestivum, J.
     Agric. Sci. Cambr., 83:43.
Gale, M. D., and Flintham, J. E., 1983, The effect of the Tom
     Thumb dwarfing gene on grain size and grain number in wheat
     Triticum aestivum, Intl. Atomic Energy Agency, Vienna, in
     press.
Gale, M. D., and Marshall, G. S., 1973, Insensitivity to gibberellin
     in dwarf wheats, Ann. Bot., 37:729.
Gale, M. D., and Marshall, G. A., 1979, A classification of the
     Norin 10 and Tom Thumb dwarfing genes in hexaploid bread
     wheat, Proc. 5th Intl. Wheat Genetics Symp., 2:995.
Gifford, R. M., 1974, A comparison of potential photosynthesis,
     productivity and yield of plant species with differing
     photosynthetic metabolism, Aust. J. Plant Physiol., 1:107.
Hardman, L. L., and Brun, W. A., 1971, Effect of atmospheric
     carbon dioxide enrichment at different developmental stages

on growth and yield components of soybeans, Crop Sci., 11:886.

Hart, R. H., Pearce, R. B., Chatterton, N. J., Carlson, G. E., Barnes, D. K., and Hanson, C. H., 1978, Alfalfa yield, specific leaf weight, CO_2 exchange rate and morphology, Crop Sci., 18:649.

Hesketh, J. D., Ogren, W. L., Hageman, M. E., and Peters, D. B., 1981, Correlations among leaf CO_2-exchange rates, areas and enzyme activities among soybean cultivars, Photosynth. Res., 2:21.

Hofner, W., and Kuhn, H., 1982, Effect of growth regulator combinations on ear development, assimilate translocation and yield in cereal crops, in: "Chemical manipulation of Crop Growth and Development, Assimilate Translocation and Yield and Creal Crops," McLaren, J. S., ed., Butterworths, London. pp. 375-390.

Hunt, L. A., 1979, Stem weight changes during grain filling in wheat from diverse sources, Proc. 5th Intl. Wheat Genetics Symp., 2:923.

Hymowitz, T., 1983, Variation in and genetics of certain antinutritional and biologically active components of soybean seed, in: "Better Crops for Feed," Ciba/Pitmans, London.

Jensen, N. F., 1978, Limits to growth in world food production, Science, 201:317.

Jong, S. K., Brewbaker, J. L., and Lee, C. H., 1982, Effects of solar radiation on the performance of maize in 41 successive monthly plantings in Hawaii, Crop Sci., 22:13.

Joppa, L. R., 1973, Agronomic characteristics of near-isogenic tall and semi-dwarf lines of Durum wheat, Crop Sci., 13:743.

Kaplan, S. L., and Koller, H. R., 1977, Leaf area and CO_2 exchange rate as determinants of the rate of vegetative growth in soybean plants, Crop Sci., 17:35.

Khan, M. A., and Tsunoda, S., 1970a, Evolutionary trends in leaf photosynthesis and related leaf characters among cultivated wheat species and its wild relatives, Jap. J. Breed., 20:133.

Khan, M. A., and Tsunoda, S., 1970b, Growth analysis of cultivated wheat species and their wild relatives with special reference to dry matter distribution among different plant organs and to leaf area expansion, Tohoku J. Agric. Res., 21:47.

King, R. W., and Chadim, H., 1983, Ear wetting and pre-harvest sprouting of wheat, in: "Third Intl. Sympos. on Pre-Harvest Sprouting in Cereals," Kruger, J. E. and La Berge, D. E., eds., Westview, Colorado. pp. 36-42.

King, R. W., Gale, M. D., and Quarrie, S. A., 1983, Effects of Norin-10 and Tom Thumb dwarfing genes on morphology, physiology and abscisic acid production in wheat, Ann. Bot., 51:201.

Konar, A., and Asana, R. D., 1975, Effect of plant competition on growth and yield of tall and semi-dwarf varieties of wheat, Indian J. Agric. Sci., 45:93.

Konzak, C. F., 1982, Evaluation and genetic analysis of semi-dwarf mutants in wheat, in: "Semi-dwarf cereal mutants and their use in cross-breeding," Intl. Atomic Energy Agency, Vienna. pp. 25-37.

Krenzer, E. G., and Moss, D. N., 1975, Carbon dioxide enrichment effects upon yield and yield components in wheat, Crop Sci., 15:71.

Kulshrestha, V. P., and Jain, H. K., 1982, Eighty years of wheat breeding in India: past selection pressures and future prospects, Z. Pflanzensücht., 89:19.

Landivar, J. A., Baker, D. N., and Jenkins, J. N., 1983a, Application of GOSSYM to genetic feasibility studies. I. Analyses of fruit abscission and yield in Okra-leaf cottons, Crop Sci., 23:497.

Landivar, J. A., Baker, D. n., and Jenkins, J. N., 1983b, Application of GOSSYM to genetic feasibility studies. II. Analyses of increasing photosynthesis, specific leaf weight and longevity of leaves in cotton, Crop Sci., 23:504.

Lavergne, D., Bismuth, E., Sarda, C., and Champigny, M. L., 1979, Physiological studies on two cultivars of *Pennisetum*: *P. americanum*, 23DB a cultivated species, and *P. mollissimum*, a wild species. I. Effects of leaf age on biochemical characteristics and activities of the enzymes associated with the photosynthetic carbon metabolism, Zeitschr. f. Pflanzenphysiol., 93:159.

Lupton, F. G. H., Oliver, R. H., and Ruckenbauer, P., 1974, An analysis of the factors determining yields in crosses between semi-dwarf and taller wheat varieties, J. Agric. Sci. Cambr., 82:483.

Lush, W. M., and Evans, L. T., 1981, The domestication and improvement of cowpeas *Vigna unguiculata L. Walp.*, Euphytica, 30:579.

Lush, W. M., and Rawson, H. M., 1979, Effects of domestication and region of origin on leaf gas exchange in cowpea *Vigna unguiculata L. Walp.*, Photosynthetica, 13:419.

MacKey, J., 1949, Genetic potentials for improved yield, in: "Proc. Workshop on Agricultural Potentiality Directed by Nutritional Needs," Rajki, S. ed., Akad. Kiado, Budapest. pp. 121-143.

McNeal, F. H., Smith, E. P., and Berg, M. A., 1974, Plant height, grain yield, and yield component relationships in spring wheat, Agron. J., 66:575.

Mahon, J. D., 1982, Field evaluation of growth and nitrogen fixation in peas selected for high and low photosynthetic CO_2 exchange, Can. J. Plant Sci., 62:5.

Makunga, O. H. D., Pearman, I., Thomas, S. M., and Thorne, G. N., 1978, Distribution of photosynthate produced before and after anthesis in tall and semi-dwarf winter wheat, as affected by nitrogen fertilizer, Ann. Appl. Biol., 88:429.

Marshall, H. G., 1982, Breeding for tolerance to heat and cold, in: "Breeding Plants for Less Favorable Environments,"

Christiansen, M. N., Lewis, C. F., eds., Wiley, New York. pp. 47-70.

Michael, G., and Seiler-Kelbitsch, H., 1972, Cytokinin content and kernel size of barley grain as affected by environmental and genetic factors, Crop Sci., 12:162.

Morgan, J., 1983, Osmo-regulation as a selection criterion for drought tolerance in wheat, Aust. J. Agric. Sci., 34:607.

Moss, D. N., and Musgrave, R. B., 1971, Photosynthesis and crop production, Adv. Agron., 23:317.

Mussell, H., and Staples, R. C., eds., 1979, "Stress Physiology in Crop Plants," Wiley, New York.

Nelson, C. J., Asay, K. H., and Horst, G. L., 1975, Relationship of leaf photosynthesis to forage yield of tall fescue, Crop Sci., 15:476.

Olugbemi, L. B., Austin, R. B., and Bingham, J., 1976, Effects of awns on the photosynthesis of wheat, *Triticum aestivum*, Ann. Appl. Biol., 84:241.

Oritani, R., Enbutsu, T., and Yoshida, R., 1979, Studies on the nitrogen metabolism in crop plants. XVI. Changes in photosynthesis and nitrogen metabolism in relation to leaf area growth of several rice varieties, Japan J. Crop Sci., 48:10.

Paleg, L. G., and Aspinall, D., 1981, "Physiology and Biochemistry of Drought Resistance in Plants," Acad. Press, Melbourne.

Passioura, J. B., 1981, The interaction between the physiology and the breeding of wheat, in: "Wheat Science - Today and Tomorrow," Evans, L. T., and Peacock, W. J., eds., Cambridge Univ. Press, Cambridge. pp. 191-201.

Patterson, F. L., Compton, L. E., Caldwell, R. M., and Schafer, J. F., 1962, Effect of awns on yield, test weight, and kernel weight of soft red winter wheats, Crop Sci., 2:199.

Patterson, F. L., and Ohm, H. W., 1975, Compensating ability of awns in soft red winter wheat, Crop Sci., 15:403.

Payne, P. I., Corfield, K. G. and Blackman, J. A., 1979, Identification of a high-molecular weight sumunit of glutenin whose presence correlates with bread-making quality in wheats of related pedigree, Theo. and Appl. Genet., 55:153.

Pearman, I., Thomas, S. M., and Thorne, G. N., 1978, Effect of nitrogen fertilizer on growth and yield of semi-dwarf and tall varieties of winter wheat, J. Agric. Sci., 91:31.

Peet, M. M., Bravo, A., Wallace, D. H., and Ozbun, J. L., 1977, Photosynthesis, stomatal resistance, and enzyme activities in relation to yield of field grown dry bean varieties, Crop Sci., 17:287.

Planchon, C., and Fesquet, J., 1982, Effect of the D genome and of selection on photosynthesis in wheat, Theo. and Appl. Genet., 61:359.

Pugsley, A. T., 1963, The inheritance of a vernalization response in Australian spring wheats, Aust. J. Agric. Sci., 14:622.

Pugsley, A. T., 1965, Inheritance of a correlated daylength response in spring wheat, Nature, 207:108.

Pugsley, A. T., 1983, The impact of plant physiology on Australian
    wheat breeding, Euphytica, 32:743-748.
Qualset, C. O., 1979, Mendelian genetics of quantitative characters
    with reference to adaptation and breeding in wheat, Proc.
    5th Intl. Wheat Genetics Symp., 2:577.
Quarrie, S. A., 1983, Genetic differences in abscisic acid physio-
    logy and their potential uses in agriculture, in: "Abscisic
    Acid," Addicott, F. T., ed., Praeger, New York. pp. 365-419.
Quisenberry, J. E., 1982, Breeding for drought resistance and
    plant water use efficiency, in: "Breeding Plants for Less
    Favorable Environments," Christiansen, M. N., Lewis, C. F.,
    eds., Wiley, New York. pp. 193-212.
Radley, M., 1970, Comparison of endogenous gibberellins and re-
    sponse to applied gibberellin of some dwarf and tall wheat
    cultivars, Planta (Berl.), 92:292.
Radley, M., 1980, Effect of abscisic acid and gibberellic acid on
    grain set in wheat, Ann. Appl. Biol., 95:409.
Raper, C. D., and Kramer, P. J., eds., 1983, "Crop Reactions to
    Water and Temperature Stresses in Humid, Temperate Climates,"
    Westview, Colorado.
Rasmusson, D. C., and Crookston, R. K., 1977, Role of multiple awns
    in determining barley yields, Crop Sci., 17:135.
Rawson, H. M., and Evans L. T., 1971, The contribution of stem
    reserves to grain development in a range of wheat cultivars
    of different height, Aust. J. Agric. Res., 22:851.
Richards, R. A., and Passioura, J. B., 1981, Seminal root morphology
    and water use of wheat. II. Genetic variation, Crop Sci.,
    21:253.
Russell, W. A., 1974, Comparative performance for maize hybrids
    representing different eras of maize breeding. Proc. 29th
    Ann. Corn and Sorghum Res. Conf., 81.
Scharen, A. L., Krupinsky, J. M., and Reid, D. A., 1983, Photosyn-
    thesis and yield of awned versus awnless isogenic lines of
    winter barley, Can. J. Plant Sci., 63:349.
Silvey, V., 1978, The contribution of new varieties to increasing
    cereal yield in England and Wales, J. Natl. Inst. Agric.
    Bot., 14:367.
Singh, B. D., Singh, R. B., Singh, R. M., Singh, Y., and Singh,
    R. P., 1978, GA3 of some wheat strains, Proc. 5th Intl.
    Wheat Genetics Symp. 1, 510-513.
Sofield, I., Wardlaw, I. F., Evans, L. T., and Zee, S. Y., 1977,
    Nitrogen, phosphorus and water contents during grain develop-
    ment and maturation in wheat, Aust. J. Plant Physiol., 4:799.
Thorne, G. N., 1982, Distribution between parts of the main shoot
    and the tillers of photosynthate produced before and after
    anthesis in the top three leaves of main shoots of Hobbit
    and Maris Huntsman winter wheat, Ann. Appl. Biol., 10:553.
Turner, N. C., and Kramer, P. J., eds., 1980, "Adaptation of Plants
    to Water and High Temperature Stress," Wiley, New York.
Vogel, O. A., Allan, R. E., and Peterson, C. T., 1963, Plant and

performance characteristics of semi-dwarf winter wheats
producing most efficiently in Eastern Washington, <u>Agron. J.</u>,
55:397.

Wattal, P. N., and Asana, R. D., 1976, Physiological analysis of
the yield of tall, semi-dwarf and dwarf cultivars of wheat
Triticum aestivum L., <u>Indian J. Plant Physiol.</u> 19:184.

Wojcieska, U., and Slusarczyk, M., 1975, The distribution of the
products of photosynthesis in the stems of long and short
strawed winter wheats, <u>Acta Agrobotanica</u>, 28:263.

Yeoh, H-H., Badger, M. R., and Watson, L., 1981, Variations in
kinetic properites of ribulose-1,5-bisphosphate carboxylase
among plants, <u>Plant Physiol.</u>, 67:1151.

QUANTITATIVE GENETIC PRINCIPLES IN PLANT BREEDING

R.J. Baker

Crop Development Centre
University of Saskatchewan
Saskatoon, Sask. S7N OWO Canada

INTRODUCTION

Quantitative genetics refers to the study of inheritance of traits whose phenotypes show more or less continuous distributions. There is no compelling reason for distinguishing among related terms such as 'biometrical genetics', 'statistical genetics', nor, in some cases, 'population genetics'. All are concerned with the study of how genes affect important characteristics, how they interact with environmental factors, and how they can be manipulated to give improved plant or animal strains.

One might ask if the understanding and application of quantitative genetic principles can be expected to enhance crop improvement efforts. If one takes the narrow view that plant breeding consists primarily of generating variability and subsequent selection of superior segregants, the answer is probably no. There is considerable latitude in some crops, and in some institutions, for plant breeding to proceed along those narrow lines. However, as time goes by, more and more plant breeders will have to 'fine-tune' their breeding programs to meet more stringent demands. An understanding of the principles of quantitative genetics is critical to the design of efficient breeding programs. Moreover, application of the methods of quantitative genetics is essential to clarify issues related to crop improvement. In the broader context, one must agree that quantitative genetics is closely allied with crop improvement.

The definition of the word 'gene' as used by quantitative geneticists appears not to have changed much since the days of

R.A. Fisher and J.B.S. Haldane, even though the science of genetics
has made tremendous progress in the intervening years. Recently, a
useful definition was developed by Richard Dawkins (1978) in his
discussion of natural selection. If his definition is translated to
the present context, one would define a 'gene' as a segment of
chromosome which is sufficiently short for it to last long enough to
function as a significant unit of artificial selection. This
definition should be close to what most plant breeders and
quantitative geneticists visualize. It seems to be practical for
both groups.

Alternate forms of a gene are referred to as alleles, though in
practice both terms, gene and allele, are often used interchangeably.
For example, the term 'gene frequency' refers to the frequency of a
particular allele at a locus.

The remainder of the genetics of quantitative genetics is quite
straightforward. Alleles at a locus segregate during meiosis.
There can be more than two alleles at one locus in a population as a
whole or in an individual of higher ploidy level. Genes on the same
chromosome will tend to be associated through successive meioses,
the closeness of the association being dependent upon their relative
proximity on the chromosome. The effect of a single gene (allele)
may be modified by another allele at the same locus (intra-locus
interaction or dominance) or by an allele at another locus
(inter-locus interaction or epistasis), or by the external
environment. Quantitative geneticists have given little thought to
the problems of extrachromosomal inheritance and such problems shall
not be considered in this paper.

Continuous variation in a quantitative trait may be due to
control by many genes or due to significant impact of non-genetic
factors upon the expression of the trait. Simply inherited traits
whose phenotypes are continuous due to environmental factors might
be considered intermediate between qualitative and quantitative.
However, since they cannot be distinguished from those traits
controlled by many genes, it is important that quantitative genetic
principles and methods apply equally well to them. Indeed, one
aspect of the study of quantitative genetics is the determination of
the number of genes and their relative effects.

The need to deal with numerous genes as well as the effect of
environment requires use of sophisticated statistical methodology.
It is difficult to itemize all possible genotypic values for more
than three or four loci and impossible to tabulate all possible
phenotypic values. Quantitative geneticists describe distributions
and relationships in terms of means, variances, covariances, regressions, correlations, path coefficients, etc. In questions concerning limits to selection, effects of sampling, and inbreeding,
researchers have developed and applied some very elegant mathematical

theory. However, there is concern that the use of statistics and mathematics in quantitative genetics, though necessary, has resulted in reluctance on the part of many plant breeders to master the basic principles of quantitative genetics. In some cases, there has been skepticism about quantitative genetic research.

Recent books by Mayo (1980) and Hallauer and Miranda (1981) discussed plant breeding in quantitative genetic terms. These authors recognized the importance of quantitative genetics in plant breeding and their books should be valuable to plant breeders and students who wish to learn how to apply quantitative genetic principles to crop improvement.

The purpose of this paper is to discuss several issues which are important in crop improvement and whose understanding and resolution depend strongly on the principles of quantitative genetics. The choice of issues to be discussed is heavily weighted toward problems in breeding self-pollinated species. However, their discussion should provide an adequate forum for showing how quantitative genetic principles can be applied to many issues of crop improvement methodology.

NUMBER OF GENES

Knowing whether or not a quantitative character is controlled by few or many genes may assist a plant breeder in planning breeding strategy. If, as Thoday and Thompson (1976) argued, most quantitative traits are such that the majority of variation can be accounted for by a few major genes, plant breeders may be able to speed initial response to selection by tracking individual genes. On the other hand, if quantitative traits are controlled by many genes (on the order of hundreds), as Comstock (1978) implied, plant breeders may have to opt for smaller short term gains in exchange for greater long term gains through recurrent selection.

The question, how many genes control a quantitative trait, is also of direct interest in the field of quantitative genetics. Recent work on selection efficiency (Crow and Kimura, 1979), choice of population sizes (Rawlings, 1979), and limits to selection (Robertson, 1970), was based on the assumption that there are many genes affecting a trait and that the contribution of any one is small relative to the phenotypic standard deviation. In contrast, some researchers such as Latter (1965) and Quereshi and Kempthorne (1968) ivestigated various aspects of selection for few genes of large effect. It is important to learn which of these two assumptions applies in particular cases.

There appear to be two basic methods for estimating the number of genes. One of them is the Castle-Wright segregation index

(Wright, 1934) which is based on the relationship between the range
and the standard deviation of genetic values. The method can easily
be modified to apply to any population with known genotypic
frequencies. It is generally known that the method gives a minimum
estimate of the number of genes. One of the difficulties with the
segregation index is the need to assume that the two parents
represent the complete range of genotypic values. In cases where
such an assumption is obviously not true, some researchers
substituted the observed phenotypic range of the segregating
population being studied. The phenotypic range, in the presence of
environmental variation, will tend to be biased upward as an
estimate of genotypic range and will therefore tend to give
estimates of gene numbers which are also biased upward. However, it
is highly unlikely that, even with as few as four or five genes, the
observed genotypic range in a finite sample will represent the true
genotypic range. Small sample sizes will therefore result in
downward biases if this method is used (Mulitze, 1983).

A closely related method is that based on observed response to
long-term selection. If selection is practiced for both higher and
lower performance, the extreme lines after a number of cycles of
recurrent selection should be a reasonable estimate of the genotypic
range. In this situaion, the initial genetic standard deviation
would bear a relationship to the genetic range that would be
dependent upon number of genes and initial gene frequencies. The
method would be sensitive to assumptions concerning the initial gene
frequencies and should probably be applied only to populations
developed by crossing two inbred lines.

The other approach to estimating numbers of genes is
represented by the methods of Wehrhahn and Allard (1965) and of
Jinks and Towey (1976). Wehrhahn and Allard (1965) developed the
inbred-backcross method for estimating numbers of genes. The method
consists of generating a number of backcross genotypes and, on the
basis of tests of inbred progeny, classifying the genotypes as being
different from or not different from the recurrent parent. The
estimated frequency of non-parental genotypes can then be compared
to genetic expectations for the appropriate number of backcrosses to
provide an estimate of the number of genes affecting the trait.

A related method is that of genotype assay (Jinks and Towey,
1976). In this method, one generates two or more grand progeny
families from each of a number of plants in a segregating
population. If the families differ, the parent is declared to be
segregating at at least one locus for the trait under study. The
frequency of heterozygous genotypes designated by this approach can
then be compared to theoretical expectations for different numbers
of genes.

Both inbred-backcross and genotype assay methods are designed for use in crosses between inbred lines and are applicable only to species that can be highly inbred. Both methods suffer from a need to evaluate a large number of genotypes in order to obtain sufficiently precise estimates of frequencies of non-parental or segregating types (Mulitze, 1983). Mulitze (1983) has also shown that both methods are very sensitive to the balance between type I and type II errors. Unfortunately, the correct choice of error rate depends in part on the actual number of genes controlling the trait.

In evaluating the genotype assay method, Mulitze (1983) concluded that the formulations given by Jinks and Towey (1976) and by Towey and Jinks (1977) pertain to progeny of assayed plants rather than to grandprogeny of assayed plants as implied by the authors.

None of these methods for estimating numbers of genes are very satisfactory. All tend to give unreasonably low estimates of numbers of genes, particularly if heritability is low. Perhaps the wisest approach to the problem of the number of genes is for plant breeders to assume in the first instance that all continuous traits are controlled by many genes. If observed distributions tend to be multimodal, if response to selection is considerably faster than anticipated, if a trait seems to be easily manipulated through backcrossing, or if major differences in the trait tend to be associated with simple inheritance of other traits, the researcher might then embark on a more energetic study of the inheritance of the trait. Subsequent study might include the Castle-Wright segregation index applied to later generations (Mulitze, 1983) as well as testcrosses and perhaps linkage studies. Only when the plant breeder is confident that he or she can identify factors that have a major influence on a quantitative trait would he or she choose to alter the strategy developed on the assumption of many genes.

An example of a quantitative trait being subject to major influence by a single gene or a group of linked genes was reported by Leisle et al. (1981). They observed a linkage between gluten strength, glume color, and presence or absence of two gliadin electrophoretic bands. Gluten strength, as measured by the micromixograph test, showed the continuous distribution of a quantitative trait. From their results, it seems quite possible to follow a gene or linked group of genes affecting gluten strength through a breeding program by rating genotypes for glume color or by carrying out gliadin electrophoresis.

EPISTASIS

The question concerning the role of epistasis in crop
improvement has two parts. Is epistatis important in the control of
quantitative traits and, if so, is epistasis apt to have a serious
impact on methods for crop improvement? An assessment of the latter
question requires formulation of response to selection in a fairly
general manner. Selection can be thought of as involving three
types of units where a unit may be a single plant or some type of
family (Baker, 1966). Selection of a selection unit is based upon
the observed phenotypic value of a genetically related criterion
unit. With mass selection, these two units are the same plant. In
family selection, they may be the same full- or half-sib family. If
selection is carried out on males but based on performance on female
siblings, the criterion and selection units are, respectively, the
female and male sibs of a family.

From each selection unit, a genetically related response unit
can be derived by direct descent. The purpose of selection is to
modify the mean genotypic values of the response units. For mass
selection, where all units are identical, individuals are selected
on the basis of their own phenotypic value and the mean genotypic
value of the selected individuals reflects response to selection.
In full-sib reciprocal recurrent selection, the selection unit is a
single plant, the criterion unit is the full-sib family produced by
crossing that plant with one from the companion population, and the
response unit is the S_1 progeny developed by selfing that plant.
Hallauer and Miranda (1981) discussed selection methods in terms of
selection units and recombination units. Their recombination unit
is equivalent to the response unit defined above.

In general, response to selection can be predicted in terms of
(a) the covariance between the genotypic values of the response
units and the phenotypic values of related criterion units, (b) the
phenotypic variance of the criterion units, and (c) the standardized
selection differential. Letting C refer to the phenotypic values of
the criterion units and R to the genotypic values of the response
units, predicted response to selection is given by

$$\text{Response} = i\ \sigma_C\ b_{RC}$$

where b_{RC} is
the regression of genotypic values of response units on phenotypic
values of criterion units and can be expressed as the ratio of the
covariance between R and C to the variance of C. The term $i\sigma_C$
represents the selection differential and can be written as the
product of the standardized selection differential and the
phenotypic standard deviation of criterion units when truncation
selection is practiced and when phenotypic values are normally
distributed. This general prediction equation is based on the
assumption that genotypic values of response units are linearly

related to phenotypic values of criterion units. The assumption
seems acceptable as long as genotypic effects and environmental
effects are not correlated.

The critical part of the equation in evaluating the importance
of epistasis concerns the covariance between criterion and response
units - the genetic part of that covariance in particular. Where
the above prediction equation holds, epistasis will be important
only if it has an important impact on the genetic covariance between
response and criterion units.

Cockerham (1963) stated that, in the absence of linkage and inbreeding, the coefficients of the various types of epistatic variance in any covariance between relatives can be determined as the
products of the coefficients of the corresponding additive and
dominance components. Thus, if a particular covariance has a coefficient k_a for the additive genetic variance and a coefficient
k_d for the dominance genetic variance, then the coefficients for
additive x additive, additive x dominant, and dominant x dominant
epistatic variances should be k_a^2, $k_a k_d$, and k_d^2 respectively.
For a series of covariances among non-inbred relatives, Cockerham
(1963) showed that k_a rarely exceeds 0.5 and the k_d rarely exceeds 0.25. Thus it would appear that the additive x additive component of epistatic variance would be expected to make the largest
contribution to covariances among relatives and its contribution
could be half as important as the contribution of additive genetic
variance itself. Cockerham's results also suggested that, in
non-inbred populations at least, contributions of trigenic and
higher order epistatic interactions to covariances should be small.

Crow and Kimura (1970) argued that epistasis was of little
importance in response to selection. They gave an example of the
impact of complementary digenic interaction on estimation of
heritability. Epistatic effects were considered to be quite large
while the effect of epistasis on heritability was considered quite
small. By ignoring epistasis, the estimate of heritability would be
expected to be in error by about 2.5 percent. They stated that it
is "these reasons, as well as the practical difficulty of measuring
epistasis, that lead the breeder to ignore epistasis".

These arguments suggest that epistasis is not apt to be an
important factor in covariances among relatives, and therefore
response to selection, in non-inbred populations. Recognizing that
response to selection in inbred populations will also depend upon
genetic covariances between relatives, one must ask whether or not
the same general conclusion can be applied to such species.
Cockerham (1963) showed that use of inbred progeny would greatly
increase the contributions of epistatic variances to various
covariances among relatives. One may even take the complementary
digenic model discussed by Crow and Kimura (1970) and determine the

impact of epistasis on the covariance between parents and progeny in
completely inbred population (e.g. single seed descent lines). In
this case, inbred progeny have the same genotype as their parents
and the parent-progeny covariance is merely the variance among
parents. Among the completely inbred progeny of a dihybrid cross,
the genotypes AABB, AAbb, aaBB, and aabb are each expected to occur
with frequency 0.25 in the absence of linkage. With complementary
interaction, their relative genotypic values would be 1, 0, 0, and
0, respectively. Additive genetic variance in this model would be
0.125 while additive x additive epistasis would be 0.0625.
Ignorning epistasis in this case would result in an error of 50
percent!

Epistasis may have much greater impact on response to selection
in an inbred crop than in a cross-fertilized crop. With inbreeding,
the coefficient for additive genetic variance in covariances between
relatives can approach unity with the implication that the
coefficients for all additive types of epistasis can also approach
unity.

The importance of epistasis in the control of quantitative
traits was considered by Hallauer and Miranda (1981) in a review of
estimates reported for maize. They concluded that attempts to
detect epistatic variances have generally not been successful but
they noted that tests based on first-order statistics invariably
show significant epistatic effects. They suggested that the
difference may be due to differences in error associated with
estimating first- and second-order statistics. Another possibile
explanation is that relatively large epistatic effects might not
lead to large epistatic variances. As stated by Crow and Kimura
(1970) and as implied by Cockerham (1963), large epistatic effects
may not have a large impact on covariances between relatives in
non-inbred populations.

Stuthman and Stucker (1975) concluded that additive x additive
epistatic variances appeared to be important components of the
genetic variance for yield of near homozygous progeny in oats.
Although their conclusion may reflect a greater contribution of
epistasis to covariances in inbred species, there is concern about
the possible impact of linkage disequilibrium on their estimates.

Epistasis can be detected by either of two basic approaches.
One can compare various covariances among relatives that have
different expected epistatic contributions (Cockerham, 1963). This
approach requires the estimation of second-order statistics with
concomitant large errors. Cockerham pointed out the advantages of
starting with inbred parents. The second approach involves the
comparison of means of different types of hybrids. First-order
statistics are used to compare hybrid means with those expected in
the absence of epistasis. Significant deviations indicate the

presence of epistasis. The triple testcross method of Kearsey and
Jinks (1968) should give a reliable test for the presence or absence
of epistatic effects.

In random mating populations, epistatic variances will have
their greatest impact at intermediate allele frequencies. At these
frequencies, digenic or higher combinations occur with frequencies
of 0.25 or lower. It is likely that these genotypes may not have
significant impact on short term response to selection. However,
this argument does not seem to apply to inbred species. Horner et
al. (1955) stated that, if a large portion of the genotypic variance
in a self-fertilizing species arises from epistasis, it may be
necessary to practice selection at higher levels of inbreeding. It
appears that epistasis may be much more critical in inbred species
and that there is relatively little reliable information on the
contributions of epistatic and non-epistatic effects to covariances
among relatives for important traits in many crop species.

LINKAGE

Linkage must certainly be a factor in the inheritance and
co-inheritance of many quantitative traits. This is particularly
true in self-pollinated crops where opportunities for recombination
are limited.

The impact of linkage on plant breeding was the subject of a
series of papers by Hanson (1959a, b). Hanson showed that large
portions of chromosomes are expected to remain intact through the
successive meioses of a typical breeding program. He concluded that
"If intermating in a self-pollinated species is at all possible, at
least one or more, and preferably four, intermating cycles should
precede the selfing generations to insure a degree of breakup of the
linkage groups and to increase the genetic recombination within the
linkage group." He further concluded that four or more parents
should be used in creating intermating populations (Hanson, 1959b).

The effects of linkage on genetic variances and covariances of
quantitative traits are well understood. Coupling linkages increase
genetic variances of traits and cause positive covariances between
traits. Repulsion linkages reduce genetic variances and cause
negative covariances. The effects are greater in magnitude as
linkage becomes tighter.

Coughtrey and Mather (1960) pointed out that, for several or
many genes, repulsion linkages will be less common among randomly
chosen individuals than will coupling linkages. For three loci, it
is possible to have all three possible pairs in coupling whereas a
maximum of two can be in repulsion. The effect becomes more
striking as the number of loci increases.

Miller and Rawlings (1967) and Meredith and Bridge (1971) investigated the effects of intermating on genetic variances and correlations in cotton, (Gossypium ssp.). They found that intermating had the expected effects of reducing the absolute magnitudes of correlations between traits and of increasing or decreasing the genetic variances for traits that were expected to be affected primarily by genes in repulsion or coupling, respectively. There is little doubt that intermating will indeed lead to greater breakup of initial linkage blocks. In considering the problem of how many crosses would have to be made in an intermating generation, Baker (1968) concluded that as few as 30 crosses among randomly chosen pairs would assure that the expected effects of intermating would not be completely negated by genetic drift.

Bos (1977) pointed out that there is not a general agreement on the need for breaking linkage blocks in a breeding program. Of those in favor of using intermating, he listed Miller and Rawlings (1967), Meredith and Bridge (1971), and Redden and Jensen (1974). One would have to include the unequivocal recommendation of Hanson (1959) and, from the tone of the article, should probably have included Baker (1968). Those of a contrary opinion include Bos himself (Box, 1977), Pederson (1974), and Stam (1977).

Bos (1977) considered theoretical expectations of dihybrid ratios in F_3, in progeny of intermated F_2, and in completely inbred progeny developed from each of those two populations. Starting with an F_1 in repulsion, he showed that the frequency of the recombinant desirable homozygote was higher in the F_3 than in the progeny of intermated F_2 plants. While these two populations are chronologically at the same level, they differ in that one has been intermated while the other has undergone an additional generation of inbreeding. This method of comparison gives the rather disturbing result that intermating has an apparently detrimental effect even in the absence of linkage between genes. If Bos had compared the progeny of the intermated F_2 and the F_2 itself, he would have obtained results that were in agreement with those from the completely inbred populations and in which there is no difference in the absence of linkage. The strength of Bos' agrument against using intermating must therefore lie in whether or not the delay of one generation or more for intermating is warranted.

Bos (1977) rightly pointed out that the maximum advantage of intermating is to increase the frequency of the recombinant double homozygote by 25 percent. However, leaving the argument at that point is somewhat unsatisfactory. For example, with five pairs of linked loci, an advantage of even 10 percent for each pair could translate into an overall advantage of over 60 percent when all five pairs are considered.

Pederson (1974) presented arguments against intermating in self-pollinated crops. Based on computer simulation of up to eight genes on each of up to three chromosomes, and considering all possible combinations of genes with the one restriction that each parent contribute the same number of plus alleles to each linkage block, Pederson made the following observations. Intermating will always be of benefit for a character controlled by loci on a single short chromosome segment. The increase in genetic variance will be on the order of 20 percent. For a character controlled by loci spread over three or four long chromosome segments, intermating will be of doubtful value because of the overriding effect of chromosome reassortment.

Stam (1977) used computer simulation to evaluate the effects of continuous intermating, intermating for one generation, and continuous selfing on response to continuous mass selection. Based on simulation of various numbers of genes, two levels of heritability, and two levels of selection intensity, Stam concluded that there was little advantage of intermating apparent in early generations, that selfing was preferred in early generations if heritability was low and that response in later generations was always greater in the intermated population. Stam (1977) further suggested that, if one generation of intermating were to be used, the optimum generation may well be the F_4 or F_5 rather than the F_2. The latter conclusion was based on theoretical considerations for two linked loci as well as simulation of 4 to 32 loci.

The prevailing quesion about linkage, particularly in breeding self-pollinated species, is whether or not repulsion linkages are important enough to warrant breaking of initial linkage blocks or whether satisfactory progress can be made from chromosome reassortment. Quantitative genetic theory shows that linkage blocks can be broken and that one generation of intermating at an optimum time can have a measureable effect. However, short term response to selection may be satisfactory with chromosome reassortment.

Linkage can have important impact in backcrossing and in the development of near-isogenic lines. Hanson (1959a) discussed the impact of linkage in backcrossed generations and showed that the segment of chromosome carried along with the one being selected would be of significant magnitude, especially with few backcrosses. It should be expected that use of donor parents with unsatisfactory quantitative attributes will often have serious impact on the performance of backcross derived lines.

PREDICTING RESPONSE

The greatest use of quantitative genetic theory in plant breeding is for predicting response to various strategies for crop

improvement. If predictions based on quantitative genetic theory
are sufficiently reliable, predicted responses to alternate
selection strategies can be used as a basis for choosing the most
efficient approach. Prediction formulae often serve to highlight
issues which are critical to the choice of crop improvement
methods.

The most general formula for response to selection was
given earlier which states that, when truncation selection is
applied and when phenotypic values are normally distributed,
response to selection is expected to be equal to the product of (i)
the standardized selection differential, (ii) the phenotypic
standard deviation of the criterion units, and (iii) the linear
regression of genotypic values of response units on the phenotypic
values of the related criterion units. This prediction is based on
the assumptions that there is a linear relationship between
genotypic values and phenotypic values and that the average
genotypic value of progeny of selected units is equal to the average
genotypic value of the selected units.

The first assumption requires only that the phenotypic values
be equal to the sum of genotypic effects and uncorrelated
environmental effects. The assumption is readily accepted in most
instances. It should be questioned when genotypes are not
properly randomized with respect to environmental factors or when
the phenotypic value is a non-linear function of genotypic and
environmental effects. In the case where the phenotype is a
non-linear function of effects, it is sometimes possible to
linearize it by appropriate transformations. In most cases, it
will not be possible to test for non-linearity.

Assuming that the progeny mean equals the mean of the selected
parents requires that genotypic values be additive and that selected
parents be random mated to produce the next generation. The first
part of this assumption can be circumvented in those cases where the
breeder can estimate the regression of additive genetic values on
phenotypic values. The latter part of the assumption can be readily
evaluated and an additional term can be included in the prediction
equation to allow for its failure. An even more generalized
approach to predicting response to selection has been proposed by
Griffing (1968).

The basic prediction formula can be modified in several ways.
One that I find very useful is to express expected response to
truncation selection as

$$\text{Response} = G_s - G_u = i\sigma_P b_{GP} = i\sigma_B r_{GP} \quad \text{and} \quad G_s = G_u + i\sigma_G r_{GP}$$

In the latter form, it is quite evident that expected response is
the product of (i) the standardized selection differential, (ii) the

correlation between genotypic and phenotypic values, and (iii) the genotypic standard deviation. Recognizing that primary interest is not in response itself, but in the mean genotypic value of the selected units, requires inclusion of a fourth component, (iv) the genotypic mean of the unselected population.

This formulation of respose to selection has advantages in pointing out some of the critical features of artificial selection. Item (iv), G_u, can be maximized by choosing a relatively well adapted starting population for the breeding program. However, this will usually be accompanied by a decreased genotypic standard deviation (item iii). Both of these items will be fixed once the starting population is chosen. The proper balance between the two conflicting items must be reached. It should be evident that item (iv) would be most important in short term selection experiments while item (iii) is more critical to long term selection experiments.

A similar balance is required between items (i) and (ii). The selection differential depends only on the number of genotypes tested and the number selected. The correlation between genotypic value and phenotypic value depends on how thoroughly each genotype is tested. Increasing the standardized selection differential through increasing the number of lines tested may be accompanied by a reduced correlation because each genotype can no longer be tested as thoroughly.

Much has been written about methods for increasing the correlation between genotypic values and phenotypic values. Experimental procedures, extent of replication within and across macro-environments, and use of highly heritable correlated traits may increase the correlation. Comparisons of alternate selection strategies often concentrate on this one factor and tend to ignore the other three factors that contribute to the performance of final selections.

It is known that the above prediction equation or its equivalent apply only to short term selection experiments. In the longer run, predictions would have to be able to account for asymmetrical responses and decreases in response rate.

One aspect of response to selection which is not fully appreciated is that of variability. Replicated selection experiments with laboratory species always show variation in response to selection. This topic has been the subject of theoretical study by Baker (1971, 1975) and Hill (1974). That selection gives variable response is beyond question. It is caused by random sampling at various stages including the selection stage itself. Response will be most variable when heritability is low and when too few individuals are selected in the selection step of the breeding program.

Variability of response to selection is critical to plant
breeders in two ways. Firstly, experiments to compare different
breeding methods should be replicated. Without replication, there
is no measure of variability and therefore no basis for declaring
that an observed difference is anything but fortuitous. There is a
possibility that theoretical arguments concerning the variability of
response may strengthen conclusions from unreplicated selection
experiments.

The second consequence of variability of response is the
possibility that sampling variability could completely destroy any
expected response from selection if population sizes are such that
variability is high. Of course, the opposite is also true -
realized response may be substantially greater than expected due to
random variability. There is little doubt that, in long-term
experiments, sampling variability will have a detrimental effect on
response to selection. Even in short-term selection experiments, it
is doubtful that a plant breeder would care to run a high risk of
obtaining no response because of random variability.

An important question in breeding inbred species is whether or
not the standard prediction formulae apply to those crops. Pederson
(1969a,b) considered the theory of response to selection in inbred
species and showed that inbreeding has an important impact on the
form of the prediction equation. Although the basic principles are
the same in inbreeding species, one has to be concerned about
covariances between plants in the generation where selection is
practiced and those in the completely inbred sub-populations that
would be derived from those plants. In random mating populations,
genotypic frequencies remain constant from generation to generation
while, in inbred species, genotypic frequencies change from
generation to generation. Thus it is necessary to have different
prediction formulae for each possible generation in which selection
might be practiced.

In recent years, several authors have been using the theory of
Kimura (1957) to investigate the probability of fixing desirable
alleles in plant breeding programs. The approach has considerable
appeal partly because it considers the effects of selection on the
individual loci influencing a quantitative trait. Response is
measured as the probability of fixing desirable alleles. Important
factors in response turn out to be initial allele frequency,
effective population size, and the selective advantage of the
desirable allele. Crow and Kimura (1979) have shown that, with
truncation selection, the selective advantage of an allele is
dependent upon the number of genes affecting a trait, its
heritability, and the selection intensity. With this paper, it
became possible to adopt what was traditionally a population

genetics or evolutionary approach to the problems of artificial selection.

Comstock (1978) and Rawlings (1979) used this general approach to discuss problems of optimum population sizes to be used in recurrent selection programs. Their conclusions were somewhat equivocal but have served to provide guidelines and have shown that definite choice of population size depends largely on the effects of alleles relative to the total phenotypic variance as well as their initial frequencies.

It is tempting to try to apply the concept of fixation of desirable alleles to the theory of breeding inbred species. In many programs, selection is carried out in crosses between two inbred lines and the result of the breeding program is an inbred line. In considering individual loci which influence a quantitative trait, it is clear that the F_1 will contain loci of three types; a) homozygous for the desirable allele because both parents carried that allele, b) heterozygous for the desirable allele because one parent carried it and the other did not, and c) homozygous for the undesirable allele because neither parent carried the desirable allele. Simple logic dictates that the best crosses will be those in which most of the loci are in the first or second class and there can be little doubt about the importance of choosing good crosses.

Selection within a cross can act only on loci of the second class, i.e. those loci which are heterozygous for the desirable allele in the F_1. The objective of the breeding program would be to increase the probability of fixation of those desirable alleles beyond the 0.5 level expected in the absence of selection. Bailey and Comstock (1976) presented some theory related to the probability of fixation of desirable alleles in inbred species. Their theory indicates that the probability of fixing the desirable allele at a segregating locus should be about 0.7 when both parents are similar in phenotype. When one parent is substantially better than the other parent, the probability of fixation of desirable alleles contributed by the better parent is considerably higher while the probability of obtaining a new inbred line which is superior to the better parent is much reduced. Bailey and Comstock (1976) had assumed that the selective advantages of desirable alleles were in the range of 0.56 to 0.82 (which they considered to be maximum for most situations) and that selection was continuous during inbreeding.

The unliklihood of obtaining inbred lines in which all the desirable alleles are concentrated led Bailey and Comstock (1976) to propose two modifications to modified pedigree selection which would improve those odds. Both procedures required some cross-pollination among selected genotypes.

Comstock (1978) suggested that one of the most significant contributions of quantitative genetics to maize breeding has been the development of a theoretical basis for comparing selection systems in terms of predicted response to selection. That statement can be extended to all economically important crops. Comstock also noted the importance of an approximate theory for the fixation of desirable alleles in enabling significant insights relative to optimum effective population sizes. Work by Bailey and Comstock (1976) showed that this same approach is useful in considering improvement in self-pollinated species.

POPULATION SIZES

One approach to the consideration of population sizes for plant breeding programs is that taken by Shebeski (1967) and Sneep (1977). These workers formulated their arguments in terms of hexaploid wheat (Triticum aestivum L. em Thell) with 21 independent loci affecting a quantitative trait. The objective of their approach was to generate and identify the one genotype that contained the desirable allele at all 21 loci. For example, Sneep (1977) pointed out that one F_2 plant in 421 would be expected to contain all 21 desirable alleles in the homozygous or heterozygous state. To be "practically certain" (i.e. probability greater than 0.98) of having an F_2 which contains at least one of the desired genotypes, one must grow an F_2 of 1684 plants. Recognizing that selection based on single F_2 plants is often unreliable, Sneep suggested testing of 1684 F_2-derived F_3 lines each consisting of a minimum of 228 plants - a total population size of 383,952 plants.

This type of reasoning leads to conclusions which indicate i) that there is a need for handling very large populations, ii) that selection should be practiced in early generations to avoid loss of desirable alleles, and iii) that single seed descent is too unreliable to be used as a method for breeding self-pollinated crops. Although the calculations used by Shebeski (1967) and Sneep (1977) are technically correct, the method errs in the choice of a practically impossible goal and in the failure to recognize that environmental influences severely limit ability to recognize superior genotypes.

Recent research on the probability of fixation of desirable alleles highlighted the importance of effective population size. In most breeding programs, effective population size is closely related to the number of individuals selected in each cycle of selection. Rawlings (1979) concluded "that an effective population size on the order of 30 would be reasonably adequate for many genetic systems." Rawlings also discussed optimum selection intensity and argued that selection intensities in the range of 1/10 to 1/4 would provide a reasonable balance between short-term and long-term objectives in a

recurrent selection program. Based on Rawlings' paper, one would
conclude that maximum population sizes of 150-300 would provide a
reasonable basis for recurrent selection.

Comstock (1978) carried out similar calculations but had
considered that it was important to be able to fix genes whose
initial frequencies were as low as 0.05 or 0.1. As a result, he
left the impression that effective population sizes on the order of
200-400 would be required to maximize response to long-term
selection. If selection intensity were 1/4, this would require
population sizes of 800-1600.

Bailey and Comstock (1976) discussed the probability of
fixation in breeding self-pollinated crops where the starting
population is a cross between two inbred lines. Because gene
frequencies are all at 0.5, effective population sizes can be quite
small. For example, one method they suggested for concentrating
desirable alleles would consist of selecting S_1 families followed
by crosses between random members of selected families. In their
example, they suggested a selection intensity of 1/10 thus requiring
10N S_1 families to be evaluated; N is the required effective
population size. They showed that, even with genes of quite small
effect, effective population sizes of 8 or 16 would give high
probabilities of fixation. This suggests that use of 80-160 S_1
lines would be reasonable if such a scheme were to be used.

From the research reported above, it seems clear that, in
recurrent selection programs, effective populations sizes should
be on the order of 10-20 if initial gene frequencies are 0.5, 30-40
if initial frequencies are 0.1 to 0.5, and 200+ for initial
frequencies less than 0.1. Moreover, these workers recommend
selection intensities of approximately 10 percent as being conducive
to optimum response. The latter conclusions agrees with findings
of Hill (1974) and Baker (1971) who found that the coefficient of
variation of response to selection tended to be minimum when
selection intensities were 10 to 20 percent.

An approach to the question of population size, particularly
in a single cycle of selection in a self-pollinated species, can be
constructed from consideration of the following prediction equation
for response to selection:

$$\text{Response} = i\ \sigma_G\ h$$

If resources are limited in that a maximum number of plots can be
grown, one has to strike a balance between allocating those
resources to testing greater numbers of lines (higher i) or to
testing each line more thoroughly (higher h). The following table
lists the optimum number of lines to test for several different
combinations of single plot heritability, number of lines to be
selected, and total number of plots to be grown. The calculations

make use of the fact that heritability of means of lines replicated
r times, h_r^2, equals $r h_1^2/(1 + h_1^2(r-1))$. The selection
intensity i is taken from Appendix Tables A and B of Falconer
(1981). This approach is similar to that due to Baker and reported
by Townley-Smith et al. (1973).

Table 1. Number of inbred lines to be tested to obtain maximum
expected response to selection

Total number of plots	Number of lines selected	Heritability (single plot)		
		0.2	0.5	0.8
5000	5	500	1250	5000
	10	556	1667	5000
	50	1000	2500	5000
3000	5	375	1000	3000
	10	429	1000	3000
	50	600	1500	3000
1000	5	167	500	1000
	10	167	500	1000
	50	333	1000	1000
100	1	20	50	100
	2	25	50	100
	5	25	100	100

The results of these calculations show quite clearly that, when
heritability is low to intermediate, it is better to expend
resources on extra replication of each line than on testing extra
lines. In fact, the results support the argument that it is likely
that testing any more than about 1000 lines in any one selection
step is inefficient in a breeding program. If one can test more
than a thousand lines, it would be more efficient to replicate each
line and thereby increase heritability.

Single seed descent is conceptually a simple method for
breeding self-pollinated crops (Brim, 1966). It consists of harvesting a single seed from each F_2 plant to produce the F_3 and
repeating the procedure until the progeny are essentially homozygous. Selection is then practiced on the homozygous lines as
suggested in the example above. Sneep (1977) dismissed the method
because of "a great loss of desirable alleles". In fact, it can be
shown that as few as 11 single seed descent lines would be sufficient to give a 99 percent probability that all 21 independent

desirable alleles are represented in one or another of the 11
plants. What is almost certainly lost is the one genotype which
contains all 21 alleles in homozygous or heterozygous condition.
Sneep (1977) has pointed out that this genotype occurs with
frequency one in 421 in the F_2 and one in 2,097,152 in completely
inbred lines.

It should be possible to develop some guidelines for a minimum
population size to be used with single seed descent. The distribution
of genotypic values in $\underline{n}$ single seed descent lines should be
equivalent to the distribution of genotypic values in a random
sample of size $\underline{n}$ taken from a large population of completely inbred
lines developed from the same cross. One should choose a sample
size of sufficient magnitude that the sample characteristics are
similar to the population characteristics at least in critical
aspects. Relatively small samples will assure that the means of the
sample and population are in close proximity. Somewhat larger
sample sizes would be required to assure that the variances of the
sample and population were similar. From a plant breeding point of
view, perhaps the most critical aspect of the sample distribution is
whether or not the most extreme value is represented. The following
table was developed to show the expected maximum value of various
sample size under the assumptions that gene are independent and of
equal effect. Expected maximum values were calculated as

$$\sum_{x=0}^{n} x \, P_x \quad \text{where}$$

$$P_x = [\sum_{i=0}^{x} {}^{n}C_i \, \tfrac{1}{2}^n]^s - [\sum_{i=0}^{x-1} {}^{n}C_i \, \tfrac{1}{2}^n]^s$$

is the probability that x is maximum value in a sample of size s;
$x=0,1,\ldots n$ where n = number of loci.

Table 2. Relative expected maximum genotypic values in various
          sample sizes

Sample size	\multicolumn{5}{c}{Number of genes}				
	5	10	20	50	100
20	0.89	0.79	0.71	0.63	0.59
50	0.96	0.84	0.75	0.66	0.61
100	0.99	0.88	0.77	0.68	0.62
500	1.00	0.94	0.83	0.71	0.65
1000	1.00	0.96	0.85	0.73	0.66
5000	1.00	1.00	0.89	0.76	0.68
∞	1.00	1.00	1.00	1.00	1.00

The data in the table show that samples of 50 or 100 should be sufficient for traits governed by five or fewer genes. The expected maximum value (i.e. the maximum value that will be observed in at least 50 percent of the samples of this size) will be close to the maximum possible with sample sizes of 100 or more. For more than 20 genes, the expected maximum even with sample sizes of 5000 is nowhere near the maximum possible value. In these cases, sample sizes of 500 or 1000 are similar to what is expected with a sample of 5000 and there is little value in going beyond those levels.

This discussion of population sizes should leave the general impression that population sizes of 500-1000 should be adequate for short-term response in self-pollinated crops and that somewhat smaller populations will be sufficient in medium to long term recurrent selection programs. It is hoped that this limited discussion will provide some impetus to other researchers to address this problem.

SELECTION AMONG CROSSES

The number of crosses used in breeding programs in inbred species varies greatly (Shebeski, 1967). This topic is the subject of much heated debate among breeders - each arguing in favor of his or her own particular choice. There seems to have been rather little written concerning objective methods for allocating resources between and within crosses in plant breeding programs. Yonezawa and Yamagata (1978), in a theoretical assessment, concluded that "number of crosses rather than size of a cross should be increased with a given total population size." Their theory is based on an implied assumption that the breeer has no a priori assessment of relative potential of different crosses.

Large numbers of crosses are used in breeding programs in some of the international plant breeding centers and in other successful breeding programs around the world. However, there are also highly successful breeding programs based on 10 to 20 crosses per year. It would be beneficial if there were an objective basis for deciding how many crosses should be made in a breeding program.

Consider that a plant breeder has a number of inbred lines which form the genetic resources for the beginning of the breeding program. A simple breeding program would be to make m crosses among random pairs and to use the modified pedigree method (Brim, 1966) to generate n inbred lines within each cross. If the product of number of crosses and number of lines within each cross has a maximum value, and if selection among crosses is based on the means of n lines within each cross, then one may ask if there is an optimum value for m, the number of crosses.

It should be clear that, as the number of crosses increases, the heritability of cross means will decrease because fewer lines can be tested within each cross. However, it should be possible to increase the selection intensity among crosses since more are being evaluated. With fewer crosses, selection intensity among crosses will likely decrease while the heritability of cross means will increase. In the latter case, it should also be possible to increase the selection intensity for within-cross selection. The optimum allocation of resources to among- and within-cross selection in this case will depend upon the genetic potential for improvement (i.e. among- and within-cross genetic variances), as well as the relative values of among- and within-cross heritabilities.

If crosses are made at random among the inbred lines included in the parental generation, and in the absence of epistasis and linkage disequilibrium, then $\sigma_{Ga} = \sigma_{Gw}$, $h_a^2 = n\ h_w^2 /(1 + (n - 1)\ h_w^2)$, and

expected response = $(i_a\ (n\ h_w^2/(1 + (n-1)\ h_w^2))^{\frac{1}{2}} + i_w\ h_w)\ \sigma_{Gw}$,

where σ_G is the genotypic standard deviation, h^2 is the heritability,

and subscripts a and w refer to among-cross and within-cross components. This formula can be used to assess different selection strategies. Below is a table of expected responses (expressed as a proportion of the within-cross genotypic standard deviation) for differing numbers of crosses, subject to the limitations that the product of the number of crosses and the number of lines is equal to 2000 and that the objective of the breeding program is to identify the best five lines within each of the best five crosses.

Table 3.   Expected responses for various numbers of crosses subject to the restriction that $\underline{mn}$ = 2000, and for selection of best five lines within each of the best five crosses

Number of crosses	Number of lines/cross	i_a	i_w	Response/σ_{Gw} $h_w^2 = 0.3$	$h_w^2 = 0.7$
5	400	0.00	2.55	1.40	2.14
10	200	0.74	2.30	1.99	2.66
20	100	1.21	2.02	2.31	2.90
50	40	1.71	1.60	2.53	3.03*
100	20	2.02	1.21	2.57*	3.01
200	10	2.30	0.74	2.47	2.87
400	5	2.55	0.00	2.11	2.45

The results, although applicable only to this specific case, do
serve to show that standard quantitative genetic theory can at least
point in the direction of an optimum number of crosses and indeed
suggests very strongly that there is an optimum for each particular
situation. While these results do not agree with the general
impression one gets by reading the paper by Yonezawa and Yamagata
(1978), they do serve to indicate that 50 or so crosses may be
required for maximum response to selection for a quantitative
trait.

This analysis assumes that one is dealing with a single
quantitative trait or at least an index based only on quantitative
traits. Any crosses in which both parents fail to contribute the
desirable allele for a simply inherited qualitative trait must be
excluded prior to such an analysis.

This limited example should serve to demonstrate the importance
of between- and within-cross selection in self-pollinated crops and
is in agreement with the recognition of the importance of between-
and within-family selection in cross-pollinated crops (see, for
example, Hallauer and Miranda, 1981). A comparison of the first row
in the table with the maximum responses (marked with asterisks)
shows that response to within cross selection only is quite
inefficient particularly at the lower heritability. It is clear
that breeders of inbred species should entertain the question of how
best to select among crosses.

Busch et al. (1974) discussed the problems of identifying
superior crosses. They evaluated F_4 and F_5 bulk populations,
random F_2 derived F_5 or F_6 lines, and mid-parent values as
possible methods for estimating cross performance. They generally
favored the use of advanced generation bulk populations. From
quantitative genetic considerations, it would seem that all three
methods should give unbiased estimates of cross means insofar as
possible inbred progeny are concerned. The differences between
methods should be primarily in terms of ease of producing the proper
material and in the precision with which it can be evaluated.

The most questionable method evaluated by Busch et al. (1974)
is that of the use of the mid-parent values to predict the mean of
all possible inbred lines that can be derived from a cross. If
additive types of epistasis are not important, the mid-parent value
should provide an unbiased estimate. The use of mid-parent value is
attractive for two reasons: i) the material for cross evaluation is
available at the beginning of the program and extra crossing and
time are not required, and ii) evaluation of p potential parents is
sufficient to select among p (p-1)/2 possible crosses. This
approach is the one used by Grafius (1965) and by Whitehouse (1968)
in selecting inbred lines for crossing. The weakness of this method
is that if epistasis is important, one of the parents may show an

epistatic effect which will occur in very low frequency in the
inbred progeny of the cross. In that case, the mid-parent value
would become a biased estimate of cross mean performance. Despite
this objection, it is highly likely that breeding programs for
inbred species could be improved considerably by more careful
evaluation of potential parents and more careful study of mid-parent
values.

This discussion of number of crosses to use in breeding inbred
species has been quite unsatisfactory. There appears to be a lack
of objective criteria for deciding upon number of crosses. The
importance of between- and within-family components of response to
selection in cross-fertilized species as well as evidence from the
application of standard quantitative genetic theory indicate that
selection among crosses is an important component of crop
improvement in inbred species. Whether or not mid-parent values can
be used extensively for selection among crosses is a matter of
debate. The proposition is attractive because of economies in
effort and time and is theoretically acceptable if additive types of
epistasis are not important.

STABILITY ANALYSIS

Genotype-environment interaction is a critical factor in crop
improvement. Questions concerning the optimum allocation of
resources to multi-experiment evaluation trials, breeding different
cultivars for specific regions, breeding for stress tolerance, and
breeding for sub-optimal conditions all depend upon the nature and
extent of genotype-environment interaction.

Comstock and Moll (1963), in an often cited paper, described
how variation due to interaction between genotypes and
macro-environments could be measured and how such measurements could
be used in objective decisions concerning allocation of resources to
replication within experiments and across macro-environments.
Their discussion of covariances of several effects in the model for
phenotypic values provides insight into choice of environments which
should maximize response to selection.

The general approach used by Comstock and Moll (1963) was one
of statistical description and provided little predictive ability.
The method fails to provide insight into the nature and causes of
interaction nor does it provide a basis for predicting performance
of particular genotypes in particular environments. Perkins and
Jinks (1968) suggested that methods based on regression of genotypic
means on environmental means "leads to more informative conclusions
and can be used to predict across generations as well as across
environments". Their approach, sometimes referred to as stability
analysis, has been applied extensively to data from testing a number

of inbred lines over a number of experiments to give a two-way
table of mean values. The method consists of dividing genotype-
environment interaction effects (as defined statistically in a
two-way table) into protions due to linear regression on environmental mean, and a residual. A similar analysis had been developed by Eberhart and Russell (1966).

Noting the similarities in the two papers, one might wonder
if they are very different. By considering experiment means as a
covariate, Eberhart and Russell (1966) were able to retain (n-2)
degrees of freedom for deviations from each of the v regression
lines. Perkins and Jinks (1968), on the other hand, recognized
that the experiment means were based on genotype values in the
first place, and ended up with (v-1)(n-2) degrees of freedom for
deviations from the v regression lines. Aside from the question
of allocation of degrees of freedom, both methods give regression
coefficients and sums of squares which are identical (Table 4).

Table 4. Comparison of two methods of stability analysis.[+]

Method 1 - Perkins and Jinks (1968)

Source of variation	df	Sum of squares	Mean square
Lines	19	8133	428.1
Environments	8	7173	896.6
Heterogeneity between regressions	19	297	15.6
Remainder	133	1824	13.7
Pooled error	957		2.6

Method 2 - Eberhard and Russell (1966)

Source of variation	df	Sum of squares	Mean square
Lines	19	8133	428.1
Environments (linear)	1	7173	7173.0
Lines x Env(linear)	19	297	15.6
Pooled deviations	140	1824	13.0
Pooled error	957		2.6

[+]Both methods besed on Tables 3 and 4 of Perkins and Jinks
   (1968) concerning height of 20 inbred lines of Nicotiana
   rustica.

Some authors have regressed the sum of the genotype-environmental interaction effect and the environmental effect on the environmental effect. This "part-whole" regression gives regression coefficients whose average value is 1.0 and coefficients of determination which are unrealistically high estimates of the amount of the interaction sums of squares explained by the regression.

A common error in stabilty analysis is that of comparing regression and deviation mean squares, rather than sums of squares, in assessing the relative importance of the two. The proper approach is to use sums of squares as in a recent paper by Tai et al. (1982). Perkins and Jinks (1968) concluded that regression accounted for a slightly larger portion of the genotype-environmental interaction than did deviations from regression. Applying the method of Tai et al. (1982) to their data (Table 4) indicates that regression accounted for only 14 percent of the interaction sum of squares.

Reflection on the whole problem of genotype-environment interaction, and regression analysis in particular, leads to the following thoughts. It appears that terms such as "stability" and "adaptability" are not clearly defined nor clearly understood. "Stress tolerance" is another term which seems to have different meanings to different people. In most genotype responses to simple environmental variables, response is decidedly non-linear. Many biologists, and even some statisticians, are not comfortable in dealing with non-linear response functions. One can think of an example where response to a particular stress follows a sigmoid curve characterized by a maximum value in the absence of stress and a sensitivity to stress. If one fits straight lines to two such curves which differ only in their maximum value, one will erroneously conclude that the two curves show a difference in sensitivity to stress.

Regression analysis of genotype-environment interaction is based on a linear model and one wonders what would happen if the underlying biology of the system was such that phenotypic values were a multiplicative function of genotypic and environmental effects. This type of effect is not uncommon in biological material and is cause for transformation prior to analysis. The observation that there is often a correlation between genotype means and the corresponding linear regressions may well reflect application of a linear model to a non-linear system. Wright (1971) showed that, if all regressions lines intersect at a common point, there will be a perfect correlation between genotype means and regression coefficients. He developed a statistical test for this possibility. His "joint regression" has one degree of freedom and turns out to be identical to Tukey's test for non-additivity in two-way tables (Tukey, 1949). Wright divided the sum of squares for regression into a portion due to joint regression and a portion due to regressions on environmental effects. Moreover, he subdivided deviations from regression into a portion due to regression on genotype effects and a residual.

The understanding of the nature of genotype-environment interaction would be improved if researchers were to concentrate on the study of responses and differences in responses. The critical issue is whether or not response curves of two or more genotypes differ sufficiently that the cultivars show a real change in rank order. Comstock (1977) noted that there has been little discussion in the literature concerning changes in the rank order of genotypes over environments.

SUMMARY

Discussion in this paper has attempted to show that the study of quantitative genetics is closely allied with the practice of plant improvement. By discussing efforts to estimate numbers of genes controlling quantitative traits as well as efforts to assess the importance of epistasis, it has been shown that techniques capable of resolving these issues are not yet available. Quantitative genetic theory has developed much more satisfactorily in dealing with problems concerning linkage and predictions of response to various selection strategies. Research relating to population sizes is currently at the forefront of quantitative genetics research. It is hoped that research on this topic will soon provide some objective guidelines for allocating resources to selection among and within crosses in inbred species. The brief discussion on stability analysis serves mainly as a caution against overly high expectations for this very popular technique.

The choice of topics has been somewhat arbitrary. Most would agree that each one is germane to a problem or issue in plant improvement. Hopefully, most will agree that quantitative genetics can help to resolve those problems or issues.

REFERENCES

Bailey, T.B., Jr., and Comstock, R.E., 1976, Linkage and the synthesis of better genotypes in self-fertilizing species, Crop Sci., 16: 363-370.
Baker, R.J., 1966, Predicted variance of response to selection, Ph.D. thesis, University of Minnesota, 72p. University Microfilms, Ann Arbor, Mich. (Diss. Abstr. 28:53-B).
Baker, R.J., 1968, Extent of intermating in self-pollinated species necessary to counteract the effects of genetic drift, Crop Sci., 8: 547-550.

Baker, R.J., 1971, Theoretical variance of response to modified pedigree selection, Can. J. Plant Sci., 51: 463-468.

Baker, R.J., 1975, Letter to the editors, Can. J. Plant Sci., 55: 355.

Bos, I., 1977, More arguments against intermating F_2 plants of a self-fertilizing crop, Euphytica, 26: 33-46.

Brim, C.A., 1966, A modified pedigree method of selection in soybeans, Crop Sci., 6: 220.

Busch, R.H., Janke, J.C. and Frohberg, R.C., 1974, Evaluation of crosses among high and low yielding parents of spring wheat (Triticum aestivum L.) and bulk prediction of line performance, Crop Sci.,14: 47-50.

Cockerham, C.C., 1963, Estimation of genetic variances, Pages 53-94, in: "Statistical Genetics and Plant Breeding," W.D. Hanson and H.F. Robinson, eds., NAS-NRC 982.

Comstock, R.E., 1977, Quantitative genetics and the design of breeding programs, Pages 705-718, in: "Proc. Int'l Conf. on Quantitative Genetics," E.Pollak, O. Kempthorne, and T.B. Bailey, Jr., eds., Iowa State University Press, Ames.

Comstock, R.E., 1978, Quantitative genetics in maize breeding, Pages 191-206, in: "Maize breeding and genetics," D.B. Walden, Ed., John Wiley and Sons, New York.

Comstock, R.E., and Moll, R.H., 1963, Genotype-environment interactions, Pages 164-196 in: "Statistical genetics and plant breeding," W.D. Hanson and H.F. Robinson, eds., NAS-NRC 982.

Coughtrey, A., and Mather, K., 1970, Interaction and gene association and dispersion in diallel crosses where gene frequencies are unequal, Heredity, 25: 79-88.

Crow, J.F., and Kimura, M., 1979, Efficiency of truncation selection, Proc. Nat. Acad. Sci.USA, 76: 396-399.

Dawkins, R. 1978, "The selfish gene," Granada Publishing Ltd., London.

Eberhart, S., and Russell, W.A., 1966, Stability parameters for comparing varieties, Crop Sci., 6: 36-40.

Falconer, D.S., 1981, "Introduction to quantitative genetics," Longman Inc., New York.

Grafius, J.E., 1965, A geometry of plant breeding, Michigan State University Res. Bull. 7.

Griffing, B., 1968, Selection in reference to biological groups. III. Generalized results of individual and group selection in terms of parent-offspring covariances, Aust.J. Biol. Sci.,21: 1171-1178.

Hallauer, A.R., and Miranda, J.B., Fo., 1981, "Quantitative genetics in maize breeding," Iowa State University Press, Ames.

Hanson, W.D., 1959a, Early generation analysis of lengths of heterozygous chromosome segments around a locus held heterozygous with backcrossing or selfing, Genetics, 44: 833-837.

Hanson, W.D., 1959b, The breakup of initial linkage blocks under selected mating systems, Genetics, 44: 857-868.

Hill, W.G., 1974, Variability of response to selection in genetic experiments, Biometrics, 30: 363-366.

Horner, T.W., Comstock, R.E. and Robinson, H.F., 1955, Non-allelic gene interactions and the interpretaton of quantitative genetic data, North Carolina Agric. Exp. Stn. Tech. Bull. 118.

Jinks, J.L., and Towey, P., 1976, Estimating the number of genes by genotype assay, Heredity, 37: 69-81.

Kearsey, M.J., and Jinks, J.L., 1968, A general method of detecting additive, dominance and epistatic variation for metrical traits. I. Theory, Heredity, 23: 403-409.

Kimura, M., 1957, Some problems of stochastic processes in genetics, Ann. Math. Stat., 28: 882-901.

Kimura, M., and Crow, J.F., 1978, Effect of overall phenotypic selection on genetic change at individual loci, Proc. Nat. Acad. Sci.USA, 75: 6168-6171.

Latter, B.D.H., 1965, The response to artificial selection due to autosomal genes of large effect. II. The effects of linkage on limits to selection in finite populations, Aust. J. Biol. Sci., 18: 1009-1023.

Leisle, D., Kosmolak, F. and Kovacs, M., 1981, Association of glume color with gluten strength and gliadin proteins in durum wheat, Can. J. Plant Sci., 61: 149-151.

Mayo, O., 1980, "The theory of plant breeding," Clarendon Press, Oxford.

Meredith, W.R., Jr., and Bridge, R.R., 1971, Breakup of linkage blocks in Gossypium hirsutum L.,Crop Sci., 11: 695-698.

Miller, P.A., and Rawlings, J.O., 1967, Breakup of initial linkage blocks through intermating in a cotton population, Crop Sci., 7: 199-204.

Mulitze, D.K., 1983, A critique of bimometrical methods for estimating the number of genes controlling quantitative traits. Ph.D. thesis, University of Saskatchewan.

Pederson, D.G., 1969a, The prediction of selection response in a self-fertilizing species. I. Individual selection, Aust. J. Biol. Sci., 22: 117-129.

Pederson, D.G., 1969b, The predictions of selection response in a self-fertilizing species. II. Family selection, Aust. J. Biol. Sci., 22: 1245-1247.

Pederson, D.G., 1974, Arguments against intermating before selection in self-fertilizing species, Theor.Appl.Genet., 45: 157-162.

Perkins, J.M., and Jinks, J.L., 1968, Environmental and genotype-environmental components of variability. III. Multiple lines and crosses, Heredity, 23: 339-356.

Quereshi, A.W., and Kempthorne, O., 1968, On the fixation of genes of large effects due to continued truncation selection in small populations of polygenic systems with linkage, Theor. Appl. Genet., 38: 249-255.

Rawlings, J.O., 1979, Long- and short-term recurrent selection in finite populations - choice of population size, Pages 201-215, in: "Proc. World Soybean Research Conference II," Raleigh.

Redden, R.J., and Jensen, N.F., 1974, Mass selection and mating systems in cereals, Crop Sci., 14: 345-350.

Robertson, A., 1970, A theory of limits in artificial selection with many linked loci, Pages 246-288, in "Mathematical topics in population genetics," K. Kojima, Ed., Springer-Verlag, New York.

Shebeski, L.H., 1967, Wheat and breeding, Pages 253-272, in: "Proc. Canadian Centennial Wheat Symp," K.F. Nelson, ed., Western Co-operative Fertilizers, Calgary.

Sneep, J., 1977, Selection for yield in early generations of self-fertilizing crops, Euphytica, 26: 27-30.

Stam, P., 1977, Selection response under random mating and under selfing in the progeny of a cross of homozygous parents, Euphytica, 26: 169-184.

Stuthman, D.D., and Stucker, R.E., 1975, Combining ability analysis of near-homozygous lines derived from a 12-parent diallel cross in oats, Crop Sci., 15: 800-803.

Thoday, J.M., and Thompson, J.N., Jr. 1976, The number of segregating genes implied by continuous variation, Genetica, 46: 335-344.

Tai, P.Y.P., Rice, E.R., Chew, V. and Miller, J.D., 1982, Phenotypic stability analysis of sugarcane cultivar performance tests, Crop Sci., 22: 1179-1184.

Towey, P., and Jinks, J.L., 1977, Alternate ways of estimating the number of genes in a polygenic system by genotype assay, Heredity, 39: 399-410.

Townley-Smith, T.F., Hurd, E.A., and McBean, D.S., 1973, Techniques in selection for yield in wheat, Pages 605-609, in: "Fourth Int'l Wheat Genetics Symp. Proc.," E.R. Sears and L.M.S. Sears, eds., Univ. of Missouri, Columbia.

Tukey, J.W., 1949, One degree of freedom for non-additivity, Biometrics, 5: 232-242.

Wehrhahn, C., and Allard, R.W., 1965, The detection and measurement of the effects of individual genes involved in the inheritance of a quantitative character in wheat, Genetics, 51: 109-119.

Whitehouse, R.N.H., 1968, An application of canonical analysis to plant breeding, Pages 61-96, in: "Proc. 5th Cong. European Assoc. for Res. on Plant Breeding," Milano.

Wright, A.J., 1971, The analysis and prediction of some two factor interactions in grass breeding, J. Agric. Res.(Camb.), 76: 301-306.

Wright, S., 1934,   The results of crosses between inbred strains of
     guinea pigs differing in number of digits,   Genetics, 19:
     537-551.
Yonezawa, K., and Yamagata, H., 1978,   On the number and size of
     cross combinations in a breeding programme on self-fertilizing
     crops,   Euphytica, 27: 113-116.

THE PATHOLOGICAL AND ENTOMOLOGICAL FRAMEWORK OF PLANT BREEDING

Arthur L. Hooker

Bioscience Director
DEKALB-PFIZER GENETICS
3100 Sycamore Road
DeKalb, IL 60115

INTRODUCTION

The purpose of this paper is to discuss gene manipulation by man in relation to disease (pathogen) and insect resistance in plants. Both disease and insect resistances are broad, complicated research areas and whole books have been devoted to each. Therefore, I can only present an overview and point out some salient features in this paper. My objective is to make it most useful to those not actively engaged in plant breeding. Furthermore, I have included only a limited number of references, mainly to serve as examples. My recent book chapter (Hooker, 1983) and other general works can be consulted for details and citations to additional research papers (Painter, 1951, 1958; Beck, 1965; Van der Plank, 1968, 1975; Pathak, 1970, 1972; Hooker, 1972a, 1984; Johnson, 1972; Maxwell et al., 1972; National Academy of Sciences, 1972; Sprague and Dahms, 1972; Nelson, 1973; Day, 1974; Gallun et al., 1975; Kennedy, 1978; Russell, 1978; Harris, 1979; Parlevliet, 1979; Maxwell and Jennings, 1980; Robinson, 1980a; Jenkins, 1981).

BIOLOGY

Nature is a complex of plants, bacteria, fungi, nematodes, insects, mites, and other pathogens or arthropods living together. There are many kinds of relationships between and among these organisms. The most common is where two organisms co-exist together in a community or ecosystem with little or no close interrelationship with each other except for some competitive effects. On the other extreme, two organisms may

live together in a symbiotic relationship where each can benefit
from the other. Of concern to man are the intermediate and
detrimental parasitic relationships where pathogens infect and
insects infest their hosts. (For ease of discussion in this
paper I am using parasite in the broad sense and regarding any
pathogen or insect that obtains its food while living on or in a
plant as a parasite.)

Considering the numerous possible contacts between and
among organisms, successful host:parasite relationships are
rare. Most plants are non-hosts to the vast majority of
organisms, including parasitic ones, which occur in nature.
While some pathogens have a wide host range and some insect
pests are omniverous, each plant species is characterized by its
own set of pathogens and insect pests that have co-evolved with
their host. However, this number is often in the 50 to 100 or
more range for any single crop species. Normally, separate sets
of genes are involved for resistance to each pathogen or insect
pest. Therefore, breeding for disease and insect resistance is
not a single objective, but one with many specific components.

New successful host:parasite relationships continue to
appear in nature. In many instances their origins are unclear.
Those best understood involve genetic changes in parasites that
allow them to extend their host range or to become sufficiently
aggressive to the point where damage becomes obvious. Often
what may appear to be "new" may simply be the introduction and
establishment of a parasite from another area of the world. In
other instances changes in crop cultural practices may favor the
increase and spread of a previously unimportant parasite and
when this happens what appears to be a new problem is
presented. Man sometimes unintentionally creates susceptibility
to a previously unimportant parasite through breeding, but the
creation of a new host, if it occurs at all, must be rare.

There are different kinds of relationships between
parasites and their hosts. Pathologists and entomologists have
coined various terms to describe them (Painter, 1951, 1958;
Hooker, 1967b; Van der Plank, 1968, 1975; Robinson, 1969, 1971;
Nelson, 1973; Day, 1974; Parlevliet and Zadoks, 1977; Gallun and
Khush, 1980; MacKenzie, 1980). In a (fully) susceptible plant,
nothing inhibits the pathogen or pest at any stage of
development throughout its full life cycle. In diseases,
pathogen development is often sub-optimal and the resistant
reaction to individual pathogens from a functional viewpoint is
either of two fundamental types, specific or general (Hooker,
1967b; Robinson, 1969; Day, 1974). In specific resistance a
differential interaction is seen when a series of different
cultivars are evaluated against a series of different parasite
biotypes. Some host cultivars are resistant to some biotypes

but are susceptible to another. Furthermore, differences
between plants is qualitative and expressed in terms of kind of
infection. Plant reaction can be separated into distinct
resistant or susceptible classes. In contrast to specific
resistance, no differential interactions are seen when a series
of different cultivars with general resistance are evaluated
against a series of different parasite biotypes. In this
situation cultivars resistant to one biotype are resistant to
all other biotypes. In general resistance differences between
plants can be qualitative but usually is quantitive and
expressed in terms of number of uninfected plants or amount of
uninfected tissue. Plant reactions grade continously from one
extreme to the other and cannot be separated into a few distinct
classes. General resistance to one parasitic species usually
does not result in resistance to another species. Specific and
general resistance can function independently or together in the
same plant.

The terms specific and general resistance are not accepted
by all pathologists nor by many entomologists. Other terms used
for specific resistance are vertical, nonuniform, differential,
race-specific, unstable, monogenic, oligogenic, and major gene
resistance. Terms used for general resistance are horizontal,
uniform, partial, non-race-specific, stable, durable, polygenic,
and others. In addition, some writers as discussed by
Parlevliet (1981) present evidence that certain host:pathogen
relationships do not fit the concept of either specific or
general resistance.

As we have seen, the biotype of the parasite must be
considered in relation to specific and general resistance. In
specific resistance where resistance is expressed against some
but not all biotypes of a single pathogen or pest, the biotypes
are usually called races, but the term pathotype is also used
(Robinson, 1969, 1980b; Russell, 1978). Races able to establish
a compatible or nearly compatible relationship with a resistant
plant are called virulent races. Those unable to establish a
compatible relationship are called avirulent races. On
susceptible plants, virulent and avirulent races produce similar
infection types. Genetic variation can exist between
individuals within a race. General resistance, as we have
noted, is expressed against all biotypes. Isolates of the
pathogen, however, can differ in aggressiveness. Aggressiveness
is also a quantitative character and among parasite isolates is
measured by the amount of infection produced on a common set of
host genotypes. New races or biotypes of parasites are a
continual problem to the plant breeder and the producers of
crops. Their appearances are natural events and where or when
they will appear are unpredictable. Parasite variation can be
dealt with realistically and should not be overemphasized.

There are effective ways of using specific resistance and much good resistance to a wide range of diseases and insects is broadly functional.

Entomologists use the terms antixenosis (nonpreference) and antibiosis (Painter, 1951; Kogan and Ortman, 1978; Gallun and Khush, 1980; Branson et al., 1983). Antixenosis is a response of the insect to the plant. With regard to antixenosis, the insect has an avoidance response in various degrees to the plant as it searches for food, oviposition sites, and/or shelter. Antibiosis, on the other hand, is the negative effect of the plant on the insect. Biological functions of the insect such as survival, development, and/or reproductionare impaired. While antixenosis and antibiosis often act together and are therefore difficult to separate and categorize both are genetic attributes of the plant and can be manipulated by man.

Tolerance is another form of useful plant reaction in agricultural terms. Tolerance has been defined as the ability of a cultivar or hybrid to endure pathogen infection or support high populations of insects without suffering as severe loss in yield or quality as do other cultivars or hybrids showing similar levels of infection or attack. Tolerance is identified as that type of resistance to the disease or the insect attack which does not have a detrimental effect on the parasite itself or the population. While it is difficult to measure, tolerance is recognized or believed to exist for several diseases and insects and is a genetic, manipulative trait of a plant (Posnette, 1969; Schafer, 1971; Maxwell and Jennings, 1980; Parlevliet, 1981).

Pathogens can also vary in their fitness to the environment (Nelson, 1973). Races are known that differ in ability to cause disease, in infection efficiency, or in sporulation efficiency depending upon temperature and other environmental factors. Obviously races of parasites with broad environmental adaptation are of most concern to the plant breeder and the grower of crops.

GENETICS

Let us now move to a brief consideration of the genetics of host:parasite interactions.

We have mentioned the host vs. non-host and pathogen vs. non-pathogen forms of variation. Little is known about the genetics of host vs. non-host reaction because these two groups of plants cannot be hybridized by conventional means and subjected to genetic study. It is interesting to speculate that "hosts" have genes that allow them to be parasitized by individual pathogens and insects rather than "non-hosts" having separate genes for resistance to the multitude of organisms that come into contact with them. If we consider man's definition of species as valid, more is known about the genetics of pathogen vs. non-pathogen than of host vs. non-host variation. For example, the fungus Helminthosporium carbonum Ullstrup is a pathogen of maize (Zea mays L.) but does not infect oats (Avena sativa L.) while H. victoriae M. and M. is a pathogen of oats but does not infect maize. These two pathogens are named differently but are morphologically similar and can be hybridized. When Nelson and Kline (1961) did this they found progeny that infected one host but not the other (parental types) but also isolates that infected both and isolates that could infect neither host. Host range was inherited as a single gene character. Studies with other Helminthosporium fungi show host range to be inherited as 1, 2, 3, and sometimes a 4 gene character (Nelson, 1970). In smut fungi (Tilletia and Ustilago spp.), non-pathogenic isolates may carry within their genomes unexpressed virulence genes corresponding to genes for resistance in cultivars of another plant species. When recombined with genes for pathogenicity, new virulent races result.

Biotechnologists might keep the concept of pathogen and non-pathogen variation and its inheritance in mind. The ability to transfer alien DNA into a host organism depends upon a suitable vector. When suitable vectors are lacking it may be possible to develop them through genetic recombination.

Studies on the inheritance of resistance leads to a better understanding of germplasm and how it may be manipulated in breeding. Such studies are desirable and very useful. However, detailed knowledge on how resistance is inherited is not needed for breeding programs to be effective.

Genes for resistance are mainly conditional genes; that is they are expressed in the plant when the plant and parasite interact under environmental conditions that normally would result in disease development or a build up in insect infestation. When the parasite is absent or does not develop, resistant plant genotypes cannot be distingished from susceptible ones except in rare situations where a known physiology or morphology confers resistance and these can be recognized in the absence of the parasite.

Biffen (1905) first reported on the inheritance of disease resistance in plants. Since that time many reports have appeared in the literature. While numerous types of genetic models have been suggested to explain resistance, they fall

under three basic types: oligogenic, polygenic, and cytoplasmic
(Hooker and Saxena, 1971; Day, 1974). The three types of
genetic systems interact in the plant to condition its reaction
to a parasite. Oligogenic and cytoplasmic resistance may be, but
is not always, specific for certain parasite races while
polygenic resistance is only known to function as a form of
general resistance.

Resistance in plants to many pathogens is oligogenic. With
oligogenic resistance a few genes are involved and the
individual effects of each gene can be measured. The genes may
function as single dominant, incompletely dominant, or as
recessive genes. Sometimes the genes are in an allelic series.
Various types of gene interaction can occur where the presence
of two or more genes are required for resistance or where one
gene modifies the expression of the other (Malm and Hooker,
1962; Hooker, 1967b; Hooker and Saxena, 1971; Day, 1974).

Genes for resistance may occur in a series of closely
linked loci. A good example of this is in maize and resistance
to the rust caused by Puccinia sorghi Schw. We first identified
what we thought were 14 alleles for resistance at the Rp locus
(Mains, 1931; Rhoades, 1935) on the short arm of chromosome 10
(Hooker and Russell, 1962; Hagan and Hooker, 1965; Wilkinson and
Hooker, 1968). These failed to segregate in populations of the
size used in conventional genetic studies and showed
differential reactions with a series of rust biotypes. The
occassional susceptible plant in test-cross populations and the
presence of flanking loci Rp5 and Rp6, 1.1 and 2.1 map units on
either side, prompted us to make a more detailed study of the Rp
locus (Saxena and Hooker, 1968; Hooker and Saxena, 1971). In
testcross progenies of up to 19,000 plants each, recombination
values ranging from 0.10 to 0.37% were observed between five
pair of alleles. No recombinations were detected between Rpd
and other alleles. Rpd is dominant to all alleles and, with the
exception of Rpd, the other alleles while dominant to the allele
rp for susceptibility show no linear order of dominance with
each other. These studies led us to believe that the terminal
end of chromosome 10 is a chromosome region with functional and
nonfunctional genes closely linked together and may have evolved
through repeated duplications and subsequent structural
modifications of individual genes.

Among the pathogens, oligogenic resistance is frequently
seen in resistance to bacteria (Corynebacterium, Erwinia,
Pseudomonas, Xanthomonas, and other spp.), viruses, nematodes
(Heterodera and Meloidogyne spp.) and to the smut, rust
(Puccinia spp.), powdery mildew (Erysiphe spp.), downy mildew
(Peronospora, Sclerospora and other spp.), anthracnose
(Colletotrichum spp.), Fusarium, Septoria and Helminthosporium

fungi. Oligogenic resistance is believed to occur in all plant
species.

In host:parasite systems where specific resistance is
involved a gene-for-gene interaction prevails. This
interaction, proposed by Flor (1956) to explain flax (<u>Linum</u>
<u>usitatissimum</u> L.) and flax rust (<u>Melampsora lini</u> (Ehrenb.)
Lev.), has now been shown to function in numerous plant diseases
(Flor, 1971; Loegering et al., 1971; Day, 1974; Lim et al.,
1974; Sidhu, 1975; Wolfe and Schwarzbach, 1975; Loegering, 1978;
Russell, 1978; Parlevliet, 1981) and some insects (Hatchett and
Gallun, 1970; Gallun, 1972; Gallun and Khush, 1980). Here, for
every gene conditioning resistance in the host, there is a
corresponding gene conditioning virulence or avirulence in the
pathogen or insect pest. Resistance results when any pair of
corresponding genes are for resistance in the plant and for
avirulence in the parasite. All other sets of corresponding
gene interactions result in compatible or nearly compatible
host:parasite interactions. Resistance in the host is usually
dominant, but can be recessive or cytoplasmic. Where it is
possible to determine, avirulence in the parasite is usually
dominant but can be recessive and perhaps cytoplasmic. When two
or more sets of corresponding genes are for resistance and
avirulence, the phenotype of the least parasite growth (greatest
resistance) prevails.

The value of specific resistance in protecting cultivars or
hybrids in agriculture clearly depends upon the absence or
rarity of corresponding genes for virulence in the pathogen or
insect in the epidemiology area where the host is to be grown.
Host cultivars having two or more effective genes for specific
resistance retain their resistance for a greater period of years
when grown in agriculture than do cultivars having only single
genes.

Polygenic resistance involves numerous genes with each
contributing a portion of the resistance; the effects of
individual genes are usually undetectable (Hooker, 1967a, 1967b;
Kim and Brewbaker, 1977; Wilcoxson, 1981; Leonard and Mundt,
1984). When estimates of gene number are made, the predicted
number involved with resistance is estimated to be lower than
that involved for more complex characters such as yield (Hughes
and Hooker, 1971). The gene action in the host is mainly
additive and the resistance is highly heritable. Where hybrids
are involved, as in maize, the reaction of the hybrid is usually
above that of the average of the two parental inbreds. I use
high numerical values to express resistance and have found the
formula $F_1 = [(2HP + LP)/3] + 0.15 (MAX-LP)$, where HP = high
parent, LP = low parent, and MAX = maximum possible score, to
quite accurately predict hybrid reaction. Polygenic resistance

is known to a wide array of pathogens (Simons, 1972). It is
believed to occur in all plant species. On the pathogen side,
few studies have been made to study the inheritance of
aggressiveness, but this presumably is also a polygenic trait.

Cytoplasmic resistance involves non-nuclear genes and is
rare (Hooker, 1974a). The best known examples are in maize
where mitochondrial membranes in cms-T cytoplasm are sensitive
to a pathotoxin produced by the fungi Helminthosporium maydis
Nisik. race T and by Phyllosticta maydis Arny and Nelson. This
susceptibility and the widespread usage of cms-T cytoplasm to
produce maize hybrids by the male-sterile method resulted in the
now famous U.S. maize blight of 1970 (Hooker, 1972b). This
disease resulted in the greatest loss due to a single disease of
a single crop in a single country in a single year of any plant
disease in all of recorded history. Also noteworthy about this
disease is that it was brought under complete control in two
years through the development and growth of disease resistant
hybrids.

The genetics of host:parasite interactions where an insect
species is the predator follow similar systems to those in plant
diseases (Athwal et al., 1971; Gallun, 1972; Gallun and Khush,
1980; Russell, 1978; Maxwell and Jennings, 1980). Among the
insects, oligogenic resistance has been observed to aphids
(Acyrthosiphon, Aphis, Macrosiphum, Rhopalosiphum and
Therioaphis spp.), greenbugs (Schizaphis spp.), midges
(Oligonychus spp.), leafhoppers (Aceratagallia, Amrasca,
Cicadulina, Dalbulus, Empoasca, Nephotettis and other spp.), and
planthoppers (Nilaparvata, Sogatella and other spp.) in numerous
crops and in maize to the European corn borer (Ostrina nubilalis
Hubner) and in barley (Hordeum vulgare L.) and wheat (Triticum
aestivum L.) to the cereal leaf beetle (Oulema melanopus L.) and
the Hessian fly (Mayetiola destructor Say). Polygenic
resistance occurs to insects such as striped rice borer (Chilo
suppressalis Walker), the sorghum shoot fly (Atherigona varia
soccata Rondani), maize earworm (Heliothis zea Boddie) and fall
army-worm (Spodoptera frugiperda Smith), striped cucumber beetle
(Acalymma vittata Fabricius), and the squash bug (Anasa tristis
De Geer). Cytoplasmic inheritance of reaction to insects is not
known.

Host tolerance and parasite fitness are both believed to be
inherited as polygenic traits. Several components of parasite
fitness are involved and each can be expressed in a quantitative
manner.

Another kind of phenotypic and genotypic variation of
interest is variation in disease symptoms which is independent
from any form of resistance mechanism. This is best illustrated

in the Helminthosporium fungi. In barley the diseases spot
blotch caused by H. sativum P.K. and B. and net blotch caused by
H. teres Sacc. have been described. These two organisms can be
hybridized. Type of symptom, whether "spot" or "net" is
inherited and segregates in the progeny from the cross. In
maize two races of H. carbonum have been described which produce
different symptoms. On certain maize inbreds, race 2 produces a
"spot" type lesion, while race 3 produces a "long-narrow-linear"
type lesion. These lesion type differences are also inherited
(Dalmacio et al., 1978). Variation in symptoms can confuse the
plant breeder and detract him from defining the main objective
and that is to develop plants that are unsatisfactory hosts to
parasites.

As more and a larger array of genes for resistance to
pathogens and insects are revealed, a greater range of options
become available to the plant breeder and the grower of crops.
Hence, studies on the inheritance of plant resistance are
beneficial. Furthermore, when newly identified genes for
resistance are incorporated into elite germplasm, they will be
used by applied plant breeders and eventually in agricultural
crops.

PARTICIPANTS

Let us now move to a consideration of gene manipulation by
man as he deals with disease and/or insect problems. This
discussion pertains mainly to world areas having a well
developed system of agricultural research and technology.
Buddenhagen (1983) discusses breeding strategies in developing
countries.

Each country has a national germplasm system where
cultivated crops are concerned. This can be viewed as a total
system involving public and private industry support. It is a
continuum starting with the collection, maintenance, evaluation,
documentation, distribution and subsequent enhancement of basic
germplasm stocks. This is followed by the breeding and release
of improved cultivars or hybrids, the production of high-quality
planting stock and its distribution to the farmer. The system
continues with the use of germplasm by farmers and others to
produce a reliable supply of high quality food and other
products for the consumer. The final result is a well fed,
clothed, and housed populace. In the United States, a National
Plant Genetic Resources Board has been appointed by the
Secretary of Agriculture to provide the Nation an overview and
to advise on the assembly, description, maintenance, and
effective utilization of plant germplasm.

While natural selection has resulted in the evolution of disease and insect resistant plants, the use of genetic resistance to control diseases and insect pests of cultivated plants is now primarily under man's control. In this section, I would like to enumerate some of the participants and to briefly describe their contributions.

The development of improved cultivars and hybrids with adequate resistance to diseases and insects is mainly the responsibility of plant breeders and their associates. In disease and insect resistance work, single individuals may perform the functions of a geneticist, entomologist, pathologist and/or breeder or two or more individuals may collaborate in the same breeding program.

Plant breeding programs are supported in various ways. Historically, breeding of agricultural crops was supported in the United States primarily by public funds administered through the State Agricultural Experiment Stations and the Agricultural Research Service, USDA. This support continues today, but considering all crops, it is declining. From early in the Nation's history, private industry has had a continued involvement in developing and merchandising improved cultivars of horticultural crops and following the advent of hybrid maize in the 1930's expanded to field crops. This expansion was accelerated by the passage of the Plant Variety Protection Act in 1970. Private breeding now encompases maize, sorghum (Sorghum bicolor (L.) Moench), sugar beets (Beta vulgaris L.), cotton (Gossypium hirsutum L.), alfalfa (Medicago sativa L.), soybeans (Glycine max (L.) Merrill), sunflowers (Helianthus annuus L.), and a wide array of flower, fruit, tree, and vegetable crops. In Europe, most plant breeding is in the hands of private industry with public funds providing for support functions. Several international organizations are also engaged in plant breeding with major emphasis on multiple resistance to diseases and insects (Khush and Coffman, 1977; Dubin and Rajaram, 1982). While much of the breeding work and support functions of such programs may be carried out at international centers, the final testing and release of new cultivars is now usually done in individual national programs.

Pathologists and entomologists are intimately involved in breeding programs for disease and insect resistance. These scientists play a significant role in determining the diseases and insect pests to consider, the production of disease inoculum or the rearing of insects, methods of pathogen inoculation or insect infestation, ways of measuring plant and pest reaction, value of resistance and the type of plant genotype to select for and advance into production where disease and/or insect reactions are concerned. Other activities also performed are

studies on insect biology, pathogen development, sources of
resistance, the incorporation of alleles for resistance into
elite germplasm suitable for seed production or for use by
breeders as parents for new cultivar or inbred line development,
and the monitoring of diseases and insects.

The breeding and release of improved cultivars and hybrids
is a complicated program. Research directors and breeders
define breeding objectives and set priorities. Breeders select
germplasm for use in new breeding starts, generate genetic
variation, observe segregating populations and select strains of
plants stable enough for evaluation. The testing program is a
large and costly component of plant breeding. Involved here are
the plant breeders, crop production experts, pathologists,
entomologists, physiologists, and those interested in quality
traits. Products from the breeding program go through a series
of preliminary to advanced tests. At the later stages of
testing, products are evaluated in numerous environments over a
wide area and, in private industry, viewed by people in seed
production and sales as well as those enumerated above. Rather
commonly, plant reaction to numerous individual diseases and
insect pests is part of the testing program. Where it can be
used international testing is a powerful technique in the
evaluation of plants for disease and insect resistance. High
resistance may not be needed, but extreme susceptibility to any
disease or insect should be avoided.

In less direct ways, others are involved in the development
and delivery of products having disease and insect resistance
into agricultural production. People in government through the
enactment and enforcement of various laws and regulations
involving seeds are also part of the germplasm system. The
importance of the Plant Variety Protection Act in the United
States has been mentioned. Various seed laws assure genetic
purity and quality seed. Large seed firms have their own
quality control laboratories to monitor seed production and the
quality of products sold to farmers. Disease and/or insect
reaction traits are not part of these standards. People
concerned with seed production and sales influence the kinds of
products that reach the farm. And finally, in free enterprise
systems, the farmer is free to select those cultivars and/or
hybrids he wishes to use on his farm.

PROGRAMS AND PRACTICES

Let us now turn to programs and practices. The optimistic
view is that most diseases and many insect pests can be
controlled within economic thresholds by plant genetic

resistance. For this to be obtainable, several research
programs and practices must be achieved (Hooker, 1977a).

Vigilance

     The first facet of any successful breeding program is to
determine objectives and be willing to change them. It is
necessary to know what diseases and insects occur in the area
served, their relative importance to crop production, and the
identity of those that can most feasibly be controlled by
genetic resistance.

     One of the objectives set forth by the first National Plant
Genetics Resources Board was the establishment of some form of
monitoring service to determine where genetic vulnerability
exists. This group presumably would know the range and types of
disease and insect resistance in the crops grown, some concept
of parasite variability and would provide national warnings to
impending disease and/or insect attack. While this is a noble
idea, it is in fact very difficult to put into practice.
Breeders often do not know the kinds of disease and/or insect
resistance or their genetics in the cultivars and/or hybrids
grown. Where cultivars and hybrids are of private origin, this
information may be regarded as proprietary and not available for
public distribution. Furthermore, cultivars and hybrids are
constantly changing as new and superior ones are developed by
breeders and put into production. Parasites can change and
these changes are largely beyond man's control. There are real
sampling and evaluation problems in the monitoring of pathogens
and insects. Finally, interpretations and predictions from
monitoring studies are difficult to make.

     Some monitoring programs are in effect. National or
international programs to survey the kinds and frequencies of
rust, powdery mildew, and smut races have been in place for many
years in several countries to assist cereal breeders.
Developments such as pheromone traps provide an excellent tool
for survey of the gypsy moth (Porthertria dispar L.) and other
insects. Following the 1970 maize blight (Hooker, 1972b) in the
U.S., the Animal and Plant Health Inspection Service (APHIS) set
up pathology positions in a series of states to monitor diseases
and pathogens. This program was largely unsuccessful and
support for it has waned. On the other hand, APHIS has a State
Cooperative Survey Program for certain insects. Many states are
participating in this computerized, growing program and
hopefully, it will provide useful information. We, as a
company, have a monitoring program for maize diseases (Smith,
1977; Smith and Kinsey, 1980). It involves the systematic
sampling of pathogens from the major U.S. maize growing areas,
their identification, their evaluation in disease tests, and the

interpretation of any significant new variation to breeders and
management. However, the usual situation is one of no organized
monitoring programs for crops and their parasites.
Pathologists, entomologists, and breeders in the course of their
normal activities become aware of new disease or insect problems
and make judgements as to future courses of action to take.
Examples of this for maize can be found in the literature
(Hooker et al., 1970; Nelson, et al., 1973; Turner and Johnson,
1980).

In each country, some form of monitoring activity is needed
to detect new problems. Only essential information needs to be
collected and this done in an effective manner and at low cost.
The information obtained should be the type that will be
valuable to those who make decisions regarding germplasm used in
breeding programs and that employed on the farm in agricultural
crops.

Problem Assessment

In problem assessment, risk situations capable of being
reduced or inadvertently established by crop plant genetic
manipulation are identified. Of concern here are potential
loss, the disease or insect problem, the pathogen, the insect,
the crop, the way the crop is grown, availability of other
control measures and potential vulnerability to disease or
insect attack if nothing is done. A study of these factors is
largely the task of both public and private plant pathologists
and entomologists. Those responsible for the administration of
research and breeders become involved when cost:benefit
estimates are made and decisions reached as to how much research
and breeding effort should be devoted to a particular problem.
When other control measures are satisfactory, the decision may
be to ignore the disease or insect in the breeding program. Even
when another control works, it may be desirable to use genetic
resistance because it is cheaper, has no undesirable effects
upon the environment, or a chemical or other control program may
be more effective when used in conjunction with partial
resistance. Breeders also need assurance that the resistance is
not negatively associated with other desirable traits. An
example here is that high resistance to stalk and root rots in
maize is often associated with low grain yield (Hooker, 1974b).

Nature and Expression of Reaction

Reliable methods are now available to inoculate plants with
most pathogens or to infest plants with many insects. Nearly
all pathogens can be either readily grown in culture or
propagated on susceptible hosts. By this means, large
quantities of inoculum can be obtained. Pathologists have also

developed means to inoculate large numbers of plants at low
cost. Some of the techniques include specifically designed
disease gardens or mechanical devices such as tractor-mounted
inoculators. The artificial rearing of insect species have
enabled entomologists to produce large numbers of egg masses or
larvae. The use of the bazooka and similar kinds of equipment
have resulted in uniform infestations of known quantities under
field conditions. Compared to when natural and sporadic
infestations were used in breeding, these advances have greatly
increased the success in selecting and breeding for insect
resistance (Dahms, 1972; Sprague and Dahms, 1972; Harris, 1979;
Maxwell and Jennings, 1980; Sutter and Branson, 1980).

While our current procedures for evaluating plant reaction
to diseases and insects are satisfactory for practical purposes
of breeding, improvements are possible and may be needed for
more critical work. Secondary spread of pathogens and
population increases for insects are important for field
evaluations. Efforts expended on improving the environment by
artificial controls producing conditions more conducive to the
growth and reproduction of parasites would be worth while. By
keeping inoculation and infestation methods simple, large
numbers of plants can be evaluated utilizing people who lack
advanced training in plant pathology or entomology. This is
important in breeding programs where many kinds of evaluations
are made and traits selected.

Various scoring methods have been developed to measure
plant reaction to disease or insect attack. As we have
mentioned, plant reaction can vary both in kind and degree.
These differences are expressed to a greater range of pathogens
than insects. Plant phenotypes need to be distinguished and if
possible, equated to the presence or absence of specific genes
and cytoplasms in the plant. Entomologists often measure the
growth and/or reproduction of the insect itself and this
information is used to determine the ability of a plant to be a
suitable host.

Improvements in rating plants for reaction to diseases and
insects are possible. Most rating scales were designed by
pathologists and entomologists to measure disease (pathogen)
development or degree of insect feeding rather than by plant
breeders to measure plant resistance. Furthermore, many of the
numerical rating scales used do not reflect gradations of plant
worth (lack of injury) comparable to the range of values within
the scale. For example, maize plants having thirty percent of
the leaf area in the lower plant canopy blighted three weeks
after anthesis will yield the same grain as those plants having
all healthy leaves. To assign the former a value of three for
30 percent blight and the latter a value of one for less than 10

percent blight with a difference of two between them is
unrealistic from an economic viewpoint. However, this rating
system is often followed because the differences in plant
reaction are apparent to those who grow crops and presumably is
of importance to them. An index system has been developed to
provide numerical values where equivalent ranges of scale values
reflect comparable differences in plant yield (Hooker, 1979;
Hooker and Perkins, 1980).

Much effort has been made to understand the histology,
physiology, and biochemistry of plant reactions to diseases and
insects (Beck, 1965; Hedin, 1977; Maxwell and Jennings, 1980;
Norris and Kogan, 1980; Bell, 1981; Giebel, 1982). At the maize
pathology and genetics laboratory at the University of Illinois,
we devoted considerable time to it ourselves (Hilu and Hooker,
1964, 1965; Hilu, 1965; Wysong and Hooker, 1966; Lim et al.,
1968, 1970; Van Dyke and Hooker, 1969; Smith, 1975; Ramundo, et
al., 1981; Raymundo and Hooker, 1982). For the most part, these
studies reveal background information useful to an understanding
of resistance but are not needed in the actual breeding for
resistance. Whether or not this information will be useful to
genetic engineering techniques applied to plants remains to be
seen.

Sources of Resistance

Locating suitable sources of resistance to pathogens and
insects is one of the important objectives in any program on the
genetics of host:parasite interactions or breeding for
resistance. It is the germplasm material upon which the program
is based. Our present practice is largely one of empirically
screening germplasm until something of interest is found. The
search then often stops and work is initiated to determine how
valuable it is in disease or insect control.

Resistance can be found in various places and its potential
usefulness can be predicted. When resistance is needed, the
first and most useful place to look for it is among and within
native cultivated germplasm. The resistance, if found, is then
present in germplasm adapted to local conditions, is readily
useable in breeding programs, and if used is not likely to bring
in undesirable traits. Adapted germplasm from other geographic
areas has frequently proven to be some of the best sources of
resistance. Non-cultivated plant species related to a crop
often have more resistance to disease and insects than do
cultivated species. Sterility, lack of chromosome pairing, and
presence of linked genes for undesirable traits are limitations
in the use of wild plant species in breeding programs.
Nevertheless, wild relatives of tobacco (<u>Nicotiana tabacum</u> L.),
tomato (<u>Lycopersicon esculentum</u> Mill.), potato (<u>Solanum</u>

tuberosum L.), sugarcane (Saccharum officinarum L.), wheat,
oats, rice (Oryza sativa L.) and other crops have contributed
valuable genes for disease resistance to cultivated species. In
species crosses, the breeder can usually transfer only small
sections of alien chromosomes to cultivated species and still
retain yield and quality. This resistance is often of the
specific type, but can be very useful. It has taken me years,
with a small amount of effort each year, to transfer resistance
to Helminthosporium turcicum Pass. from Tripsacum floridanum
Porter ex. Vasey to maize. The resistance is now in inbred
lines where hybrid performance is competitive in yield and other
agronomic traits with current commercial hybrids. The
resistance segregates as a single dominant gene and has been
given the symbol Ht3 (Hooker and Perkins, 1980).

We now know there are gene centers or gene pools located in
various parts of the world that are particularly rich in genes
for resistance to specific pathogens or insects (Leppik, 1970).
Part of this is related to primary and secondary gene centers of
the crop plant and part is related to the fact that strains of
the pathogen in the breeder's epidemiology area have never been
exposed to plants having the genes for resistance present in the
exotic varieties. Hence, most, if not all, of the pathogen
population is avirulent and thus the resistance is effective.
Sources of resistance are also often found where host and
parasite have been associated with each other for long periods
of time. Here, resistant plant types have greater survival
value, and alleles for resistance increase in frequency in the
plant population.

It is quite important to know something about how
resistance is expressed, or may be expressed, when one looks for
sources of resistance. Sometimes, new forms of resistance not
previously recognized are discovered. An example of this is the
chlorotic-lesion form of resistance with reduced sporulation to
Helminthosporium turcicum, the fungus causing northern leaf
blight of maize (Hooker, 1961). This resistance has been widely
used and worth an undeterminable amount of money in the control
of this disease. It presumably escaped attention because the
usual rating scales were based on amount of infection and
applied late in the season. One is looking for alleles for
resistance regardless of how they function or are inherited. As
alleles for resistance, whether partial or more complete, are
located, they can be combined in various ways to produce plant
genotypes with the most effective resistance combined with other
traits. After this is done, the usefulness of the resistance in
agriculture can be assessed.

It is desirable to know something about the breeding
potential of various sources of resistance. Some of this comes

through observation of material in the breeding program, but special studies can be made. A workable procedure, after a number of sources of resistance have been located, is to cross them to a few susceptible but elite cultivars or inbreds. An examination of the frequency and types of segregates in backcross, F2 or more advanced generations can reveal sources that have the best breeding potential for both resistance and other traits. In my work, I have frequently experienced situations where two or more genotypes appear equally good per se as sources of resistance, but one proves to be distinctly superior in hybrid combination. The intercrossing of different sources of resistance can also reveal genes for resistance or modifier genes not seen in original sources. Occasionally, superior forms of resistance result when intermediate forms are hybridized.

My experience with maize and its diseases is that an intensive search for sources of resistance will reveal a large number and range of types (Hooker and LeRoux, 1957; Hooker, 1961, 1962, 1977c, 1978; Hooker et al., 1964; Perkins and Hooker, 1981). Furthermore, when subjected to a genetic study, a wide array of genes, cytoplasms, and interacting systems for resistance will be revealed (Hooker and Russell, 1962; Malm and Hooker, 1962; Hooker, 1963, 1967a, 1972b, 1977b; Lee et al., 1963; Hagan and Hooker, 1965; Saxena and Hooker, 1968; Wilkinson and Hooker, 1968; Hooker and Saxena, 1971; Hughes and Hooker, 1971; Smith and Hooker, 1973; Lim and Hooker, 1976; Hooker and Perkins, 1980; Hooker and Tsung, 1980; Carson and Hooker, 1981, 1982). There is no reason to expect that other plant species and their parasites would be different.

There are extensive germplasm collections in the world that have not adequately been evaluated for disease and insect resistance. The available genes for various types and components of resistance await identification. Furthermore, alleles for different components of resistance should be combined in various ways in elite germplasm and the effects measured. There is a real need for programs of germplasm enhancement where more alleles for resistance from unadapted sources are incorporated into elite germplasm used by breeders. Each applied breeder cannot afford to use unproductive but resistant sources of germplasm and hope to make progress in developing cultivars and hybrids superior to those now in cultivation. However, a few people devoting major attention to enhancement efforts can be of great value to many others.

The previous work can be and is done in both public and private research programs. Scientists in the public sector should release newly discovered resistant germplasm as soon as its worth has been proven. Breeders now have ways of speeding

breeding programs and the objective of all research should be to
see that it is used as soon as feasible. It is here that much
more work needs to be done. The opportunity of locating new
alleles for resistance is great. The real value comes through
the incorporation of the various types of alleles for resistance
to the numerous diseases and insects of a crop into the
germplasm pool now being cycled and recycled by breeders.

Genetics

     Genetic studies of host:parasite interactions are usually
made by breeders, entomologists or pathologists in association
with plant breeding programs. Plant genetic studies are done by
the above three groups of scientists whereas studies of insect
variation or of pathogen variation are usually made by
entomologists or pathologists, respectively. The latter studies
are of greatest use when the entomologists and pathologists have
a good comprehension of the kinds of germplasm in use in current
breeding programs and in agriculture. While we now know a great
deal about the genetics of host:parasite interactions, more
information continues to be revealed (Ellingboe, 1981; Kiyosawa,
1982; Parlevliet, 1983).

Breeding for Resistance

     Methods of breeding for resistance to pathogens and insects
are similar to those for other traits. Of course, specialized
environments (disease or insect attack) are needed so that genes
and cytoplasms for resistance will be expressed and proper
selections made in the breeding program. In addition, genetic
variation in the insect or pathogen may be an important
consideration in the breeding and testing program.

     While multidisciplinary, cooperative, and interactive
relationships are needed to breed for disease and/or insect
resistance, all breeding programs ultimately end up being
centered around certain portions of germplasm. This segment of
germplasm is in fact an individual breeding effort where one
person often makes most of the decisions. It, in essence, is
this person's program and he is responsible for it. Because of
selection pressure, programs that have disease and/or insect
resistance as the primary objective have been the most
successful when breeding for these traits is the goal. However,
products from such programs must also possess other needed
attributes to be released and eventually used in agriculture.
Again because of selection pressure, when other breeding
objectives predominate, progress in disease and insect
resistance is less evident. Total performance counts, and it is
difficult to get each of all traits within desired ranges.

There are several facets to plant breeding programs.

Plant breeders use various methods to generate variation. The most widely used involves hybridization between selected plants of contrasting types within a species. These parents are usually elite cultivars or inbred lines. It is by this means that genes for general resistance and some genes for specific resistance are manipulated. Artificial gene mutations may be attempted (Simons, 1979) and wide crosses involving exotic strains or wild species which may include unique chromosome manipulation or embryo rescue techniques are sometimes used. Here, genes for specific resistance are primarily manipulated. The newer techniques of biotechnology show promise and will be explored as an aid to plant breeding (Kosuge et al., 1983; Thanutong et al., 1983). Somaclonal variation following growth in tissue culture and plant regeneration is now operational for several crops (Brettell and Ingram, 1979; Fujiwara, 1982; Larkin and Scowcroft, 1983). Resistance to disease has been achieved via tissue culture. Genetic engineering involves cloning genes, their modification,insertion into the plant, and incorporation with expression in the plant genome are exciting possibilities for the future. Some attempts are now underway to clone genes for disease resistance.

Backcross breeding is commonly used for disease or insect resistance when the resistance is simply inherited and the source is a poorly adapted type. By this means, most of the traits of a productive but susceptible genotype can be recovered with the allele(s) for resistance substituted into it.

Plant populations can be improved through recurrent selection. This breeding procedure is most often used in cross-pollinated species. The frequency of alleles for resistance increases in the population under improvement resulting in a more resistant population and a higher probability of obtaining resistant lines when plants within the population are selected and selfed.

In cultivar or inbred line development, selection is imposed during the selfing generations. Maximum variation is seen during the early selfing generations, and selection for disease resistance is often done at this time. Large numbers of plants can be easily inoculated and because heritability of disease reaction is quite high, individual plant selections are effective. Selection for insect resistance may also be done during early generations but is often delayed. The reason for this is that insect infestations are more costly to make and because of escape, individual plant reactions are less reliable than for many diseases. Therefore, selections are made among individual progeny rows where genetic variation within rows is less and among rows greater. After genetic uniformity is achieved, cultivars or inbred lines are subjected to comprehensive evaluations for numerous traits. These are described for disease and insect resistance in the following section or elsewhere in the symposium for other traits.

Breeders have found that it is easier to breed for resistance to some parasites than others. In general, those that attack actively growing tissue or have specialized feeding habits are easier to control than those that attack senescent or dormant tissue or have unspecialized feeding habits. Among the pathogens, resistance to the viruses, most bacteria, vascular pathogens (Fusarium and Verticillium spp.), rusts, powdery mildews, downy mildews, Helminthosporium fungi, smuts, and certain nematodes is quite readily attained in most plant species. Among the insects, breeding for resistance to aphids, greenbugs, plant hoppers, and certain leaf chewing insects has been the most successful.

Disease and Insect Reactions

Products from the breeding program need to be evaluated for reaction to diseases and insects before they are released into agriculture. This task should be the responsibility of trained pathologists and entomologists, and the reaction ratings should be obtained objectively and without bias. These data are heavily used by those who decide what products are to be advanced through the testing program and eventually released for sale, by those who determine where a product will be sold, and by those who advise growers which products to plant. Breeders, of course, also use this information as they select germplasm for new and continued use in their breeding programs.

Because of legal ramifications, it has become increasingly difficult for private industry to give numerical ratings for disease or insect reactions or to advertise a product as resistant even when all objective data supports this fact. Various types of exclusions are included in advertisements or attached to the seed container. This limits legal warranty. Nevertheless, much good information on disease and insect reactions, as part of product descriptions, do become available to the farmer. It would be helpful if all products of the same plant species were described in a uniform and meaningful way. Growers would then be able to make better comparisons among cultivars on hybrids for disease or insect reactions.

Public researchers or extension advisors often try to help farmers by reporting their observations of cultivar or hybrid reactions to naturally occurring diseases or insects. The value of this practice is questioned when the data are imperfectly obtained or when only part of the available genotypes evaluated. It often happens that the genotypes most widely grown will encounter the pathogen or pest, be observed as susceptible and so listed, while a less widely grown genotype may escape contact with the pathogen or insect and be mistakenly listed as resistant. Productive genotypes damaged by the parasite might also be more satisfactory to the grower than a less damaged genotype with lower yield potential.

While time and effort would be involved, it should be possible to objectively evaluate plant genotypes and to place them into broad categories such as generally safe, slightly vulnerable, moderately vulnerable, vulnerable, or highly vulnerable to specific pathogens or insects when grown in known cropping situations.

Gene and Cytoplasm Management

Much has been written about different strategies for deploying genes and cytoplasms for resistance in agriculture (Hooker, 1967b, 1977a, 1983; Van der Plank, 1968; Browning and Frey, 1969; Pathak, 1970; Watson, 1970; Robinson, 1971, 1973; Knott, 1972; Simons, 1972; Frey et al., 1973; Nelson, 1973; Khush and Coffman, 1977; Parlevliet and Zadoks, 1977; Parlevliet, 1979, 1983; Gallun and Khush, 1980; Johnson, 1981; Dubin and Rajaram, 1982; Lewin, 1982). Many of these concepts are relative to pathogens and a few insects that are genetically capable of matching host genes for resistance with corresponding virulence genes so that the resistance becomes ineffective. This phenomena does not apply to the vast majority of disease and insect problems. The "probability" of this happening in agriculture must also be considered as well as the "possibility." For example, the probablity of matching virulence is much higher in the potato late blight fungus (Phytophthora infestans (Mont.) DBy.) than in the Helminthosporium fungi. The goal should always be resistance that functions adequately in an agricultural situation and for the life of the cultivar or hybrid.

Various schemes for gene and cytoplasm management in crop plants have been suggested for pathogens. The following are some of the proposals: 1) avoid using only single-gene resistance, 2) greater use of polygene general resistance, 3) use of single genes in multi-line cultivars where each line has a different gene, 4) zonal deployment of genes based upon the epidemiology of the pathogen from early to later maturing cultivars or from one geographic region to another, 5) diversity of genotypes on farms within a region, 6) diversity of genotypes within a region over time so that each successive cultivar is unlike the previous, 7) pyramiding a collection of single genes for specific resistance in the same cultivar, and 8) use of complex resistance. There may be others. The value and limitations of these schemes have been discussed elsewhere (Hooker, 1977a, 1983; Parlevliet, 1981). While general resistance is of utmost importance and should be used to the greatest extent possible compatible with all other objectives, some breeders may now be giving too little attention to specific resistance. They gain comfort in claiming they are working with general resistance and cannot be blamed for resistance that fails. It follows that much good resistance is not being used and the level of general resistance selected can be too low to provide adequate protection. In actual practice, general resistance and placing several effective genes for specific resistance in the same genotype is the objective of successful breeders for disease resistance. Other schemes are not used or fall into place without plan.

More attention continues to be given to insect resistance in breeding programs. The range of insect species considered is expected to widen. Major genes for high resistance to insects have been identified in numerous crops and when compared to pathogens are used more extensively in agricultural crops than are polygenic systems. Insect biotypes with corresponding genes for virulence may be selected from insect populations and when this happens or is predicted to happen, cultivars with a new effective gene for resistance are bred and released. Entomologists and breeders have found this to be an effective way to cope with insect variation.

Cultivar and Hybrid Release for Production

When results are in from the testing program, decisions are made on which cultivar or hybrid to release. In U.S. public institutions, the common practice is for the plant breeder to present the results of the testing program before a Plant Variety Committee and a decision is made regarding release. In private industry, management plays a large role. People in sales and production as well as in research are involved. In addition to product superiority, factors such as farmer acceptance, market penetration, production profitability, and potential sales volume are considered. Not all superior products developed through breeding programs reach the farm.

Disease and insect reactions are considered in the decision to release products from plant breeding programs. Extreme susceptibility to any common disease or insect will usually prevent release unless the genotype is outstanding in other respects and an effective alternative means of disease or insect control is known. When all current cultivars or hybrids are susceptible, a genotype with good resistance but only average performance in other respects may be released because of its unique resistance. However, the usual pattern is for newly released cultivars or hybrids to have an adequate degree but not necessarily a high degree of resistance to several diseases and insects. In some countries, minimum standards have been established, and only products that meet these standards for disease reaction are put on recommended lists (Doodson, 1982). We do not have these requirements in the U.S.

Seed Production

With the exception of farmers who may save their own seed of self-pollinated crops, seed production in the United States is now commercialized and in the hands of private industry. Breeders seed is the most pure and is maintained under the control of the public or private plant breeder. It is used to produce foundation seed. With public cultivars, foundation seed is produced by an organization set up for that purpose and is available to private industry to produce certified seed for sale to farmers. Not all seed sold by private industry is certified. Seed of public inbreds is also produced by foundation seed organizations and sold to private industry for use in producing hybrid seed. Within private industry, foundation seed is produced in a special unit set up within the company to serve this function. Such seed is available only to the company itself. Hybrid seed and seed of cultivars is sold to farmers and others who grow plants or crops. With some industry firms where a processed product such as dried, canned, or frozen fruits or vegetables is sold, the firm may support a plant breeding program and the seed is used by the growers of the processed product but is never sold to others.

Sale of Seed

Seed may be sold directly by the company producing the seed or merchandised through some form of distributors. Within large firms, special sales organizations are established to advertise and market seed. This is a professional group with sales organized into regions and down to districts. However, when this type of sales organization is in place, the final salesman is often a farmer dealer or a local person known and respected by the farmers of the community.

Use by Farmers

Farmers and other growers of field and vegetable crops have come to expect all new cultivars and hybrids to have resistance to the common diseases in his area. This is an obligation that must be met by breeders. To a lesser extent, this is also true for some insects. Resistance is less used in fruit, ornamental,

or tree crops. Some of these are perennial crops or possess
unique quality characteristics. The range of diseases and
insects controlled by resistance continues to expand for each
crop. This trend is expected to continue.

Of the various alternatives offered to him by the breeder
and the seed industry, the farmer selects specific cultivars and
hybrids for growth on his farm. With most farmers, small
amounts of a new cultivar or hybrid are grown and observed first
hand. Only after a satisfactory performance has been
demonstrated on his farm, do progressive farmers grow a large
percentage of any one cultivar or hybrid. However, by this
means, only a few cultivars or hybrids prove themselves to be
superior, and they become widely grown.

Consumers

Consumers are the main beneficiaries from the growth by
farmers of superior performing, disease and insect resistant
cultivars and hybrids. A stable amount of high quality food is
available to them at low cost. When all works well, the
populace is well fed, clothed, and sheltered.

ACKNOWLEDGMENT

The author acknowledges Eldon Ortman and Daniel Palmer for
helpful discussions and critical reading of the manuscript and
Ms. Andrea Cochrane for typing the manuscript.

REFERENCES

Athwal, D. S., Pathak, M. D., Bacalangco, E. H., and Pura, C. D., 1971, Genetics of resistance to brown planthoppers and green leafhoppers in Oryza sativa L., Crop Sci., 11:747.

Beck, S. D., 1965, Resistance of plants to insects, Ann. Rev. Entomol., 10:207.

Bell, A. A., 1981, Biochemical mechanisms of disease resistance, Ann. Rev. Plant Physiol., 32:21.

Biffen, R. H., 1905, Mendel's laws of inheritance and wheat breeding, Jour. Agr. Sci., 1:4.

Branson, T. F., Welch, V. A., Sutter, G. R., and Fisher, J. R., 1983, Resistance to larvae of Diabrotica virgifera virgifera in three experimental maize hybrids, Environ. Entomol., 12:1509.

Brettell, R. I. S., and Ingram, D. S., 1979, Tissue culture in production of novel disease-resistant crop plants, Biol. Rev. 54:329.

Browning, J. A., and Frey, K. J., 1969, Multiline cultivars as a means of disease control, Ann. Rev. Phytopathol., 7:355.

Buddenhagen, I. W., 1983, Breeding strategies for stress and disease resistance in developing countries, Ann. Rev. Phytopathol. 21:385.

Carson, M. L., and Hooker, A. L., 1981, Inheritance of resistance to stalk rot of corn caused by Colletotrichum graminicola, Phytopathology, 71:1190.

Carson, M. L., and Hooker, A. L., 1982, Reciprocal translocation testcross analysis of genes for anthracnose stalk rot resistance in a corn inbred line, Phytopathology, 72:175.

Dahms, R. G., 1972, Techniques in the evaluation and development of host-plant resistance, Jour. Environ. Qual., 1:254.

Dalmacio, S. C., Mac Kenzie, D. R., and Nelson, R. R., 1979, Heritability of differences in virulence between races 2 and 3 of Cochliobolus carbonum, Philipp. Phytopathol., 15:47.

Day, P. R., 1974, "Genetics of Host-Parasite Interaction," Freeman, San Francisco.

Doodson, J. K., 1982, Disease testing at NIAB, Britain's unique agricultural institute, Plant Dis., 66:875.

Dubin, H. J., and Rajaram, S., 1982, The CIMMYT's international approach to breeding disease-resistant wheat, Plant Dis., 66:967.

Ellingboe, A. H., 1981, Changing concepts in host-pathogen genetics, Ann. Rev. Phytopathol., 19:125.

Flor, H. H., 1956, The complementary genic systems in flax and flax rust, Advan. Genet., 8:29.

Flor, H. H., 1971, Current status of the gene-for-gene concept, Ann. Rev. Phytopathol., 9:275.

Frey, K. J., Browning, J. A., and Simons, M. D., 1973, Management of host resistance genes to control diseases, Z. PflKrankh.

PflSchutz., 80:160.

Fujiwara, A. ed., 1982, " Plant Tissue Culture 1982," Maruzen, Tokyo.

Gallun, R. L., 1972, Genetic interrelationships between host plants and insects, Jour. Environ. Qual., 1:259.

Gallun, R. L., and Khush, G. S., 1980, Genetic factors affecting expression and stability of resistance, in: "Breeding Plants Resistant to Insects," F. G. Maxwell and P. R. Jennings, eds., Wiley, New York.

Gallun, R. L., Starks, K. J., and Guthrie, W. D., 1975, Plant resistance to insects attacking cereals, Ann. Rev. Entomol., 20:337.

Giebel, J., 1982, Mechanism of resistance to plant nematodes, Ann. Rev. Phytopathol., 20:257.

Hagan, W. L., and Hooker, A. L., 1965, Genetics of reaction to Puccinia sorghi in eleven corn inbred lines from Central and South America, Phytopathology, 55:193.

Harris, M. K. ed., 1979, "Biology and Breeding for Resistance to Arthropods and Pathogens in Agricultural Plants," Texas A & M University, College Station.

Hatchett, J. H., and Gallun, R. L., 1970, Genetics of the ability of the Hessian fly, Mayetiola destructor, to survive on wheats having different genes for resistance, Ann. Entomol. Soc. Amer., 63:1400.

Hedin, P. A., ed., 1977, "Host Plant Resistance to Pests," American Chemical Society, Washington.

Hilu, H. M., 1965, Host-pathogen relationships of Puccinia sorghi in nearly isogenic resistant and susceptible seedling corn, Phytopathology, 55:563.

Hilu, H. M., and Hooker, A. L., 1964, Host-pathogen relationship of Helminthosporium turcicum in resistant and susceptible corn seedlings, Phytopathology, 54:570.

Hilu, H. M., and Hooker, A. L., 1965, Localized infection by Helminthosporium turcicum on corn leaves, Phytopathology, 55:189.

Hooker, A. L., 1961, A new type of resistance in corn to Helminthosporium turcicum, Plant Dis. Reptr., 45:780.

Hooker, A. L., 1962, Additional sources of resistance to Puccinia sorghi in the United States, Plant Dis. Reptr., 46:14.

Hooker, A. L., 1963, Monogenic resistance in Zea mays L. to Helminthosporium turcicum, Crop Sci., 3:381.

Hooker, A. L. 1967a, Inheritance of mature plant resistance to rust, Phytopathology, 57:815.

Hooker, A. L., 1967b, The genetics and expression of resistance in plants to rusts of the genus Puccinia, Ann. Rev. Phytopathol., 5:163.

Hooker, A. L., 1972a, Breeding and testing for disease reaction, McGraw-Hill Encyclopedia of Science and Technology, 3rd ed., 10:371.

Hooker, A. L., 1972b, Southern leaf blight of corn-present status and future prospects, Jour. Environ. Qual., 1:244.

Hooker, A. L., 1974a, Cytoplasmic susceptibility in plant disease, Ann. Rev. Phytopathol., 12:167.

Hooker, A. L., 1974b, Selection for stalk rot resistance in corn and its effect on grain yield, Genetika, 6:27.

Hooker, A. L., 1977a, A plant pathologist's view of germplasm evaluation and utilization, Crop Sci., 17:689.

Hooker, A. L., 1977b, A second major gene locus in corn for chlorotic-lesion resistance to Helminthosporium turcicum, Crop Sci., 17:132.

Hooker, A. L., 1977c, Stalk rot resistance through gamete selection from exotic maize, Maydica, 22:173.

Hooker, A. L., 1978, Additional sources of monogenic resistance in corn to Helminthosporium turcicum, Crop Sci., 18:787.

Hooker, A. L., 1979, Estimating disease losses based on the amount of healthy leaf tissue during the plant reproductive period, Genetika, 11:181.

Hooker, A. L., 1983, Breeding to control pests, in: "Crop Breeding," D. R. Wood, ed., American Society of Agronomy, Madison.

Hooker, A. L., 1984, Corn and sorghum rusts, in: "The Cereal Rusts, Vol. II," A. P. Roelfs and W. R. Bushnell, eds., Academic Press, New York.

Hooker, A. L., and LeRoux, P. M., 1957, Sources of protoplasmic resistance to Puccinia sorghi in corn, Phytopathology, 47:187.

Hooker, A. L., and Russell, W. A., 1962, Inheritance of resistance to Puccinia sorghi in six corn inbred lines, Phytopathology, 52:122.

Hooker, A. L., and Perkins, J. M., 1980, Helminthosporium leaf blights of corn-the state of the art, Amer. Seed Trade Assn., Annual Corn and Sorghum Industry Research Conference, 35:68.

Hooker, A. L., and Saxena, K. M. S., 1971, Genetics of disease resistance in plants, Ann. Rev. Genetics, 5:407.

Hooker, A. L., Smith, D. R., Lim, S. M., and Beckett, J. B., 1970, Reaction of corn seedlings with male-sterile cytoplasm to Helminthosporium maydis, Plant Dis. Reptr., 54:708.

Hooker, A. L., and Tsung, Y. K., 1980, Relationship of dominant genes in corn for chlorotic lesion resistance to Helminthosporium turcicum, Plant Dis., 64:387.

Hooker, A. L., Hilu, H. M., Wilkinson, D. R., and Van Dyke, C. G., 1964, Additional sources of chlorotic lesion resistance to Helminthosporium turcicum in corn, Plant Dis. Reptr., 48:777.

Hooker, A. L., Nelson, R. R., and Hilu, H. M., 1965, Avirulence of Helminthosporium turcicum on monogenic resistant corn, Phytopathology, 55:462.

Hughes, G. R., and Hooker, A. L., 1971, Gene action conditioning resistance to northern leaf blight in maize, Crop Sci., 11:180.

Jenkins, J. N., 1981, Breeding for insect resistance, in: "Plant

Breeding II," K. J. Frey, ed., Iowa State University Press, Ames.

Johnson, H. W., 1972, Development of crop resistance to disease and nematodes, Jour. Environ. Qual., 1:23.

Johnson, R., 1981, Durable resistance: definition of, genetic control, and attainment in plant breeding, Phytopathology, 71:567.

Kennedy, G. G., 1978, Recent advances in insect resistance of vegetable and fruit crops in North America: 1966-1977, Entomol. Soc. Am. Bull., 24:375.

Khush, G. S., and Coffman, W. R., 1977, Genetic evaluation and utilization (GEU) program. The rice improvement program of the International Rice Research Institute, Theor. Appl. Genet., 51:97.

Kim, S. K., and Brewbaker, J. L., 1977, Inheritance of general resistance in maize to Puccinia sorghi Schw., Crop Sci., 17:456.

Kiyosawa, S., 1982, Genetics and epidemiological modeling of breakdown of plant disease resistance, Ann. Rev. Phytopathol., 20:93.

Knott, D. R., 1972, Using race-specific resistance to manage the evolution of plant pathogens, Jour. Environ. Qual., 1:227.

Kogan, M., and Ortman, E. F., 1978, antixenosis-a new term proposed to define Painter's "nonpreference" modality of resistance, Entomol. Soc. Am. Bull., 24:175.

Kosuge, T., Meredith, C. P., and Hollaender, A., 1983, "Genetic Engineering of Plants," Plenum Press, New York.

Larkin, P. J., and Scowcroft, W. R., 1983, Somaclonal variation and crop improvement, in: "Genetic Engineering of Plants," T. Kosuge, C. P. Meredith and A. Hollaender, eds., Plenum Press, New York.

Lee, B. H., Hooker, A. L., Russell, W. A., Dickson, J. G., and Flangas, A. L., 1963, Genetic relationships of alleles on chromosome 10 for resistance to Puccinia sorghi in 11 corn lines, Crop Sci., 3:24.

Leppik, E. E., 1970, Gene centers of plants as sources of disease resistance, Ann. Rev. Phytopathol., 8:323.

Lewin, R., 1982, Never ending race for genetic variants, Science, 218:877.

Lim, S. M., and Hooker, A. L., 1976, Estimates of combining ability for resistance to Helminthosporium maydis race 0 in a maize population, Maydica, 21:121.

Lim, S. M., Hooker, A. L., and Paxton, J. D., 1970, Isolation of phytoalexins from corn with monogenic resistance to Helminthosporium turcicum, Phytopathology, 60:1071.

Lim, S. M., Kinsey, J. G., and Hooker, A. L., 1974, Inheritance of virulence in Helminthosporium turcicum to monogenic resistant corn, Phytopathology, 64:1150.

Lim, S. M., Paxton, J. D., and Hooker, A. L., 1968, Phytoalexin

production in corn resistant to Helminthosporium turcicum,
Phytopathology, 58:720.

Leonard, K. J., and Mundt, C. C., 1984, Methods for estimating
epidemiological effects of quantitative resistance to plant
diseases, Theor. Appl. Genet., 67:219.

Loegering, W. Q., 1978, Current concepts in interorganismal
genetics, Ann. Rev. Phytopathol., 16:309.

Loegering, W. Q., McIntosh, R. A., and Burton, C. H., 1971, Computer
analysis of disease data to derive hypothetical genotypes for
reaction of host varieties to pathogens, Can. Jour. Genet.
Cytol., 13:742.

Mac Kenzie, D. R., 1980, The problem of variable pests, in: "Breeding
Plants Resistant to Insects," F. G. Maxwell and P. R. Jennings,
eds., Wiley, New York.

Mains, E. B., 1931, Inheritance of resistance to rust, Puccinia
sorghi, in maize, Jour. Agr. Res., 43:419.

Malm, N. R., and Hooker, A. L., 1962, Resistance to rust, Puccinia
sorghi Schw., conditioned by recessive genes in two corn inbred
lines, Crop Sci., 2:145.

Maxwell, F. G., and Jennings, P. R. eds, 1980, "Breeding Plants
Resistant to Insects," Wiley, New York.

Maxwell, F. G., Jenkins, J. N., and Perrott, W. J., 1972, Resistance
of plants to insects, Adv. Agron., 24:187.

National Academy of Sciences, 1972, Genetic vulnerability of major
crops, Washington.

Nelson, R. R., 1970, Genes for pathogenicity in Cochliobolus
carbonum, Phytopathology, 60:1335.

Nelson, R. R. ed., 1973, "Breeding Plants for Disease Resistance,
Concepts and Applications." Penn. State Univ. Press, University
Park.

Nelson, R. R., and Kline, D. M., 1961, The pathogenicity of certain
species of Helminthosporium to species of the gramineae, Plant
Dis. Reptr., 45:644.

Nelson, R. R., Blanco, M., Dalmacio, S., and Moore, B. S., 1973, A
new race of Helminthosporium carbonum on corn, Plant Dis.
Reptr., 57:822.

Norris, D. M., and Kogan, M., 1980, Biochemical and morphological
bases of resistance, in: "Breeding Plants Resistant to Insects,"
F. G. Maxwell and P. R. Jennings, eds., Wiley, New York.

Painter, R. H., 1951, "Insect Resistance in Crop Plants," MacMillan,
New York.

Painter, R. H., 1958, Resistance of plants to insects, Ann. Rev.
Entomol., 3:267.

Parlevliet, J. E., 1979, Components of resistance that reduce the
rate of epidemic development, Ann. Rev. Phytopathol., 17:203.

Parlevliet, J. E., 1981, Disease resistance in plants and its
consequenses for plant breeding, in: "Plant Breeding II," K. J.
Frey, ed., Iowa State University Press, Ames.

Parlevliet, J. E., 1983, Rare-specific resistance and cultivar-specific virulence in the barley-leaf rust pathosystem and their consequences for the breeding of leaf rust resistant barley, Euphytica, 32:367.

Parlevliet, J. E., and Zadoks, J. C., 1977, The integrated concept of disease resistance; a new view including horizontal and vertical resistance in plants, Euphytica, 26:5.

Pathak, M. D., 1970, Genetics of plants in pest management, in: "Concepts of Pest Management," R. L. Rabb and F. E. Guthrie, eds.,North Carolina State University, Raleigh.

Pathak, M. D., 1972, Resistance to insect pests in rice varieties, in: "Rice Breeding," International Rice Research Institute, Los Baños.

Perkins, J. M., and Hooker, A. L., 1981, Reactions of eighty-four sources of chlorotic lesion resistance in corn to three biotypes of Helminthosporium turcicrum, Plant Dis., 65:502.

Posnette, A. F., 1969, Tolerance of virus infection in crop plants, Rev. Appl. Mycol., 48:113.

Raymundo, A. D., and Hooker, A. L., 1982, Single and combined effects of monogenic and polygenic resistance on certain components of northern corn leaf blight development, Phytopathology, 72:99.

Raymundo, A. D., Hooker, A. L., and Perkins, J. M., 1981, Effect of gene HtN on the development of northern corn leaf blight epidemics, Plant Dis., 65:327.

Rhoades, V. H., 1935, The location of a gene for disease resistance in maize, Proc. Natl. Acad. Sci. USA, 21:243.

Robinson, R. A., 1969, Disease resistance termonology, Rev. Appl. Mycol., 48:593

Robinson, R. A., 1971, Vertical resistance, Rev. Pl. Path., 50:233.

Robinson, R. A., 1973, Horizontal resistance, Rev. Pl. Path., 52:483.

Robinson, R. A., 1980a, New concepts in breeding for disease resistance, Ann. Rev. Phytopathol., 18:189.

Robinson, R. A., 1980b, The pathosystem concept, in: "Breeding Plants Resistant to Insects," F. G. Maxwell and P. R. Jennings, eds., Wiley, New York.

Russell, G. E., 1978, "Plant Breeding for Pest and Disease Resistance," Butterworths, London-Boston.

Saxena, K. M. S., and Hooker, A. L., 1968, On the structure of a gene for disease resistance in maize, Proc. Natl. Acad. Sci. USA, 61:1300.

Schafer, J. F., 1971, Tolerance to plant disease, Ann. Rev. Phytopathol., 9:235.

Sidhu, G., 1975, Gene-for-gene relationships in plant parasitic systems, Sci. Prog., Oxf., 62:467.

Simons, M. D., 1972, Polygenic resistance to plant disease and its use in breeding resistant cultivars, Jour. Environ. Qual., 1:232.

Simons, M. D., 1979, Modification of host-parasite interactions through artificial mutagenesis, Ann. Rev. Phytopathol., 17:75.

Smith, D. R., 1975, Expression of monogenic chlorotic-lesion resistance to Helminthosporium maydis in corn, Phytopathology 65:1160.

Smith, D. R., 1977, Monitoring corn pathogens, Amer. Seed Trade Assn., Annual Corn and Sorghum Industry Research Conference, 32:106.

Smith, D. R., and Hooker, A. L., 1973, Monogenic chlorotic-lesion resistance in corn to Helminthosporium maydis, Crop Sci., 13:330.

Smith, D. R., and Kinsey, J. G., 1980, Further physiologic specialization in Helminthosporium turcicum, Plant Dis., 64:779.

Smith, D. R., Hooker, A. L., and Lim, S. M., 1970, Physiologic races of Helminthosporium maydis, Plant Dis. Reptr., 54:819.

Sprague, G. F., and Dahms, R. G., 1972, Development of crop resistance to insects, Jour. Environ. Qual., 1:28.

Sutter, G. R., and Branson, T. F., 1980, A procedure for artificially infesting field plots with corn rootworm eggs, Jour. Econ. Entomol., 73:135.

Thanutong, P., Furusawa, I., and Yamamoto, M., 1983, Resistant tobacco plants from protoplast-derived calluses selected for their resistance to Pseudomonas and Alternaria toxins, Theor. Appl. Genet., 66:209.

Turner, M. T., and Johnson, E. R., 1980, Race of Helminthosporium turcicum not controlled by Ht genetic resistance in corn in the American corn belt, Plant Dis., 64:216.

Van der Plank, J. E., 1968, "Disease Resistance in Plants," Academic Press, New York.

Van der Plank, J. E., 1975, "Principles of Plant Infection," Academic Press, New York.

Van Dyke, C. G., and Hooker, A. L., 1969, Ultrastructure of host and parasite in interactions of Zea mays with Puccinia sorghi, Phytopathology, 59:1934.

Watson, I. A., 1970, Changes in virulence and population shifts in plant pathogens, Ann. Rev. Phytopathol., 8:209.

Wilcoxson, R. D., 1981, Genetics of slow rusting in cereals, Phytopathology, 71:989.

Wilkinson, D. R., and Hooker, A. L., 1968, Genetics of reaction to Puccinia sorghi in ten corn inbred lines from Africa and Europe, Phytopathology, 58:605.

Wolfe, M. S., and Schwarzback, E., 1975, The use of virulence analysis in cereal mildews, Phytopath. Z., 82:297.

Wysong, D. S., and Hooker, A. L., 1966, Relation of soluble solids content and pith condition to Diplodia stalk rot in corn hybrids, Phytopathology, 56:26.

THE GENOMIC SYSTEM OF CLASSIFICATION AS A GUIDE TO INTERGENERIC

HYBRIDIZATION WITH THE PERENNIAL TRITICEAE

Douglas R. Dewey

U. S. Department of Agriculture
Agricultural Research Service
Utah State University-UMC 63
Logan, UT 84322

INTRODUCTION

Of the approximately 325 species in the tribe Triticeae
(= Hordeeae), about 250 are perennials that include many of the
world's important forage grasses. Although more than 75% of the
species in the Triticeae are perennials, they have received far less
attention from cytogeneticists and plant breeders than have the
annuals, which include three major cereal crops--wheat, barley, and
rye. In addition to being important in their own right as forages,
the perennials form a vast genetic reservoir that might be used to
improve the annual cereals.

Hybridization between annual and perennial Triticeae species
has been a relatively common plant-breeding practice since the early
1930's, especially in the U.S.S.R. (Tsitsin, 1960; 1975). However,
only a few perennial species were involved in those early programs,
whose goals usually were to transfer disease resistance or the
perennial habit to the annuals. Annual cereal X perennial grass
hybridization remained at a more or less static level until the
1970's when advances in hybridization techniques (Kruse, 1973),
embryo culturing (Murashige, 1974), and control of homoeologous
pairing (Riley, 1974) stimulated a renewed interest in wide hybridization in the Triticeae. Over the past 10 years I have noticed
heightened interest and activity in hybridization between annual and
perennial species of Triticeae as evidenced by a substantial
increase in requests by cereal breeders and cytogeneticists for
seeds from the U.S. Living Collection of Perennial Triticeae
Grasses, which I curate (Dewey, 1977).

If forage and cereal breeders are to use the perennial
Triticeae effectively, they must understand the cytogenetic characteristics and biological relationships of these grasses. The purposes of this paper are to: 1) summarize the current information
concerning the genomic and phylogenetic relationships among the
perennials, 2) relate genomic and phylogenetic relationships to a
new system of classification, and 3) provide genomic information
that might assist perennial grass breeders and wheat, barley, and
rye breeders in using the perennials in intergeneric hybridization
programs.

GENERA OF THE TRIBE TRITICEAE

The generic makeup of the Triticeae has varied widely since the
time of Linnaeus (1753) who named seven genera--Aegilops, Elymus,
Hordeum, Lolium, Nardus, Secale, and Triticum--that have at one time
or another been included in the Triticeae. All of the Linnaean
genera except Lolium and Nardus have remained in the Triticeae.
Bentham (1881) placed all grasses with a simple spike into the
Triticeae (as Hordeeae), and he recognized 12 genera; but only six
of those--Agropyron, Asperella, Elymus, Hordeum, Secale, and
Triticum--are bona fide members of the Triticeae (Table 1). Bentham
absorbed Aegilops into Triticum, a decision that has support in some
genetic and taxonomic circles in North America (Bowden, 1959; Sears,
1975). The Bentham treatment is the basis of many of the traditional treatments followed in floras of Europe, North America, and
the British Commonwealth, including the treatment by A. S. Hitchcock
in his Manual of the Grasses of the United States (Hitchcock,
1951).
The Nevski (1933) treatment of the Triticeae (as Hordeeae),
which encompassed 25 genera (Table 1), differed substantially from
the traditional Bentham point of view and earned Nevski the dubious
reputation, in some circles, as a taxonomic splitter. Nevertheless,
many of Nevski's generic concepts have come into rather widespread
use. His treatment of Agropyron as a small genus consisting of just
the crested wheatgrasses is now generally accepted worldwide. Two
of his genera, Taeniatherum Nevski and Psathyrostachys Nevski, are
also accepted worldwide. Chinese agrostologists (Keng, 1965)
accepted much of Nevski's work, which is still the basis of the
Triticeae treatments in China (Guo and Wang, 1981). Some of
Nevski's generic concepts and proposals for the Triticeae were
ill-advised; but his insights into the biology of the Triticeae were
remarkable when considered in terms of information available at that
time and his age (he died at age 31).

Pilger (1954) and Tzvelev (1973, 1976) based their treatments
of the Triticeae on Nevski's concepts, but they corrected some of
Nevski's nomenclatural errors (e.g. Nevski chose the wrong type

species for Elymus) and realigned, merged, or eliminated some of the genera on the basis of new information. Today, Tzvelev's (1976) Poaceae URSS is more or less the standard treatment used in the U.S.S.R. and surrounding countries.

The Triticeae is treated by four authors--A. Melderis, C. J. Humphries, T. G. Tutin, and S. A. Heathcote--in the 1980 Vol. 5 of Flora Europaea. These authors incorporated some of Nevski's generic concepts into the traditional Bentham treatment and recognized 14 genera within the geographic confines of Europe (Table 1).

The most recent and certainly the most unconventional treatment is that of Löve (1982, 1984), who recognized 38 genera entirely on the basis of genomic relationships (Table 1). His taxonomic philosophy, with which I generally concur (Dewey, 1982; 1983), is that a system of classification should reflect phylogeny and biological relationship. Few, if any, will question the value of genomic relationships, as determined by genome analysis, as indicators of phylogeny and biological closeness. Yet many disagreements arise with respect to how genomic data should be applied to taxonomy, at least at the generic level. Löve (1982) equated a genus to a group of species that are based on a given genome or a particular combination of genomes. For example, Crithodium = A-genome species, Sitopsis = B-genome species, Gigachilon = AB-genome species, and Triticum = ABD-genome species. Some may feel that the genomic system of classification as practiced by Löve results in an unnecessary proliferation of small or monotypic genera. However, we should not encumber ourselves with preconceived notions of how many genera should constitute a tribe or how many species should constitute a genus. Those numbers should be determined by the biological facts, which will differ from tribe to tribe and genus to genus.

Adoption of the genomic system of classification will remove much, but not all, of the subjective judgment associated with circumscribing a genus. The genomic system requires identification of the various genomes in the tribe and then building genera around the genomes. However, genomes are not discrete genetic units; but as with all biological traits, they form a continuum from complete homology to partial homology (homoeology) to nonhomology. So some subjectivity is required in deciding when related genomes are sufficiently different to warrant separate designations. Nevertheless, Triticeae cytogeneticists are in general agreement on the genomes found in the tribe.

The genomic system of classification is not without its drawbacks, which relate primarily to the difficulty of constructing easily used taxonomic keys based on gross morphology. Even the strongest proponents of the genomic system admit that taxonomic keys

Table 1.   Genera included in the tribe Triticeae by various authors.*

Bentham (1881)	Nevski (1933)	Hitchcock (1951)	Pilger (1954)
Agropyrum	Aegilops	Aegilops	Aegilops
Asperella	Agropyron	Agropyron	Agropyron
Elymus	Aneurolepidium	Elymus	Amblyopyrum
Hordeum	Anthosachne	Hordeum	Crithopsis
Kralika	Asperella	Hystrix	Dasypyrum
Lepturus	Brachypodium	Lolium	Elymus
Lolium	Clinelymus	Monerma	Eremopyrum
Nardus	Critesion	Parapholis	Henrardia
Oropetium	Crithopsis	Scribneria	Heteranthelium
Psilurus	Cuviera	Secale	Hordelymus
Secale	Elymus	Sitanion	Hordeum
Triticum	Elytrigia	Triticum	Hystrix
	Eremopyrum		Leymus
	Haynaldia		Malacurus
	Heteranthelium		Psathyrostachys
	Hordeum		Secale
	Malacurus		Sitanion
	Psathyrostachys		Taeniatherum
	Roegneria		
	Secale		
	Sitanion		
	Taeniatherum		
	Terrella		
	Trachynia		
	Triticum		

Table 1. (continued)

Keng (1965)	Tzvelev (1976)	Melderis et al. (1980)	Löve (1984)
Aegilops	Aegilops	Aegilops	Aegilemma
Agropyron	Agropyron	Agropyron	Aegilonearum
Aneurolepidium	Amblyopyrum	Crithopsis	Aegilopoides
Asperella	Dasypyrum	Dasypyrum	Aegilops
Clinelymus	Elymus	Elymus	Agropyron
Elymus	Elytrigia	Eremopyrum	Amblyopyrum
Elytrigia	Eremopyrum	Festucopsis	Australopyrum
Eremopyrum	Henrardia	Hordelymus	Chennapyrum
Hordeum	Heteranthelium	Hordeum	Comopyrum
Lepturus	Hordelymus	Leymus	Critesion
Lolium	Hordeum	Psathyrostachys	Crithodium
Parapholis	Hystrix	Secale	Crithopsis
Psathyrostachys	Leymus	Taeniatherum	Cylindropyrum
Roegneria	Psathyrostachys	Triticum	Dasypyrum
Secale	Secale		Eigopyrum
Triticum	Taeniatherum		Elymus
	Triticum		Elytrigia
			Eremopyrum
			Festucopsis
			Gastropyrum
			Gigachilon
			Henrardia
			Heteranthelium
			Hordelymus
			Hordeum
			Kiharapyrum
			Leymus
			Lophopyrum
			Orrhopygium
			Pascopyrum
			Patropyrum
			Psathyrostachys
			Pseudoroegneria
			Secale
			Sitopsis
			Taeniatherum
			Thinopyrum
			Triticum

*The treatments are not necessarily comparable because all do not cover the tribe worldwide and most authors have worked from different geographical and historical perspectives.

will always be less convenient than those based on traditional
morphologically oriented taxonomy. However, taxonomic keys can and
are being constructed to accommodate the genomic system (Barkworth
et al., 1983). In my mind, the advantages of a taxonomic system
that closely reflects biological relationships far outweigh the
inconvenience associated with less easily used taxonomic keys.

The antithesis of the genomic system of classification of the
Triticeae as advocated by Löve, with 38 genera, is the proposal of
Krause (1898) and Stebbins (1956) that calls for uniting all of the
genera into one. This proposal does nothing to solve the taxonomic
problems associated with the Triticeae, but simply moves them to a
subgeneric level where they are obscured because they all carry the
same generic name. Combining all species of Triticeae into one
genus is the biological equivalent of "sweeping the problem under
the rug."

APPLICATION OF THE GENOMIC SYSTEM OF CLASSIFICATION TO THE PERENNIAL TRITICEAE

The conditions that have led to much of the taxonomic confusion
in the Triticeae--many polyploids, different types of polyploidy,
and extensive hybridization between species--are the very conditions
that make genome analysis such an effective means of bringing order
out of apparent chaos. Genome analysis has proved its effectiveness
in unraveling the complex genomic and phylogenetic relationships in
Triticum and its closely related annual genera (Kihara, 1963). The
same procedures of genome analysis that have served so well in
Triticum and other annuals can work equally well in the perennials.

None of the perennial Triticeae has been studied as intensely
as Triticum and the other annual cereals in Hordeum and Secale, yet
a significant amount of genomic data are now available on many of
the perennials. Löve (1982, 1984) has recently converted the
genomic data into a comprehensive worldwide taxonomic treatment of
both annuals and perennials, but many information gaps remain in the
perennials. Consequently, a certain amount of extrapolation and
interpolation is necessary in circumscribing the perennial genera.
When I feel that the information gap is too large, I defer a
taxonomic decision. This is why I do not at this time consider
Hordelymus, Festucopsis, and Australopyrum, all of which are small
perennial genera. My concept and handling of Elytrigia, Lophopyrum,
and Thinopyrum are quite different from Löve's, and those
differences are described later in the manuscript. The nine
perennial genera (Table 2) treated in this manuscript account for
over 95% of the perennials in the tribe. Secale is not included
though it contains one perennial, S. montanum. Critesion (=Hordeum
pro parte) is included even though it contains a sizeable number of
annuals.

Table 2. The perennial genera of the Triticeae tribe based on genome content.

Genus	Type species	Genome(s)	Approx. no. species	Chrom. no. 2n
Agropyron	A. cristatum	P	10	14,28,42
Pseudoroegneria	P. strigosa	S	15	14,28
Psathyrostachys	P. lanuginosa	N	10	14
Critesion	C. jubatum	H	30	14,28,42
Thinopyrum	T. junceum	J-E	20	14,28,42,56,70
Elytrigia	E. repens	SX	5	42,56
Elymus	E. sibiricus	SHY	150	28,42,56
Leymus	L. arenarius	JN	30	28,42,56,70,84
Pascopyrum	P. smithii	SHJN	1	56

Uniformity in genome symbols facilitates an understanding of genomic relationships, but no unified system has been in use until Löve (1982) proposed genome symbols for the entire tribe, annuals and perennials. I am accepting Löve's genome designations even though it means changing some symbols that I have used for many years. Hopefully, others will do the same. Some changes may be necessary as the genomes become better defined. I use the genome designations X and Y to indicate unspecified genomes of unknown origin.

The principles involved in circumscribing a genus on genomic grounds are simple and straightforward and they involve a three-step process: 1) determine the genomic constitution of the type species of the genus, 2) bring all taxa with the same basic genome or combination of genomes into that genus, and 3) exclude from that genus all taxa that do not have the same basic genome or combination of genomes as the type species. For example, the type species of Agropyron is A. cristatum, whose genome is designated as P. Agropyron then consists only of those species with just the P genome. All species that contain genomes other than the P genome are excluded from Agropyron. When these procedures are applied to traditional Agropyron, it becomes a small genus of fewer than 10 species in contrast to its traditional definition as a large genus with more than 100 species.

As simple as the genomic system of classification may seem, it requires the accumulation of a vast amount of cytogenetic data from interspecific and intergeneric hybrids. A great deal of cytogenetic data has been acquired on the perennials and their hybrids over the past 30 years, but some critical data are still missing. For instance, the genome formulas of two type species, Elytrigia repens (2n=42) and Leymus arenarius (2n=56), are not known with certainty. The diploid source of the Y genome, which is found in many polyploid species of Elymus from the Far East, is not known. The genomic relationships of the Australian perennials and the Mediterranean Festucopsis with the remainder of the tribe are obscure. The genome content of Critesion is poorly defined. In spite of the stated and unstated deficiencies, it is timely and appropriate to implement the genomic system of classification, which is far better than any current taxonomic classification in reflecting biological relationships. Some adjustments will obviously be necessary as more genomic data are gathered; however, the basic structure of the genomic system of classification, as presented herein, will remain intact.

Although I subscribe in principle to Löve's (1982) genomic classification and nomenclature as they apply to the annual genera, I use conventional nomenclature when referring to species of Triticum, Hordeum, and Secale to make the discussion less confusing to the reader. My purpose here is to introduce the genomic system of classification and nomenclature for the perennial genera and not get the reader entangled with the proposed taxonomic changes in the annuals. Throughout the paper, I will apply the genomically based nomenclature to the perennials, even when referring to previously published studies that use traditional nomenclature. To equate new and traditional nomenclature, the reader is referred to the Appendix, which contains the names and authorities of all perennial taxa cited in the text.

AGROPYRON GAERTNER - 1770
Name derivation: From Greek agros = field, and pyros = wheat
Type species: Agropyron cristatum (L.) Gaertn.
Genome: P

Circumscription and Description

Agropyron is the best known, but possibly the least understood, genus of the perennial Triticeae. A clear perception of the past and present limits of Agropyron is basic to any discussion involving the perennial Triticeae. Traditionally, Agropyron has been the largest genus of the tribe with well over 100 species, encompassing almost all of the perennials with one spikelet per node. Number of spikelets per node is a convenient character in a taxonomic key; but, unfortunately, it is a poor measure of biological relationship. When defined as perennial species with one spikelet per node, Agropyron is a biologically diverse complex whose member species differ widely with respect to morphology, mode of reproduction, genome content, ecological adaptation, and geographical distribution. Because of the biological diversity of traditional Agropyron, statements that characterize the genus as a whole are usually not possible. Consequently, the traditional definition of Agropyron is so broad that it does not have much utility.

When defined in terms of only those species with the P genome (designated in my previous publications as the C genome), Agropyron becomes a small genus of no more than 10 species, which constitute what is known as the "crested wheatgrass complex." The most common species of this complex are A. cristatum, A. desertorum, and A. fragile (Appendix). Most agrostologists now accept the narrow genomic definition of Agropyron, but they have come to that definition by conventional taxonomic means. Nevski (1933) was the first modern botanist to define Agropyron in its restricted sense, and he did this without the benefit of cytogenetic information. Tzvelev (1973) reaffirmed the narrow definition of Agropyron and published it in the now widely used Poaceae URSS (Tzvelev, 1976). Melderis (1978) also defined Agropyron in the narrow sense, and that treatment is now part of Flora Europaea (Melderis et al., 1980). Only in North America, where the Hitchcock (1951) and Bowden (1965) grass treatments prevail, is Agropyron still thought of in its broad traditional sense.

Agropyron species occur at three ploidy levels--2n=14, 28, and 42 (Dewey and Asay, 1975)--with the tetraploids being the most common (Table 3). Frequently occurring quadrivalents at metaphase I in the tetraploids and frequent hexavalents in the hexaploids indicate that the polyploid taxa are autoploid or near autoploid (Dewey, 1969). Others feel that the polyploid taxa are better described as segmental alloploids (Schulz-Schaeffer et al., 1963), but the differences in the two points of view are largely semantic. Hybridization between the various crested wheatgrass taxa, representing all ploidy levels, leaves no doubt that Agropyron is founded on one basic genome (Asay and Dewey, 1979) and can be treated as one gene pool (Asay and Dewey, 1983). Hereafter, diploid, tetraploid, and hexaploid Agropyron will be represented genomically as PP, PPPP, and PPPPPP, respectively, even though some modifications of the basic P genome have occurred.

The crested wheatgrasses are cross-pollinating, caespitose, long-lived perennials, typically with pectinate (comb-like) spikes; although spike types range from broad-pectinate (A. cristatum) to narrow-linear (A. fragile). These grasses are native to inland Europe and Asia, but they are now widely grown in the western half of North America on arid rangelands. Species of Agropyron are noted for their tolerance to drought and cold; they have moderate tolerance to salinity.

Table 3. Chromosome numbers of 279 species and subspecies of Triticeae grasses from 9 genera. Data compiled primarily from Tzvelev (1976), Löve (1984), Jacobsen and Bothmer (1981), and Dewey (published and unpublished).

Genus	Genome	No. species or subspecies						Total
		2n=14	2n=28	2n=42	2n=56	2n=70	2n=84	
Agropyron	P	5	13	1				19
Pseudoroegneria	S	9	10					19
Psathyrostachys	N	6						6
Critesion	H	22	11	7				40
Thinopyrum	J-E	3	9	8	2	1		23
Elytrigia	SX			9	3			12
Elymus	SHY		92	27	4			123
Leymus	JN		25	1	8	1	1	36
Pascopyrum	SHJN				1			1
Total		45	160	53	18	2	1	279
% of Grand Total		16	57	19	6	0.7	0.4	

Intergeneric Hybridization

Agropyron is more or less genetically isolated from other genera in the Triticeae. With two possible exceptions (Dewey, 1963a; R. C. Wang and C. Hsiao, unpublished), the P genome is apparently not found intact in any species outside of Agropyron, and intergeneric hybridization is rare. It is this high level of genetic isolation that has convinced most agrostologists that Agropyron should be treated as a separate genus regardless of how the remainder of the perennial Triticeae might be handled.

Agropyron (P) - Pseudoroegneria (S). Interpretation of chromosome pairing in these hybrids is often aided by size differences between the Agropyron (large) and Pseudoroegneria (small) chromosomes. Chromosomes of tetraploid P. spicata (SSSS) paired only rarely with chromosomes of diploid A. cristatum (PP) in their triploid hybrids (PSS) (Dewey, 1964a). All chromosome pairing in hybrids (PPSS) of A. desertorum (PPPP) and P. spicata (SSSS) was attributed to autosyndesis (Dewey, 1967b). Attempts to stabilize a new fertile species from the PPSS hybrids have been hindered by continued meiotic irregularities and sterility in advanced generations. Problems of this nature simply mean that the tetraploid parent species are not strict autoploids. A better strategy for producing meiotically regular and fertile PPSS amphiploids is to hybridize a PP diploid with an SS diploid and then double the PS hybrid. This procedure insures that complete and balanced P and S genomes are brought together in a PPSS amphiploid. Such hybrids have been made at Logan of diploid A. cristatum (PP) X diploid P. libanotica (SS) and P. stipifolia (SS), but the amphiploids have not been produced. Because Agropyron and Pseudoroegneria contain highly desirable and economically important range grasses, continued efforts should be made to produce stable intergeneric amphiploids.

Agropyron (P) - Psathyrostachys (N). No reported hybrids.

Agropyron (P) - Critesion (H). No reported hybrids.

Agropyron (P) - Thinopyrum (J-E). Some homologies may exist between the P genome of Agropyron and one of the genomes of T. intermedium. An average of 2.9 and a maximum of five trivalents per cell in pentaploid hybrids (PPE_1E_2X) of tetraploid A. desertorum (PPPP) X hexaploid T. intermedium ($E_1E_1E_2E_2XX$) suggests that the P genome is similar to the X genome of T. intermedium (Dewey, 1963a). That suggestion is reinforced by the nature of chromosome pairing in hexaploid A. cristatum (PPPPPP) X T. intermedium ($E_1E_1E_2E_2XX$) hybrids ($PPPE_1E_2X$), which produced up to three quadrivalents per cell and averaged 1.1 (Dewey, 1963b). The Agropyron X T. intermedium hybrids have not been advanced beyond the F_1

generation, but the opportunity for gene exchange between <u>Agropyron</u> and <u>T. intermedium</u> appears to be possible.

Agropyron (P) - Elytrigia (SX). Crosses between tetraploid <u>A. desertorum</u> (PPPP) and hexaploid <u>E. repens</u> ($S_1S_1S_2S_2XX$) produced pentaploid hybrids (PPS_1S_2X) that formed up to 14 bivalents at metaphase I (Dewey, 1961). Size differences between <u>A. desertorum</u> (large) and <u>E. repens</u> (small) chromosomes made it possible to conclude that almost all pairing was autosyndetic and that little or no genetic exchange occurred between the P genome of <u>Agropyron</u> and the S or X genomes of <u>E. repens</u>. The same conclusion was reached on the basis of chromosome pairing in diploid <u>A. cristatum</u> (PP) X <u>E. repens</u> ($S_1S_1S_2S_2XX$) hybrids (PS_1S_2X) (Dewey, 1964b). Seventy-chromosome amphiploids ($PPPPS_1S_1S_2S_2XX$) and 56-chromosome amphiploids ($PP_1S_1S_1S_2S_2XX$) are meiotically unstable and relatively sterile even after four or more sexual generations (K. H. Asay, unpublished). The prospects of introgressing genes from <u>Agropyron</u> to <u>E. repens</u> or vice versa are poor, as are the chances of producing immediately useful amphiploids containing the P, S, and X genomes.

Agropyron (P) - Elymus (SHY). Only two intergeneric hybrids are known that combine the P genome of <u>Agropyron</u> and the SH genomes of <u>Elymus</u>. Boyle and Holmgren (1968) observed meiosis in a natural hybrid (PSH) of <u>A. cristatum</u> (PP) and <u>Elymus trachycaulus</u> (SSHH) and concluded that gene exchange probably did not occur between the parental genomes. However, mean chromosome associations of 9.2^I, 8.6^{II}, 0.5^{III} and 0.03^{IV} in a PPSH hybrid between <u>A. desertorum</u> (PPPP) and <u>E. trachycaulus</u> (SSHH) led Napier and Walton (1982) to conclude that the chomosomes of the S, H, and P genomes may have sufficient segmental homology to enable some pairing in hybrids, at least in certain genetic backgrounds. Gene exchange will be difficult between <u>Agropyron</u> and <u>Elymus</u>, and it is probably not a practical breeding objective. Synthesis of PPSSHH amphiploids may be the best means of combining the P, S, and H genomes.

Agropyron (P) - Leymus (JN). No reported hybrids.

Agropyron (P) - Pascopyrum (SHJN). No reported hybrids.

Agropyron (P) - Triticum (ABD). No reported hybrids. Early attempts to hybridize <u>A. cristatum</u> with <u>Triticum</u> failed (White, 1940; Smith, 1942). Even recent attempts to hybridize <u>Agropyron</u> and <u>Triticum</u>, using the latest techniques to facilitate crossing, have also failed (A. Mujeeb-Kazi, pers. comm.). The closest thing to an <u>Agropyron</u> - <u>Triticum</u> hybrid is a successful cross of <u>T. aestivum</u> (AABBDD) X an amphiploid of <u>A. desertorum</u> X <u>Elytrigia repens</u> ($PPPPS_1S_1S_2S_2XX$) whose meiosis has not yet been reported (Mujeeb-Kazi et al., 1984). Even if intergeneric hybrids can be obtained, gene transfer from <u>Agropyron</u> to the cereals may not be

possible. I suspect that the entire arsenal of techniques to manipulate alien genetic material will be needed before Agropyron genes
can be incorporated into the genetic background of Triticum. The
rewards, if any, of such a program are more or less unpredictable;
but the challenge of the unknown will probably attract curious
scientists.

Agropyron (P) - Hordeum (I). No reported hybrids. Attempts by
White (1940) and Smith (1942) to hybridize A. cristatum with cultivated barley were unsuccessful.

Agropyron (P) - Secale (R). The only documented hybrid between
Secale and Agropyron was made by Krasniuk (1935). This difficult-
to-make hybrid was obtained in 1932 and apparently has not been
repeated since that time. The cross involved S. cereale (RR) X A.
cristatum (PPPP). Favorsky (1935) analyzed the hybrid (PPR) cytologically and observed seven bivalents at metaphase I. He
interpreted this pairing to mean that the S. cereale chromosomes
were more or less homologous with seven of the A. cristatum chromosomes. This is an obvious misinterpretation because we know that
chromosomes of tetraploid Agropyron pair autosyndetically in intergeneric hybrids (Dewey, 1967b). The Secale X Agropyron hybrid was
completely sterile, and efforts to backcross it were unsuccessful.
Further references to this hybrid do not appear in the literature,
and presumably it has been lost.

PSEUDOROEGNERIA LÖVE - 1980
Name derivation: From Greek pseudo = false, and
                          Roegneria = a genus described by Nevski
Type species: Pseudoroegneria strigosa (M. Bieb.) Löve
Genome: S

Circumscription and Description

     Pseudoroegneria is a newly erected genus (Löve, 1980a) con-
sisting of about 15 species that are built around one genome
designated S. This genome should not be confused with the genome in
species of Aegilops section Sitopsis, whose genome is also
represented by the letter S (Kihara, 1949). Löve (1982) designated
the Sitopsis genome as B; I use his genome designations because they
encompass the entire tribe. The Pseudoroegneria grasses are
caespitose, long-anthered, and cross-pollinating perennials; they
have narrow, linear spikes with single, distantly spaced spikelets,
with or without awns. The most commonly known species in North
America is P. spicata (bluebunch wheatgrass), a common native on
western rangeland. All species of Pseudoroegneria were previously
encompassed in traditional Agropyron and then in Elytrigia
(Appendix). Although Löve (1980a) was the first to treat
Pseudoroegneria as a genus, others (Nevski, 1934; Tzvelev, 1976)

recognized its species as a biological unit and placed them in separate sections of traditional Agropyron or Elytrigia. I have previously followed Tzvelev's (1976) treatment of Elytrigia (with Pseudoroegneria as a section) but with the proviso that Elytrigia would some time in the future require further partitioning (Dewey, 1983a). In my opinion, the time has come to partition Elytrigia sensu Tzvelev into several genera. Those who accept the genomic definition of genera, will have no difficulty in accepting Pseudoroegneria because it is genomically parallel to Agropyron in that it is based on a single genome.

Although the basic S genome is the sole genome found in the 15 species of Pseudoroegneria, it is found in combination with other genomes in more than 150 additional species (Table 2). The S genome is a component of most, if not all, species in three polyploid genera--Elymus, Elytrigia, and Pascopyrum. All species with the S genome, alone or in combination with others, form a group more or less equivalent to Elymus sensu Melderis (1978). I find that such a large and biologically diverse genus as Elymus sensu Melderis is unmanageable and serves no useful purpose.

Pseudoroegneria consists of about equal numbers of diploid and tetraploid taxa (Table 3). Hybrids between the diploid species have almost complete chromosome pairing, but with high or complete sterility, indicating different versions of the same basic genome in each diploid, i.e. S_1S_1, S_2S_2 etc. (Stebbins & Pun, 1953a). Some of the Pseudoroegneria species (P. stipifolia and P. spicata) have diploid and tetraploid races, and the tetraploids behave cyto-logically as autoploids or near-autoploids, which I represent ge-nomically as SSSS (Dewey, 1975a). Other tetraploids (P. geniculata) may be best described as segmental alloploids ($S_1S_1S_2S_2$), with most of the chromosomes pairing as bivalents (C. Hsiao, unpublished). The chromosome number and genome content of all species of Pseudoroegneria have not been determined, so it may be necessary to move certain species into or out of Pseudoroegneria as more cytogenetic information is obtained.

Pseudoroegneria is a genus of the Northern Hemisphere, with its species occurring on open rocky hillsides from the Middle East and Transcaucasia across Central Asia and Northern China to Western North America. These grasses are exceptionally drought tolerant and have excellent forage quality.

Intergeneric Hybridization

Because the S genome is found in Pseudoroegneria, Elymus, Elytrigia, and Pascopyrum, intergeneric hybridization might be expected among species in these four genera. Those expectations are met except for the combination of Pseudoroegneria - Pascopyrum, which has not been reported.

Pseudoroegneria (S) - Agropyron (P). See Agropyron - Pseudoroegneria.

Pseudoroegneria (S) - Psathyrostachys (N). No reported hybrids.

Pseudoroegneria (S) - Critesion (H). These intergeneric hybrids are rarer than expected. Most Elymus species have a genome formula of SSHH, with the S genome coming from Pseudoroegneria and the H genome from Critesion (Dewey, 1974b). If Elymus arose in that manner, one would expect to see SH hybrids occurring naturally. If such hybrids exist, they have not been reported. We have succeeded in producing only one artificial SH intergeneric hybrid and an SSHH amphiploid from a cross between diploid P. spicata (SS), and the diploid C. brevisubulatum (HH) (D. R. Dewey, unpublished). Synthetic Pseudoroegneria - Critesion polyploid hybrids include P. spicata (SS) X C. brachyantherum ($H_1H_1H_2H_2$) and P. spicata (SS) X C. arizonicum ($H_1H_1H_2H_2H_3H_3$) (Dewey, 1971a). Additional efforts should be made to hybridize Pseudoroegneria and Critesion, especially at the diploid level, and produce SSHH amphiploids to test the hypothesis that Elymus arose in this fashion. If the hypothesis is verified, a large scale Pseudoroegneria - Critesion hybridization program should be undertaken to synthesize new species of Elymus, some of which could be useful forage grasses.

Pseudoroegneria (S) - Thinopyrum (J-E). Only one hybrid of this combination is known. Dvorak (1981) hybridized diploid P. stipifolia (SS) with tetraploid T. scirpeum ($E_1E_1E_2E_2$) to produce triploid hybrids (SE_1E_2). Mean chromosome associations of 7.8^I, 5.9^{II}, and 0.41^{III} were attributed primarily to autosyndetic pairing between the E genomes. Stebbins and Pun (1953a) speculated that the E and S genomes might be variations of the same basic genome. That hypothesis is refuted by Dvorak's (1981) data, which show that the E and S genomes are distinctly different. Gene exchange between the E and S genomes is not expected.

Pseudoroegneria (S) - Elytrigia (SX). This hybrid combination has distinct plant-breeding possibilities. Hybrids between tetraploid P. spicata (SSSS) and hexaploid E. repens ($S_1S_1S_2S_2XX$) produced pentaploid hybrids of the type SSS_1S_2X (Dewey, 1967a). The F_1 hybrids were surprisingly fertile and backcrossed readily to E. repens. Through repeated backcrossing and selection, the caespitose growth habit of P. spicata was transferred to E. repens (Dewey, 1976). Inasmuch as an aggressive spreading habit is the major objection to E. repens, development of a nonrhizomatous or mildly rhizomatous strain of E. repens is a significant improvement. Two germplasm releases have been made of this intergeneric hybrid derivative (Asay and Dewey, 1981).

Intergeneric hybrids between tetraploid P. stipifolia (SSSS) and hexaploid E. repens ($S_1S_1S_2S_2XX$) (Dewey, 1970d) are genomically

similar to the hybrids of P. spicata and E. repens and the two
hybrid populations can be bred separately or combined. Both types
of breeding programs are in progress at Logan. Natural hybridization between Pseudoroegneria and Elytrigia has not been reported,
but it may occur and remain undetected. Elytrigia repens is such a
variable species that fertile or partially fertile Pseudoroegneria -
Elytrigia hybrid derivatives may be thought to be part of the
natural variation found in E. repens.

Pseudoroegneria (S) - Elymus (SH). An exceptionally large
number of artificial hybrids have been produced between diploid
species of Pseudoroegneria (SS) and tetraploid species of Elymus
(SSHH) to produce SSH triploids (Dewey, 1982; Löve and Connor,
1982). Any of the several Pseudoroegneria diploids (SS) can probably be hybridized with any of the more than 75 Elymus tetraploids
(SSHH), usually without embryo rescue procedures. Most of the triploid (SSH) hybrids are highly sterile; nevertheless, genes can
probably be introgressed between Elymus and Pseudoroegneria without
too much difficulty by simple backcrossing. The awned character of
P. spicata (SS) was transferred to E. lanceolatus (SSHH) after just
one backcross (Dewey, 1970a). Occasional P. spicata plants in
nature have rhizomes, and it is almost certain that the rhizomatous
trait came from E. lanceolatus. Although Pseudoroegneria - Elymus
hybridization and introgression are possible, most grass breeders
have not yet come close to exhausting intraspecific and intrageneric
variation, so they find little incentive to enter into an intergeneric breeding program.

Synthesis of $S_1S_1S_2S_2HH$ amphiploids from S_1S_2H hybrids is a
viable plant breeding option. Several, and possibly most, of the 27
hexaploid Elymus taxa (Table 3) have an $S_1S_1S_2S_2HH$ genome formula.
At Logan, we have synthesized more than 10 amphiploids of this
nature, and each is an incipient species. One amphiploid hybrid, P.
libanotica X E. caninus, proved to be equivalent to a naturally
occurring hexaploid, E. transchyranus (Dewey, 1972c). Another
amphiploid, E. canadensis X P. libanotica, is more vigorous than
either parent species and deserves the attention of grass breeders
(Dewey, 1974a).

Pseudoroegneria (S) - Leymus (JN). No reported hybrids.

Pseudoroegneria (S) - Pascopyrum (SHJN). No reported hybrids.

Pseudoroegneria (S) - Triticum (ABD). Mujeeb-Kazi et al.
(1984) have produced the only verifiable hybrids of Pseudoroegneria
with Triticum. The hybrid of tetraploid P. stipifolia (SSSS) with
T. aestivum (AABBDD) was very difficult to obtain, as indicated by
an embryo recovery of 0.5%. The same authors also succeeded in
hybridizing P. geniculata (possibly $S_1S_1S_2S_2$) and T. aestivum.
Because the identification and genome constitution of the P.
geniculata parent are uncertain, judgment must be deferred as to

whether this is indeed a Pseudoroegneria - Triticum hybrid.
Cytological analyses have yet to be reported on these two apparent
Pseudoroegneria - Triticum hybrids, so the relationship of the S
genome of Pseudoroegneria to the A, B, and D genomes of Triticum
cannot be stated unequivocally. Nevertheless, it seems safe to
assume that the relationship is remote and that plant breeders will
find it difficult or even impossible to introgress genes from the S
genome into Triticum.

Pseudoroegneria (S) - Hordeum (I). No reported hybrids.

Pseudoroegneria (S) - Secale (R). No reported hybrids.

PSATHYROSTACHYS NEVSKI - 1934
Name derivation: From Greek psathyros = brittle, and
                      stachys = spike
Type species: Psathyrostachys lanuginosa (Trin.) Nevski
Genome: N

Circumscription and Description

     Psathyrostachys is one of the most easily circumscribed and
least controversial genera in the Triticeae. It is a small
genus with no more than 10 species (Table 2). These grasses are
caespitose, long-anthered, cross-pollinating, long-lived perennials
with multiple spikelets per node, one or more florets per spikelet,
subulate glumes, and a fragile rachis. The genus was erected by
Nevski (1934) and consisted of species that had previously been in
traditional Elymus or traditional Hordeum (Appendix). Today,
agrostologists are almost unanimous in their recognition and
treatment of Psathyrostachys (Keng, 1965; Tzvelev, 1976; Melderis et
al., 1980). The genomic definition of Psathyrostachys as species
containing the N genome is identical to the conventional taxonomic
definition accepted throughout Europe and Asia. Unfortunately,
Psathyrostachys is not generally recognized in North America, where
it is still placed in traditional Elymus (Bowden, 1964).

     Chromosome counts have been made on about half of the
Psathyrostachys taxa, all of which are diploid (Table 3). If the
remainder of Psathyrostachys species prove to be diploid, it will be
the only completely diploid genus in the perennial Triticeae. The
Psathyrostachys genome is designated with the letter N, although for
many years I used the letter J (Dewey, 1970b; 1982). At present, it
can only be assumed that all Psathyrostachys species contain the N
genome. Chromosome pairing in interspecific F_1 hybrids show that
P. juncea and P. fragilis have the N genome, but with sufficient
modifications in each species to lead to almost complete sterility
in the hybrids (Dewey and Hsiao, 1983). Though the N genome is the
exclusive genome of the no more than 10 Psathyrostachys species, it

occurs in combination with the J genome of Thinopyrum in another 30
species that constitute the genus Leymus (Table 2). The N genome is
also one of the four genomes, SHJN, that constitute Pascopyrum.

Psathyrostachys is a genus of the Eurasian interior, where its
species grow on rocky open slopes and steppes from the Middle East
and European Russia across Central Asia to Northern China. Only one
species, P. juncea (Russian wildrye), has gained importance as a
forage grass. This species is tolerant to alkalinity and to drought
and has good forage quality. Russian wildrye is now widely used to
revegetate arid rangelands in North America.

Intergeneric Hybridization

Psathyrostachys (N) - Agropyron (P). No reported hybrids.

Psathyrostachys (N) - Pseudoroegneria (S). No reported
hybrids.

Psathyrostachys (N) - Critesion (H). Jacobsen and Bothmer
(1981) have been exceptionally successful in hybridizing P. fragilis
with different species of Critesion. However, the description and
meiotic behavior of these intergeneric hybrids have not been published. At Logan, we have obtained hybrids between diploid P.
juncea (NN) and diploid C. brevisubulatum (HH) (R. C. Wang,
unpublished). The NH diploids are vegetatively weak plants with
minimal chromosome pairing, illustrating the integrity of the
parental genomes. If NNHH amphiploids can be produced, they will be
unique without counterparts in nature.

Psathyrostachys (N) - Thinopyrum (J-E). No reported hybrids.
Inasmuch as Psathyrostachys and Thinopyrum are the putative donors
of the J and N genomes of Leymus, the two genera should hybridize.
Opportunities for natural hybridization are minimal or nonexistent
because the two genera rarely if ever occupy the same site. I am
unaware of controlled crosses between species of the two genera, and
this should be a high priority project, which would test the
proposed origin of Leymus.

Psathyrostachys (N) - Elytrigia (SX). No reported hybrids.

Psathyrostachys (N) - Elymus (SHY). The only known hybrid is
one plant from a cross of P. juncea (NN) with Elymus scribneri
(SSHH) (Dewey, 1967c). Chromosome pairing was negligible in the
hybrids, showing the distinctness of the S, H, and N genomes. The
hybrid was vegetatively weak and totally sterile. Doubling the
chromosome complement of the hybrid would produce an SSHHNN amphiploid, which would be novel because SSHHNN allohexaploids apparently
do not occur in nature.

Psathyrostachys (N) - Leymus (JN). Because Psathyrostachys and Leymus share a common genome, N, hybridization between species of the two genera is expected. Intergeneric triploid hybrids (J_1N_1N) have been produced by crossing P. juncea (NN) with allotetraploid $(J_1J_1N_1N_1)$ L. cinereus, L. triticoides, L. innovatus, L. salinus, L. multicaulis, L. racemosus, and L. secalinus (Dewey, 1970b; 1972d). Psathyrostachys juncea is the only species that has been used extensively in crossing, but similar results would be expected from other Psathyrostachys species. The triploid hybrids (J_1N_1N) frequently form seven univalents (J_1) and seven bivalents (N_1N), thus opening the way to genetic exchange between the N genomes of Psathyrostachys and Leymus. Psathyrostachys - Leymus hybrids are highly sterile, but introgression through backcrossing can probably be achieved. Inasmuch as the germplasm resource of P. juncea is very limited, introgression of genes from Leymus may be a worthy objective. Amphiploids $(J_1J_1N_1N_1NN)$ have been produced from four different hybrids, but none of the amphiploids show promise as a forage grass.

Psathyrostachys (N) - Pascopyrum (SHJN). No reported hybrids.

Psathyrostachys (N) - Triticum (ABD). No reported hybrids. Rather extensive efforts to hybridize P. juncea with T. aestivum have failed (A. Mujeeb-Kazi, pers. comm.).

Psathyrostachys (N) - Hordeum (I). The only successful hybrid has been made by Jacobsen and Bothmer (1981) who crossed P. fragilis (NN) with H. vulgare (II). The results of their studies are yet to be published.

Psathyrostachys (N) - Secale (R). No reported hybrids.

CRITESION RAFINESQUE - 1819
Name derivation: From Greek crithe = barley
Type species: Critesion jubatum (L.) Nevski
Genome: H

Circumscription and Description

Critesion is one of the more difficult, complex, and controversial genera to deal with taxonomically because it is genomically heterogeneous. As defined herein, Critesion is a genus of about 30 species, all of which had previously been in traditional Hordeum (Appendix). Rafinesque (1819) established Critesion as a segregate genus of Hordeum with one species, C. jubatum. Nevski (1934) perpetuated Rafinesque's concept of Critesion as a genus with one species, C. jubatum. Löve and Löve (1975) expanded Critesion by including the perennial species of traditional Hordeum section Stenostachys. Löve (1980a, b) enlarged Critesion further by encompassing all of traditional Hordeum except cultivated barley, H. vulgare, and its wild large seeded close relatives--H. spontaneum, H. deficiens, H. agriocrithon et al.

Löve's (1982) separation of Critesion from Hordeum is generally consistent with cytogenetic data, which show little or no homology between the genome of the H. vulgare complex and the genome(s) of the remainder of traditional Hordeum (Morrison and Rajhathy, 1959; Rajhathy and Morrison, 1959). Löve included Hordeum bulbosum in Critesion; however, H. bulbosum has considerable genomic homology with H. vulgare (Kasha and Sadasaviah, 1971; Bothmer et al., 1983), so I have elected to leave H. bulbosum in Hordeum. By my definition, Hordeum consists of the H. vulgare complex plus H. bulbosum, with a genome designation of I. Critesion then takes in the remainder of the species of traditional Hordeum, whose genome designation is H.

All species of Critesion are caespitose grasses with dense spikes, three spikelets per node, one floret per spikelet, and a fragile rachis. Beyond those morphological similarities, Critesion species differ widely with respect to important biological traits. Most are perennials (both short- and long-lived); some are annuals, and a few are biennials (Rajhathy et al., 1964). Most species of Critesion are small-anthered and self-fertilizing; however, taxa of the C. brevisubulatum complex are long-anthered and cross-pollinating (Bothmer, 1979; Dewey, 1979).

The genomic situation in Critesion is poorly understood. The type species, C. jubatum, is a segmental allotetraploid that is represented genomically as $H_1H_1H_2H_2$ (Starks and Tai, 1974). Some other Critesion species (C. californicum, C. compressum, C. stenostachys, C. muticum, and C. brachyantherum) also have the H genome (Starks and Tai, 1974; Rajhathy and Morrison, 1961; Hunziker et al., 1973). Other Critesion species or species-complexes (C. secalinum, C. brevisubulatum, C. pusillum, and C. murinum) may have genomes with few or no homologies with those of the C. jubatum complex. If the genomic system of classification is followed strictly, species with distinctly different genomes must be placed in different genera, meaning that Critesion may need to be partitioned into several more genera. Until more genomic information is available, the genome designation H will be used for all of Critesion. In addition to being the genomic foundation of Critesion, the H genome(s) is a component of most of the approximately 150 species of Elymus (Dewey, 1982). Thus, the H genome rivals or even exceeds the S genome in biological and geographical distribution.

Critesion is a genus rich in diploids, accounting for almost half of the diploid taxa in the nine perennial genera of the tribe (Table 3). The polyploids cover the full spectrum from strict autoploidy to strict alloploidy. The taxa of the C. brevisubulatum

complex form an autoploid series of diploids, tetraploids, and hexaploids (Dewey, 1979). The polyploids of the C. jubatum complex are
best described as segmental alloploids (Wagenaar, 1959; Starks and
Tai, 1974). The C. murinum complex consists of diploids, tetraploids, and hexaploids; and the polyploid taxa behave cytologically
as alloploids (Rajhathy and Morrison, 1962). Conclusions concerning
the nature of polyploidy in Critesion must be drawn cautiously
because of the apparent presence of genes that promote asynapsis
(Wagenaar, 1960) and suppress homoeologous pairing (Rajhathy et al.
1964; Starks and Tai, 1974; Subrahmanyam, 1978).

Species of Critesion are distributed widely in temperate and
subarctic regions of the Northern Hemisphere and South America.
They occupy coastal as well as inland continental sites and grow at
altitudes from below sea level to 4500 meters. These grasses are
adapted to mesic sites, often saline or alkaline meadows. Some of
the Critesion species, especially the annuals, are weedy and often
move into disturbed waste places. None of the species is sufficiently important as a forage grass to warrant the attention of
grass breeders.

Intergeneric Hybridization

Because the H genome of Critesion has such a wide biological,
ecological, and geographical distributions, opportunities for
intergeneric hybridization may be higher for Critesion than for any
other genus in the Triticeae. Species of Critesion have hybridized,
either naturally or artificially, with at least nine other genera in
the Triticeae including the cereal genera--Triticum, Hordeum, and
Secale. A comprehensive treatment of intergeneric hybrids in
Critesion and Hordeum is given by Fedak (1984).

Critesion (H) - Agropyron (P). No reported hybrids.

Critesion (H) - Pseudoroegneria (S). See Pseudoroegneria -
Critesion.

Critesion (H) - Psathyrostachys (N). See Psathyrostachys -
Critesion.

Critesion (H) - Thinopyrum (J-E). No verified hybrids.

Critision (H) - Elytrigia (SX). Natural hybrids between C.
secalinum and A. repens are rare, yet they have been given the
hybrid binomial X Agrohordeum langei (Hansen, 1959a). Kerguelen
(1975) gave the parentage of the natural hybrid X Agrohordeum rouxii
as Elytrigia pycnantha X Critesion secalinum, but that conclusion
must be verified. The only synthetic intergeneric hybrid of this
combination involved C. secalinum (2n=28) and E. repens (2n=42)
(Cauderon and Saigne, 1961). The authors interpreted the level of

chromosome pairing in the hybrids (12.5 bivalents) to mean that one
of the C. secalinum genomes is closely related to one genome of E.
repens. This interpretation needs reevaluation because there is
nothing in the morphology of the two species to suggest such a close
relationship.

Critesion (H) - Elymus (SHY). Inasmuch as Critesion is the
putative donor of the H genome to more than 100 SH-genome species of
Elymus, intergeneric hybridization and introgression between
Critesion and Elymus is anticipated. Hybrids between taxa of the C.
jubatum complex ($H_1H_1H_2H_2$) and the E. trachycaulus complex (SSHH)
are so common in nature that some were thought to be good species,
e.g. the so-called Elymus macounii is a common hybrid of C. jubatum
and E. trachycaulus (Boyle and Holmgren, 1955). Other natural hybrids with the same basic Critesion - Elymus parentage include X
Agrohordeum pilosilemma, X Elyhordeum dakotense, X Elyhordeum
stebbinsianum, X Elyhordeum montanense, X Sitordeum californicum
(Bowden, 1967), X Elyhordeum arcuatum (Mitchell and Hodgson, 1968),
and X Elyhordeum kolymense (Probatova and Sokolovskaya, 1982).
Artificial hybrids have been produced with the same general parentage as the natural hybrids cited above (Stebbins et al., 1946;
Bowden, 1957; Dewey, 1968). Inability to distinguish between auto-
and allosyndetic pairing in the hybrids makes it impossible to conclude how much homology exists between the H genome of C. jubatum
and the H genome of Elymus species. Although introgression between
C. jubatum and Elymus is probable, it has not been conclusively
demonstrated.

More definitive conclusions concerning Critesion - Elymus
genome homologies come from triploid hybrids of C. bogdanii (HH)
with E. canadensis (SSHH) and E. elymoides (SSHH) (Dewey, 1971b).
An average of more than five bivalents per metaphase-I cell in both
hybrids led to the conclusion that tetraploid Elymus had a modified
H genome from Critesion in addition to an S genome from
Pseudoroegneria. The question still remains as to whether more than
one of the several Critesion genomes entered into the origin of
Elymus. A series of Critesion - Elymus gayanus hybrids (Jacobsen
and Bothmer, 1981) may help define Critesion - Elymus genomic relations, but cytological data have not been published on those
hybrids.

Löve and Connor (1982) reported on hybrids between annual C.
marinum (2n=14) and two perennial species of Elymus from New
Zealand, E. enysii (2n=28) and E. scabrus (2n=42). The level of
chromosome pairing in both hybrids suggested that a genome similar
to the H genome of C. marinum is found in E. enysii and E. scabrus.

Critesion (H) - Leymus (JN). Natural hybridization apparently
occurs between C. jubatum ($H_1H_1H_2H_2$) and tetraploid species of
Leymus (JJNN) in parts of Canada and Alaska. Intergeneric hybrids

of this parentage include X Elyhordeum dutillyanum, X Elyhordeum littorale, and X Elyhordeum piperi (Bowden, 1967). Only one hybrid, X Elyhordeum littorale, has been analyzed cytologically by Hodgson and Mitchell (1965) who concluded that L. mollis and H. brachyantherum are genetically isolated and do not share a common genome.

Critesion (H) - Pascopyrum (SHJN). No reported hybrids.

Critesion (H) - Triticum (ABD). Hybridization between Critesion and Triticum is a recent development that has generated considerable plant-breeding interest. Few, if any, homologies exist between the Critesion and Triticum genomes as evidenced by extremely low levels of pairing in hybrids of C. bogdanii (HH) X T. timopheevi (AAGG) (Kimber and Sallee, 1976), C. chilense (HH) X T. turgidum (AABB) (Martin and Laguna, 1980), C. chilense (HH) X T. aestivum (AABBDD) (Martin and Chapman, 1977), C. chilense (HH) X Aegilops squarrosa (DD) (Martin, 1983), and C. pusillum (HH) X T. aestivum, (Finch and Bennett, 1980). Hybrids of H. chilense with six tetraploid taxa of Triticum--T. timopheevi, T. dicoccoides, T. georgicum, T. carthlicum, T. turanicum, and T. polonicum--had almost no chromosome pairing and reinforced the conclusion that C. chilense chromosomes do not pair with those of Triticum (Padilla and Martin, 1983b). The prospects of gene exchange between Critesion and Triticum are remote, but production of amphiploids and addition lines offer a means of bringing whole or partial genomes together. Amphiploids have been produced from the C. bogdanii X T. timopheevi hybrid (Kimber, 1979), the C. chilense X T. turgidum hybrid (Martin and Laguna, 1982), and the C. chilense X T. aestivum hybrid (Chapman and Miller, 1978), which have some promise as new cereal crops. Addition lines for each of the seven C. chilense chromosomes have been produced (Miller et al., 1982).

Critesion (H) - Hordeum (I). By the traditional taxonomic definition of Hordeum, the Critesion - Hordeum hybrids are interspecific; however under the genomic definition, these are intergeneric hybrids. Although cross-compatibility between H. vulgare and taxa of the C. jubatum complex is reasonably good, lack of chromosome pairing (less than 1 bivalent per cell) in the hybrids indicate no homology between the genomes of the two groups of species (Rajhathy and Morrison, 1959). On the basis of chromosome pairing in C. depressum X H. vulgare hybrids, Morrison and Rajhathy (1959) concluded that the two species have no genomes in common. On similar grounds, Fedak (1982) concluded that there was no obvious homology between the genomes of hexaploid C. parodii and diploid H. vulgare. Finch and Bennett (1980) likewise found no evidence of appreciable homology between H. vulgare chromosomes and those of C. murinum (2n=28 and 42) and C. procerum (2n=42). The most comprehensive Critesion - Hordeum hybridization project is that of Bothmer et al. (1983) who obtained hybrids of H. vulgare with C. jubatum, C.

lechleri, C. procerum, C. arizonicum, C. marinum, C. roshevitzii, C. brevisubulatum, C. capense, C. brachyantherum, and C. depressum. Chromosome pairing was low in all hybrids, usually less than two chiasmata per cell, indicating little or no homology between the genomes Critesion and Hordeum.

Cauderon (1956) found practically no homology between the chromosomes of Hordeum bulbosum and C. secalinum. Padilla and Martin (1983a) found no homology between the chromosomes of diploid C. chilense and diploid H. bulbosum and they concluded that the two species are phylogenetically distinct. The cytogenetic data support the concept of separating Hordeum and Critesion. The genetic distance between Hordeum and Critesion indicates transfer of genes from Critesion to cultivated barley will be difficult.

Critesion (H) - Secale (R). The first Critesion - Secale hybrid was made 40 years ago when Brink et al. (1944) hybridized C. jubatum ($H_1H_1H_2H_2$) with S. cereale (RR) with the aid of embryo culture. The authors concluded that there was little homology between the parental genomes. Wagenaar (1959) repeated the C. jubatum X S. cereale hybrid and demonstrated conclusively, on the basis of chromosome size differences between the parent species, that Critesion and Secale chromosomes rarely pair. Chromosome size differences also aided Finch and Bennett (1980) in concluding that C. chilense (2n=14) and C. jubatum (2n=28) chromosomes were not homologous with S. cereale (2n=14) and S. africanum (2n=14) chromosomes. To date, genes have not been introgressed between Critesion and Secale, and prospects appear dim for such transfers in the future.

THINOPYRUM LÖVE - 1980
Name derivation: From Greek thino = a combining form of this,
                  the shore, and pyros = wheat
Type species: Thinopyrum junceum (L.) Löve
Genome: J-E

Circumscription and Description

Thinopyrum is genus erected recently by Löve (1980b) who included in it only six species of the former Agropyron junceum complex. On genomic grounds, I find it necessary to expand the genus to about 20 species from Thinopyrum, Lophopyrum, and part of Elytrigia. Thinopyrum sensu Löve consists of species based on the J genome (Löve, 1982). Lophopyrum sensu Löve is a genus of about 10 species, with L. elongatum as the type species and the E genome as the basic genome (Löve, 1982). However, the J genome of Thinopyrum and the E genome of Lophopyrum are so close that the two genomes and the two genera should be combined. Hybrids between tetraploid T. junceiforme ($J_1J_1J_2J_2$) and diploid T. elongatum (EE) had a maximum

of 7III per cell and averaged 2.8III (Cauderon and Saigne, 1961).
The high frequency of trivalents means that the J and E genomes are
so closely related that they should be treated as variations of the
same genome.

Karyotypes of diploid T. bessarabicum (JJ) and diploid T.
elongatum (EE) are very similar (Heneen and Runemark, 1972a; b),
again supporting the treatment of the two genomes as one. Dvorak
(1981) favored combining the J and E genome designations under the
letter E. I retain the designation J for both genomes because J is
the older of the two genome designations, being applied as early as
1940 (Östergren, 1940a). In different parts of the paper, this
genome may be represented as J, E, or J-E.

Some of the taxa of Elytrigia sensu Löve are genomically closer
to Thinopyrum sensu Dewey than they are to E. repens ($S_1S_1S_2S_2XX$),
the type species of Elytrigia. These taxa are those of the
"intermediate wheatgrass complex," which will be treated hereafter
as Thinopyrum intermedium. Cauderon (1958) has shown that T.
intermedium has one or more genomes in common with T. elongatum and
T. junceum but no genomes in common with E. repens. Chromosome
pairing was almost complete in octoploid hybrids of hexaploid T.
intermedium X decaploid T. ponticum and the hybrids were very
fertile (Lyubimova, 1970). Certainly, T. intermedium and T.
ponticum must reside in the same genus.

As treated herein, Thinopyrum consists of all of Thinopyrum
sensu Löve, all of Lophopyrum sensu Löve, plus those species of
Elytrigia closely related to T. intermedium (Appendix). At least
for now, I am leaving species such as E. pungens and E. pycnantha
with E. repens in Elytrigia. Some genomic compromises have been
made in defining the limits of Thinopyrum. Although the J-E genome
is the core of the genus, other genomes may occur in certain complex
polyploid species of Thinopyrum. When genomic relationships are
better understood, some adjustments to the generic boundaries of
Thinopyrum may be required.

Thinopyrum as here circumscribed consists of three species
complexes--T. junceum, T. elongatum, and T. intermedium--which are
given sectional status in the genus (Appendix). Species in the T.
junceum complex (T. junceum, T. bessarabicum, T. junceiforme, T.
distichum et al.) are self-fertilizing maritime grasses, usually
rhizomatous, and with a fragile rachis. The T. elongatum complex
(T. elongatum, T. curvifolium, T. ponticum, T. scirpeum et al.) are
caespitose, self- or cross-pollinating grasses of coastal areas as
well as saline inland sites. The grasses of the T. intermedium
complex (T. intermedium, T. gentryi, T. podperae et al.) are usually
rhizomatous and cross-pollinating; they are adapted to the more
favorable inland sites that are not highly saline.

Thinopyrum consists of diploids, 2n=14, (T. bessarabicum, T. elongatum), segmental allotetraploids, 2n=28, (T. junceiforme, T. scirpeum, T. distichum, and T. curvifolium) segmental allohexaploids, 2n=42, (T. junceum), complex segmental octoploids, 2n=56, (T. runemarkii), and decaploids, 2n=70, (T. ponticum) (Table 3). Much remains to be done in defining the type of polyploidy and genomic relationships in Thinopyrum.

The nomenclature of diploid (2n=14) T. elongatum and its decaploid (2n=70) relative known in North America as tall wheatgrass is badly confused and must be clarified. The specific epithet elongatum has been applied to both the diploid and decaploid taxa, which are really very different. The diploid is a small plant, less than 50 cm tall; it is self-fertilizing and grows on sea coasts. The decaploid is a large, coarse grass with culms over 100 cm tall; it is cross-pollinating and usually grows on inland sites. The solution to this nomenclatural problem lies with the type specimen, which is located in Vienna, Austria. Jan Dvorak (pers. comm.) has examined the type specimen of T. elongatum, and he is confident that the type specimen is the diploid, which must then carry the name elongatum. The specific epithet ponticum has been correctly applied to the 70-chromosome species known in North America as tall wheatgrass (Holub, 1973). The epithet turcicum was applied by McGuire (1983) to the 56-chromosome race, which is morphologically similar to the 70-chromosome ponticum. I concur with Melderis (1978) who considers the two chromosome races (2n=56, 70) as infraspecific taxa.

Thinopyrum is a genus of Europe, the Middle East, and Central Asia, except for T. distichum, which is native to South Africa. Species of the T. junceum complex grow along the coastlines of Europe from the Baltic Sea and the Mediterranean Sea to the North Sea. The T. elongatum complex grows around the perimeter of the Mediterranean Sea and inland in the Middle East and European Russia. The grasses of the T. intermedium complex grow on inland sites from Europe and the Middle East and throughout Central Asia.

Intergeneric Hybridization

Thinopyrum has been the perennial Triticeae genus of greatest interest to wheat breeders since the 1930's when N. V. Tsitsin and his Soviet colleagues first demonstrated that T. ponticum (2n=70), T. intermedium (2n=42), and T. junceum (2n=42) hybridized readily with various species of Triticum (Tsitsin, 1960). Other Thinopyrum - Triticum hybridization programs were also launched in the 1930's in Canada (Armstrong, 1936; Peto, 1936) and in the U.S. (Smith, 1942; 1943). Some wheat breeders had great expectations for transferring the perennial habit, disease resistance, drought tolerance, and salt tolerance from Thinopyrum to Triticum. Interest in

Thinopyrum - Triticum hybridization waned when the initial high
expectations were not realized. In the last 10 years, renewed
interest has been shown in Thinopyrum - Triticum hybridization
because of new techniques of transferring genes between species.
Sharma and Gill (1983a) have prepared a recent and comprehensive
compilation of intergeneric hybrids involving Triticum, and the
reader is directed to that publication for a more complete picture
of the status of Triticum - Thinopyrum hybridization.

Thinopyrum (J-E) - Agropyron (P). See Agropyron -
Thinopyrum.

Thinopyrum (J-E) - Pseudoroegneria (S). See Pseudoroegneria
- Thinopyrum.

Thinopyrum (J-E) - Psathyrostachys (N). No hybrids of this
combination have been reported, but a hybridization program
should be launched to test the hypothesis that Thinopyrum (JJ)
and Psathyrostachys (NN) are the progenitors of Leymus (JJNN). A
direct test of this hypothesis would be to hybridize diploid T.
bessarabicum with any of the diploid species of Psathyrostachys
and make a JJNN amphiploid, which would then be hybridized with a
JJNN species of Leymus. Another approach is to hybridize T.
bessarabicum (JJ) with a tetraploid Leymus (JJNN) to produce triploid hybrids (JJN), which would be analyzed cytologically to
determine homology of the Thinopyrum genome with the J genome of
Leymus. If Thinopyrum and Psathyrostachys are the ancestral diploid
parents of Leymus, their geographic distributions must have been
different from what they are today because Thinopyrum and
Psathyrostachys are not sympatric and have no opportunity to
hybridize naturally.

Thinopyrum (J-E) - Critesion (H). No verified hybrids. See
Critesion - Thinopyrum.

Thinopyrum (J-E) - Elymus (SHY). The only known Thinopyrum -
Elymus hybrid combination comes from a cross of T. intermedium
and E. trachycaulus (Napier and Walton, 1983). Six hybrids (five
pentaploids and one septaploid) were obtained from about 800
florets with the aid of embryo culture. The pentaploid (2n=35)
hybrid averaged 19.6^I, 6.9^{II}, 0.43^{III}, and 0.1^{IV} per cell. The
septaploid (2n=49) hybrid, which was the consequence of an unreduced E. trachycaulus gamete (n=28) and a reduced T. intermedium
gamete (n=21), averaged 8.1^I, 20.3^{II}, and 0.05^{III}. If T.
intermedium is represented as $E_1E_1E_2E_2XX$ and E. trachycaulus as
SSHH, the pentaploid and septaploid hybrids would be E_1E_2XSH and
E_1E_2XSSHH, respectively. Virtually all of the chromosome pairing
in both hybrids can be attributed to autosyndesis. The genome
formulas previously assigned to T. intermedium and E.

trachycaulus adequately account for the amount and kind of pairing observed in the hybrids.

Thinopyrum (J-E) - Leymus (JN). Inasmuch as the J genome of Thinopyrum is apparently found in Leymus (JN), one might expect to find these intergeneric hybrids in coastal areas of the North Sea where Thinopyrum and L. arenarius grow together. Hybrids between tetraploid T. junceiforme and octoploid L. arenarius are indeed so common that they are mentioned in most floras of Scandanavia and Northern Europe (Holmberg, 1926; Hubbard, 1968; Melderis et al., 1980). Genomic relations between the parent species are difficult to establish because the parents and hybrids are complex polyploids and autosyndetic pairing cannot be distinguished from allosyndetic pairing. Nevertheless, the frequency of the hybrids suggests that Thinopyrum and Leymus share a common genome.

Petrova (1970) obtained three F_1 hybrids and an amphiploid between diploid T. elongatum (EE) and tetraploid L. mollis (JJNN). The hybrids (EJN) formed up to six bivalents per metaphase-I cell and averaged 3.6^{II}. Petrova attributed the pairing to autosyndesis between the J and N genomes and concluded that the T. elongatum genome was not homologous with either genome of L. mollis. However, there is no evidence that the J and N genomes can pair, and I attribute the pairing in Petrova's hybrids to be allosyndetic pairing between the E genome of T. elongatum and the J genome of L. mollis. Occasional quadrivalents in Petrova's amphiploid hybrid (EEJJNN) suggests to me that the E and J genomes have considerable homology.

Thinopyrum (J-E) - Pascopyrum (SHJN). No reported hybrids. The J genome of Thinopyrum apparently occurs in Pascopyrum, so intergeneric hybridization should be possible. The two genera are not sympatric, making natural hybridization impossible. Artificial hybrids do not occur, but an effort should be made to hybridize diploid T. bessarabicum (JJ) with octoploid P. smithii (SSHHJJXX) to test for the presence of the J genome in P. smithii.

Thinopyrum (J-E) - Elytrigia (SX). Natural hybridization is relatively common between Thinopyrum and certain species of Elytrigia. According to Kerguelen (1975) some of the more common hybrid taxa (traditional nomenclature) and their putative parentage (revised nomenclature) are: Agropyron X acutum (= Thinopyrum junceum X Elytrigia pycnantha), Agropyron X duvalli (= Thinopyrum junceiforme X Elytrigia pycnantha), Agropyron X obtusiusculum (= Thinopyrum junceiforme X Elytrigia pycnantha), and Agropyron X littoreum (= Thinopyrum junceiforme X Elytrigia repens).

The correct names and exact parentage of the above hybrids are open to some question (Hansen, 1959a,b; Hubbard, 1968; Tzvelev, 1976), but the important issue is that <u>Thinopyrum</u> - <u>Elytrigia</u> natural hybridization is quite common.

Östergren (1940a), Cauderon and Saigne (1961), and Heneen (1963b) studied chromosome pairing in <u>T</u>. junceiforme ($J_1J_1J_2J_2$) X <u>E</u>. <u>repens</u> ($S_1S_1S_2S_2XX$) and their hybrids ($J_1J_2S_1S_2X$), which averaged from 9.7 to 11.8 bivalents per cell. Bivalent pairing in the hybrids is best interpreted as the consequence of auto-syndesis, J_1-J_2 pairing and S_1-S_2 pairing. Cauderon and Saigne (1961) and Heneen (1963b) concluded that <u>T</u>. junceiforme and <u>E</u>. <u>repens</u> had no genomes in common.

Cauderon (1958) studied chromosome pairing in hybrids of hexaploid <u>T</u>. <u>junceum</u> with hexaploid <u>E</u>. <u>pycnantha</u>, hexaploid <u>T</u>. intermedium, and octoploid <u>E</u>. <u>pungens</u>. All hybrids had most of their chromosomes paired, and multivalent associations were common. However, high ploidy levels and the inability to distinguish autosyndesis from allosyndensis made interpretation of chromosome pairing in these hybrids very difficult and rather tentative. From a compilation of several studies, Cauderon (1966) concluded that <u>T</u>. <u>junceum</u> (2n=42) and <u>E</u>. <u>pungens</u> (2n=56) had the E genome in common; <u>T</u>. <u>junceum</u> and <u>T</u>. <u>intermedium</u> (2n=42) also had the E genome in common; whereas <u>T</u>. <u>junceum</u> and <u>E</u>. <u>pycnantna</u> (2n=42) and <u>T</u>. <u>junceum</u> and <u>E</u>. <u>repens</u> (2n=42) had no genomes in common. The genomic relationships among these complex polyploids are still uncertain. Analysis of chromosome pairing in polyhaploids of these species or in their hybrids with diploid <u>T</u>. <u>bessarabicum</u> will be needed to clarify genomic relationships between <u>Thinopyrum</u> and <u>Elytrigia</u>.

Crosses between tetraploid <u>T</u>. <u>curvifolium</u> ($E_1E_1E_2E_2$) and hexaploid <u>E</u>. <u>repens</u> ($S_1S_1S_2S_2XX$) produced 35-chromosome hybrids ($E_1E_2S_1S_2X$) that averaged 14.8^I, 9.7^{II}, 0.2^{III}, and 0.03^{IV} (Dewey, 1980b). Most, if not all, pairing was the consequence of auto-syndetic pairing of E_1-E_2 and S_1-S_2. The hybrids were treated with colchicine to form 70-chromosome amphiploids ($E_1E_1E_2E_2S_1S_1S_2S_2XX$), which were meiotically irregular but quite fertile. The <u>T</u>. curvifolium - <u>E</u>. <u>repens</u> amphiploid was hybridized with an <u>Agropyron</u> <u>desertorum</u> - <u>E</u>. <u>repens</u> amphiploid ($PPPPS_1S_1S_2S_2XX$) to produce exceptionally vigorous and moderately fertile 70-chromosome progeny of the type $PPE_1E_2S_1S_1S_2S_2XX$. The three-genus amphiploids seem to have considerable potential as forage grasses, and they should be subjected to further improvement by grass breeders.

Thinopyrum section Thinopyrum (J) - Triticum (ABD). Östergren (1940b) hybridized tetraploid <u>T</u>. junceiforme ($J_1J_1J_2J_2$) with tetraploid <u>Triticum turgidum</u> (AABB) and observed 18.4^I and 4.8^{II} in the hybrids (J_1J_2AB). Most or all of the pairing occurred between the J genomes contributed by <u>T</u>. <u>junceiforme</u>. The lack of homology between <u>Thinopyrum</u> and <u>Triticum</u> genomes was indicated in <u>T</u>. aestivum

(AABBDD) X $\underline{T.\ bessarabicum}$ (JJ) hybrids (ABDJ), which averaged 27.6^I and 0.2^{II} (Alonso and Kimber, (1980). Obviously, introgression of genes from the J genome into Triticum will not come easily.

Pienaar (1981) was somewhat more optimistic concerning the possibility of gene exchange between Thinopyrum and Triticum. He hybridized tetraploid Thinopyrum distichum (probably a J-genome segmental alloploid) with tetraploid Triticum durum and observed an average of 14.1^I, 4.8^{II}, 0.4^{III}, 0.8^{IV}, 0.01^V, and 0.01^{VI}. A similar level of pairing was observed in Triticum aestivum X Triticum distichum pentaploid hybrids. The multivalent associations in the hybrids suggest some homology between the Thinopyrum and Triticum chromosomes. If so, it may be possible to transfer genes from Thinopyrum distichum to Triticum by meiotic crossing-over facilitated by genes that promote homoeologous pairing.

Pienaar (1981) succeeded in producing amphiploids (2n=56 and 2n=70) of both Triticum X Thinopyrum hybrids. The amphiploids will probably have little direct use as cereal crops, yet they may serve as the means of producing addition lines or even substitution lines of Triticum with increased protein content, larger kernels, and greater salt tolerance. The anticipated benefits of Triticum with added or substituted Thinopyrum chromosomes are all speculative. The only certainty is that the procedures to achieve the hoped-for results will be long and tedious.

Thinopyrum section Lophopyrum (E) - Triticum (ABD). Early hybrids between Triticum and Thinopyrum section Lophopyrum involved Thinopyrum ponticum (2n=70) with Triticum aestivum (2n=42) and Triticum dicococcum (2n=28) to produce 56- and 49-chromosome hybrids, respectively (Peto, 1936; Cugnac and Simonet, 1953). About 75% of the chromosomes paired as bivalents or multivalents in the hybrids. Most of the pairing was attributed to autosyndesis of the T. ponticum chromosomes, but some pairing was thought to occur between T. ponticum and Triticum chromosomes. Subsequent studies have failed to fully resolve the question of homology between T. ponticum and T. aestivum genomes (Dvorak, 1976). Nevertheless, individual chromosomes from decaploid T. ponticum have been shown to be homoeologous with Triticum aestivum chromosomes in group 6 (Johnson and Kimber, 1967) and group 7 (Knott et al., 1977).

The use of diploid Thinopyrum elongatum in hybrids with Triticum aestivum can provide more definitive answers to genomic relationships between Thinopyrum section Lophopyrum and Triticum because the possibility of autosyndetic pairing among Thinopyrum chromosomes is eliminated. Formation of up to five bivalents per cell in T. elongatum X Aegilops squarrosa (the source of the D genome in Triticum) suggests that the E genome of Thinopyrum is homoeologous with the D genome of Triticum (Dvorak, 1971). Through an elaborate system of addition and substitution lines, Dvorak (1980) was able to show that five of the seven T. elongatum chromosomes had well-defined homoeologies with Triticum chromosomes.

The homoeology of T. ponticum and T. elongatum chromosomes with those of Triticum is close enough to make whole chromosome substitutions or additions possible. Single gene transfers between Triticum and Thinopyrum are also possible, especially when the diploidizing Ph gene is absent or when assisted by irradiation (Sharma and Knott, 1966). Cauderon (1979) summarized the characters that had been introgressed from T. ponticum into Triticum. Most of the transfers involved simply inherited disease resistance and usually required irradiation and/or deletion of the Ph gene. Though not spectacular, these achievements are not insignificant. Transfer of quantitatively inherited traits will be much more difficult and may not be possible with existing techniques.

Amphiploid hybrids have not been obtained between Thinopyrum ponticum (2n=70) and Triticum aestivum (2n=42) or any other species of Triticum. Even if such amphiploids could be obtained, they would have such high chromosome numbers (2n=84 to 112) that meiotic instability, sterility, and chromosome loss would preclude their stabilization. Amphiploids have been obtained between Thinopyrum elongatum (2n=14) and Triticum turgidum (2n=28), Triticum timopheevi (2n=28), and T. aestivum (2n=42) (Jenkins and Mochizuki, 1957; Jenkins, 1958). Hexaploid amphiploids have also been produced between T. elongatum (2n=14) and Triticum durum (2n=28), and between T. elongatum (2n=14) and Aegilops squarrosa (2n=14) (Evans, 1964). None of the amphiploids are being used directly as cereal or forage crops, but they have proven useful as sources of addition or substitution lines (Dvorak and Knott, 1974).

Thinopyrum section Trichophorae (J-E) - Triticum (ABD). Taxa of the Thinopyrum intermedium complex are the most widely used perennial Triticeae grasses in crosses with Triticum. Hybrids are easily obtained without treatment of stigmas with plant hormones or embryo rescue procedures (Armstrong, 1936; Smith, 1942; 1943). Knobloch (1968) compiled the following list of hybrids: Thinopyrum intermedium s. lat. X Triticum aestivum, T. persicum, T. polonicum, T. durum, T. sphaerococcum, T. dicoccoides, T. dicoccum, T. monococcum, T. turgidum, T. compactum, T. pyramidale, T. timopheevi, T. macha, T. aegilopoides, T. carthlicum, T. orientale, and T. spelta. Aegilops species that have been hybridized with Thinopyrum intermedium include A. longissima, A. variabilis, A. juvenalis, A. speltoides, A. crassa, A. triaristata, A. triuncialis, A. caudata, and A. cylindrica (Knobloch, 1968). Even though some of the above Triticum and Aegilops taxa may not be valid species or some names may be synonyms, it is evident that controlled hybridization between Thinopyrum intermedium and Triticum aestivum and its close

relatives is widespread and common. Nevertheless, natural
Thinopyrum - Triticum hybrids have not been reported.

The ease with which Thinopyrum intermedium hybridizes with
Triticum and the level of chromosome pairing in the F_1's (8 to 12
bivalents per cell) led early Soviet workers to conclude that many
chromosomes and even whole genomes of T. intermedium and Triticum
were homologous (Veruschkine, 1936). Subsequent studies give more
conservative estimates of genome homologies. Chromosome pairing in
Thinopyrum intermedium ($E_1E_1E_2E_2XX$) X Triticum durum (AABB), T.
dicoccoides (AABB), and T. aestivum (AABBDD) averaged 4.8^{II} to
6.2^{II}, which caused Peto (1936) to conclude that T. intermedium
had a genome partially homologous with the A or B genome of
Triticum. At that time, Peto was unaware that hexaploid T.
intermedium is a segmental autoalloploid whose E genomes can pair
autosyndetically in polyhaploids (Dewey, 1962) or in intergeneric
hybrids (Stebbins and Pun, 1953b). When the autosyndetic pairing
capability of T. intermedium chromosomes is taken into account, only
a few homologies are indicated between the chromosomes of T.
intermedium and Triticum.

Mean chromosome pairing in hybrids of Thinopyrum intermedium
with various tetraploid and hexaploid species of Triticum usually
falls in the range of 4 to 8 bivalents plus 0 to 1 trivalent per
cell. Love and Suneson (1945) concluded that there was very little
homology between T. intermedium and Triticum chromosomes. Cauderon
(1958) drew that same conclusion. Matsumura et al. (1958) believed
that there was considerable homology between one T. intermedium
genome and the B genome of Triticum, but that conclusion is
questionable.

Pairing is so limited between Thinopyrum intermedium and
Triticum chromosomes that gene introgression is unlikely unless
aided by genes that promote homoeologous pairing, a procedure used
successfully by Cauderon and Ryan (1974) and Wang et al. (1977).
Whole chromosome substitutions are possible, but they usually carry
unfavorable genes or upset the genetic balance (Cauderon, 1958;
1979). Transfer of quantitatively inherited traits from T.
intermedium to Triticum has not been successful.

Thinopyrum intermedium - Triticum amphiploids have received a
great deal of attention, especially as a means of producing a perennial wheat or a large seeded forage grass (Tsitsin, 1975).
Unfortunately, the amphiploids have not lived up to the expectations
of their developers. Complete amphiploids of T. intermedium and
Triticum are meiotically unstable and not very fertile (Cauderon,
1958; 1979). Partial amphiploids, those with just one T.
intermedium genome in combination with the full Triticum complement,
have shown more promise than complete amphiploids (Cauderon, 1979).

Thinopyrum (J-E) - Hordeum (I). No reported hybrids.

Thinopyrum (J-E) - Secale (R). Thinopyrum intermedium is the only species of Thinopyrum that has been hybridized with Secale, and that cross has been made several times, the first being made by V. F. Lyubimova before 1937 (Lyubimova, 1973). Interpretation of chromosome pairing is aided in these hybrids by chromosome size differences (Secale chromosomes are large and Thinopyrum chromosomes are small), which allow the investigator to distingush between auto- and allosyndetic pairing. Stebbins and Pun (1953b) observed averages of 12.1^I, 6.3^{II}, and 1.1^{III} at metaphase I in tetraploid hybrids of S. cereale (RR) and T. intermedium ($E_1E_1E_2E_2XX$). They attributed virtually all of the pairing, including trivalents, to autosyndesis among the T. intermedium chromosomes. Only rarely did a Thinopyrum chromosome pair with a Secale chromosome or two Secale chromosomes pair together. Stebbins and Pun concluded that T. intermedium contained two closely homologous genomes plus a third distinctly different genome and that none of the genomes of T. intermedium was homologous with the Secale genome. Gaul (1953) studied the same hybrid combination, S. cereale X T. intermedium, and arrived at the same conclusions as Stebbins and Pun.

Zennyozi (1963) added to the list of Thinopyrum - Secale hybrids by successfully crossing S. africanum, S. montanum, and S. kuprijanovii with T. intermedium. Chromosome pairing was similar in all hybrids, with averages ranging from 12.4 to 12.6^I, 6.4 to 6.5^{II}, 0.81 to 0.84^{III}, and $0.03-0.06^{IV}$ per cell. Heteromorphic bivalents consisting of Secale and Thinopyrum chromosomes were rare, as were Secale - Secale bivalents. Chromosome pairing in all Thinopyrum - Secale hybrids is consistent with genome formulas of $E_1E_1E_2E_2XX$ for T. intermedium and R_1R_1, R_2R_2 etc. for the various diploid Secale species.

ELYTRIGIA DESVAUX - 1810
Name derivation: A combination of Elymus and Triticum
Type species: Elytrigia repens (L.) Nevski
Genomes: SX

Circumscription and Description

Elytrigia is difficult to define genomically because the genus, consisting of about five species, is made up entirely of complex polyploids whose genomic constitutions are uncertain. These taxa, with a few exceptions, constitute the strongly rhizimatous, long-anthered, and cross-pollinating species previously included in traditional Agropyron (Appendix). First established in 1810, Elytrigia gained little recognition until Nevski (1933) separated it again from traditional Agropyron. However, Nevski (1934) almost

immediately reversed himself and relegated _Elytrigia_ to sectional
status within _Agropyron_. Tzvelev (1973, 1976) is the agrostologist
responsible for popularizing the concept of _Elytrigia_ as a genus in
modern taxonomy. _Elytrigia_ sensu Tzvelev is a genus of about 30
species, which includes both rhizomatous and caespitose cross-
pollinating grasses previously in traditional _Agropyron_, except the
crested wheatgrasses. Melderis (1978) absorbed _Elytrigia_ into
Elymus where the caespitose species were placed in section
Caespitosae and the rhizomatous species went into section _Elytrigia_.
In recent years, _Elytrigia_ sensu Tzvelev has gained some acceptance
in North America (Dvorak, 1981; McGuire, 1983), but most continue
Hitchcock's (1951) traditional usage of _Agropyron_ s. lat. At
present, _Elytrigia_ is treated in different terms in Europe, Asia,
and North America.

The genomic definition of _Elytrigia_ is much narrower than traditional definitions. The genomic system of classification brings
all taxa with the same or similar genome constitution as E. _repens_,
the type species, into _Elytrigia_ and excludes all others. _Elytrigia_
repens is a segmental autoallohexploid ($S_1S_1S_2S_2XX$) whose S
genomes are derived from _Pseudoroegneria_ and the X genome is of
undetermined origin (Dewey, 1976). _Elytrigia repens_ is an extremely
variable species that is partitioned into several subspecies
(Melderis et al., 1980; Tzvelev, 1976; Löve, 1984), which are presumably genomically comparable. _Elytrigia elongatiformis_ is a 56-
chromosome taxon that has the full chromosome complement of E.
repens plus another genome (Dewey, 1980c). Beyond E. _repens_ and E.
elongatiformis, the species composition of _Elytrigia_ is tentative.
Nevski (1934) placed E. _lolioides_ in the same series as E.
elongatiformis, so E. _lolioides_ is almost surely a legitimate member
of _Elytrigia_. Inclusion of the European species, E. _pycnantha_ and
E. _pungens_, in _Elytrigia_ is debatable; but at present they seem to
fit better in _Elytrigia_ than in _Thinopyrum_, the closest related
genus. My reasons for excluding the _Thinopyrum intermedium_ complex
from _Elytrigia_ are given in the previous section dealing with
Thinopyrum.

Elytrigia consists exclusively of polyploid species with 42 or
56 chromosomes. The hexaploids are the most common (Table 3). None
is a strict alloploid nor a strict autoploid. These species are
best described as complex segmental autoalloploids.

Elytrigia is a genus of Europe and Asia, with several of its
members (E. _pungens_ and E. _pycnantha_) being restricted to coastal
and interior regions of Western Europe. _Elytrigia repens_ has an
extensive natural distribution throughout much of Europe and Asia,
and it has become established in many parts of North America and
other temperate regions of the world, where it is often considered
to be a weedy species.

Intergeneric Hybridization

Elytrigia, a small genus with respect to number of species, is one of the more universal genera because of the almost worldwide distribution of E. repens. This wide distribution puts E. repens in contact with many other Triticeae species and provides an opportunity for natural hybridization. A greater proportion of natural intergeneric hybrids occur with Elytrigia (usually E. repens) than with any other genus in the tribe.

Elytrigia (SX) - Agropyron (P). See Agropyron - Elytrigia.

Elytrigia (SX) - Pseudoroegneria (S). See Pseudoroegneria - Elytrigia.

Elytrigia (SX) - Psathyrostachys (N). No reported hybrids.

Elytrigia (SX) - Critesion (H). See Critesion - Elytrigia.

Elytrigia (SX) - Thinopyrum (J-E). See Thinopyrum - Elytrigia.

Elytrigia (SX) - Elymus (SHY). Elytrigia and Elymus share the S genome of Pseudoroegneria, so there is a genomic basis for Elytrigia - Elymus intergeneric hybridization. Species of the two genera occasionally occupy the same site, providing an opportunity for natural hybridization. Nevertheless, Elytrigia - Elymus natural hybrids are rare and none is completely documented. Ullman (1936) cited a report but gave no data on a natural hybrid between Elytrigia repens and Elymus caninus from the Island of Funen, Denmark. Lepage (1952) observed what he considered to be hybrids of Elytrigia repens with Elymus canadensis from Alaska, and he applied the name X Agroelymus hodgsonii to the putative hybrid. Pohl (1962) reported that natural hybridization occurred between Elytrigia repens and Elymus trachycaulus in Iowa and that at least some specimens identified as Agropyron pseudorepens were probably hybrids. Some of the Elytrigia repens X Elymus trachycaulus hybrids cited by Pohl were partially fertile, suggesting natural introgression between the two taxa. Considerable hybridization seems to occur between Elytrigia repens and E. arizonicus (SSHH) in the Pinaleno Mountains of Arizona, but the putative hybrids have not been examined cytologically (G. L. Pyrah, pers. comm.).

Elytrigia repens and Elymus lanceolatus crossed readily (12 seeds from 26 florets) and chromosome pairing averaged 6.8^I, 12.5^{II}, 1.1^{III}, 0.01^{IV}, and 0.01^V in the F_1 hybrids (Dewey, 1965). Pairing was attributed to a combination of auto- and allosyndesis, with one of the S genomes of Elymus lanceolatus pairing with one genome of Elytrigia repens. In view of what we know now about the genomic constitutions of the parent species, the original

interpretation may be only partially correct. Hybrids between E.
repens $(S_1S_1S_2S_2XX)$ and E. lanceolatus (SSHH) are represented
genomically as SS_1S_2XH. To account for the observed pairing, up
to 14^{II} per cell, one must assume considerable homology between
the X genome of E. repens and the H genome of E. lanceolatus.
Cauderon (1958) concluded the E. repens contained an H (as Z) genome
similar to one of those of Critesion secalinum. However, the ease
with which E. repens hybridizes with Thinopyrum species (see
Thinopyrum - Elytrigia hybrids) indicates that the X genome of E,
repens may be related to the J-E genome of Thinopyrum. Further work
must be done to identify the X genome of E. repens.

Controlled hybrids between Elytrigia repens and a New Zealand
species, Elymus enysii (2n=28), have done little to clarify the
genomic relationships between Elytrigia and Elymus. The hybrids
formed 3 to 5 closed bivalents per cell, which were interpreted to
mean that E. repens and E. enysii shared a common genome (Löve and
Connor, 1982). However, I suspect that the bivalents were the
consequence of autosyndetic pairing between the S_1-S_2 genomes of
E. repens $(S_1S_1S_2S_2XX)$ and that the two species have no genomes in
common. Hybrids of Elymus scabrus (2n=42) X E. repens formed up to
7 bivalents per cell (Löve and Connor, 1982). The authors interpreted the pairing as evidence of genomic homologies between the
parental species, but I interpret it as evidence of lack of homology. Furthermore, the nature of reported chromosome pairing in
hybrids of E. enysii and E. scabrus with Pseudoroegneria spicata and
Critesion marinum (Löve and Connor, 1982) is difficult, if not
impossible, to reconcile with the genomic content of E. repens.
Obviously, additional work is needed to clarify apparent contradic-
tions involving the genomic composition of the Australian-New
Zealand species of Triticeae.

Elytrigia (SX) - Leymus (JN). It is uncertain whether
Elytrigia (SX) and Leymus (JN) share a common genome, yet natural
intergeneric hybrids have been reported on several occasions. The
most commonly reported hybrid involves E. repens and L. arenarius
(Nevski, 1934; Elven, 1981), which carries the hybrid binomial X
Elymotrigia bergrothii (Hansen, 1959a). These hybrids, which are
completely sterile, occur on the coastal regions of Scandinavia.
Nevski (1934) described natural hybrids of E. repens and L.
secalinus from the Lena River region of Siberia. Natural hybrids
from western North America once thought to be between E. repens and
L. mollis (= X Agroelymus adamsii) have since been identified as
hybrids between Elymus trachycaulus and Leymus mollis (= X
Agroelymus jamesensis) (Bowden, 1967).

The only artificial Elytrigia - Leymus hybrid is from the cross
of E. repens X L. angustus (Dewey, 1972b). Leymus angustus (2n=84)
is a complex polyploid that contains the J and N genomes, but it
cannot be assigned a specific genome formula. The E. repens X L.
angustus hybrids (2n=63) averaged 24.3^{I} and 19.3^{II} at

metaphase I, and all pairing was presumed to be autosyndetic. However, genome analysis in such complex high polyploids is of questionable value.

Identification of the X genome in E. repens should be a high priority project because the circumscription of the genus on genomic grounds depends on knowing the source of that genome. It is vitally important to know whether the X genome is related to the J genome of Thinopyrum, as I think it must be, or if it is related to a Hordeum genome as suggested by Cauderon et al. (1961). Hybrids of E. repens X diploid Thinopyrum bessarabicum (JJ) and E. repens X diploid Critesion (HH) should be useful in resolving the identity of the X genome.

Elytrigia (SX) - Pascopyrum (SHJN). The only known intergeneric hybrid of this combination is E. repens ($S_1S_1S_2S_2XX$) X P. smithii (SSHHJJNN) (D. R. Dewey, unpublished). Five hybrids were obtained in 1976, but they have not yet been analyzed cytologically.

Elytrigia (SX) - Triticum (ABD). When defined in its restricted genomic sense, Elytrigia has been hybridized only rarely with Triticum. Armstrong (1936) and Smith (1943) were unsuccessful in crossing E. repens and E. pungens, respectively, with tetraploid and hexaploid Triticum. Tsitsin and Gruzdeva (1959) commented that E. repens hybridized with Triticum only with great difficulty. They proposed using Thinopyrum intermedium as a bridging species between E. repens and Triticum, but nothing further has been reported on this project since the original study was reported in 1959.

Cauderon (1958) failed to obtain hybrids of hexaploid Elytrigia pycnantha or E. repens with tetraploid and hexaploid Triticum. Her crosses of octoploid E. pungens with diploid and hexaploid Triticum were unsuccessful; however, the crosses of E. pungens with tetraploid T. timopheevi and T. dicoccum were successful. Recently, Mujeeb-Kazi et al. (1984) were able to obtain hybrids of E. repens (2n=42) and E. pungens (2n=56) with T. aestivum (2n=42), but the chromosome pairing data are not yet available.

Hexaploid hybrids of T. timopheevi (AAGG) and T. dicoccum (AABB) with E. pungens averaged 16.0 to 24.8^I, 7.6 to 11.6^{II}, 0.12 to 0.86^{III}, and 0.01 to 0.06^{IV} per metaphase-I cell (Cauderon, 1958). Cauderon explained all of the chromosome pairing on the basis of autosyndesis among E. pungens chromosomes. She assigned the genome formula $K_1K_1K_2K_2E_4E_4E_5E_5$ to E. pungens, where E is the genome of diploid Thinopyrum elongatum and K is an unspecified genome. If E. pungens contains the basic E genome, geneticists should be able to introgress genes from E. pungens to Triticum in the same manner as from Thinopyrum elongatum and Thinopyrum intermedium to Triticum. However, Triticum - E. pungens amphiploids have not been produced and genes have not been introgressed from E. pungens into Triticum. The genome constitution of octoploid E. pungens, as well as that of the other complex polyploid species of Elytrigia, needs to be defined more explicitly.

Elytrigia (SX) - Hordeum (I). No reported hybrids.

Elytrigia (SX) - Secale (R). No reported hybrids.

ELYMUS LINNAEUS - 1753
Name derivation: From Greek elymos, a species of millet
Type species: Elymus sibiricus L.
Genomes: SHY

Circumscription and Description

Elymus is by far the largest genus of the Triticeae when defined on the basis of genome content. It contains approximately 150 species, which is five times larger than the next largest genus (Table 2). These grasses are relatively short-lived perennials, and most of them are caespitose, small-anthered, and self-pollinating. Traditionally, Elymus has been defined as those species with multiple spikelets per node coupled with multiple florets per spikelet. Although such a definition is admirably suited to taxonomic keys, it simply has little biological relevance.

Elymus has a long and varied history, going back to its establishment by Linnaeus (1753). During the past 50 years, Elymus has been treated in vastly different terms by different authors (Nevski, 1934; Gould, 1947; Hitchcock, 1951; Bowden, 1964; Tzvelev, 1973; Melderis, 1978; Löve, 1982). Today, the Hitchcock (1951) definition of Elymus (multiple spikelets per node and multiple florets per spikelet) still prevails in North America. In Poaceae URSS (Tzvelev, 1976) Elymus consists of caespitose and self-fertilizing species regardless of number of spikelets per node; this definition approaches the genomic classification. Elymus in Flora Europaea (Melderis et al., 1980) consists of Elymus sensu Tzvelev plus Elytrigia sensu Tzvelev, which makes it an even larger genus and certainly much more diverse than Elymus as defined genomically. Melderis and McClintock (1983) recently reaffirmed the unusually broad definition of Elymus found in Flora Europaea.

About 75% of the Elymus species are tetraploids (Table 3). The type species, Elymus sibiricus, is an allotetraploid that contains a genome derived from Pseudoroegneria (SS) and a genome from Critesion (HH) (Dewey, 1974b). Many Elymus tetraploids from Europe (E. caninus, E. alaskanus), Central Asia (E. fibrosus, E. mutabilis), North America (E. canadensis, E. trachycaulus, E. lanceolatus, E.

glaucus) and South America (E. tilcarensis, E. agropyroides) are
also SSHH allotetraploids, but each has its own variation of the S
and H genomes. East Asian species such as E. ciliaris have a Y
(unknown origin) genome in place of the H genome, and those tetraploids are represented genomically as SSYY (Liu and Dewey, 1983).
One species, Elymus drobovii, has a genome formula of SSHHYY (Dewey,
1980a). The genome structure and constitution of many New Zealand-
Australian species of Elymus is still uncertain. After the genomic
relationships are fully understood in Elymus, several new genera may
need to be constructed. Hexaploids are relatively common in Elymus
(Table 3) and most are segmental autoalloploids of the type
$S_1S_1S_2S_2HH$ or $SSH_1H_1H_2H_2$. Octoploids, presumably of the type
$S_1S_1S_2S_2H_1H_1H_2H_2$ are rare in Elymus, accounting for less than 5%
of the total.

Elymus is the most widely distributed genus in the Triticeae,
with species occurring naturally in Europe, Asia, North America,
South America, and New Zealand-Australia. These grasses are not as
drought or salt tolerant as many of the other perennial Triticeae,
but they have good forage quality and are highly productive under
favorable climatic conditions. The only report of apomixis in the
Triticeae comes from Elymus scabrus (Hair, 1956). Disease
resistance, wide adaptation, and apomixis are some characteristics
that make Elymus a genus of interest to breeders of annual Triticeae
cereal crops.

Intergeneric Hybrids

Because Elymus is such a large and ubiquitous genus, the
opportunity for natural hybridization is greater for Elymus
than for other Triticeae genera. Although most Elymus species
are self-fertilizing, they can outcross with surprising frequency
to other self-fertilizing species. Whenever self-fertilizing E.
trachycaulus and self-fertilizing Critesion jubatum grow on the
same site, intergeneric hybrids are almost always present
(Wagenaar, 1960). The genome content of Elymus (SHY) also
contributes to its opportunity for intergeneric hybridization.
The S and H genomes of Elymus are found alone or together in
Pseudoroegneria (S), Critesion (H), Elytrigia (SX), and
Pascopyrum (SHJN), thus providing a genomic basis for
intergeneric hybridization with those four genera. Elymus can
also hybridize with genera that do not carry the S, H, or Y
genomes--Agropyron (P), Psathyrostachys (N), Leymus (JN), and
Thinopyrum (J-E).

Elymus (SHY) - Agropyron(P). See Agropyron - Elymus.

Elymus (SHY) - Pseuderoegneria (S). See Pseudoroegneria -
Elymus.

Elymus (SHY) - Psathyrostachys (N). See Psathyrostachys - Elymus.

Elymus (SHY) - Critesion (H). See Critesion - Elymus

Elymus (SHY) - Thinopyrum (J-E). See Thinopyrum - Elymus

Elymus (SHY) - Elytrigia (SX). See Elytrigia - Elymus

Elymus (SHY) - Leymus (JN). The frequency of Elymus - Leymus hybridization is something of a surprise inasmuch as the two genera have no genomes in common. Natural hybridization of JN-genome species, especially Leymus mollis and L. innovatus, with SH-genome species of the Elymus trachycaulus complex occurs quite frequently in Canada and Alaska. Intergeneric hybrids of this nature have been given the following hybrid binomials (based on traditional generic names): X Agroelymus colvillensis, X Agroelymus hirtiflorus, X Agroelymus turneri, X Agroelymus jamesensis, and X Agroelymus ungavensis (Bowden, 1967; Lepage, 1965). Genomically similar hybrids occur in East Asia under the name X Leymotrix ajanensis (= Leymus ajanensis X Hystrix sibirica) (Probatova and Kharkevitch, 1982).

The following artificial Elymus - Leymus hybrids have been produced: E. stebbinsii X L. condensatus (Stebbins and Walters, 1949); E. canadensis X L. triticoides and E. canadensis X L. secalinus (Dewey, 1970c). These artificial hybrids have been analyzed cytologically and so have the following natural hybrids: E. trachycaulus - L. mollis (Bowden, 1959b); E. elymoides - L. cinereus (Dewey and Holmgren, 1962); and E. trachycaulus - L. innovatus (Sadasiviah and Weijer, 1981). All of the above are tetraploid (2n=28) hybrids that can be represented genomically as SHJN. Average chromosome associations at metaphase I in the various hybrids ranged from 21.3 to 26.8^I, 0.6 to 3.1^{II}, and 0.00 to 0.15^{III}. The small amount of pairing indicates residual homologies between the different genomes, but it is not possible to specify which genomes are involved in the pairing. The low level of pairing in this whole series of hybrids is consistent with the hypothesis that the four genomes are distinctly different and warrant separate designations.

Sakamoto (1971) hybridized hexaploid E. tsukushiensis with tetraploid L. mollis and obtained pentaploid hybrids that averaged 30.9^I and 2.0^{II}. All bivalents were loosely connected rods and Sakamoto concluded that none of the five genomes represented in the hybrids were homologous. Elymus tsukushiensis is an allohexaploid with three distinctly different genomes (Sakamoto, 1964), which I tentatively represent as SHY; Leymus mollis has the JN genomes, and the hybrids would be represented as SHYJN. The insignificant amount of pairing in the pentaploid SHYJN hybrids demonstrates lack of homology among the five genomes.

Elymus(SHY) - Pascopyrum (SHJN). Hybridization between Elymus and Pascopyrum is expected because of the phylogenetic relationship; however, such hybrids are rare. The only documented hybrid is E. canadensis X P. smithii (Dewey, 1970c). The hybrids are represented genomically as SS_1HH_1JN. If the SH genomes of Elymus were homologous with the SH genomes found in Pascopyrum, one would expect 14^I and 14^{II} in the hybrids. Observed chromosome associations were 13.37^I and 14.31^{II}, or very near the expectation. About half of the bivalents were rings, indicating close homologies; nevertheless, the hybrid was totally sterile.

Elymus (SHY) - Triticum (ABD). All early attempts to hybridize Elymus and Triticum failed (Armstrong, 1936; Smith, 1943; Cauderon, 1958). Only in recent years have such crosses been successful, and they usually required post-pollination treatment of the florets with gibberellic acid and embryo culturing on artificial media. Two centers in North America--Kansas State University, Manhattan, KS; and CIMMYT, El Batan, Mexico-- have been especially active and unusually successful in producing Elymus - Triticum hybrids.

Sharma and Gill (1983b) produced hybrids of T. aestivum (AABBDD) with E. trachycaulus (SSHH), E. yezoensis (SSYY), and E. ciliaris (SSYY). Chromosome pairing was negligible (less than one bivalent/cell) in the two hybrids involving E. trachycaulus and E. yezoensis with Triticum, indicating few if any homologies between the Elymus genomes (SHY) and those of Triticum (ABD). Pairing was considerably higher in the E. ciliaris X T. aestivum hybrid (4.25^{II} and 0.11^{III}) even though it is genomically equivalent to the E. yezoensis X T. aestivum hybrid, which averaged 0.73^{II}. Apparently the genetic complement of E. ciliaris has the effect of neutralizing the Ph gene in T. aestivum and allowing some pairing among the A, B, and D genomes. Attempts to produce amphiploids from the hybrids were unsuccessful; however, backcross progenies were obtained by pollinating the F_1's with T. aestivum pollen. Presumably, all of the Elymus chromosomes will be eliminated in a few generations. Introgression between Elymus and Triticum is improbable unless it can be induced by irradiation or some means of promoting nonhomologous chromosome pairing.

Mujeeb-Kazi and Bernard (1982) reported success in obtaining a remarkable number of Elymus - Triticum hybrids including E. trachycaulus (SSHH) X T. aestivum (AABBDD), E. fibrosus (SSHH) X T. aestivum (AABBDD), E. fibrosus X T. turgidum (AABB), E. dahuricus (SSHHYY?) X T. aestivum, and E. canadensis (SSHH) X T. aestivum. The meiotic data have not been published on these interesting hybrids, but more complete insights into Elymus - Triticum genomic relationships will certainly be gained when the data are published.

Transfer of apomixis from E. scabrus to Triticum has been an unrealized goal of wheat geneticists for many years. I have

provided seed of E. scabrus to a number of laboratories around the
world for the purpose of hybridizing this apomict with Triticum.
Reports of successful crosses of E. scabrus and Triticum have not
been forthcoming. The genomic constitution of E. scabrus is uncertain, yet it seems to be part of the SHY-genome complex. The recent
report of hybrids with E. scabrus with Elytrigia repens, Critesion
marinum, Pseudoroegneria spicata, and Elymus stewartii (Löve and
Connor, 1982) is encouraging because it is the first demonstration
of intergeneric hybridization for E. scabrus. A project has been
recently initiated at Logan, Utah, to extend the hybridization
limits of E. scabrus to Triticum (J. G. Carman, pers. comm.). Even
if the hybrids can be obtained, that may be the simplest aspect of
the intended transfer of apomixis from E. scabrus to Triticum.

Elymus (SHY) - Hordeum (I). Hybrids (SHI) of E. canadensis
(SSHH) X H. vulgare (II) had almost no chromosome pairing at metaphase I (0.81 chiasmata per cell) (Mujeeb-Kazi and Rodriguez, 1982).
These data show that the H genome of E. canadensis is not comparable
to the I genome of H. vulgare and other cultivated barleys. The
nonhomology of the H genome, which originates in Critesion, with the
I genome supports the taxonomic separation of Critesion and Hordeum.
Hybrids (SH H I) of H. vulgare (II) with a hexaploid Elymus,
presumably E. patagonicus ($SSH_1H_1H_2H_2$), averaged 19.6^I, 2.6^{II},
0.8^{III}, 0.14^{IV} (Mujeeb-Kazi and Rodriguez, 1980). The bivalent
associations in the hybrids can be accounted for by autosyndesis
among the H genomes of E. patagonicus. The cause of the multivalent
associations need to be studied further, but I suspect that they are
also the consequence of autosyndesis within the E. patagonicus
complement that has undergone some structural changes.

Elymus (SHY) - Secale (R). No reported hybrids.

LEYMUS HOCHSTETTER - 1848
Name derivation: An anagram of Elymus
Type species: Leymus arenarius (L.) Hochst.
Genomes: JN

Circumscription and Description

     Leymus is a genus of about 30 species, all of which had previ-
ously been in Elymus (Appendix). The genomic definition of
Leymus corresponds fully with the more conventional definition based
on morphology. These species usually have multiple spikelets per
node and are for the most part rhizomatous, long-anthered, and
cross-pollinating. Hochstetter (1848) recognized that Elymus
arenarius differed substantially from the other species in Linnaeus´
Elymus, and he erected the genus Leymus to accommodate L. arenarius
and similar taxa. Little attention was given to Hochstetter´s work
until Pilger (1947) restored the concept of Leymus as a genus, a

view subsequently endorsed by Tzvelev (1960). Leymus is now
recognized as a genus in Poaceae URSS (Tzvelev, 1976) and in Flora
Europaea (Melderis et al., 1980); it is even gaining recognition in
North America (Barkworth et al., 1983).

Leymus, a polyploid genus (Table 3), is based on variations of
two basic genomes, the N genome of Psathyrostachys and the J genome
of Thinopyrum. Genome analysis in many hybrids between
Psathyrostachys juncea (NN) and tetraploid species of Leymus has
shown that Leymus contains the N genome (previously designated as J)
(Dewey, 1972a). Until recently, the second genome of Leymus was
designated X, indicating a genome of unknown origin (Dewey, 1982).
Löve (pers. comm.) suggested on morphological grounds that the J
genome of Thinopyrum was the X genome of Leymus. Karyotype
analyses of T. bessarabicum (JJ) and several tetraploid Leymus
species also point to Thinopyrum as the source of the X genome (C.
Hsiao, unpublished). Consequently, tetraploid Leymus can be
represented genomically as JJNN. Tetraploid taxa of Leymus appear
to be strict allotetraploids. Presumably, the taxa at higher ploidy
levels (2n=42 to 84) contain only the J and N genomes that have
undergone duplication and some modification. Multivalents are seen
at metaphase I in octoploid L. cinereus and dodecaploid L. angustus
(Dewey, 1972b), but their frequency is much less than expected from
true autoalloploids. Genes promoting bivalent pairing appear to be
operating in the high polyploid taxa of Leymus, including octoploid
L. arenarius (2n=56), the type species (Heneen, 1963a).

Species of Leymus are long-lived perennials that are distributed from the coastal regions of the North Sea (L. arenarius) across
Central Asia (L. racemosus, L. angustus, L. secalinus) to East Asia
(L. chinensis) into Alaska (L. mollis) and western North America (L.
innovatus, L. triticoides, L. cinereus). These grasses often grow
on saline or alkaline sites. Their tolerance to salinity probably
comes from the Thinopyrum genome (J), and tolerance to drought and
alkalinity comes from the Psathyrostachys genome (N). Several of
the species (L. racemosus, L. mollis, L. arenarius) have very large
seeds; nevertheless, all species of Leymus have generally poor
seedling vigor and are rather difficult to establish. Their
perennial habit, large seeds, tolerance to salinity and alkalinity,
and resistance to disease have made some species of Leymus
attractive to cereal breeders.

Leymus (JN) - Agropyron (P). No reported hybrids.

Leymus (JN) - Pseudoroegneria (S). No reported hybrids.

Leymus (JN) - Psathyrostachys (N). See Psathyrostachys -
Leymus.

Leymus (JN) - Critesion (H). See Critesion - Leymus.

Leymus (JN) - Thinopyrum (J-E). See Thinopyrum - Leymus.

Leymus (JN) - Elytrigia (SX). See Elytrigia - Leymus.

Leymus (JN) - Elymus (SHY). See Elymus - Leymus.

Leymus (JN) - Pascopyrum (SHJN). As one of the putative ancestral genera of Pascopyrum, Leymus should hybridize with Pascopyrum.
Contrary to expectations, Leymus - Pascopyrum hybridization is
either very rare or it is difficult to recognize. Leymus
triticoides, as one of the immediate parents of P. smithii, would be
expected to hybridize naturally with P. smithii wherever the two
species occupy the same or adjacent sites, but natural hybrids have
not been reported. Stebbins (1956) spoke of L. triticoides and P.
smithii hybrids in which chromosome pairing was complete and the
hybrids were fertile. This report is incorrect, probably because of
misidentified plant material, because P. smithii is always octoploid
(2n=56) and L. triticoides is tetraploid (2n=28) and they could not
produce hybrids with complete pairing. Although modest attempts to
hybridize L. triticoides and P. smithii at Logan have failed, I am
confident that the two species can be hybridized.

At Logan we have obtained a large number of hybrids from
unemasculated crosses of P. smithii (2n=56) X L. angustus (2n=84)
(D. R. Dewey, unpublished). The hybrids have been verified cytologically, but their chromosome pairing has not been analyzed. It
is doubtful that definitive results concerning genomic relationships
could be obtained from the hybrids because of their high chromosome
number (2n=70).

Leymus (JN) - Triticum (ABD). Next to Thinopyrum, Leymus is
the genus of perennial Triticeae of greatest interest to wheat,
barley, and rye breeders. Petrova (1960) stated that wide hybridization of Leymus with Triticum, Hordeum, and Secale was initiated
by N. V. Tsitsin in the early 1940's with the intent of incorporating the drought tolerance, salt tolerance, disease resistance,
and high number of seeds/spike of Leymus into the cereals. The
first Leymus - Triticum hybrids involved octoploid L. arenarius
(2n=56) and tetraploid L. racemosus (2n=28) crossed with T.
aestivum (2n=42), T. compactum (2n=42), and T. durum (2n=28)
(Petrova, 1960). Tetraploid L. mollis (2n=28) was brought into the
Soviet Leymus - Triticum breeding program in the mid 1950's.

Petrova (1960) studied meiosis in tetraploid hybrids (ABJN) of
T. carthlicum (AABB) X L. racemosus (JJNN) and observed little or no
chromosome pairing between the Leymus and Triticum genomes. On the
basis of pairing in hybrids of T. durum and L. racemosus, she concluded that Triticum and Leymus contained distinctly different
genomes. These conclusions have been substantiated in the more
recent studies of Mujeeb-Kazi and Rodriguez (1981).

Because introgression of individual genes or blocks of genes from Leymus to Triticum seemed improbable, Soviet geneticists turned to amphiploidy as a means of combining the genomes of Triticum and Leymus (Tsitsin and Petrova, 1976). Amphiploids with chromosome numbers of 2n=56, 70, 84, and 98 were produced from hybrids of L. arenarius, L. mollis, and L. racemosus with T. aestivum, T. dicoccocum, T. durum, and T. carthlicum. Sterility problems precluded the use of the complete amphiploids as cereal crops. Partial amphiploids (2n=42) with the A and B genomes of T. durum plus one genome (probably J) from L. mollis are stable and highly productive. The partial amphiploids are characterized by the annual habit, large spikes, many seeds per spikelet, high protein (21 to 24%) and satisfactory baking quality. Late maturity and weak straw prevent their immediate use under cultivation. The partial amphiploids are now being used in crosses with hexaploid Triticum (Tsitsin and Petrova, 1976).

A T. aestivum (AABBDD) X L. racemosus (JJNN) hybrid (ABDJN) was obtained from the CIMMYT wide hybridization program in 1976 (Mujeeb-Kazi and Rodriguez, 1981). The hybrid averaged 32.8^I and 1.02^{II}, which confirms previous conclusions (Petrova, 1960) that the J and N genomes are not homologous with each other or with the A, B, or D genomes of Triticum. Mujeeb-Kazi and Rodriguez were unable to make an amphiploid from the hybrid, but they were able to obtain a backcross plant with the genome formula of AABBDDJN. They hoped to use the backcross plant as the starting point for obtaining addition lines that carry J or N chromosomes.

The first successful Triticum - Leymus hybridizations involved only the large-seeded, self-fertilizing species of Leymus from section Leymus, i.e., L. arenarius, L. mollis, and L. racemosus. Early attempts to hybridize small-seeded, cross-pollinating species of Leymus--L. triticoides, L. ambiguus, L. condensatus--with Triticum failed (Smith, 1942). However, in very recent times Mujeeb-Kazi et al. (1984) have obtained hybrids of Triticum aestivum with small-seeded, cross-pollinating species, L. triticoides (2n=28) and L. cinereus (2n=28), and also with L. angustus (2n=84), a species with relatively large seeds and moderate self-fertility. Inasmuch as all species of Leymus contain only the J and N genomes, it should be possible to hybridize any species of Leymus with Triticum if the proper hybridization and embryo rescue techniques are employed.

Leymus (JN) - Hordeum (I). Attempts to hybridize L. racemosus and other Leymus species with Hordeum were made as early as 1933, but without success (Bakhteyev and Darevskaya, 1960). Smith (1942) repeated the Hordeum - Leymus hybridization effort on an expanded scale, but no hybrids were obtained. Not until 1943, did Bakhteyev and Darevskaya (1960) finally succeed in hybridizing H. vulgare (II) with L. racemosus (JJNN), and they applied the hybrid binomial X Elymordeum zizinii in honor of their colleague N. V. Tzitsin.

The intergeneric hybrid was totally sterile and did not respond to attempts to produce an amphiploid.

Meiotic data that have a bearing on possible homologies between the J and N genomes of _Leymus_ and the I genome of barley are inconclusive. According to Ahokas (1970) some bivalents were observed in the _Hordeum_ _vulgare_ X _L. raceomosus_ hybrid produced by Bakhteyev and Darevskaya. Schooler and Anderson (1980) reported on a reputed hybrid of autotetraploid _H. vulgare_ (IIII) X _L. mollis_ (JJNN) that seemed to have substantial allosyndetic pairing between the _Hordeum_ and _Leymus_ genomes. Having read their paper very carefully, I can only conclude that they were dealing with a self of _Hordeum_ _vulgare_ and not a _Hordeum_ X _Leymus_ hybrid. Ahokas (1970) hybridized _H. vulgare_ with octoploid _L. arenarius_, but chromosome pairing data were not reported. The issue of homology or nonhomology between _Hordeum_ and _Leymus_ genomes is still unresolved, but I have no reason to believe that they have any significant homology.

Leymus (JN) - Secale (R). _Leymus_ does not hybridize readily with _Secale_. Smith (1942) was unsuccessful in crossing _S. cereale_ with _L. ambiguus_, _L. condensatus_, and _L. racemosus_. A natural pentaploid (2n=35) hybrid ($J_1J_2N_1N_2R$) of octoploid (2n=56) _L. arenarius_ ($J_1J_1J_2J_2N_1N_1N_2N_2$) and diploid (2n=14) _S. cereale_ (RR) was described by G. Östergren in southern Sweden and analyzed cytologically by Heneen (1963a). The hybrid usually formed 7^I and 14^{II} at metaphase I. Interpretation of chromosome pairing was aided by size differences between _Secale_ (large) and _Leymus_ (small) chromosomes. All pairing was attributed to autosyndesis between the J_1-J_2 and N_1-N_2 genomes.

PASCOPYRUM LÖVE - 1980
Name derivation: From Latin _pascuum_ = pasture,
                  and Greek _pyros_ = wheat
Type species: _Pascopyrum smithii_ (Rydb.) Löve
Genomes: SHJN

Circumscription and Description

Pascopyrum is a monotypic genus erected by Löve, (1980a) to accommodate a genomically unique octoploid species (_Agropyron smithii_) known commonly as western wheatgrass. This species is a strongly rhizomatous, long-anthered, cross-pollinating grass whose spikes often contain both single and double spikelets per node. Because of its rhizomatous habit, _P. smithii_ was placed in _Elytrigia_ by Nevski (1933), and Chinese agrostologists have also followed that course (Keng, 1965). _Pascopyrum smithii_ is mentioned only

incidentally in Flora Europaea, but under the name Elymus smithii
(Melderis et al., 1980). General usage in North America continues
to be Agropyron smithii.

Pascopyrum smithii is always octoploid (2n=56) (Gillett and
Senn, 1960). Reports of 2n=28 for P. smithii (Carnahan and Hill,
1961) apparently come from misidentified collections of Elymus
lanceolatus, a tetraploid species that is sometimes confused with P.
smithii. This is the only true allooctoploid species in the
Triticeae, and it apparently is the product of hybridization between
Leymus triticoides (JJNN) and Elymus lanceolatus (SSHH) (Dewey,
1975b). The immediate and ultimate sources of the SHJN genomes of
Pascopyrum are outlined in Fig. 1.

Pascopyrum is a North American genus; its sole member, P.
smithii, is widely distributed throughout western North America
(Hitchcock, 1951). It is one of the dominant grasses of the
Northern Great Plains and the Intermountain Region. Pascopyrum
smithii is more or less the morphological and ecological composite
of its putative parents, Leymus triticoides and Elymus lanceolatus.
It is adapted to heavy saline-alkaline soils, reflecting the input
of the JJNN genomes. The impact of the SSHH genomes is less obvious
but those genomes certainly contribute to the forage qualities of P.
smithii, making it a premier forage grass.

Intergeneric Hybrization

Pascopyrum has been a participant in fewer intergeneric hybrids
than any other perennial genus of the Triticeae. One reason for
such few hybrids is that Pascopyrum has only one species, whose
natural distribution is limited to western North America. A second
reason is that P. smithii flowers 2 to 3 weeks later than most
species of Triticeae, which precludes much natural crossing and
makes controlled crossing inconvenient. Factors favoring
intergeneric hybridization with Pascopyrum include its
cross-pollinating habit, high self-sterility, and its own hybrid
background. Being based on four genomes, SHJN, should make
Pascopyrum susceptible to hybridization with its diploid and
tetraploid progenitors. For unknown reasons, hybrids of Pascopyrum
with Pseudoroegneria (S), Critesion (H), Thinopyrum (J),
Psathyrostachys (N), Elymus (SHY), and Leymus (JN) are rare or
nonexistent. Furthermore, Pascopyrum is the only perennial genus of
Triticeae that has not been hybridized with Triticum, Hordeum, or
Secale. The report of a P. smithii X T. aestivum hybrid (Sharma
and Gill, 1981) involved Elymus trachycaulus misidentified as P.
smithii. If genetic isolation is an attribute used in the
definition of a genus, Pascopyrum certainly qualifies as a genus.

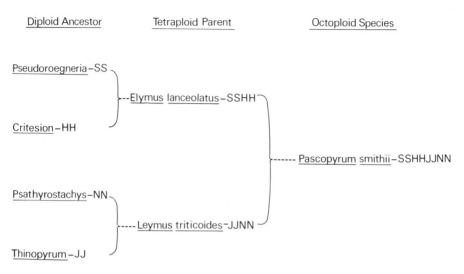

Fig. 1. Origin of <u>Elymus lanceolatus</u>, <u>Leymus triticoides</u>, and <u>Pascopyrum smithii</u>.

Pascopyrum (SHJN) - Agropyron (P). No reported hybrids.

Pascopyrum (SHJN) - Pseudoroegneria (S). No reported hybrids.

Pascopyrum (SHJN) - Psathyrostachys (N). No reported hybrids.

Pascopyrum (SHJN) - Critesion (H). No reported hybrids.

Pascopyrum (SHJN) - Thinopyrum (J-E). No reported hybrids.

Pascopyrum (SHJN) - Elytrigia (SX). See <u>Elytrigia - Pascopyrum</u>.

Pascopyrum (SHJN) - Elymus (SHY). See <u>Elymus - Pascopyrum</u>.

Pascopyrum (SHJN) - Leymus (JN). See <u>Leymus - Pascopyrum</u>.

Pascopyrum (SHJN) - Triticum (ABD). No reported hybrids. White (1940) attempted this cross, but without success. The <u>P. smithii</u> X <u>T. aestivum</u> hybrid reported by Sharma and Gill (1981) was actually <u>E. trachycaulus</u> X <u>T. aestivum</u> (Sharma and Gill, 1983a).

Pascopyrum (SHJN) - Hordeum (I). No reported hybrids.

Pascopyrum (SHJN) - Secale (R). No reported hybrids.

SUMMARY AND CONCLUSIONS

Only during the past 30 years has much attention been given to the cytogenetics of the perennial Triticeae. Considerable cytogenetic data are now available; but most of it is not very detailed nor sophisticated, and much remains to be done. Some indication of what has been accomplished and what remains to be done comes from comparison of the number of species and subspecies whose chromosome numbers were known in 1947, 1961, and 1984. In a comprehensive review paper, Myers (1947) listed chromosome numbers for 66 taxa of perennial Triticeae. In an updated version of the Myers review, Carnahan and Hill (1961) added another 91 taxa to the list, or a total of 157 species and subspecies with known chromosome numbers. In 1984, the total is 279 (Table 3), or an increase since 1961 of 122 additional taxa with reported chromsosome numbers. In Löve's (1984) conspectus, more than 150 perennial species or subspecies remain without a chromosome count. Obviously, many years' work lay ahead just to get the most rudimentary cytogenetic data on many of those grasses.

The depth of cytogenetic information varies widely from genus to genus. Agropyron, because of its economic importance, probably has the most complete cytogenetic data, which includes chromosome numbers, karyotypes, B-chromosome behavior, and type of polyploidy. Less than half of the species of Pseudoroegneria have chromosome counts, and the nature of polyploidy in some of the tetraploids is not yet known. Chromosome numbers are known for only half of the Psathyrostachys species, yet we assume that the entire genus is diploid. Chromosome counts are available for most of the Critesion species, but a great deal of additional information is needed to define the genomic relationships within this genomically variable genus.

Essentially all of the taxa of Thinopyrum have been observed by cytogeneticists because these are the species of greatest interest to wheat breeders. Chromosome numbers are known for these taxa, but the nature of polyploidy and genomic relationships among the high polyploids are poorly understood. Although Elytrigia contains only about five species, all are complex polyploids whose genomic constitutions and interrelationships are still being unraveled. Chromosome numbers are known for about 80% of the 30 species of Leymus, and there is little doubt that Leymus contains only the J and N genomes. The genomic structure of some of the high polyploids of Leymus is the main area of cytogenetic uncertainty in this genus.

Chromosome number, genomic content, and genomic structure are all
known for Pascopyrum.

Certain elements of the genus Elymus are well known from a
cytogenetic standpoint, but others are entirely unknown. At least
half of the 150 species of Elymus do not have chromosome counts. On
the other hand, species such as those of E. canadensis complex have
been involved in numerous cytogenetic studies and their genomic
structure and constitution are known with surety. The greatest
information void in Elymus is in those species from China, which has
more than 50 species of Elymus (including Roegneria) (Keng, 1965).
Many of these species contain a Y genome of unknown origin. Recent
reports from China indicate that a Y-genome diploid has been found
(Yen Chi, pers. comm.). This same genome may occur in the
Australian species of Elymus. After the Chinese species of Elymus
and other Y-genome taxa have been described cytogenetically, it may
be desirable to partition Elymus into two or more genera.

Although cytogenetic information on the perennial Triticeae is
by no means complete, sufficient information is available to construct a useful system of classification based on genomic relationships. The genomic system of classification is a significant
departure from conventional taxonomy, yet the two systems have a
good deal in common. Agropyron, Psathyrostachys, and Leymus are
essentially the same under the genomic system and the more conventional treatments of Tzvelev (1976) and Melderis et al. (1980). The
other genera of the genomic system are usually equivalent to one or
more sections or series within conventional genera, so the differences in the system often relate to the taxonomic level rather than
to entirely different groupings of taxa.

The taxonomic treatment of Tzvelev (1976) in Poaceae URSS is
the conventional treatment that comes closest to the genomic
classification. In fact, the two treatments agree in most respects.
Tzvelev's Elytrigia is a combination of Pseudoroegneria, Elytrigia,
and Thinopyrum; and he treats Critesion under Hordeum. Other than
that, the two treatments are almost identical.

The taxonomic treatment of Melderis et al. (1980) in Flora
Europaea differs from the genomic system largely with respect to the
handling of Elymus, but that one difference is enormous. Elymus
sensu Melderis (1978) becomes a huge genus that combines traditional
Elymus with Elytrigia sensu Tzvelev. As such, it is a morphologically, ecologically, reproductively, and genomically heterogenous
assemblage of species that defies logic. If anything, Elymus sensu
Melderis is even more genetically diverse and unmanageable than traditional Agropyron. Needless to say, I think that the Melderis
treatment of Elymus is not much more useful than the proposal of
Stebbins (1956) to combine all Triticeae species into one genus.

Genomic relationships as reflected in the genomic system of classification should guide grass breeders who are interested in wide-hybridization breeding. Knowing that Elytrigia contains the S genome of Pseudoroegneria explains how the caespitose habit of P. spicata was introgressed into E. repens. Pseudoroegneria X Elytrigia hybridization should be emphasized now that we know the genomic relationships between the two genera. Synthesis of new species of Elymus should be possible through hybridization of Pseudoroegneria diploids with Critesion diploids followed by induced amphiploidy. Gene introgression between Psathyrostachys and Leymus is a distinct possibility in view of the common N genome that they share. Synthesis of new strains of Pascopyrum smithii can be expected from hybridization between Leymus and Elymus followed by chromosome doubling. The need and value of intergeneric hybridization breeding in forage grasses are debatable issues because grass breeders have not yet used much of the genetic variability within a species or within a genus. However, wide hybridization programs in these grasses are easily justifiable when the benefits of basic cytogenetic studies are combined with the applied benefits in the form of new gene combinations or entirely new species of use to plant breeders.

Wheat, barley, and rye breeders should find the genomic system of classification of value in their search for genes outside of their own genera. In the past, wheat breeders have looked to traditional Agropyron for genes to introgress into Triticum. The literature is replete with statements to the effect that Agropyron is the perennial genus most closely related to Triticum. Comments of that sort are meaningless, misleading, or even incorrect unless they are followed by qualifying statements that specify which part of traditional Agropyron is being referred to. The genomic system of classification promotes more precise thinking and communication.

The J-E genome of Thinopyrum has been the genome most widely exploited by wheat breeders, and it will probably continue to be the most useful genome because it is closest to the Triticum genomes, as evidenced by the success of substitution lines (Dvorak, 1980). Wheat breeders should look to all species of Thinopyrum--not only T. intermedium, T. elongatum, and T. ponticum--for genes to be introgressed into Triticum. The J genome of JN species of Leymus is the same basic genome as the J-E found in Thinopyrum, so Leymus might be as valuable to wheat breeders as Thinopyrum.

The J genome of Thinopyrum and Leymus can be used by wheat breeders as a source of individual genes introgressed by ph-assisted crossing over, by substitution of whole chromosomes, by addition of whole chromosomes, or by addition of an entire genome. So far, only rather simply inherited traits have been introgressed. The transfer of quantitatively inherited characters such as tolerance to environmental stress will be difficult or impossible without the simultaneous transfer of unwanted genetic material.

The H genome of Critesion may be as closely related to the genome of barley as the J genome of Thinopyrum and Leymus is to the genomes of wheat. However, the flexibility for introgressing genes, chromosomes, or genomes from Critesion to Hordeum is restricted by the diploid condition of barley as contrasted to genomically buffered polyploid wheat. Consequently, intergeneric hybridization may be less useful in barley breeding than in wheat breeding.

Introgression of genetic material from the P, S, H, and N genomes of Agropyron, Pseudoroegneria, Critesion, Psathyrostachys, and Elymus to wheat will almost surely prove to be more difficult than movement of genes from the J genome of Thinopyrum or Leymus. The homoeology, if any, of the P, S, H, and N genomes of the perennials with the A, B, and D genomes of Triticum may be so remote that introgression through crossing over between homoeologs is not a practical plant breeding option. Addition of chromosome pairs from the P, S, H, and N genomes to wheat may be possible without unduly upsetting the gene balance in polyploid wheat. The prospects for substituting chromosomes are even poorer than making chromosome additions.

The addition of whole P, S, H, or N genomes to those of Triticum, Hordeum, or Secale through hybridization and induced amphiploidy may be the best means of combining the genomes of the distantly related perennials with those of the annuals. Diploid species of Agropyron, Pseudoroegneria, Critesion, and Psathyrostachys should be used in these crosses to produce amphiploids with no more than 56 chromosomes, which seems to be the maximum number for a successful cereal crop. Amphiploids with 42 chromosomes are probably preferable to those with 56 chromosomes. The hexaploid amphiploids of Triticum timopheevi X Critesion bogdanii (Kimber and Sallee, 1976) and Critesion chilense X Triticum turgidum (Martin and Laguna, 1980) are examples of what can be accomplished by combining whole genomes of diploid perennials with the genomes of tetraploid annuals.

LITERATURE CITED

Ahokas, H., 1970, Some artificial intergeneric hybrids in the
        Triticeae, Ann. Bot. Fenn., 7:182-192.
Alonzo, L. C.. and Kimber, G., 1980, A hybrid between diploid
        Agropyron junceum and Triticum aestivum, Cereal Res.
        Commun., 8:355-358.
Armstrong, J. M., 1936, Hybridization of Triticum and Agropyron. I.
        Crossing results and description of the first generation
        hybrids, Canad. J. Res., 14-C:190-202.

Asay, K. H., and Dewey, D. R., 1979, Bridging ploidy differences in crested wheatgrass with hexaploid X diploid hybrids, Crop Sci., 19:519-523.

Asay, K. H., and Dewey, D. R., 1981, Registration of Agropyron repens X A. spicatum germplasms RS-1 and RS-2, Crop Sci., 21:351.

Asay, K. H., and Dewey, D. R., 1983, Pooling the genetic resources of the crested wheatgrass species-complex, Proc. XIV Int. Grassland Congr., Pages 124-127, Westview Press, Boulder.

Bakhteyev, F. Kh., and Darevskaya, E. M., 1960, Wide hybridization in the genus Hordeum L., Pages 242-253, in: "Wide Hybridization in Plants," N. V. Tsitsin, ed., Israel Program for Sci. Transl., Jerusalem.

Barkworth, M. E., Dewey, D. R., and Atkins, R. E., 1983, New generic concepts in the Triticeae (Gramineae) of the Intermountain Region: Key and comments, Great Basin Naturalist, 43:561-572.

Bentham, G., 1881, Notes on Gramineae, Tribe XIII. Hordeeae, J. Linn. Soc., Bot., 19:129-134.

Bothmer, R., 1979, Revision of the Asiatic taxa of Hordeum sect. Stenostachys, Bot. Tidsskr., 74:117-146.

Bothmer, R., Flink, J., Jacobsen, N., Kotimäki, and Landström, T., 1983, Interspecific hybridization with cultivated barley (Hordeum vulgare L.), Hereditas 99:219-244.

Bowden, W. M., 1957, Natural and artificial X Elymordeum hybrids, Canad. J. Bot., 36:101-123.

Bowden, W. M., 1959a, The taxonomy and nomenclature of the wheats, barleys, and ryes and their wild relatives, Canad. J. Bot., 37:657-684.

Bowden, W. M., 1959b, Chromosome numbers and taxonomic notes on northern grasses. I. Tribe Triticeae, Canad. J. Bot., 37:1143-1151.

Bowden, W. M., 1964, Cytotaxonomy of the species and interspecific hybrids of the genus Elymus in Canada and neighboring areas, Canad. J. Bot., 42:547-601.

Bowden, W. M., 1965, Cytotaxonomy of the species and interspecific hybrids of the genus Agropyron in Canada and neighboring areas, Canad. J. Bot., 43:1421-1448.

Bowden, W. M., 1967, Taxonomy of intergeneric hybrids of the tribe Triticeae from North America, Canad. J. Bot., 45:711-724.

Boyle, W. S., and Holmgren, A. H., 1955, A cytogenetic study of natural and controlled hybrids between Agropyron trachycaulum and Hordeum jubatum, Genetics, 40:539-545.

Boyle, W. S., and Holmgren, A. H., 1968, A cytotaxonomic study of a natural hybrid between Agropyron cristatum and A. subsecundum, Madroño, 19:277-280.

Brink, R. A., Cooper, D. C., and Ausherman, L. E., 1944, A hybrid between Hordeum jubatum and Secale cereale reared from an artificially cultivated embryo, J. Heredity, 35:66-75.

Carnahan, H. L., and Hill, H. D., 1961, Cytology and genetics of
    forage grasses, Bot. Rev., 27:1-162.
Cauderon, Y, 1956, Étude de l'hybride F₁ entre Hordeum bulbosum
    L. et H. secalinum Schreb., Ann. Amelior. Pl., 1956:307-317.
Cauderon, Y, 1958, Étude cytogénétique des Agropyron francais
    et de leurs hybrids avec les blés, Ann. Amelior. Pl.,
    1958:389-567.
Cauderon, Y., 1966, Genome analysis in the genus Agropyron,
    Hereditas, Suppl. Vol. 2:218-234.
Cauderon, Y., 1979, Use of Agropyron species for wheat improvement,
    Proc. Conf. Broadening Genetic Base of Crops, Pages 129-139,
    Wageningen.
Cauderon, Y., and Ryan, G., 1974, Aegilops speltoides promotion of
    homoeologous pairing in one Triticum aestivum X Agropyron
    intermedium derivative, Wheat Inform. Serv., 39:1-5.
Cauderon, Y, and Saigne, B., 1961, New interspecific and inter-
    generic hybrids involving Agropyrum, Wheat Inform. Serv.,
    12:13-14.
Chapman, V., and Miller, T. E., 1978, The amphiploid of Hordeum
    chilense X Triticum aestivum, Cereal Res. Commun.,
    6:351-352.
Cugnac, A. de., and Simonet, M., 1953, Étude cytogénétique de
    quelques hybrides intergénériques d'Agropyrum,
    Agroelymus, Agrohordeum, et Agrotriticum, Ann. Amelior. Pl.,
    1953:433-452.
Dewey, D. R., 1961, Hybrids between Agropyron repens and Agropyron
    desertorum, J. Heredity, 52:13-21.
Dewey, D. R., 1962, The genome structure of intermediate wheatgrass,
    J. Heredity, 53:282-290.
Dewey, D. R., 1963a, Cytology and morphology of a synthetic
    Agropyron trichophorum X Agropyron desertorum hybrid, Amer.
    J. Bot., 50:522-562.
Dewey, D. R., 1963b, Morphology and cytology of synthetic hybrids of
    Agropyron trichophorum X Agropyron cristatum, Amer. J. Bot.,
    51:1028-1034.
Dewey, D. R., 1964a, Synthetic hybrids of New World and Old World
    Agropyrons. I. Tetraploid Agropyron spicatum X diploid
    Agropyron cristatum, Amer. J. Bot. 51:763-769.
Dewey, D. R., 1964b, Genome analysis of Agropyron repens X Agropyron
    cristatum synthetic hybrids, Amer. J. Bot., 51:1062-1068.
Dewey, D. R., 1965, Synthetic hybrids of New World and Old World
    Agropyrons. II. Agropyron riparium X Agropyron repens, Amer.
    J. Bot., 52:1039-1045.
Dewey, D. R., 1967a, Synthetic hybrids of New World and Old World
    Agropyrons. III. Agropyron repens X tetraploid Agropyron
    spicatum, Amer. J. Bot., 54:93-98.
Dewey, D. R., 1967b, Synthetic hybrids of New World and Old World
    Agropyrons. IV. Tetraploid Agropyron spicatum f. inerme X
    tetraploid Agropyron desertorum, Amer. J. Bot., 54:403-409.

Dewey, D. R., 1967c, Synthetic hybrids of Agropyron scribneri X Elymus junceus, Bull. Torrey Bot. Club, 94:388-395.

Dewey, D. R., 1968, Synthetic hybrids among Hordeum brachyantherum, Agropyron scribneri, and Agropyron latiglume, Bull. Torrey Bot. Club, 95:454-464.

Dewey, D. R., 1969, Hybrids between tetraploid and hexaploid crested wheatgrasses, Crop Sci., 9:787-791.

Dewey, D. R., 1970a, The origin of Agropyron albicans, Amer. J. Bot. 57:12-18.

Dewey, D. R., 1970b, Genome relations among diploid Elymus junceus and certain tetraploid and octoploid Elymus species, Amer. J. Bot., 57:633-639.

Dewey, D. R., 1970c, Genome relations among Elymus canadensis, Elymus triticoides, Elymus dasystachys, and Agropyron smithii, Amer. J. Bot., 57:861-866.

Dewey, D. R., 1970d, Cytogenetics of Agropyron stipaefolium and its hybrids with Agropyron repens, Bull. Torrey Bot. Club, 97:315-320.

Dewey, D. R., 1971a, Genome relations among Agropyron spicatum, A. scribneri, Hordeum brachyantherum, and H. arizonicum, Bull. Torrey Bot. Club, 98:208-206.

Dewey, D. R., 1971b, Synthetic hybrids of Hordeum bogdanii with Elymus canadensis and Sitanion hystrix, Amer. J. Bot., 58:902-908.

Dewey, D. R., 1972a, Cytogenetic and genomic relationships of Elymus giganteus with E. dasystachys and E. junceus, Bull. Torrey Bot. Club, 99:77-83.

Dewey, D. R., 1972b, Cytogenetics of Elymus angustus and its hybrids with Elymus giganteus, Elymus cinereus, and Agropyron repens, Bot. Gaz., 133:57-64.

Dewey, D. R., 1972c, The origin of Agropyron leptourum, Amer. J. Bot., 59:836-842.

Dewey, D. R., 1972d, Genome analysis of diploid Elymus junceus and five tetraploid Elymus species, Bot. Gaz., 133:415-420.

Dewey, D. R., 1974a, Hybrids and induced amphiploids of Elymus canadensis X Agropyron libanoticum, Amer. J. Bot., 61:181-187.

Dewey, D. R., 1974b, Cytogenetics of Elymus sibiricus and its hybrids with Agropyron tauri, Elymus canadensis, and Agropyron caninum, Bot. Gaz., 135:80-87.

Dewey, D. R., 1975a, Genome relations of diploid Agropyron libanoticum with diploid and autotetraploid Agropyron stipifolium, Bot. Gaz., 136:116-121.

Dewey, D. R., The origin of Agropyron smithii, Amer. J. Bot., 62:524-530.

Dewey, D. R., 1976, Derivation of a new forage grass from Agropyron repens X Agropyron spicatum hybrids, Crop Sci., 16:175-180.

Dewey, D. R., 1977, U.S. living collection of perennial Triticeae grasses, Pages 69-72, in: "Systematic Collections of the Agricultural Research Service," U.S.D.A. Misc. Publ., 1343.

Dewey, D. R., 1979, The *Hordeum violaceum* complex of Iran, *Amer. J. Bot.*, 66:166-172.

Dewey, D. R., 1980a, Cytogenetics of *Agropyron drobovii* and five of its interspecific hybrids, *Bot. Gaz.*, 141:469-478.

Dewey, D. R., 1980b, Hybrids and induced amphiploids of *Agropyron repens* X *A. curvifolium*, *Crop Sci.*, 20:473-478.

Dewey, D. R., 1980c, Morphological, cytological, and taxonomic relationships between *Agropyron repens* and *Agropyron elongatiforme*, *Syst. Bot.*, 5:61-70.

Dewey, D. R., 1982, Genomic and phylogenetic relationships among North American perennial Triticeae grasses, Pages 51-80, *in*: "Grasses and Grasslands," J. E. Estes et al., eds., Univ. of Oklahoma Press, Norman.

Dewey, D. R., 1983a, Historical and current taxonomic perspectives of *Agropyron*, *Elymus*, and related genera, *Crop Sci.*, 23:637-642.

Dewey, D. R., 1983b, New nomenclatural combinations in the North American perennial Triticeae (Gramineae), *Brittonia*, 35:30-33.

Dewey, D. R., and Holmgren, A. H., 1962, Natural hybrids of *Elymus cinereus* X *Sitanion hystrix*, *Bull. Torrey Bot. Club*, 89:217-228.

Dewey, D. R., and Asay, K. H., 1975, The crested wheatgrasses of Iran, *Crop Sci.*, 15:844-849.

Dewey, D. R., and Hsiao, C., 1983, A cytogenetic basis for transferring Russian wildrye from *Elymus* to *Psathyrostachys*, *Crop Sci.*, 23:123-126.

Dvorak, J., 1971, Hybrids between diploid *Agropyron elongatum* and *Aegilops squarrosa*, *Canad. J. Genet. Cytol.*, 13:90-94.

Dvorak, J., 1976, The cytogenetic structure of a 56-chromosome derivative from a cross between *Triticum aestivum* and *Agropyron elongatum* (2n=70), *Canad. J. Genet. Cytol.*, 18:271-279.

Dvorak, J., 1980, Homology between *Agropyron elongatum* chromosomes and *Triticum aestivum* chromosomes, *Canad. J. Genet. Cytol.*, 22:237-259.

Dvorak, J., 1981, Genome relationships among *Elytrigia* (= *Agropyron*) *elongata*, *E. stipifolia*, "*E. elongata* 4x," *E. caespitosa*, *E. intermedia*, and "*E. elongata* 10x," *Canad. J. Genet. Cytol.*, 23:481-492.

Dvorak, J., and Knott, D. R., 1974, Disomic and ditelosomic additions of diploid *Agropyron elongatum* chromosomes to *Triticum aestivum*, *Canad. J. Genet. Cytol.*, 16:399-417.

Elven, R., 1981, A hybrid between *Elytrigia* and *Elymus*, *Blyttia*, 39:115-120.

Evans, L. E., 1964, Genome construction within the Triticinae. I. The synthesis of hexaploid (2n=42) having chromosomes of *Agropyron* and *Aegilops* in addition to the A and B genomes of *Triticum durum*, *Canad. J. Genet. Cytol.*, 6:19-28.

Favorsky, N. V., 1935, On the reduction division in the hybrid of
    Secale cereale L. X *Agropyrum cristatum* (L.) Gaertn. in
    connection with the question of its sterility, Socialistic
    Grain Farming 1:115-125.
Fedak, G., 1983, Hybrids between *Hordeum parodii* and *H. vulgare*,
    Canad. J. Genet. Cytol., 25:101-103.
Fedak, G., 1984, Wide crosses in *Hordeum*, in: "Barley Monograph,"
    Rasmusson, D. C., ed., American Society of Agronomy, Madison,
    In press.
Finch, R. A., and Bennett, M. D., 1980, Mitotic and meiotic
    chromosome behaviour in new hybrids of *Hordeum* with *Triticum*
    and *Secale*, Heredity, 44:201-209.
Gaul, H., 1953, Genomonalytische untersuchungen bei *Triticum* X
    Agropyron intermedium unter berücksichtigung von *Secale*
    cereale X *A. intermedium*, Z. Indukt. Abstammungs-
    Vererbungsl., 85:505-546.
Gillett, J. M., and Senn, H. A., 1960, Cytotaxonomy and
    infraspecific variation of *Agropyron smithii* Rydb., Canad. J.
    Bot., 38:747-760.
Gou, P. C., and Wang, S. J., 1981, Research on the evolution of the
    inflorescence and the generic relationships of the Triticeae
    in China, Acta Bot. Nor-Occ. Sinica, 1:12-19.
Gould, F. W., 1947, Nomenclatural changes in *Elymus* with a key to
    the Californian species, Madroño, 9:120-128.
Hair, J. B., 1956, Subsexual reproduction in *Agropyron*, Heredity,
    10:129-160.
Hansen, A., 1959a, Die Gras-Hybriden in der Flora Frankreichs Kritik
    und Ergänzungen, Jard. Bot. Etat., 29:61-68.
Hansen, A., 1959b, Die *Elytrigia*-Arten und-Hybriden on der
    polnischen Ostseeküste, Frag. Flor. Geobot., 5:181-189.
Heneen, W. K., 1963a, Cytology of the intergeneric hybrid *Elymus*
    arenarius X *Secale cereale*, Hereditas, 49:61-77.
Heneen, W. K., 1963b, Meiosis in the interspecific hybrid *Elymus*
    farctus X *E. repens* (= *Agropyron junceum* X *A. repens*),
    Hereditas, 49:107-118.
Heneen, W. K., and Runemark, H., 1972a, Chromosomal polymorphism in
    isolated populations of *Elymus* (*Agropyron*) in the Aegean. I.
    Elymus striatulus sp. nov., Bot. Not., 125:419-429.
Heneen, W. K., and Runemark, H., 1972b, Cytology of the *Elymus*
    (*Agropyron*) *elongatus* complex, Hereditas, 70:155-164.
Hitchcock, A. S., 1951, Tribe 3. Hordeae, Pages 230-280, in: "Manual
    of the Grasses of the United States," U.S.D.A. Misc. Publ.
    200, Second edition revised by Agnes Chase, U.S. Govt.
    Printing Office, Washington, D. C.
Hochstetter, C. F., 1848, Nachträglicher Commentar zu meiner
    Abhandlung: "Aufbau der Graspflanze etc.," Flora, No.
    7:105-119. Regensburg.
Hodgson, H. J., and Mitchell, W. W., 1965, A new *Elymordeum* hybrid
    from Alaska, Canad. J. Bot., 43:1355-1358.

Holmberg, O. R., 1926, Genus 91. _Agropyron_, Pages 268-279, _in_:
        "Skandinaviens Flora, Vol. 2," P. A. Norstedt and Söners,
        Forlag, Stockholm.
Holub, J., 1973, New names in Phanerogamae 2, _Folia Geobot._
        Phytotax. (Praha), 8:171.
Hubbard, C. E., 1968, Grasses, a guide to their structure, identifi-
        cation, uses, and distribution in the British Isles, Penguin
        Books, Middlesex.
Hunziker, J. H., Naranjo, C. A., and Zeiger, E., 1973, Las
        relationes evolutivas entre _Hordeum compressum_ y otras
        especies diploides Americanas afines, _Kurtziana_, 7:7-26.
Jacobsen, N., and Bothmer, R., 1981, Interspecific hybridization in
        the genus _Hordeum_ L., Proc. 4th Int. Barley Genet. Symp.,
        Pages 710-715.
Jenkins, B. C., 1958, Research report No. 3 (1956-57) of the Samuel
        Rosner Chair in Agronomy, 52 pages, University of Manitoba,
        Winnipeg.
Jenkins, B. C., and Mochizuki, A., 1957, A new amphiploid from a
        cross between _Triticum durum_ and _Agropyron elongatum_ (2n=14),
        Wheat Inform. Serv., 5:15.
Johnson, R., and Kimber, G., 1967, Homoeologous pairing of a chromo-
        some from _Agropyron elongatum_ with those of _Triticum aestivum_
        and _Aegilops speltoides_, _Genet. Res. Camb._, 10:63-71.
Kasha, K., and Sadasaviah, R. S., 1971, Genome relationships between
        Hordeum vulgare L. and _H. bulbosum_ L., _Chromosoma_ (Berlin),
        35:264-287.
Keng, Y. L., 1965, Tribus 7. Hordeae Bentham, Pages 340-451, _in_:
        "Flora Illustrata Plantarum Primarum Sinicarum," Keng, Y. L.,
        ed., Scientific Publishing Co., Peking.
Kerguelen, M., 1975, Les Gramineae (Poaceae) de la flore francaise.
        Essai de mise au point taxonomique et nomenclature, _Lejeunia_
        nouvelle Serie, 75:1-344.
Kihara, H. 1949, Genomanalyse bei _Triticum_ and _Aegilops_. IX.
        Systematischer Aufbau der Gattung _Aegilops_ auf
        genomanalytischer Grundlage, Cytologia, 14:135-144.
Kihara, H., 1963, Interspecific relationship in _Triticum_ and
        Aegilops, _Rep. Kihara Inst. Biol. Res._, 15:1-12.
Kimber, G., 1979, Amphiploids as a genetic resource in the
        Triticeae, _Indian J. Genet. Pl. Breed._, 39:133-137.
Kimber, G., and Sallee, P. J., 1976, A hybrid between _Triticum_
        timopheevi and _Hordeum bogdanii_, _Cereal Res. Commun._,
        4:33-37.
Knobloch, I. W., 1968, A check list of crosses in the Gramineae,
        Mimeographed publication (170 pages). I. W. Knobloch,
        East Lansing.
Knott, D. R., Dvorak, J., and Nanda, J. S., 1977, The transfer to
        wheat and homoeology of an _Agropyron elongatum_ chromosome
        carrying resistance to stem rust, _Canad. J. Genet. Cytol._,
        19:75-79.

Krasniuk, A. A., 1935, The hybrid Secale cereale X Agropyron
    cristatum, Socialistic Grain Farming, 1:106-114.
Krause, E. H. L., 1898, Floristische Notizen II. Gräser, Bot.
    Centralbl., 73:337-343.
Kruse, A., 1973, Hordeum X Triticum hybrids, Hereditas, 73:157-161.
Lepage, E., 1952, Études sur quelques plantes americaines. II.
    Hybrides intergénériques: Agrohordeum et Agroelymus,
    Naturaliste Canad., 79:241-266.
Lepage, E., 1965, Revision genéalogique de quelques X Agroelymus,
    Naturaliste Canad., 92:205-216.
Linnaeus, C., 1753, Species plantarum, Laurentii Salvii, Holmiae.
Liu, C. W., and Dewey, D. R., 1983, The genome constitution of
    Elymus fedtschenkoi, Acta Genet. Sinica, 10:20-27.
Löve, A., 1980a, IOPB chromosome number reports. LXVI. Poaceae-
    Triticeae-Americanae, Taxon, 29:163-169.
Löve, A., 1980b, Chromosome number reports. LXVII. Poaceae-
    -Triticeae, Taxon 29:350-351.
Löve, A., 1982, Generic evolution of the wheatgrasses, Biol.
    Zentralbl., 101:199-212.
Löve, A., 1984, Conspectus of the Triticeae, Feddes Repert., 95: In
    press.
Löve, A., and Connor, H. E., 1982, Relationships and taxonomy of New
    Zealand wheatgrasses, New Zealand J. Bot., 20:169-186.
Löve, A., and Löve, D., 1975, Nomenclatural notes on arctic plants,
    Bot. Not., 128:497-523.
Love, R. M., and Suneson, C. A., 1945, Cytogenetics of certain
    Triticum-Agropyron hybrids and their fertile derivatives,
    Amer. J. Bot., 32:451-456.
Lyubimova, V. F., 1970, Cytogenetic investigations of hybrids
    obtained from crossing Agropyron glaucum Roem. et Schult.
    with Agropyron elongatum (Host) P.B., Genetika, 6:5-14.
Lyubimova, V. F., 1973, Wheat-quackgrass-rye hybrids: A cytogenetic
    study, Genetika, 9:5-16.
Martin, A., 1983, Cytology and morphology of the hybrid Hordeum
    chilense X Aegilops squarrosa, J. Heredity, 74:487.
Martin, A., and Chapman, V., 1977, A hybrid between Hordeum chilense
    and Triticum aestivum, Cereal Res. Commun., 5:365-368.
Martin, A., and Laguna, E. S., 1980, A hybrid between Hordeum
    chilense and Triticum turgidum, Cereal Res. Commun.,
    8:349-353.
Martin, A., and Laguna, E. S., 1982, Cytology and morphology of the
    amphiploid Hordeum chilense X Triticum turgidum conv. durum,
    Euphytica, 31:261-267.
Matsumura, S., Muramatsu, M., and Sakamoto, S., 1958, Genome
    analysis in Agropyron a genus related to Triticum, Rep.
    Kihara Inst. Biol. Res., 9:8-16.
McGuire, P. E., 1983, Elytrigia turcica sp. nova, an octoploid
    species of the E. elongata complex, Folia Geobot. Phytotax.,
    18:107-109.

Melderis, A., 1978, Taxonomic notes on the tribe Triticeae
    (Gramineae), with special reference to the genera Elymus L.
    sensu lato, and Agropyron Gaertner sensu lato, Linn. Soc.
    Bot., 76:369-384.
Melderis, A., Humphries, C. J., Tutin, T. G., and Heathcote, S. A.,
    1980, Tribe Triticeae Dumort. Pages 190-206, in "Flora
    Europaea," Volume 5, Tutin, T. G., et al., eds., Cambridge
    University Press, Cambridge.
Melderis, A., and McClintock, D., 1983, The genera Elymus L. and
    Leymus Hochst. in Britain, Watsonia, 14:391-395.
Miller, T. E., Reader, S. M., and Chapman, V., 1982, The addition of
    Hordeum chilense chromosomes to wheat, in: "Induced
    Variability in Plant Breeding," Pages 79-81, Int. Symp. on
    Mutation and Polyploidy, Wageningen.
Mitchell, W. W., and Hodgson, H. J., 1968, Hybridization within the
    Triticeae of Alaska: A new X Elyhordeum and comments,
    Rhodora, 70:467-473.
Morrison, J. W., and Rajhathy, T., 1959, Cytogenetic studies in the
    genus Hordeum. III. Pairing in some interspecific and
    intergeneric hybrids, Canad. J. Genet. Cytol., 1:65-77.
Mujeeb-Kazi, A., and Bernard, M., 1982, Somatic chromosome
    variations in backcross-1 progenies from intergeneric hybrids
    involving some Triticeae, Cereal Res. Commun., 10:41-45.
Mujeeb-Kazi, A., and Rodriguez, R., 1980, Some intergeneric hybrids
    in the Triticeae, Cereal Res. Commun., 8:469-475.
Mujeeb-Kazi, A., and Rodriguez, 1981, An intergeneric hybrid of
    Triticum aestivum X Elymus giganteus, J. Heredity,
    72:253-256.
Mujeeb-Kazi, A., and Rodriguez, 1982, Cytogenetics of hybrids of
    Elymus canadensis X Hordeum vulgare, J. Heredity, 73:77-79.
Mujeeb-Kazi, A., Roldan, S., and Miranda, J. L., 1984, Intergeneric
    hybrids of Triticum aestivum L. with several Agropyron and
    Elymus species, Cereal Res. Commun., 12: In press.
Murashige, T., 1974, Plant propagation through tissue cultures,
    Annual Rev. Pl. Physiol, 25:135-166.
Myers, W. M., 1947, Cytology and genetics of forage grasses, Bot.
    Rev., 13:319-421.
Napier, K. V., and Walton, P. D., 1982, A hybrid between Agropyron
    trachycaulum and A. desertorum, Crop Sci., 22:657-660.
Napier, K. V., and Walton, P. D., 1983, Hybrids between Agropyron
    trachycaulum and A. intermedium, Euphytica, 32:231-239.
Nevski, S. A., 1933, Agrostologische Studien. IV. Uber das System
    der Tribe Hordeeae Benth., Trudy Bot. Inst. Akad. Nauk SSSR,
    Ser. 1, Fl. Sist. Vyss. Rast., Pages 9-32.
Nevski, S. A., 1934, Tribe XIV. Hordeae Benth., Pages 469-579, in:
    "Flora of the U.S.S.R.," Vol. II, Komarov, V. L., ed., Israel
    Program for Sci. Transl., Jerusalem.
Östergren, G., 1940a, Cytology of Agropyron junceum, A. repens and
    their spontaneous hybrids, Hereditas, 26:305-316.

Östergren, G., 1940b, A hybrid between Triticum turgidum and Agropyron junceum, Hereditas, 26:395-398.

Padilla, J. A., and Martin, A., 1983b, Morphology and cytology of Hordeum chilense X H. bulbosum hybrids, Theor. Appl. Genet., 65:353-355.

Padilla, J. A., and Martin, A., 1983b, New hybrids between Hordeum chilense and tetraploid wheats, Cereal Res. Commun., 11:5-7.

Peto, F. H., 1936, Hybridization of Triticum and Agropyron. II. Cytology of the male parents and F$_1$ generation, Canad. J. Res., 14-C:203-214.

Petrova, K. A., 1960, Hybridization between wheat and Elymus, Pages 226-237, in: "Wide Hybridization in Plants," Tsitsin, N. V., ed., Israel Program for Sci. Transl., Jerusalem.

Petrova, K. A., 1970, Morphological and cytological investigation of Agropyron elongatum (Host) P.B. 2n=14 X Elymus mollis Trin. 2n=28; F$_1$ hybrids and amphidiploids, in: "Otdalen. Gibridiz. i. Poliploidiya," Pages 158-176, U.S.S.R. Nauk, Moscow.

Pienaar, R. de V., 1981, Genome relationships in wheat X Agropyron distichum (Thunb.) Beauv. hybrids, Z. Pflanzenzücht, 87:193-212.

Pilger, R., 1947, Additamenta agrostologica. I. Triticeae (Hordeeae), Bot. Jahrb. Syst., 74:1-13.

Pilger, R., 1954, Das System der Gramineae, Bot. Jahrb. Syst., 76:281-384.

Pohl, R. W., 1962, Agropyron hybrids and the status of Agropyron repens, Rhodora, 64:143-147.

Probatova, N. S., and Kharkevitch, S. S., 1982, New taxa of the Poaceae from the Khabarovsk Region, Bot. Zhurn., (Leningrad), 67:1408-1404.

Probatova, N. S., and Sokolovskaya, A. P., 1982, Synopsis of chromosome numbers in Poaceae from the Soviet Far East, Bot. Zhurn., (Leningrad), 67:62-70.

Rajhathy, T., and Morrison, J. W., 1959, Cytogenetic studies in the genus Hordeum. IV. Hybrids of H. jubatum, H. brachyantherum, H. vulgare, and a hexaploid Hordeum sp., Canad. J. Genet. Cytol., 1:124-132.

Rajhathy, T., and Morrison, J. W., 1961, Cytogenetic studies in the genus Hordeum. V. H. jubatum and the New World species, Canad. J. Genet. Cytol., 3:378-390.

Rajhathy, T., and Morrison, J. W., 1962, Cytogenetic studies in the genus Hordeum. VI. The murinum-complex, Canad. J. Genet. Cytol., 4:240-247.

Rajhathy, T., Morrison, J. W., and Symko, S., 1964, Interspecific and intergeneric hybrids in Hordeum, Proc. 1st Int. Barley Genet. Symp., Pages 195-212.

Rafinesque, C. S., 1819, Prodrome des nouveaux genres de plantes observés en 1817 et 1818 dans l'intérieur des États-Unis d'Américque, J. Phys. Chim. Hist. Nat. Arts, 89:96-104.

Riley, R., 1974, Cytogenetics of chromosome pairing in wheat, Genetics, 78:193-203.

Runemark, H., and Heneen, W. K., 1968, Elymus and Agropyron, a problem of generic delimitation, Bot. Not., 121:51-79.

Sadasiviah, R. S., and Weijer, J., 1981, Cytogenetics of some natural intergeneric hybrids between Elymus and Agropyron species, Canad. J. Genet. Cytol., 23:131-140.

Sakamoto, S., 1964, Cytogenetic studies in the tribe Triticeae. 1. A polyhaploid plant of Agropyron tsukushiense var. transiens Ohwi found in a state of nature, Jap. J. Genet., 39:393-400.

Sakamoto, S., 1971, An intergeneric hybrid between Agropyron tsukushiense and Elymus mollis, Annual Rep. Natl. Inst. Genet., 21:97.

Schooler, A. B., and Anderson, M. K., 1980, Behavior of intergeneric hybrids between Hordeum vulgare (4x) and Elymus mollis type, Cytologia, 45:157-162.

Schulz-Schaeffer, J., Allderdice, P. W., and Creel, G. C., 1963, Segmental alloploidy in tetraploid and hexaploid Agropyron species of the crested wheatgrass complex (Section Agropyron), Crop Sci., 3:525-530.

Sears, E. R., 1975, The wheats and their relatives, Pages 59-91, in: "Handbook of Genetics, Vol 2," King, R. C., ed., Plenum Press, New York.

Sharma, H. C., and Gill, B. S., 1981, New hybrids between Agropyron and wheat. I. A. ciliare X wheat and A. smithii X wheat, Wheat Inform. Serv., 52:19-22.

Sharma, H. C., and Gill B. S., 1983a, Current status of wide hybridization in wheat, Euphytica, 32:17-31.

Sharma, H. C., and Gill, B. S., 1983b, New hybrids between Agropyron and wheat. 2. Production, morphology, and cytogenetic analysis of F_1 hybrids and backcross derivatives, Theor. Appl. Genet., 66:111-121.

Sharma, D., and Knott, D. R., 1966, The transfer of leaf-rust resistance from Agropyron to Triticum by irradiation, Canad. J. Genet. Cytol., 8:137-143.

Smith, D. C., 1942, Intergeneric hybridization of cereals and other grasses, J. Agric. Res., 64:33-47.

Smith, D. C., 1943, Intergeneric hybridization of Triticum and other grasses, principally Agropyron, J. Heredity, 34:219-224.

Starks, G. D., and Tai, W., 1974, Genome analysis of Hordeum jubatum and H. compressum, Canad. J. Genet. Cytol., 16:663-668.

Stebbins, G. L., Jr., 1956, Taxonomy and the evolution of genera with special reference to the family Gramineae, Evolution, 10:235-245.

Stebbins, G. L., and Pun, F. T., 1953a, Artificial and natural hybrids in the Gramineae, tribe Hordeae. V. Diploid hybrids of Agropyron, Amer. J. Bot., 40:444-449.

Stebbins, G. L., and Pun, F. T., 1953b, Artificial and natural hybrids in the Gramineae, tribe Hordeae. VI. Chromosome pairing in Secale cereale X Agropyron intermedium and the problem of genome homologies in the Triticinae, Genetics, 38:600-608.

Stebbins, G. L., Valencia, J. I., and Valencia, R. M., 1946,
    Artificial and natural hybrids in the Gramineae, tribe
    Hordeae. II. Agropyron, Elymus, and Hordeum, Amer. J. Bot.,
    33:579-586.
Stebbins, G. L., and Walters, M. S., 1949, Artifcial and natural
    hybrids in the Gramineae, tribe Hordeae. III. Hybrids
    involving Elymus condensatus and E. triticoides, Amer. J.
    Bot., 36:291-301.
Subrahmanyam, N. C., 1978, Meiosis in polyhaploid Hordeum:
    Hemizygous ineffective control of diploid-like behavior in a
    hexaploid?, Chromosoma (Berlin), 66:185-192.
Tsitsin, N. V., 1960, The significance of wide hybridization in the
    evolution and production of new species and forms of plants
    and animals, Pages 2-30, in: "Wide Hybridization in Plants,"
    Tsitsin, N. V., ed., Israel Program for Sci. Transl.,
    Jerusalem.
Tsitsin, N. V., 1975, Origin of new species and forms of plants,
    Proc. 12th Int. Bot. Congr., Pages 3-10.
Tsitsin, N. V., and Gruzdeva, E. D., 1959, Hybrids of Agropyron
    glaucum Roem. et Schult. X A. repens (L.) P. B. Bull. Princ.
    Jand. Bot., 33:53-60.
Tsitsin, N. V., and Petrova, K. A., 1976, Forty-two-chromosome
    wheat-wildrye amphiploids, Dokl. Akad. Nauk SSSR,
    228:1215-1218.
Tzvelev, N. N., 1960, De speciebus nonnulis novis vel minus cognitis
    e Pamir, Bot Mater. Gerb. Inst. Komarova Akad. Nauk SSSR,
    20:413-439.
Tzvelev, N. N., 1973, Conspectus specierum tribus Triticeae Dum.
    familiae Poaceae in Flora URSS, Novit. Syst. Plantarium
    Vascularium, 10:19-59.
Tzvelev, N. N., 1976, Tribe 3. Triticeae Dum., Pages 105-206, in:
    "Poaceae URSS," Nauka Publishing House, Leningrad.
Ullmann, W., 1936, Natural and artificial hybridization of grass
    species and genera, Herbage Reviews, (1936):105-142.
Veruschkine, S. M., 1936, The main lines of work with Triticum-
    Agropyron hybrids at the Saratov Station, Selekc. & Semenov,
    8:23-35.
Wagenaar, E. B., 1959, Intergeneric hybrids between Hordeum jubatum
    L. and Secale cereale L., J. Heredity, 50:195-202.
Wagenaar, E. B., 1960, The cytology of three hybrids involving
    Hordeum jubatum L.: The chiasma distributions and the
    occurrence of pseudo ring-bivalents in genetically induced
    asynapsis, Canad. J. Bot., 38:69-85.
Wang, R. C., Liang, G. H., and Heyne, E. G., 1977, Effectiveness of
    ph gene in inducing homoeologous chromosome pairing in
    Agrotricum, Theor Appl. Genet., 51:139-142.
White, W. J., 1940, Intergeneric crosses between Triticum and
    Agropyron, Sci. Agric., 21:198-232.
Zennyozi, A., 1963, F1 hybrids between four species of Secale and
    Agropyron intermedium, Wheat Inform. Serv., 15,16:30-31.

APPENDIX

Genomically based nomenclature and traditional nomenclature and common synonyms of perennial Triticeae species referred to in the text.

Genomically based nomenclature	Traditional nomenclature and common synonyms
AGROPYRON Gaertner (P genome)	
Agropyron cristatum (L.) Gaertner s. lat. (See Tzvelev, 1976, for subspecies)	Agropyron cristatum (L.) Gaertner
Agropyron desertorum (Fisch. ex Link) Schultes	Agropyron desertorum (Fisch. ex Link) Schultes
Agropyron fragile (Roth) Candargy	Agropyron fragile (Roth) Candargy
	Agropyron sibiricum (Willd.) Beauvois
PSEUDOROEGNERIA Löve (S genome)	
Pseudoroegneria geniculata (Trin.) Löve s. lat. (See Tzvelev, 1976, for subspecies)	Agropyron geniculatum (Trin.) C. Koch
	Elytrigia geniculata (Trin.) Nevski
Pseudoroegneria libanotica (Hackel) D. R. Dewey comb. nov., based on Agropyron libanoticum Hackel, 1904, Allgem. Bot. Zeitschr. 10:21	Agropyron libanoticum Hackel
	Elytrigia libanotica (Hack.) Holub
	Elymus libanoticus (Hack.) Melderis
Pseudoroegneria spicata (Pursh) Löve	Agropyron spicatum (Pursh) Scribner & Smith
	Elytrigia spicata (Pursh) D. R. Dewey
	Elymus spicatus (Pursh) Gould
Pseudoroegneria stipifolia (Czern. ex Nevski) Löve	Agropyron stipifolium Czernajew ex Nevski
	Elytrigia stipifolia (Czern. ex Nevski) Nevski
Pseudoroegneria strigosa (M. Bieb.) Löve s. lat. (See Tzvelev 1976 for subspecies)	Agropyron strigosum (M. Bieb.) Boissier
	Elytrigia strigosa (M. Bieb.) Nevski
Pseudoroegneria tauri (Boiss. & Bal.) Löve	Agropyron tauri Boissier & Balansa
	Elytrigia tauri (Boiss. & Bal.) Tzvelev

Genomically based nomenclature	Traditional nomenclature and common synonyms
PSATHYROSTACHYS Nevski (N genome)	
Psathyrostachys fragilis (Boiss.) Nevski	Hordeum fragile Boissier
	Elymus fragilis (Boiss.) Grisebach
Psathyrostachys juncea (Fisch.) Nevski	Elymus junceus Fischer
Psathyrostachys lanuginosa (Trin.) Nevski	Elymus lanuginosus Trinius
	Hordeum lanuginosum (Trin.) Schenk
CRITESION Rafinesque (H genome)	
*Critesion arizonicum (Covas) Löve	Hordeum arizonicum Covas
Critesion bogdanii (Wilensky) Löve	Hordeum bogdanii Wilensky
Critesion brachyantherum (Nevski) Barkworth & D. R. Dewey	Hordeum brachyantherum Nevski
Critesion brevisubulatum (Trin.) Löve (See Bothmer, 1979, for subspecies)	Hordeum brevisubulatum (Trin.) Link
Critesion californicum (Covas & Stebbins) Löve	Hordeum californicum Covas & Stebbins
Critesion capense (Thunb.) Löve	Hordeum capense Thunberg
*Critesion chilense (Roem. & Schult.) Löve	Hordeum chilense Roemer & Schultes
**Critesion depressum (Scribn. & Smith) Löve	Hordeum depressum (Scribn. & Smith) Rydberg
Critesion glaucum (Steud.) Löve	Hordeum glaucum Steudel
Critesion jubatum (L.) Nevski	Hordeum jubatum Linnaeus
*Critesion lechleri (Steud.) Löve	Hordeum lechleri (Steud.) Schenk
*Critesion marinum (Hudson) Löve	Hordeum marinum Hudson
Critesion murinum (L.) Löve	Hordeum murinum Linnaeus
Critesion muticum (K. Presl.) Löve	Hordeum muticum K. Presl
Critesion parodii (Covas) Löve	Hordeum parodii Covas
*Critesion procerum (Nevski) Löve	Hordeum procerum Nevski
Critesion pusillum (Nutt.) Löve	Hordeum pusillum Nuttall
Critesion secalinum (Schreb.) Löve	Hordeum secalinum Schreber
Critesion stenostachys (Godron) Löve	Hordeum stenostachys Godron
*=Annuals	

Genomically based nomenclature	Traditional nomenclature and common synonyms
THINOPYRUM Löve	(J-E genome)
Section Thinopyrum	
Thinopyrum bessarabicum (Savul & Rayss) Löve	Agropyron bessarabicum Savul. & Rayss Elytrigia bessarabica (Savul. & Rayss) Dubovik Elymus striatulus Runemark
Thinopyrum distichum (Thunb.) Löve	Agropyron distichum (Thunb.) P. Beauvois Elytrigia disticha (Thunb.) Prokudin ex Löve Elymus distichus (Thunb.) Melderis
Thinopyrum junceiforme (Löve & Löve) Löve	Agropyron junceiforme Löve & Löve Elytrigia junceiformis Löve & Löve Agropyron junceum ssp. boreoatlanticum Simonet & Guinochet
Thinopyrum junceum (L.) Löve	Agropyron junceum (L.) P. Beauvois Agropyron junceum ssp. mediterraneum Simonet Elytrigia juncea (L.) Nevski Elymus farctus (Viv.) Runemark ex Melderis Elymus diae Runemark
Thinopyrum runemarkii Löve	

Section Lophopyrum (Löve) D. R. Dewey comb. nov. based on Agropyron subgenus Elytrigia section Holopyron series Elongatae Nevski, 1934, Flora U.S.S.R., Vol. II:647 Synonym: Lophopyrum (Nevski) Löve sect. Lophopyrum

Thinopyrum curvifolium (Lange) D. R. Dewey, comb. nov., based on Agropyron curvifolium Lange, 1860, Pug. Impr. Hisp.:55	Agropyron curvifolium Lange Elytrigia curvifolia (Lange) Holub Elymus curvifolius (Lange) Melderis Lophopyrum curvifolium (Lange) Löve
Thinopyrum elongatum (Host) D. R. Dewey, comb. nov., based on Triticum elongatum Host, 1802, Gram. Austr. 2:18	Agropyron elongatum (Host) Beauvois Elytrigia elongata (Host) Nevski Elymus elongatus (Host) Runemark Lophopyrum elongatum (Host) Löve

Genomically based nomenclature	Traditional nomenclature and common synonyms
Thinopyrum ponticum (Podp.) Barkworth & D. R. Dewey	Agropyron elongatum ssp. ruthenicum Beldie Elytrigia pontica (Podp.) Holub Elymus elongatus ssp. ponticus (Podp.) Melderis Lophopyrum ponticum (Podp.) Löve
Thinopyrum scirpeum (K. Presl) D. R. Dewey, comb. nov., based on Agropyron scirpeum K. Presl, 1826, Fl. Sic.:49	Agropyron elongatum ssp. scirpeum (K. Presl) Ciferri & Giacomini Elytrigia scirpea (K. Presl) Holub Lophopyrum scirpeum (K. Presl) Löve
Section Trichophorae (Nevski) D. R. Dewey, comb. nov. based on Agropyron subgenus Elytrigia section Holopyron series Trichophorae Nevski, 1934, Flora U.S.S.R. Vol. II:648	
Thinopyrum gentryi (Melderis) D. R. Dewey, comb. nov., based on Agropyron gentryi Melderis, 1970, Flora Iranica 70:165	Agropyron gentryi Melderis Elytrigia gentryi (Melderis) Tzvelev Elytrigia intermedia ssp. gentryi (Melderis) Löve
Thinopyrum intermedium (Host) Barkworth & D. R. Dewey (See Tzvelev, 1976, for subspecies)	Agropyron intermedium (Host) Beauvois Elytrigia intermedia (Host) Nevski Elymus hispidus (Opiz) Melderis Agropyron glaucum (Desf. ex DC.) Roemer & Schultes
Thinopyrum podperae (Nabelek) D. R. Dewey, comb. nov., based on Agropyron podperae Nabelek, 1929, Publ. Fac. Sci. Univ. Masaryk, Brno, 3:24	Agropyron podperae Nabelek Elytrigia podperae (Nabelek) Holub Elytrigia intermedia ssp. podperae (Nabelek) Löve

Genomically based nomenclature	Traditional nomenclature and common synonyms
ELYTRIGIA Desvaux (SX genomes)	
Elytrigia elongatiformis (Drob.) Nevski	*Agropyron elongatiforme* Drobov
	Elytrigia repens ssp. *elongatiformis* (Drob.) Tzvelev
Elytrigia lolioides (Kar. & Kir.) Nevski	*Agropyron lolioides* (Kar. & Kir.) Candargy
	Elytrigia ciliolata (Nevski) Nevski
	Elymus lolioides (Kar. & Kir.) Melderis
Elytrigia pungens (Pers.) Tutin	*Agropyron pungens* (Pers.) Roemer & Schultes
(See Melderis, 1978, for subspecies)	*Elymus pungens* (Pers.) Melderis
	Agropyron campestre Grenier & Godron
Elytrigia pycnantha (Godr.) Löve	*Agropyron pycnanthum* (Godr.) Grenier
	Elymus pycnanthus (Godr.) Melderis
	Agropyron litorale Dumortier
Elytrigia repens (L.) Nevski	*Agropyron repens* (L.) P. Beauvois
(See Melderis, 1978, for subspecies)	*Elymus repens* (L.) Gould
ELYMUS Linnaeus (SHY genomes)	
Elymus agropyroides K. Presl	*Elymus agropyroides* K. Presl
Elymus alaskanus (Scribn. & Merr.) Löve	*Agropyron alaskanum* Scribner & Merrill
(See Löve, 1984 for subspecies)	*Roegneria borealis* (Turcz.) Nevski
Elymus asiaticus Löve	*Hystrix sibirica* (Trautv.) Kuntze
Elymus canadensis Linnaeus	*Elymus canadensis* Linnaeus
Elymus caninus (L.) Linnaeus	*Agropyron caninum* (L.) P. Beauvois
	Roegneria canina (L.) Nevski
Elymus ciliaris (Trin.) Tzvelev	*Agropyron ciliare* (Trin.) Franchet
	Roegneria ciliaris (Trin.) Nevski
Elymus drobovii (Nevski) Tzvelev	*Agropyron drobovii* Nevski
	Roegneria drobovii (Nevski) Nevski

Genomically based nomenclature	Traditional nomenclature and common synonyms
Elymus elymoides (Rafin.) Swezey	Elymus sitanion Shultes
	Elymus hystrix (Nutt.) M. E. Jones
	Sitanion hystrix (Nutt.) J. G. Smith
Elymus enysii (Kirk) Löve & Connor	Agropyron enysii Kirk
Elymus fibrosus (Schrenk) Tzvelev	Agropyron fibrosum (Schrenk) Candargy
	Roegneria fibrosa (Schrenk) Nevski
Elymus gayanus Desvaux	Elymus gayanus Desvaux
Elymus glaucus Buckley	Elymus glaucus Buckley
	Clinelymus glaucus (Buckl.) Nevski
Elymus lanceolatus (Scribn. & Smith) Gould	Agropyron dasystachyum (Hook.) Scribner
(See Dewey, 1983b, for subspecies)	Elytrigia dasystachya (Hook.) Löve & Löve
Elymus mutabilis (Drob.) Tzvelev	Agropyron mutabilis Drobov
(See Tzvelev 1976 for subspecies)	Roegneria mutabilis (Drob.) Hylander
Elymus patagonicus Spegazzini	Elymus patagonicus Spegazzini
Elymus scabrus (R. Br.) Löve	Agropyron scabrum R. Brown
Elymus scribneri (Vasey) M. E. Jones	Agropyron scribneri Vasey
Elymus sibiricus Linnaeus	Elymus sibiricus Linnaeus
	Clinelymus sibiricus (L.) Nevski
Elymus stebbinsii Gould	Agropyron parishii Scribner & Smith
	Elytrigia parishii (Scribn. & Smith) D. R. Dewey
Elymus tilcarensis (J. H. Hunziker) Löve	Agropyron tilcarense J. H. Hunziker
Elymus trachycaulus (Link) Gould ex Shinners	Agropyron trachycaulum (Link) Malte ex H. F. Lewis
(See Löve 1984 for subspecies)	Agropyron pauciflorum (Schwein.) A. Hitchccock
	Roegneria pauciflora (Schwein.) Hylander
	Elymus pauciflorus (Schwein.) Gould
	Agropyron tenerum Vasey
Elymus transhyrcanus (Nevski) Tzvelev	Agropyron transhyrcanum (Nevski) Bondarenko
	Roegneria transhyrcana Nevski
	Agropyron leptourum (Nevski) Grossheim
	Roegneria leptoura Nevski
	Elymus stewartii Löve

Genomically based nomenclature	Traditional nomenclature and common synonyms
Elymus tsukushiensis Honda	Agropyron tsukushiense (Honda) Ohwi
	Clinelymus tsukushiensis (Honda) Honda
Elymus yezoensis Honda	Agropyron yezoense Honda

LEYMUS Hochstetter (JN genomes)

Genomically based nomenclature	Traditional nomenclature and common synonyms
Leymus ajanensis (V. Vassil.) Tzvelev	Elymus ajanensis (V. Vassil.) Voroschilov
	Elymus interior Hultén
Leymus ambiguus (Vasey & Scribn.) D. R. Dewey	Elymus ambiguus Vasey & Scribner
Leymus angustus (Trin.) Pilger	Elymus angustus Trinius
	Aneurolepidium angustum (Trin.) Nevski
Leymus arenarius (L.) Hochstetter	Elymus arenarius Linnaeus
Leymus chinensis (Trin.) Tzvelev	Elymus chinensis (Trin.) Keng
	Agropyron chinense (Trin.) Ohwi
	Aneurolepidium chinense (Trin.) Kitagawa
	Elymus pseudoagropyron (Trin. ex. Griseb.) Turczaninov
Leymus cinereus (Scrib. & Merr.) Löve	Elymus cinereus Scribner & Merrill
Leymus condensatus (K. Presl) Löve	Elymus condensatus K. Presl
	Aneurolepidium condensatum (K. Presl) Nevski
Leymus innovatus (Beal) Pilger	Elymus innovatus Beal
Leymus mollis (Trin.) Pilger	Elymus mollis Trinius
Leymus salinus (M. E. Jones) Löve	Elymus salinus M. E. Jones
Leymus secalinus (Georgi) Tzvelev	Elymus dasystachys Trinius
(See Tzvelev, 1976 for subspecies)	Aneurolepidium dasystachys (Trin.) Nevski
	Elymus dasystachys (Trin.) Pilger
	Elymus secalinus (Georgi) Bobrov
	Aneurolepidium secalinum (Georgi) Kitagawa
Leymus triticoides (Buckl.) Pilger	Elymus triticoides Buckley

Genomically based nomenclature

Traditional nomenclature and common synonyms

PASCOPYRUM Löve (SHJN genomes)

Pascopyrum smithii (Rydb.) Löve

Agropyron smithii Rydberg
Elymus smithii (Rydb.) Gould
Elytrigia smithii (Rydb.) Nevski

EVOLUTIONARY RELATIONSHIPS AND THEIR INFLUENCE ON PLANT BREEDING

Gordon Kimber

Department of Agronomy
University of Missouri
Columbia, Missouri 65211

Clues providing insights into the evolutionary past may be found as patterns hidden in the morphology, cytology and biochemistry of today's species. The commonality of the structure and function of DNA in almost all species is such a pattern. The ubiquitous distribution of DNA provides, in itself, the most convincing argument for the validity of evolution, and at the same time, the basis for the subtle changes in genotype that fuel the processes of evolution.

Clearly, all methods employed for the investigation of the evolutionary past measure, either directly or indirectly, the similarity of the DNA of two or more species. The taxonomist in placing two species in the same genus or two genera in the same tribe or family is making judgements based on indirect measures of the DNA of the forms involved. It is assumed, in this example, that forms which differ but little from each other in their morphological characteristics also differ from each other by trivial changes in their DNA.

The erection of specific or generic rank on the basis of morphology is, perhaps, several orders removed from direct measurements of the similarity or dissimilarity of the various DNAs involved. Furthermore, changes at a single locus can produce such distinct phenotypes that specific rank could be established on the basis of change of a single nucleotide. Similar arguments can be made for other methods of investigating the evolutionary past.

*Contribution from the Missouri Agricultural Experiment Station. Journal Series Number 9413.

The ability to identify and order the nucleotide pairs in a gene provides the opportunity for the study of the evolution of a single gene with remarkable precision. The sequencing of the individual nucleotides throughout the entire length of the nuclear DNA of eukaryotic species could be the definitive method for determining evolutionary relationships. The elegance and precision of such techniques provides the most detailed insight into the arrangement of the building blocks of evolution, yet there are two types of associated difficulty. First, even the unequivocal characterization of a single gene currently represents a substantial labor. The effort of sequencing perhaps three billion nucleotide pairs in a single higher organism is immense. Second, even if such a task were feasible, the intractable problem of assigning evolutionary significance to the observed differences between the sequences of two individuals renders the undertaking an impractical exercise.

Just as nucleotide sequencing has associated advantages and difficulties as an evolutionary tool, so do all other methods which have been employed for the determination of species relationships. In general it is expected that the method which compares the greatest proportion of DNA will, on average, be the most reliable. Methods which are based on the comparison of a single locus or, at the best, a few loci will generally provide unequivocal but often conflicting results. Table 1 lists methods of determining evolutionary relationships in descending order of the proportion of the DNA compared. The actual order of some of the entries is arbitrary and may need alteration in specific cases. For example, the investigation of chromosome pairing in hybrids is a valuable method of determining species relationships in allopolyploid series but of limited value when only diploids are available for hybridization. Similarly, banding studies are practicable when there are many and distinct bands, but they are of less use when there are few bands.

Table 1. Methods for determining evolutionary relationships in
          arbitrary descending order of the proportion of the
          nuclear DNA which is compared.

> Total nucleotide sequencing
> Chromosome pairing in hybrids
> Comparative morphology
> Chromosome banding
> Karyotype analysis
> DNA hybridization
> DNA amount
> Protein electrophoresis
> Chromatography
> Immunological techniques
> Restriction enzyme analysis

Understanding the evolutionary relationships of cultivated
species has both intrinsic interest and possible practical value.
These two aspects are interrelated, for the theoretical studies
can provide the basis for the logical implementation of techniques
designed to allow the introduction of genetic material from related,
but wild species. In addition, the recognition of whether a cultivated species is diploid, auto- or allopolyploid profoundly influences the breeding methodology and also the expectations of improvement. For simplicity, diploids, auto- and allopolyploids will be
considered separately.

The determination of the species relationships of diploid
species is accompanied by difficulties not attendant to similar
studies involving polyploids. For example, chromosome pairing in
diploid hybrids cannot be considered a reliable indication of relationship. For chromosome pairing in hybrids to be a reliable measure of species relationships, there must be competition for synap-
sic partners in order that preferential pairing affinity may be
determined. If there is considerable chromosome pairing in an F_1
diploid hybrid it cannot be assumed that the genomes of the diploids
are identical or even closely related. The A- and D- genome diploid
cottons, for example, are geographically isolated and morphologically distinct, yet the hybrid between them (Endrizzi and Phillips,
1960) had a mean of 5.90 bivalents per cell and two cells of the
269 observed had 12 bivalents and two univalents. The absence of
chromosome pairing in diploid hybrids may be a better indication of
the divergence of the species than is frequent pairing an indication of a close relationship.

It is probable that the morphological characteristics of
diploids provide the best measure of their evolutionary relationships since, apart from the effects of single genes of profound
effect, the morphology will be the result of the interaction of
many loci. Consequently it will represent an examination of the
greatest proportion of the DNA possible. The order of the methods
of determining genomic relationships listed in Table 1, therefore,
should be altered, with chromosome pairing in hybrids being placed
much lower in the list.

The interpretation of the genomic relationships of diploid
species is greatly enhanced when there are associated allopolyploid
forms with which they can be hybridized. When only diploid species
are known in a group, the determination of evolutionary relationships can be tautological. The production of a hybrid is considered an indication of close relationships while the failure of hybrids
to occur is taken to show divergence. However, hybrid production
is affected by many factors that have little, if any, evolutionary
significance. This lack of clarity results in an often unclear
connection between evolutionary relationships and the breeding
methodology of diploid species.

The description of the introduction of alien genetical material into cultivated diploid species is also confused by semantics.
If it is possible to cross a cultivated diploid species with a wild
diploid relative and derive a fertile F_1, then the wild species
cannot be considered alien. Consequently there are few clear examples of the deliberate introduction of alien variation into cultivated diploid species, and even fewer where there is potential
commercial gain. The increase in heterosis of several *Zea mays*
hybrids to which small chromosomal segments of teosinte and *Tripsacum* had been added (Cohen et al., 1981) is probably the only such
case in a diploid, major crop plant.

The effects of evolutionary relationships on breeding methodology at the diploid level are clearly limited by the small number
of options available for cytogenetical manipulation in diploids
and also the difficulties associated with unequivocal determination
of genomic similarity. However, various possibilities do exist and
they can be turned to practical advantage. The anthropomorphic
"Compilospecies concept" of Harland and de Wet (1963) is a recognition of the ability of some species to assimilate genetical material during their evolution from other, related species. It is
suggested, for example, that *Z. mays* may have no obvious ancestor
because it is a compilospecies and has consumed its ancestral forms.
Thus, such a species would probably be more amenable to the introduction of genetical material from related species than other
species which did not fit the compilospecies concept. The introduction of genetical material to cultivated diploids may not be
limited to transfers from other diploids. Vardi (1973) has demonstrated experimentally the transfer of characters from cultivated
tetraploid wheats to wild diploid species. Therefore, the deliberate transfer of genetical material down a polyploid series to cultivated diploid forms is possible. This type of manipulation may
be expected to be more productive if the diploid is genomically
similar or identical to one of the genomes in the polyploid.

The evolutionary relationships of autopolyploids are usually
obvious; consequently the relationships will have little direct
effect on plant breeding methodology. Autopolyploidy *per se*
affects breeding methodology profoundly. Reductions of self or
cross fertility are common, complex segregation patterns are observed, the recovery of homozygous genotypes is more difficult,
and several commercial autopolyploid crops are vegetatively
propagated.

Mitigation of the undesirable effects of autopolyploidy occur
naturally and are attempted artificially. Approximation to regular
bivalent formation may be observed in the autohexaploid *Phleum
pratense* (Nordenskiold, 1945) while many autoploid species seem to
evolve towards some heterozygosity (Bingham, 1980). Doyle (1979a,
1979b, 1980), Gaul and Friedt (1975) and Sybenga (1973) have all

described attempts at the allopolyploidization of autopolyploids.

Consideration of the evolutionary relationships of autoploids in breeding methodology seems to be limited to the suggestion of Alonso and Kimber (1983c) regarding the recognition of patterns of preferential chromosome pairing as a method of recognizing the progress of diploidization. First, an autotetraploid tester should be made from a diploid line with good agronomic characteristics. Hybrids between this line and a series of inbred diploids must then be made. The meiotic chromosome pairing of this series of auto-triploids should be analysed and values of the relative pairing affinity (Alonso and Kimber, 1981), should be calculated. The diploid line involved in the triploid hybrid that gives the highest value of relative affinity should be hybridized with the diploid used to produce the autotetraploid tester and an autotetraploid should be made from this hybrid. It is expected that this autotetraploid would tend to have regular bivalent formation and good fertility. Alternatively, the selected diploid line may be hybridized with the triploid involving the same selection. Kimber and Alonso (1981) have pointed out that the 2:2 pattern of pairing expected in such a derived autotetraploid will result in a low frequency of multivalents, even when the value of the relative effinity is as low as 0.8. Consequently it is expected that substantial progress towards diploidization may be made.

Almost by definition, the evolutionary relationships of allo-polyploids are clearly recognized. Consequently, their influence on plant breeding is easier to document and comprehend. The generally regular chromosome pairing and segregation resulting from the differential affinity, whether genetically enhanced or not, of the chromosomes of the divergent genomes of allopolyploids leads to the conventional breeding methodology of intraspecific hybridization, selfing and selection. The recognition of evolutionary connections mainly affects the techniques for the manipulation of chromosomes, chromosome segments or individual loci which may be introduced from alien, but related, species.

The order of methods for determining species relationships in Table 1 is, perhaps, most applicable to allopolyploid species. The conceptual limitations on the interpretation of the data and the impracticality of sequencing nucleotides for the entire nuclear genome of a sufficiently large number of replicates of the putative parents and their derived amphiploids, places chromosome pairing as the most practical and, arguably, the most reliable method of determining evolutionary relationships in allopolyploid series.

The determination of evolutionary relationships on the basis of chromosome pairing in hybrids was first accomplished by Rosenberg (1909) in the genus *Drosera*. The method, termed "genomic analysis", was elegantly and exhaustively employed by Kihara and

his colleagues (see review by Lilienfeld, 1951) and led to an essentially correct picture of the evolutionary and species relationships
in the genus *Triticum* as defined by Morris and Sears (1967). The
generality of the technique has led to its adoption in studies of
many other allopolyploid groups with little modification.

The ideal process of genomic analysis is the production of
triploid hybrids between allotetraploids and their putative diploid
ancestors. The formation of the basic number of bivalents in the
triploid hybrid is taken as an indication of genomic homology.
Thus, this form of genomic analysis is based essentially on measurements of the total amount of chromosome pairing per cell. The determination of genomic homology becomes more difficult when there
are not exactly the basic number of bivalents, when multivalents
are observed, or when tetraploid or high ploidy hybrids are analysed. Generally, reductions in total pairing are assumed to indicate some differentiation of otherwise identical genomes, multivalents are taken to demonstrate residual homology or translocation
heterozygosity. High polyploid hybrids are more equivocal.

Recently, numerical methods for the analysis of chromosome
pairing in hybrids have been developed (Kimber, Alonso and Sallee,
1981; Alonso and Kimber, 1981; Kimber and Alonso, 1981; Espinasse
and Kimber, 1981; Kimber and Alonso, 1983; Alonso and Kimber, 1983a,
b, c, d). The essential features of these techniques are the calculation of a measure of how often the chromosomes pair (mean arm-
pairing frequency, c) and a measure of the similarity of two, or
more, of the genomes present (relative affinity, x). The value of
c is obtained from the frequencies of the observed meiotic figures
(Alonso and Kimber, 1981) and it ranges from 0.0 when there is no
chromosome pairing at all, to 1.0 when every possible arm is paired
in every cell. The expected meiotic figures are calculated on the
basis of various assumptions of synapsis, chiasmata formation and
the relative affinity of the genomes present (Alonso and Kimber,
1981). The relative affinity (x) ranges from 0.5, when all the
genomes are equally related to each other, to 1.0 when two, or more,
genomes pair to the exclusion of all other genomes. These assumptions and definitions result in various models of chromosome pairing
at increasing levels of ploidy. At the triploid level two models
are expected: one, the 3:0 model where all three genomes are equally
related and, two, the 2:1 model where two of the genomes are more
closely related to each other and equally, and more distantly related to the third. As a consequence of similar assumptions at the
tetraploid level, four models are considered, and they are designated 4:0, 2:2, 2:1:1 and 3:1; while at the pentaploid level the
six models are designated 5:0, 2:2:1, 2:1:1:1, 3:2, 3:1:1 and 4:1.
The optimum value of the relative affinity is calculated (by minimum sums of squares of differences between observed and expected
pairing) for each of the appropriate models. The model which fits
the observed data best, its associated value of x, and the observed

value of c, is taken to describe the evolutionary relationships of
the species involved.

Numerical genomic analysis is not simply a quantification of
genomic analysis based on the amount of chromosome pairing, but
represents a measure of the importance of the pattern of chromosome
pairing in recognizing evolutionary relationships. The patterns
of chromosome pairing in hybrids can be represented by the diagrams
of Figure 1. It is difficult, if not impossible, to represent the
equality of the relative affinity of the most distantly related
chromosomes in some models, and of the most closely related chromosomes in others, by lines of equal length. The designation of the
model (3:1, 2:1 etc) and the value of x determine the pattern of
chromosome pairing; while the value of c is a measure of the amount
of chromosome pairing. The hybrids represented by the diagrams of
Figure 2 would both have identical values of x, but the value of c
would be larger in a) and in b). Thus, the pattern of pairing
would be the same but the amount of synapsis would differ. Similar
figures could be constructed for any of the other models.

Studies of chromosome pairing in hybrids involving species of
the genus *Triticum* have led to the reassignment of some genome
symbols and to the recognition of a clear need for the production
of new hybrids to allow the investigation of various incomplete or
dubious genomic designations (Kimber, 1983a). The recognition of
patterns of chromosome pairing as an indication of evolutionary relationships seems to work in species other than wheat relatives
(Alonso and Kimber, 1981; Kimber and Alonso, 1981; Espinasse and
Kimber, 1981). Further, the recognition of the patterns of chromosome pairing and measurement of the value of c now allows a logical
choice of the method of introduction of alien variation (Kimber,
1983b), and thus the evolutionary relationships of the species has
the most influence on the breeding methodology that is employed.

Perhaps the best-documented evolutionary relationships are to
be found in the wheat group, and their influence on technique selection for the introduction of alien variation will be taken as typical. Three groups of hybrids can be recognized. First, tetraploid
hybrids with very low values of c, irrespective of the value of x.
In most of the hybrids of this type the alien species will be
genomically very remote from wheat and induced homoeologous chromosome pairing will not allow sufficiently frequent recombination
between the chromosomes of the alien genome and the wheat genomes
to be of practical value. For example, Kimber and Alonso (1981)
have demonstrated the 3:1 pairing pattern for the hybrid between
wheat and rye (*Secale cereale*) when chromosome 5B was missing. It
is probable that the group of three genomes was the A, B and D from
wheat and the isolated genome was the rye. Even though the value of
c was raised to 0.375 in this hybrid, the value of x was 0.963 and
thus little pairing would occur between the wheat and rye chromosomes.

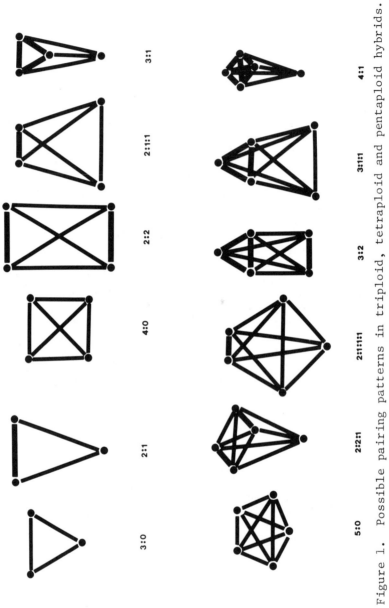

Figure 1. Possible pairing patterns in triploid, tetraploid and pentaploid hybrids.

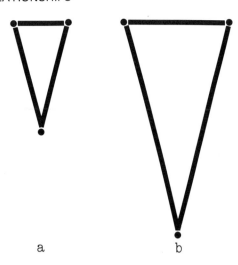

Figure 2.   The pairing patterns in two triploid hybrids with the
             same value of relative affinity but a) with a high
             value of mean arm-pairing frequency and b) with a low
             value of mean arm-pairing frequency.

The recognition of this very diverse evolutionary relationship leads
to the conclusion that the most practical method for the introduction of alien variation from a hybrid of this type is either by
irradiation of, or centromeric break and fusion in, derivatives of
its amphiploid. The work of Sears (1956, 1972), Driscoll and
Jensen (1963), Kimber (1971) and Lukasewski and Gustafson (1983)
may be quoted as typical of the various published techniques.

     Second, tetraploid and pentaploid hybrids with low values of
c which, if homoeologous synapsis is induced, fit those models
which would indicate frequent chromosome pairing between the wheat
and alien chromosomes. The series of hybrids between _T. cestivum_
and _T. kotschyi_ with various levels of chromosome pairing described
by Sears (1982) would be typical. This set of hybrids is particularly interesting as the best-fit model (3:2) is the same, independently of the level of chromosome pairing (Kimber, 1983b). While
this may be fortuitous at the lowest levels of pairing, the regu-
larity of the fit to the same model at higher levels of pairing
has significance both for the introduction of alien variation and
in understanding the processes regulating chromosome pairing. The
fit to the 3:2 model cannot indicate which genomes are pairing in
this pattern. However, Alonso and Kimber (1983b) have shown, using
chromosomes marked by being telocentric, that the relationship between the B and S genomes is similar to that between the A and D
genomes in tetraploid hybrids involving S-genome diploids. It may
be reasonable to assume similar relationships in the hybrid with

T. kotschyi, which is genomically ABDUS. Two equally probable
patterns of pairing can be envisioned: ADU:BS or BSU:AD. The first
pattern would have a greater probability for the introduction of
alien variation, for the two alien genomes (S and U) are in different clusters. The induced recombinant "Compair" produced by Riley,
Chapman and Johnson (1968) may be quoted as a typical product of
this type of hybrid.

Third, those hybrids in which the high values of x and c of
the best-fit model indicate true homology. Here the introduction
of the alien variation is presumably the simplest. It is not essential to irradiate or to induce homoeologous recombination, as normal
recombination alone will allow the recovery of the desirable genotype. Several possible practical techniques allowing the introduction of desirable genetic material are available for utilization
in this type of situation. Pentaploid hybrids between a hexaploid
and its tetraploid progenitor are often fertile and may be crossed
directly back to the hexaploid. Here the genome found only in the
hexaploid will not be reorganized, as it is hemizygous in the hybrid, and recombination will be limited to the genomes introduced
from the tetraploid species. Avivi (1979) has suggested the possible introduction of the high grain-protein content of some tetraploid wheats into bread wheat through this type of hybrid. The
production of a tetraploid hybrid between the hexaploid and one of
its diploid progenitors opens the possibility of at least three
variations on the process of introduction. First, the hybrid can
be colchicined and an amphiploid produced. This may be selfed or
back-crossed to the hexaploid. Only the introduced diploid genome
will be free to recombine; the other two should be homozygous.
Second, if the diploid parent carried the genome found only in the
hexaploid, then the hybrid could be backcrossed with the tetraploid
progenitor. The production of an unreduced gamete in the hybrid
would allow a hexaploid to be reconstituted without the necessity
of colchicine treatment. Worstell (1982) has described this method
for the introduction of alien variation from tetraploid into hexaploid wheat. All three genomes would undergo recombination in this
type of crossing program. Third, the hybrid may be pollinated by
the hexaploid parent. If the two homologous genomes in the hybrid
were to pair and segregate normally and most of the remaining
chromosomes were to be recovered in the gamete, then an almost
hexaploid derivative would be obtained. Alonso and Kimber (1983d)
have described this type of crossing program for hybrids between
T. aestivum and *T. tauschii*. In this case only the D genome would
be recombined, with the first back-cross derivative having 11/12
of the genotype of the recurrent parent.

The ability to perceive and measure the evolutionary relationships of the wild relatives of cultivated species provides the
basis for influencing breeding methodology and for the development
of techniques for the introduction of alien variation. The poten-

tial tends to increase, in order, from tautological limitations in
the case of cultivated diploids, through some possibilities of
diploidization of autopolyploids, to the logical choice of various
methods for the introduction of alien variation to allopolyploids.
The theoretical and practical importance of recognizing evolutionary relationships are related. The ability to look back on the
events of the evolutionary past is a fascinating intellectual exercise; the practical application of the conclusions that may be
drawn is an activity of potential human and commercial significance.
James (1981) points out that the total cost to world agriculture
of losses caused by disease is some $50 billion per annum. The
prevention of worse losses or the reduction of the current figure
by plant breeding methodology based on an understanding of evolutionary relationships could save many from starvation and provide
a financial reward for all.

SUMMARY

The evolutionary relationships of plant species may be measured in various ways. The choice of method depends, to some degree, on the ploidy relationships of the species being investigated.
In general, the best method is that which compares the largest proportion of the nuclear DNA. The recognition of evolutionary relationships can provide a basis for both the normal breeding methodology and for the selection of techniques for the introduction of
alien variation. The identification of evolutionary relationships
in diploids, auto- and allopolyploids is discussed, and the effect
of these relationships on plant breeding methodology is described.

LITERATURE CITED

Alonso, L. C., and Kimber, G., 1981, The analysis of meiosis in
    hybrids. II. Triploid hybrids, Can. J. Genet. Cytol., 23:221-
    234.
Alonso, L. C., and Kimber, G., 1983a, A study of genomic relation-
    ships in wheat based on telocentric chromosome pairing. I,
    Z. Pflanzenzüchtg., 90:23-31.
Alonso, L. C., and Kimber, G., 1983b, A study of genomic relation-
    ships in wheat based on telocentric chromosome pairing. II,
    Z. Pflanzenzüchtg., 90:273-284.
Alonso, L. C., and Kimber, G., 1983c, Preferential chromosome pair-
    ing in trisomics, Z. Pflanzenzüchtg., In press.
Alonso, L. C., and Kimber, G., 1983d, A new method for the intro-
    duction of alien variation in wheat, Z. Pflanzenzüchtg., In
    press.
Avivi, L., 1979, Utilization of Triticum dicoccoides for the im-
    provement of grain protein quality and quantity in cultivated
    wheats, Monografia Genetica Agraria, 4:27-38.
Bingham, E. T., 1980, Maximizing heterozygosity in autopolyploids,
    in: "Polyploidy: Biological Relevance," W. H. Lewis, Ed.,

Plenum, New York, Pp. 471-489.

Cohen, J. I., Galinat, W. C., and Pasupuleti, C. V., 1981, Evaluation of exotic germplasm for crop improvement, Agr. Ags., p. 81.

Doyle, G. G., 1979a, The allotetraploidization of maize. Part 1: the physical basis - differential pairing affinity, Theor. Appl. Genet., 54:103-112.

Doyle, G. G., 1979b, The allotetraploidization of maize. Part 2: the physical basis - the cytogenetics of segmental allotetraploids, Theor. Appl. Genet., 54:161-168.

Doyle, G. G., 1980, The allotetraplodization of maize, Part 3: Gene segregation in trisomic heterozygotes, Theor. Appl. Genet., 61:103-112.

Driscoll, C. J., and Jensen, N. F., 1964, A genetic method for detecting intergeneric translocations, Genetics, 48:458-468.

Endrizzi, J. E., and Phillips, L. L., 1960, A hybrid between *Gossypium arboreum* L. and *G. raimondii* Ulb, Can. J. Genet. Cytol., 2:311-319.

Espinasse, A., and Kimber, G., 1981, The analysis of meiosis in hybrids. IV. Pentaploid hybrids, Can. J. Genet. Cytol., 23: 627-638.

Gaul, H., and Friedt, W., 1975, Progress in the diploidization of autotetraploid barley, Barley Genetics III, Proc. Third Int. Barley Genet. Symp., Garching, Pp. 378-387.

James, C., 1981, The cost of disease to world agriculture, Seed Sci. Tech., 9:679-685.

Lukaszewski, A. J., and Gustafson, J. P., 1983, Translocation and modifications of chromosomes in Triticale x Wheat hybrids, Theor. Appl. Genet., 64:239-248.

Kimber, G., 1971, The design of a method, using ionising radiation, for the introduction of alien variation into wheat, Indian J. Genet. Pl. Breed., 31:580-584.

Kimber, G., 1983a, Genome analysis in the genus *Triticum*, Proc. 6th Int. Wheat Genet. Symp., Kyoto, Japan, In Press.

Kimber, G., 1983b, Technique selection for the introduction of alien variation in wheat, Z. Pflanzenzüchtg., In press.

Kimber, G., and Alonso, L. C., 1981, The analysis of meiosis in hybrids, III. Tetraploid hybrids, Can. J. Genet. Cytol., 23: 235-254.

Kimber, G., and Alonso, L. C., 1983, The pairing of telocentric chromosomes in triploids and trisomics, Genetics, In press.

Kimber, G., Alonso, L. C., and Sallee, P. J., 1981, The analysis of meiosis in hybrids. I. Aneuploid hybrids, Can. J. Genet. Cytol., 23:209-219.

Lilienfeld, F. A., 1951, H. Kihara: Genome-analysis in *Triticum* and Aegilops, concluding review, Cytologia, 16:101-123.

Morris, R., and Sears, E. R., 1967, The cytogenetics of wheat and its relatives, in: "Wheat and wheat improvement," Reitz, L. P., and Quisenberry, K. S., Eds., Amer. Soc. Agron. Monogr., Madison, Wisconsin, Pp. 19-87.

Nordenskiold, H., 1945, Cytogenetic studies in the genus *Phleum*, Acta. Agri. Suecana, 1:1-138.

Rosenberg, O., 1909, Cytologische und morphologische Studien an *Drosera longifolia* x *rotundifolia*, Kungl. Sv. Vetensk. Acad. Handl., 43:1-64.

Riley, R., Chapman, V., and Johnson, R., 1968, The incorporation of alien disease resistance in wheat by genetic interference with the regulation of meiotic chromosome synapsis, Genet. Res. Camb., 12:199-219.

Sears, E. R., 1956, The transfer of leaf-rust resistance from *Aegilops umbellulata* to wheat, Brookhaven Symp. Biol., 9:1-22.

Sears, E. R., 1972, Chromosome engineering in wheat, Stadler Genet. Symp., G. P. Redei, Ed., Univ. Missouri, 4:25-38.

Sears, E. R., 1982, A wheat mutation conditioning an intermediate level of homoeologous chromosome pairing, Can. J. Genet. Cytol., 24:715-719.

Sybenga, J., 1973, Allopolyploidization of autoploids. 2. Manipulation of the chromosome pairing enzyme, Euphytica, 22:433-444.

Worstell, J. V., 1982, High seed-protein from wild emmer wheat: methods for introduction and chromosome substitution, Ph.D. Thesis, Univ. Missouri-Columbia.

MUTATIONS IN WHEAT THAT RAISE THE LEVEL OF MEIOTIC CHROMOSOME PAIRING*

E. R. Sears

Department of Agronomy
University of Missouri
Columbia, Missouri 65211

That chromosome 5B of wheat suppresses the pairing of homoeologous (related) chromosomes was discovered by Okamoto (1957) and Sears and Okamoto (1958) in Missouri and independently by Riley and Chapman (1958) in England. This discovery answered the question why the chromosomes of the three genomes of wheat are unable to pair with each other even though they are genetically so closely related that extra dosage of the right chromosome can compensate substantially for the complete absence of a particular chromosome of either of the other two genomes. Also, the discovery of the 5B effect suggested a way of inducing the chromosomes of wheat to pair with those of other species and genera; namely, by deleting 5B or the gene concerned.

Although the possibility already existed of inducing homoeologous pairing in wheat by incorporating the chromosomes of *Triticum (Aegilops) speltoides* or by using nullisomic-5B tetrasomic-5D (or -5A), it appeared that a mutation of the gene concerned would be of substantial value. Accordingly, Okamoto (1966) attempted to produce such a mutation. He x-rayed emasculated spikes, pollinated them with rye, and observed the resulting hybrid plants cytologically. Although 3.4% of the hybrids had high pairing, presumably as the result of deletion of the 5B pairing gene, the high-pairing plants were so nearly sterile that he was unable to recover any of the mutations. In 1966 I repeated Okamoto's experiment and obtained a similar rate of mutation (Sears, 1977). I was also unable to recover any of the several

*Contribution from the Missouri Agricultural Experiment Station.
 Journal Series Number 9490.

mutations produced. In retrospect, both Okamoto and I were
probably fortunate, for it is likely that all of the mutations
involved long deficiencies, in each case including, besides the
pairing gene, a gene or genes essential for normal male fertility.

A mutation (numbered 10/13) that conditions an increased
level of homoeologous pairing was induced with ethyl methanesulfonate
(EMS) by Wall et al. (1971a). Two lines of evidence indicated that
it was a mutation of the pairing suppressor on the long arm of
chromosome 5B: (1) The 5B gene was the only pairing suppressor
known at the time, and (2) results obtained from monosomic analysis
appeared to confirm its location in 5B (Wall et al., 1971b). The
mutation was designated *ph* (for *pairing homoeologous*). The level
of pairing conditioned by the 10/13 mutation was substantially
lower than that caused by deficiency for chromosome 5B. Either the
EMS treatment had only partially inactivated the pairing gene, or
there are two loci involved, only one of which was inactivated.

Because 10/13 does not condition the high pairing level desired
for use in transferring alien genes to wheat, an experiment was
undertaken (Sears, 1977) to induce a deficiency for the *Ph* gene or
genes. The experiment involved x-irradiating pollen and using it
on emasculated spikes of plants that had a genetically marked 5B
monosome. The short arm of this monosome was largely or wholly
replaced by a segment of rye chromosome 5R carrying the gene *Hp*
(*Hairy peduncle*). Thus, those offspring that received a maternal
5B could be identified by their hairy necks, and they were discarded.
The remainder had either a paternal, irradiated 5B or none at all.
The latter were identifiable as aberrant seedlings and were
discarded, and the rest were grown to flowering. Those that were
male-fertile were used as males in crosses with *T. kotschyi (Ae.
variabilis)*. Many were infertile, presumably because of loss of
essential fertility genes located on chromosome 5B. Only one high-
pairing mutation was obtained in the 910 irradiated gametes tested
(Sears, 1977). It was given the designation *ph1b*. Also obtained
was a mutation to an intermediate level of pairing, which proved to
be located on the short arm of 3D and was designated *ph2* (Sears, 1982).

When a cross was made to combine 10/13 and *ph1b*, and the F_1
was tested in a cross with *T. kotschyi*, there was segregation for
pairing level, 31 being high-pairing, 7 intermediate-pairing, and
9 low-pairing. This was a reasonable fit to the 2:1:1 ratio
expected if the two mutations were at non-linked loci and if high
pairing were epistatic to lower pairing. The two loci could be on
chromosome 5B, 50 or more units apart, or they could be on different
chromosomes, contrary to the previous evidence.

There was other evidence that there might be two pairing
suppressors on 5BL: Mello-Sampayo (1972) had transferred a terminal
segment of 5DL to 5BL, presumably substituting it for a homoeologous

segment of 5BL, and observed a reduction in the effect of the arm on pairing. This suggested that a gene responsible for part of the *Ph1* effect had been removed and replaced by a less-potent 5D homoeoallele.

Another test of the 10/13 mutation was made, consisting of a cross of 10/13 with monoisosomic 5BL, the selection of a monosomic F_1 plant, and a cross of this plant (as male) with *T. kotschyi*. Of the eight offspring examined at MI, two were low-pairing. This was not possible if the gene (or genes) concerned was on chromosome 5B; all should have been intermediate or, if the 5B monosome had failed to be transmitted, high-pairing.

Subsequent to the production of the 10/13 mutation, Mello-Sampayo (1967, 1968, 1971), Upadhya and Swaminathan (1967), Driscoll (1973), and Mello-Sampayo and Canas (1973) discovered additional suppressors of homoeologous pairing, substantially less effective than *Ph1*, on the short arms of chromosomes 3A and 3D, and on chromosome 4D; and a deficiency for the 3DS locus was recovered in my mutation experiment, as mentioned above. Since the level of pairing associated with this mutation was about the same as with 10/13, the two were crossed and the F_1 pollinated by *T. kotschyi*. No segregation was found, the entire hybrid population of 20 showing intermediate pairing.

These experiments left very little room for doubt that the 10/13 mutation involved the *Ph2* locus rather than *Ph1*. The only escape from this conclusion was if the 10/13 line had become contaminated by *ph2*, so that the hybrids supposedly carrying 10/13 actually had *ph2*. Fortunately, this possibility could easily be checked, for *ph2* involves such a long deficiency that chromosomes carrying it do not pair with telo-3DS (Sears, 1982). Therefore, it was only necessary to go back to the 10/13 x *ph2* F_1, grow a few F_2 plants, pollinate each with ditelosomic-3DS, and look among the offspring for regular pairing of the telosome. When this was done, one plant proved to be segregating for pairing *vs.* non-pairing with the telosome, and two of the three others were either homozygous or heterozygous for pairing (all examined offspring—two and one, respectively—showing pairing with the telosome). Thus there is no longer any question that 10/13 involves the *Ph2* locus on 3DS, and it should be designated *ph2b*. From a practical point of view, it should be a more useful mutation than *ph2a* (formerly *ph2*). Because it involves either an intragenic change or a relatively short deficiency, it has a less detrimental effect on vigor and fertility than does *ph2a*.

Since the just-mentioned plants with a heteromorphic bivalent were heterozygous for the *ph2b* mutation, they constituted suitable material for a test of linkage between *ph2b* and the centromere. Accordingly, two of them were pollinated by *T. kotschyi*, and 123

hybrid offspring were classified for level of pairing (Table 1).
There were 3 crossovers and 117 non-crossovers, with the remaining
3 hybrids being deficient for the chromosome concerned and thus
non-classifiable. The rate of crossing over was 3/120, or 2.5%.

Linkage of *ph1b* with the centromere was also determined,
using a plant (kindly provided by Dr. Carla Ceoloni) which was
heteromorphic for a 5BL telosome carrying *Ph1* and a complete 5B
chromosome with *ph1b*. Of 110 offspring from a cross with *T.
kotschyi*, only one was a crossover (Table 1). Thus the rate of
recombination was 0.9%.

Another high-pairing mutation was induced by irradiation of a
tetraploid cultivar (Giorgi, 1978). This mutation is evidently a
short deficiency for *Ph1*, located near the middle of the 5BL arm
(Dvořák and Chen, pers. comm.). Such a location is not incompatible
with the close linkage found for *ph1b*, for there is evidence that
proximal crossing-over is infrequent when a telosome is involved
(Sears, 1972; Dvôrák and Chen, 1984).

Table 1. Classification of offspring from crosses of *Ph2ph2b* and
Ph1ph1b x *Triticum kotschyi*, in which *Ph2* and *Ph1* were
located on telosomes and the mutant alleles on complete
chromosomes.

Locus concerned	Level of pairing	Type of chromosome present		
		Complete	Telocentric	None
Ph2	Low	0	58	0
	Inter-mediate	59	3	3
Ph1	Low	1	56	0
	High	53	0	0

The close linkage of *ph1b* to the 5B centromere should make it
possible to use the centromere as a fairly reliable marker for the
gene. It will first be necessary, however, to recover a 5BL telosome carrying *ph1b*. Unfortunately, the only recombinant obtained
in the linkage experiment was one in which *Ph1* had been transferred
to the complete chromosome, rather than *ph1b* to the telosome.
However, it should not be difficult to recover a 5BL telosome
among offspring of monosomic 5B carrying *ph1b*.

As noted in a previous publication (Sears, 1982), an inter-mediate-pairing mutation such as *ph2a* or *ph2b* should sometimes be useful for inducing the transfer of alien genes to wheat. This might be the case when the desired gene was carried by an alien chromosome so closely related to its wheat homoeologues as to be able to pair with them at a level lower than that provided by *ph1b*. Under the influence of *ph2a* or *ph2b* the alien chromosome might then pair and recombine with its wheat homoeologues while a negligible amount of recombination was occurring among the A-, B-, and D-genome chromosomes themselves.

If a pairing level lower than that induced by *ph2a* or *ph2b*, but higher than the control level, were desired, it could be pro-vided by deficiency for the pairing suppressor on 3A or 4D. However, no effort to induce a mutation of either of these genes is known to be under way.

No interest has been shown in mutations of any of the several genes that promote pairing in wheat. Simple deficiency for one of these loci would lead to a reduced pairing level, which might be useful in amphiploids or other lines in which homoeologous pairing would otherwise occur and give rise to chromosome aberrations.

Although it would be of value to have a mutation, or combina-tion of mutations, that induced a higher level of pairing than *ph1b* does, the prospects for this are not good. Mello-Sampayo and Canas (1973) were unable, by adding 3DS deficiency, to increase the pairing in wheat X *T. sharonense* over that caused by 5BL deficiency alone. Riley and Law (1965), however, found that the highest level of pairing in 5B-deficient hybrids of wheat with various diploid relatives was achieved when *T. tripsacoides* (Ae. *mutica*) or *T. speltoides* was involved. Probably the pairing-promotive effect of these two species slightly augmented the effect of 5B deficiency. It appears that there is a ceiling for homoeo-logous pairing, and that *Ph1* deficiency is sufficient, or almost sufficient, to reach that level. Nevertheless, there remains a possibility that more pairing of wheat chromosomes with distantly related alien chromosomes, such as rye, might be obtained by combining a *ph2* mutation with *ph1b*.

LITERATURE CITED

Dvořák, J., and Chen, K.-Y., 1984, Distribution of nonstructural variation between wheat cultivars along chromosome arm 6Bp: Evidence from the linkage map and physical map of the arm, Genetics, 106:325-333.
Driscoll, C. J., 1973, Minor genes affecting homoeologous pairing in hybrids between wheat and related genera, Genetics, 74:s566.
Giorgi, B., 1978, A homoeologous pairing mutant isolated in *Triticum*

durum cv. Cappelli, Mut. Breed. Newsl., 11:4-5.

Mello-Sampayo, T., 1967, Emparelhamento cromosómico na neiose de híbridos pentaploides de trigo, Protug. Acta Biol. (ser. A), 10:109-128.

Mello-Sampayo, T., 1968, Homoeologous chromosome pairing in pentaploid hybrids of wheat, Proc. 3rd Int. Wheat Genet. Symp., Canberra, pp. 179-184.

Mello-Sampayo, T., 1971, Genetic regulation of meiotic chromosome pairing by chromosome 3D of *Triticum aestivum*, Nature New Biol., 230:22-23.

Mello-Sampayo, T., 1972, Compensated monosomic 5B-trisomic 5A plants in tetraploid wheat, Can. J. Genet. Cytol., 14:463-475.

Mello-Sampayo, T., and Canas, A. P., 1973, Suppressors of meiotic chromosome pairing in common wheat, in: "Proc. 4th Int. Wheat Genet. Symp.," Sears, E. R. and Sears, L. M. S., eds., Columbia, pp. 709-713.

Okamoto, M., 1957, Asynaptic effect of chromosome V, Wheat Inf. Serv., 5:6.

Okamoto, M., 1966, Studies on chromosome 5B effects in wheat, Proc. 2nd Int. Wheat Genet. Symp., 1963 (Hereditas Suppl. 2):409-417.

Riley, R., and Chapman, V., 1958, Genetic control of the cytologically diploid behaviour of hexaploid wheat, Nature, 182: 713-715.

Riley, R., and Law, C. N., 1965, Genetic variation in chromosome pairing, Adv. Genet., 13:57-114.

Sears, E. R., 1972, Reduced proximal crossing-over in telocentric chromosomes of wheat, Genet. Iber., 24:233-239.

Sears, E. R., 1977, An induced mutant with homoeologous pairing in common wheat, Can. J. Genet. Cytol., 19:585-593.

Sears, E. R., 1982, A wheat mutation conditioning an intermediate level of homoeologous chromosome pairing, Can. J. Genet. Cytol. 24:715-719.

Sears, E. R., and Okamoto, M., 1958, Intergenomic chromosome relationships in hexaploid wheat, Proc. X Int. Congr. Genet., Montreal, 2:258-259.

Upadhya, M. D., and Swaminathan, M. S., 1967, Mechanisms regulating chromosome pairing in *Triticum*, Biol. Zentralbl., 86 (Suppl.): 239-255.

Wall, A. M., Riley, R., and Chapman, V., 1971a, Wheat mutants permitting homoeologous meiotic chromosome pairing, Genet. Res., 18:311-328.

Wall, A. M., Riley, R., and Gale, M. D., 1971b, The position of a locus on chromosome 5B of *Triticum aestivum* affecting homoeologous meiotic chromosome pairing, Genet. Res., 18:329-339.

CHROMOSOME MANIPULATION IN PLANT BREEDING:

PROGRESS AND PROSPECTS

Ralph Riley   Colin N. Law

Agricultural and Food   Plant Breeding Institute
Research Council   Cambridge, CB2 2LQ
London, WlN 6DT   England
England

SUMMARY

Cytogeneticists and breeders have for long engaged in the
intraspecific manipulation of chromosomes using aneuploids in
genetic analysis and monosomics and nullisomics to transfer intact
pairs of chromosomes between parental genotypes. Similarly intact
pairs of chromosomes have been transferred from donor to recipient
cross compatible species to make alien chromosome addition or substitution lines. Much research effort has been applied to increasing the precision with which the manipulated chromosome can be
identified including chromosome structural distinction by gross
morphology, by banding or by molecular probing. In well explored
groups, such as the wheat group, biochemical genetic markers can be
exploited in chromosome recognition. Emphasis is placed on the
precise match of the manipulation undertaken, particularly in the
introduction of alien variation, to be optimisation of the phenotypic expression being sought. Finally, a grave limitation is
placed on the fullest exploitation of chromosome manipulation as
well as on plant genetic engineering by the limited knowledge available on the processes of gene expression in plants and on the intermediary metabolism of plant processes.

INTRODUCTION

Occasionally in genetic studies or in plant breeding there is
advantage in manipulating not single genes but large arrays of
associated genes. Such arrays may consist of all of the genes in
a set of chromosomes, or on a single chromosome, or on a substantial-

segment of chromosome. The experimental manipulation of entire
sets of chromosomes may be of three kinds:

1. The reduction from the normal chromosome number (2*n*) to the
haploid number (*n*).
2. The increase, within a species, from the normal diploid number
(2*n*) to a higher multiple, such as triploid (3*n*) or tetraploid (4*n*)
or analogous changes when the initial number is polyploid. Thus
in many parts of the world synthetic autoriploid varieties of
sugar beet (*Beta vulgaris*) are used and autotetraploid Italian
ryegrass (*Lolium italicum*) and red clover (*Trifolium pratense*) are
cultivated.
3. The combination in a synthetic allopolyploid, or amphiploid,
of the full complements of chromosomes of two distinct species.

   Examples of the use of synthetic amphiploids in agriculture
or in variety production are the so-called "hybrid ryegrass" and
synthetic *Brassica napus* (oilseed rape). Hybrid ryegrass is a
tetraploid combining the chromosome sets of pernnial ryegrass and
Italian ryegrass. Hybrid ryegrass has the rapid early growth and
high nutritive value of Italian ryegrass and the persistency of
perennial ryegrass and greater genetic stability than is found in
the interspecific hybrid at the diploid level (Breese et al., 1975).
B. napus (2*n*=38) is the natural allotetraploid derived from *B.
campestris* (2*n*=20) and *B. oleracea* (2*n*=18). By resynthesizing *B.
napus* from its parents and hybridizing the natural and synthetic
tetraploid it has been possible to introduce useful genetic variation from the diploid species into the tetraploid (Olsson, 1963).

ANEUPLOIDS

   The manipulation of intact single chromosomes in plant breeding
and in relevant research makes use of plants that have chromosome
numbers different from those that are normal in the species. Such
plants are *aneuploid* because they do not have the good, or *euploid*
number. The following are the designations given to the range of
aneuploids most commonly used in research where the euploids is
designated 2*n*:

Nullisomic - 2*n*-2 lacks both chromosomes of one homologous pair
Monosomic - 2*n*-1 lacks one chromosome
Trisomic   - 2*n*+1 has one chromosome present three times
Tetrasomic - 2*n*+2 has one chromosome present four times

   Thus in bread wheat (*Triticum aestivum*), in which the euploid
is 2*n*=42, nullisomics are 2*n*-2=40, monosomics are 2*n*-1=41, trisomics
are 2*n*+1=43, and tetrasomics are 2*n*+2=44.

Fully effective research work depends upon the availability of a complete set of a particular type of aneuploid. That is where stocks or lines have been constructed and indentified, with each chromosome, in turn, present in a particular aneuploid condition.

Complete sets of monosomics are available in the following crop species:

Wheat	*(T. aestivum)*	$2n=42$, with 21 monosomic lines
Oats	*(Avena sativa)*	$2n=42$, with 21 monosomic lines
Tobacco	*(Nicotiana tabacum)*	$2n=48$, with 24 monosomic lines

Complete sets of trisomics are available in the following crop species:

Wheat	*(T. aestivum)*	$2n=42$, with 21 trisomic lines
Oats	*(Av. sativa)*	$2n=42$, with 21 trisomic lines
Tomato	*(Lycopersicon esculentum)*	$2n=24$, with 21 trisomic lines
Spinach	*(Spinacia oleracea)*	$2n=12$, with 6 trisomic lines
Chilli	*(Capsicum annuum)*	$2n=24$, with 12 trisomic lines
Rye	*(Secale cereale)*	$2n=14$, with 7 trisomic lines
Rice	*(Oryza sativa)*	$2n=24$, with 12 trisomic lines
Sorghum	*(Sorghum vulgare)*	$2n=20$, with 10 trisomic lines
Barley	*(Hordum vulgare)*	$2n=14$, with 7 trisomic lines

Access to an extensive list of references to plant aneuploids is provided in Khush (1973).

ANEUPLOIDS IN GENETIC STUDIES

Both trisomics and monosomics have been used to provide information on the chromosomal location of genes of significance in varietal synthesis. Such information is of importance in enabling breeders to determine the likelihood of synthesizing genotypes with desired combinations of beneficial alleles.

The determination of which chromosome in the complement carries a particular genetic locus depends upon the distorted segregation that occurs for allelic differences when the chromosome with the relevant locus is in either the monosomic or the trisomic condition.

When a chromosome is in the monosomic condition it is not transmitted from meiosis to 50% of the gametes and therefore the genes that it carries do not segregate according to the ratios expected of chromosomes in the disomic conditions. In practice, therefore, if the chromosomal location is sought of a dominant allele in a euploid genotype, this genotype is hybridized with the set of lines monosomic, in turn, for every chromosome of the complement and carrying a

recessive allele at the same locus. The F_2 generation will segregate
with the expected ratio of 3 dominant : 1 recessive phenotype from
monosomic hybrids derived from all of the monosomic parental lines
except for that which was monosomic for the chromosome carrying the
dominant allele. In such hybrids the dominant allele will be in the
hemizygous condition in the F_1 generation, so the recessive phenotype
will not be displayed in the F_2 generation except perhaps in the rare
nullisomic progeny.

Alternatively, if the chromosome location of a recessive allele
is sought in an unknown genotype, a set of monosomic F_1 hybrids is
again created by hybridization with the parental lines carrying the
dominant allele and monosomic, in turn, for every chromosome of the
complement. The recessive phenotype will be displayed by the F_1
monosomic for the chromosome carrying the recessive allele while the
remainder of the monosomic F_1s will display the dominant phenotype.

More complex analyses are required where two or more loci are
involved in the determination of a character. Nevertheless the
procedures are relatively straightforward and valuable guidance to
breeders has arisen from genetic analyses based on monosomics.
Detailed descriptions of the procedures of aneuploid analysis are
given by Sears (1953) and Khush (1973).

Trisomics have been widely used to ascertain the chromosomal
location of genes in the provision of information for breeders on the
genetic structure of crop species. The methodology is analogous to
that involved when monosomic hybrids are used, except that trisomic
hybrids are produced. Where the locus under study is on the trisomic
chromosome there is distortion from the usual disomic segregation in
the back-cross or F_2 generation (for references see Khush, 1973).
Further refinement of gene location has been developed by making use
of telocentric chromosomes, which are misdivision products from which
one arm has been deleted. They arise, often in monosomics, by
aberrant division at meiosis of unpaired chromosomes. The extent of
the disassociation in segregation of a particular allele from the
telocentric or the complete homologous chromosome permits the estimation of linkage between the relevant locus and the centromere
(Sears, 1962).

INTERVARIETAL CHROMOSOME SUBSTITUTION

A technique which makes use of aneuploids has been elaborated ve
effectively in bread wheat, initially for breeding research and now
for application to variety production. Known as intervarietal chromosome substitution, the technique results in the transfer of a single
unchanged pair of chromosomes of a donor variety into a recipient
variety where it replaces the homologous pair. This is achieved by
using monosomic lines of the recipient variety as the recurrent seed

parent so that the chromosome to be replaced is absent from most egg
cells. When this recurrent monosomic parent is pollinated by the
donor parental variety, or by one of the monosomic back-cross progeny,
a euhaploid pollen grain will usually achieve fertilization. Therefore, the resulting monosomic offspring will have the donor monosomic
chromosome but will lack the recipient homologue while the background
chromosomes will, according to the stage of the back-crossing
programme, increasingly revert to the gene content of the recipient
variety. Following the completion of back crossing, the final monosomic derivative is allowed to self pollinate and this leads to the
segregation of a small proportion of euploid progeny in which the
substituted donor chromosome is in the disomic condition while the
background chromosomes will have reverted largely to their genetic
status in the recipient variety. This procedure is being used at the
Plant Breeding Institute, Cambridge, England, to introduce into
developing varieties chromosome $5B^S/7B^S$ from the variety Bersee
(Johnson, 1976). This chromosome carries several genes that apparently
determine durable resistance in mature plants to stripe (yellow) rust
(Puccinia striiformis). The use of inter-varietal chromosome substitution ensures that the several genes determining rust resistance will
be held together in new varieties.

     An additional procedure that can be employed in the refinement
of established intervarietal substitution lines permits recombination
to occur in the substituted chromosome. The disomic substitution
line is hybridized with the recipient variety. The resulting hybrids
are heterozygous only in the substituted chromosome whereas all the
other, background chromosomes are homozygous. Among the derivatives
of such hybrids selection can be practised for recombination in the
substituted chromosome and thus, in breeding work, its precise gene
content can be adjusted (Law and Worland, 1973).

ALIEN GENETIC VARIATION

     In some breeding programmes desirable genetic variation for
particular characters is not available within the crop species
concerned. It is then necessary to seek it by induced mutation or by
genes from other species. This is readily possible when homologous
chromosomes of the parental species will pair at meiosis and recombine.
However, when there is little or no meiotic pairing and recombination
between chromosomes derived from the two parents and amphiploids
combining the parental genomes are not useful, an alternative is to
attempt to incorporate in the recipient crop species some limited,
perhaps single, genetic attribute of a donor species.

     There has been a progressive evolution of the cytogenetic method-
ology aimed at achieving this objective and more work in this field
has concentrated on wheat, and its relatives, than on any other crop
group. In this discussion, therefore, the principles will be discussed
using wheat as a model.

The first step in the limitation of the amount of genetic
material transferred to wheat from a related alien source involves
the construction of so-called "alien chromosome addition lines."
In these a single pair of chromosomes from another species is added
to the full complement of wheat chromosomes. Plant material of this
kind has been most extensively studied in the wheat-rye combination
(initially by O'Mara, 1940, 1951). In the process the wheat-rye
amphiploid triticale (*Triticosecale*) is constructed and then back
crossed to the wheat species parent. The back crossing programme is
continued, using wheat as the recurrent parent, and during the process
the rye chromosome content of the derivatives is diluted until only
one rye chromosome remains, while the full complement of wheat
chromosomes is retained. Such monosomic addition plants are then
allowed to self-pollinate and among the resulting progeny those plants
are selected which have the alien chromosome in the disomic condition.
In these, since the alien chromosome is paired, meiosis is regular and
the addition condition is true breeding and stable.

Of course, from any combination of species, as many distinct
addition lines can be created as there are chromosomes in the haploid
complement of the donor species. In the case of lines with the
addition of *Secale*, *Aegilops* or *Agropyron* chromosomes to wheat, the
addition lines differ phenotypically from the recipient wheat
varieties. However, while some beneficial changes of phenotype occur
there are also usually disadvantageous modifications, particularly
reduced yield. As a result, as yet, no alien chromosome addition
line has been used as a commercial variety.

The next step, after the introduction of alien genetic variation
by the addition of single chromosome pairs, involves the replacement
of a pair of chromosomes of the recipient species by an alien pair.
This can be achieved in wheat because its polyploid status permits
marked deviations into aneuploidy. Derivatives of the procedure are
called "disomic chromosome substitution lines." They are achieved
by hybridizing wheat monosomics with disomic addition lines to
produce plants in which a wheat chromosome and the alien chromosome
are simultaneously monosomic. The complete replacement of the wheat
monosomic may be attained in the next generation either by selfing
the doubly monosomic plants or two generations later following the
hybridization of the doubly monosomic individuals with the disomic
addition line.

Disomic substitution lines frequently display attributes of the
phenotype conditioned by genes on the alien chromosome. They are
moreover frequently more fertile and more stable from generation to
generation than disomic addition lines. It is consequently not
surprising that wheat varieties have been employed commercially which
carry an alien chromosome pair substituting for a wheat pair.
The rye-chromosome substitution varieties of wheat that have been

used most widely all have chromosome IR for IB substitutions of
a IBL/IRS translocation. Varieties with this constitution, which
it should be said arose fortuitously from hybridization programmes
using rye ·or triticale as a parent, include Orlando, Zorba, Weique,
Neuzucht, Riebesel 74/51 (Zeller, 1973) and Clement.

It is important to emphasize that alien chromosomes can only
be efficiently substituted for recipient chromosomes to which they
are related genetically. Thus the substituted and substituting
chromosomes will probably have evolved from the same chromosome of
the common ancestor of the donor and recipient species. Much of
their gene content may therefore be similar in activity, linkage
order and dispersion among non-informational nucleotide sequences.

Despite the success of disomic substitution lines, in general,
the amount of alien genetic material introduced into such genotypes
is too extensive for the retention of the genotypic balance of the
organism. This probably arises because, despite their similarity,
there has been no adjustment, by selection, of the gene contents to
permit efficient interactions between the alien chromosome and the
chromosomes of the recipient species. This leads to the conclusion
that the introduced alien chromosomal material should be limited, as
far as possible, to only that restricted segment necessary to incorporate in the recipient the single desired phenotypic modification
from the donor. E. R. Sears of the University of Missouri, USA,
had the earliest and clearest insight into this problem and sought
to solve it by the use of induced translocation (Sears, 1956). No
doubt he will be reviewing the present state of this technology
elsewhere in this Stadler Symposium.

PRECISION IN IDENTIFICATION

The effectiveness of chromosome manipulation in either practical
plant breeding or in cytogenetic research resides in the certainty
that can be applied to tracing, through the generations of breeding,
that the structure of the plant genome accords with the expections
of the programme. Such certainty is necessary if at the end-point
of a process of genotypic manipulation the experiemntalist can have
confidence that the new genotype is that predicted at the outset.

From the earliest applications of manipulative technologies
chromosome structural markers have been used to ensure that chromosomes which could be so recognized were indeed in the required condition. Chromosome size, arm ratio or satellite status have enabled
a few chromosomes to be recognized and traced through pedigrees.
Experimental improvement in this process came initially from the
work of Ernie Sears and subsequent co-workers in wheat, who recognized that the high frequency of mis-division of monosomic chromo-

somes to give rise to telocentrics and isochromosomes could be used
to provide structural markers. This culminated in the production
of the Missouri set of telocentric lines marking every arm of every
wheat chromosome. The introduction of one or two of these lines
into a breeding programme enabled the unequivocal identification
of particular members of the chromosome complement throughout a
series of sexual generations.

Continuing to refer to techniques applied with wheat as a model
for the systems used to trace chromosomes, reference must now be
made to "banding" procedures of the type exploited in mouse by
Pradue and Gall (1970). Similar Giemsa C-banding has been used in
wheat by Gill and Kimber (1974), Iordansky et al. (1978) and Van
Niekert and Pienaar (1983), while Linde-Laursen (1975) employed
these banding procedures on barley. An alternative, so called,
N-banding procedure was successfully used in wheat by Gerlach (1977).
All of these banding methods reveal the linear differentiation of
regions of the chromosome that are heterochromatic and separate
normal-staining euchromatic regions. The patterns of banding along
the chromosome may be distinctive and permit the recognition of
individual members of the karyotype. However, some chromosome
pairs are not distinctive even when a combination of banding and
other morphological features is employed. There also appears to be
some difficulty created by the inadequate repeatability of banding
results between different laboratores (Van Niekerk and Pienaar, 1983)
so that, while in the hands of individual research workers, banding
may enable unequivocal recognition of some chromosomes, transfers
of methodology and recognizable plant material between laboratories
may be uncertain.

A developing methodology in marking chromosomes lies in the use
of *in situ* hybridization of the DNA of chromosomes in cytological
preparations with tritium labelled cRNA transcribed by *E. coli* RNA
polymerase from chimaeric plasmids incorporating plant DNA as described by Bedbrook et al. (1980) and Jones and Flavell (1982) and
used by several other groups. So far, in wheat, molecular probes
of this kind have always consisted of segments of sequences that
are highly repeated in the wheat genome. Consequently, they have
not been specific markers for any individual chromosome pair and
therefore of somewhat limited value in tracing particular chromosomes.
Although, in some cases Hutchinson and Lonsdale (1982), the probes
were preferential in labelling most intensively the chromosomes of
a particular genome of the allopolyploid. Until probes are available made from single or limited sequence plant DNA their value
will be restricted. However, we must look to the future when probes
of this type and of known information content can be employed to
mark specific chromosomes and indeed specific genetic loci. At that
time structural marking will have reached its ultimate.

Finally, in relation to markers for chromosomes, attention must
be paid briefly to the phenotype effects of their genes. From the

earliest study of aneuploids in wheat considerable use was made of the modification to plant morphology caused by the changed dosage of specific chromosomes, often resulting from the effects of recognised genetic loci (Sears, 1954). This use of genetic markers has subsequently been supplemented by markers based on disease resistance/ susceptibility differences and latterly particularly by the use of biochemical markers(Hart, 1982). From Table 1 it will be seen that more than 90 loci are marked by enzyme structural genes detectable by relatively routine, though tedious, zymogram methodology and that most chromosomes of the wheat complement of 21 pairs possess at least one such marker. To these may be added the biochemical markers for glutenin (Glu-1) production on the long arms of homoeologous group 1 chromosomes markers for gliadin production on the short arms of the same chromosomes (Gli-1) and the other gliadin genes (Gli-2) distally placed on the short arms of the homoeologous group 6 chromosomes (Payne et al., 1983). In wheat and in other plant groups the new precision in the recovery of desired genotypes is being created by the availability of biochemical markers such as these and the prospects for an even greater supply of such tools must be high.

RESULTS FROM CHROMOSOME MANIPULATION IN WHEAT

As part of an extensive study of the genetics of grain protein production in wheat Law et al. (1983) have studied lines carrying single chromosomes of Ae. umbellulata either added to wheat or substituted for their wheat homoeologues. Particular interest has attached to chromosome 1U from Ae. umbellulata because of the presence on the chromosomes of homoeologous group 1 of the Glu-1 and the Gli-1 loci. The incorporation of chromosome 1U indeed caused the presence of a form of high-molecular-weight subunit of glutenin and of a form of gliadin not normally present in wheat as revealed by electrophoresis. In order to determine whether the introduction of the Ae. umbellulata alleles for Glu-1 and Gli-1 modified the baking quality of the flour, lines carrying 1U were tested by the sodium dodecyl sulfate (SDS) test of Axford et al. (1979) (Table 2).

Chromosome 1U had some effect on total protein content but most strikingly in the line in which it substitutes for chromosome 1A of wheat there is a marked increase in SDS volume. This indicates a potentially very valuable improvement in bread-making quality of the flour of this line compared with Chinese Spring, the recipient variety. Unfortunately, this is accompanied by a massive reduction in yield so that the substitution line cannot itself be used commercially. Instead, interference is being practised to remove the restriction on recombination between homoeologous chromosomes which is exercised by the Ph allele on chromosome 5B. In this way the beneficial contribution to baking quality of 1U may be associated with the yield benefits provided by 1A. Finally, the study of this system shows that a quite specific substitution is necessary to improve the geno-

Table 1: Enzymatic marker genes in wheat for chromosome identification (personal communication from M. D. Gale and C. C. Ainsworth, Plant Breeding Institute, Cambridge).

ENZYME	GENE/GENE SET	CHROMOSOME	MAP LOCATION	REFERENCE
Acid Phosphatase	Acph 2 to 6 & 8	4Aβ, 4BS, 4DL		13,17
Alcohol dehydrogenase	Adh-1 set	4Aα, 4BL, 4DS		12,13,17
	Adh-A1	4Aα	23.1% proximal to Rht-1	32
	Adh-2	5AL, 5BL, 5DL		20
Aminopeptidase	Amp-1 set	6AS, 6BS, 6DS		13,17
α-Amylase (malt)	α-Amy-1 set	6AL, 6BL, 6DL		29,30
	α-Amy-B1	6BL	19.4% from centromere	29
	α-Amy-D1	6DL	11.9% from centromere	29
	α-Amy2	6BL	20.6% from α-Amy-B1	29
α-Amylase (green)	α-Amy-2 set	7AL, 7BL, 7DL		29,30
	α-Amy-B2	7BL	6% from centromere	11
β-Amylase	β-Amy-1 set	4Aβ, 4DL		1,21
	β-Amy-D1	4DL	>35% from centromere	1
	β-Amy-2 set	5AL, 5BL		1
	β-Amy-A2	5AL	2.3% proximal to B1	1
Endopeptidase	Ep-1 set	7AL, 7BL, 7DL		17
Esterase	Est-1 set	3AL, 3BL, 3DL		3,25,27
	Est-2 set	6AL, 6BL, 6DL		27

ENZYME	GENE/GENE SET	CHROMOSOME	MAP LOCATION	REFERENCE
Glucose-6-phosphate dehydrogenase	G-6-pd	groups 1 and 6		7
Glucose phosphate isomerase	Gpi-1 set	1AS, 1BS. 1DS		15
	Gpi-D1	1DS	34.5% proximal to Gli-D1 36.2% distal to Glu-D1	9
Glutamate oxaloacetate transaminase	Got-1 set	6AS, 6BS, 6DS		14
	Got-2 set	6AL, 6BL, 6DL		14
	Got-3 set	3AL, 3BL, 3DL		14
	Got-E3	3DL	4.3% from centromere	18
Hexokinase	Hk-1 set	1BS, 1DS		2
	Hk-2	3BS		2
Lipopurothionins	Pur-1 set	1AL, 1BL, 1DL		10
Lipoxygenase	Lpx-1 set	4Aα, 4BL, 4DS		7
	Lpx-2 set	5AL, 5BL, 5DL		7
Malate dehydrogenase	Mdh-1 set	1AL, 1BL, 1DL		5
Peroxidase	Per 1, Per 3	7AS, 7DS		6,22,23
	Per 2	4BL		22,23
	Per-4 set	1BS, 1DS		4
Phosphoglucomutase	Pgm-1	4BL		7,35
6-Phosphogluconate	6-Pgd-1	6AL, 6BL		19

(continued)

Table 1 continued.

ENZYME	GENE/GENE SET	CHROMOSOME	MAP LOCATION	REFERENCE
dehydrogenase	6-Pgd-2	7BL		19
Phosphodiesterase	Pde-1 set	3AS, 3BS, 3DS		33,34
Shikimate dehydrogenase	Skdh-1 set	5AS, 5BS, 5DS		24,28
Triose phosphate isomerase	Tpi-1 set	3AS, 3BS, 3DS		31

FOOTNOTE TO TABLE 1

(1) Ainsworth, C. C., Gale, M. D., and Baird, S., 1983, The genetics of β-amylase isozymes in wheat. I. Allelic variations among hexaploid varieties and intrachromosomal gene locations, Theor. Appl. Genet., 66:39-49.

(2) Ainsworth, C. C., 1984, The genetic control of hexokinase isozymes in wheat, Genet. Res. Camb., 42:219-227.

(3) Ainsworth, C. C., Gale, M. D., and Johnson, H. M., 1984, The genetic control of grain esterases in hexaploid wheat. I. Allelic variation, Theor. Appl. Genet., In press.

(4) Ainsworth, C. C., Johnson, H. M., Jackson, E. A., Miller, T. E., and Gale, M. D., 1984, The chromosomal locations of leaf peroxidase genes in hexaploid wheat, rye and barley, In preparation.

(5) Benito, C., and Salinas, J., 1983, The chromosomal location of malate dehydrogenase isozymes in hexaploid wheat (Triticum aestivum L.), Theor. Appl. Genet., 64:255-258.

(6) Benito, C., and Perez de la Vega, M., 1979, The chromosomal location of peroxidase isozymes of the wheat kernel, Theor. Appl. Genet., 55:73-76.

(7) Beniot, C., 1982, Chromosomal location of isozyme markers in hexaploid wheat and rye, 4th Int. Congr. Isozymes, Texas, Abstract.

(8) Chojecki, A. J. S., and Gale, M. D., 1982, Genetic control of glucose phosphate isomerase in wheat and related species, Heredity, 49:337-347.

(9) Chojecki, A. J. S., Gale, M. D., Holt, L. M., and Payne, P. I., 1983, The intrachromosomal mapping of a glucose phosphate isomerase structural gene, using allelic variation among stocks of Chinese Spring wheat, Genet. Res. Camb., 41:221-226.

(10) Fernandex de Caleya, R., Hernandez-Lucas, C., Carbonaro, P., and Garcia-Olmedo, F., 1976, Gene expression in alloploids: genetic control of lipopurothionins in wheat, Genetics, 83:687-699.

(11) Gale, M. D., Lae, C. N., Chojecki, A. J., and Kempton, R. A., 1983, Genetic control of α-amylase production in wheat, Theor. Appl. Genet., 64:309-316.

(12) Hart, G. E., 1970, Evidence for triplicate genes for alcohol dehydrogenase in hexaploid wheat, Proc. Nat. Acad. Sci. U.S.A., 66:1136-1141.

(13) Hart, G. E., 1973, Homoeologous gene evolution in hexaploid wheat, Proc. 4th Int. Wheat Gen. Symp., Sears, E. R. and Sears, L. M. S., eds., pp. 805-810.

(14) Hart, G. E., 1975, Glutamate oxaloacetate transminase of Triticum: Evidence for multiple systems of triplicate structural genes in hexaploid wheat, pp. 637-657, in: "Isozymes: Developmental Genetics, Vol. 3," Markert, C. K., ed., Academic Press, New York.

(15) Hart, G. E., 1979, Evidence for a triplicate set of glucose-

phosphate isomerase structural genes in hexaploid wheat, Biochem. Genet., 17:585-598.

(16) Hart, G. E., personal communication.

(17) Hart, G. E., and Langston, P. J., 1977, Chromosomal location and evolution of isozyme structural genes in hexaploid wheat, Heredity, 39:263-277.

(18) Hart, G. E., McMillin, D. E., and Sears, E. R., 1976, Determination of the chromosomal location of a glutamate oxaloacetate transaminase structural gene using Tritcium-Agropyron translocations, Genetics, 83:49-61.

(19) Hsam, S. L. K., Zeller, F. J., and Huber, W., 1982, Genetic control of 6-phosphogluconate dehydrogenase (6-PGD) isozymes in cultivated wheat and rye, Theor. Appl. Genet., 62:317-320.

(20) Jaaska, V., 1979, NADP-dependent aromatic alcohol dehydrogenase in polyploid wheats and their diploid relatives. On the origin and phyolgeny of polyploid wheat, Theor. Appl. Genet., 53:209-217.

(21) Joudrier, P., and Cauderon, Y., 1976, Localization chromosomique de genes controlant la synthese de certains constituants β-amylasique du grain de ble tendre, Comptes Rendua Ac. Sc. Paris D., 181:115-118.

(22) Kobrehel, K., 1978, Identification of chromosome segments controlling the synthesis of peroxidases in wheat seeds and in transfer lines with Agropyron elongatum, Can. J. Bot., 56:1091-1094.

(23) Kobrehel, K., and Fiellet, P., 1975, Identification of genomes and chromosomes involved in peroxidase synthesis of wheat seeds, Can. J. Bot., 53:2336-2344.

(24) Koebner, R. M. D., and Shepherd, K. W., 1982, Shikimate dehydrogenase - a biochemical marker for group 5 chromosomes in the Triticinae, Genet. Res. Camb., 41:209-213.

(25) Kostova, R., Vladova, R., and Vorkova, M., 1980, Electrophoretic investigation of some enzyme systems in ditelocentric lines of Triticum aestivum L. cv. Chinese Spring. I. The esterase enzyme system, Genetics and Plant Breeding, 13:77-88.

(26) May, C. E., Vickery, R. S., and Driscoll, C. J., 1973, Gene control in hexaploid wheat, Proc. 4th Int. Wheat Gen. Symp., Sears, E. R., and Sears, L. M. S., eds., pp. 843-850.

(27) Nakai, Y., 1973, Isozyme variation in *Aegilops* and *Triticum*. II. Esterase and acid phosphatase isozymes studied by the gel isoelectro focusing method, Seiken Ziho, 24:45-73.

(28) Neuman, P. R., and Hart, G. E., 1982, Genetics of shikimate dehydrogenase in hexaploid wheat, 4th Int. Congr. Isozymes, Texas, In press.

(29) Nishikawa, K., Furata, Y., Hina, Y., and Yamada, T., 1981, Genetic studies of α-amylase isozymes in wheat. IV. Genetic analyses in hexaploid wheat, Jap. J. Genet., 56:385-345.

(30) Nishikawa, K., and Nobuhara, M., 1971, Genetic studies of α-amylase isozymes in wheat. I. Location of genes and variation in tetra- and hexaploid wheat, Jap. J. Genet., 46:345-353.

(31) Pietro, M. E., and Hart, G. E., 1982, Genetic Control of Triosephosphate isomerase in hexaploid wheat, 4th Int. Congr. on Isozymes, Texas.

(32) Suseelan, K. N., Rao, M. V. P., Bhatia, C. R., and Rao, I. R., 1982, Genetic control of Triosephosphate isomerase in hexaploid wheat, 4th Int. Congr. on Isozymes, Texas.

(33) Wolf, G., and Lerch, B., 1973, Genome analysis in the *Triticinae* using isoenzymes of phosphodiesterase, Proc. 4th Int. Wheat Genet. Symp., Sears, E. R., and Sears, L. M. S., eds., pp. 885-889.

(34) Wolf, G., Rimpau, J. and Lelley, T., 1977, Localization of structural and regulatory genes for phosphodiesterase in wheat (*Triticum aestivum*), Genetics, 86:597-605.

(35) Youssefian, S., 1982, The localization of the phosphoglucomutase structural gene in hexaploid wheat and its relatives, M. Phil. Dissertation, Cambridge University.

Table 2.   Yield, percentage protein and SDS sedimentation
           volume for the substitution and addition of the
           1U chromosome of *Ae. umbellulata* into the variety
           Chinese Spring (Law et al., 1983)

Line	Chromosome constitution	Yield/ Plant (% CS)	% Protein (=Nx5.7)	SDS (%CS)
Chinese Spring (CS)	1A 1B 1D	16.14	18.09	55.38
CS 1A (1U)	1U 1B 1D	8.24(51)	18.31	74.31(134)
CS 1B (1U)	1A 1U 1D	8.79(54)	18.78	50.56(91)
CS 1D (1U)	1A 1B 1U	8.41(52)	19.78	38.25(69)
CS addition 1U	1A 1B 1D 1U	12.32(76)	18.20	57.25(103)

type since chromosome 1A apparently makes the least contribution to
baking quality. 1U can most usefully be substituted for 1A while
chromosomes 1B and 1D – with their more useful effects – are retained.

There is potential benefit, in terms of increased grain protein
content, from the substitution into wheat of chromosome 2M of *Ae.
comosa*. Riley et al., (1968) initially isolated this chromosome in
a wheat addition line and demonstrated that it carries the gene Yr-8
giving resistance to certain races of stripe rust. Subsequently
Law et al., (1977) showed that the addition of homoeologous group 2
chromosomes to wheat from *Ae. comosa, Ae. umbellulata and S. montanum*,
or their substitution for wheat homoeologues, affects grain protein
content. Chromosome 2M has the most marked effect in increasing
protein content and, in the substitution line in which it replaces
chromosome 2D – the least potent wheat chromosome – there is markedly
higher protein content and little effect on yield (Table 3).

The segment of chromosome 2M increasing grain protein content
has been transferred to the homoeologous region of wheat chromosome
2D of the varieties Maris Widgeon and Rothwell Perdix (Law et al.,
1983). In all tests with these lines the grain protein increased by
between one quarter per cent and one per cent without any significant
difference in grain yield although the unmodified Rothwell Perdix

lost yield to a greater extent than the manipulated quivalent due to stripe rust.

Table 3. Mean yields, percentage proteins and yield components of lines in which chromosome 2M of *Ae. comosa* has been substituted for either chromosome 2A, 2B, or 2D in the variety Chinese Spring. (Law et al., 1983)

Genotypes	Yield per plant	% protein (=Nx5.7)	250 Grain weight	Grain Number per Ear	Tiller Number
Chinese Spring 2A/2M	5.0	18.92***	7.2**	21***	9.9***
Chinese Spring 2B/2M	6.2	18.81***	7.5*	24***	8.8**
Chinese Spring 2D/2M	6.9	21.09***	8.4***	22***	9.3**
Chinese Spring	7.3	16.82	6.9	36	7.5

Significantly different from Chinese Spring   * P 0.05-0.01
                                                    ** P 0.01-0.001
                                                    *** P<0.001

Clearly these findings have practical potential, so attention has been given to assessing the consequence of introducing a segment of chromosome 2M into a high yielding commercial wheat genotype. A chromosome combining 2M with each of its wheat homoeologoues, in turn, has been put into an essentially Maris Hobbit genotypic background following crossing between Chinese Spring and Maris Hobbit and two backcrosses to Hobbit. Comparisons of these lines in field trials were made in 1982 (Table 4). As with the whole chromosome substitution lines in Chinese Spring there is a marked increase in protein content and with the least reduction of yield when the 2M segment substitutes for the 2D segment. There is also a relationship in this, and in other samples, between increased protein content and increased grain size. There is also a negative correlation between grain number per ear and grain protein content. Perhaps the crucial general principle that is being re-emphasized in this work is the specifity with which desirable phenotypic improvement in

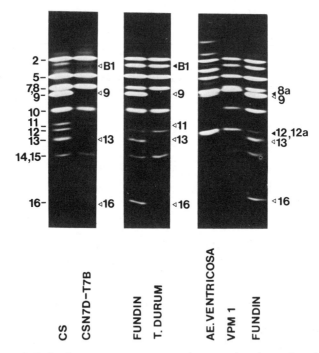

Fig. 1.   α–AMY-2 phenotypes, comparisons showing the isozymes
          controlled by α–Amy-D2 in bread wheat and the exchange
          in VPM 1 where the α–Amy-D2a allele is replaced by
          α–Amy-D2b of *Ae. ventricosa*.

provided. In this instance it occurs only when 2M replaces 2D
wholly or in part:

Table 4. Mean yield, percentage proteins and yield
          components of advanced backcross lines in
          which chromosome 2M of *Ae. comosa* has been
          substituted for either chromosome 2A, 2B
          or 2D in the variety Hobbit 'sib'.

Genotypes	Number of Lines	Yield per Plant	% Protein (=Nx5.7)	250 Grain Weight	Grain Number per Ear	Tiller Number
Hobbit's2A/2M	15	16.8**	14.88**	11.9	28***	12.6**
Hobbit's2B/2M	13	15.6**	15.33**	11.5	27***	12.4**
Hobbit's2D/2M	13	22.4	14.42	14.8***	30***	12.8**
Hobbit's'	1	25.7	13.91	11.9	48	11.2

Significantly different from Hobbit's   ** P 0.01-0.001
                                        *** P 0.001

    A recent example of the use of enzymatic markers is provided
by the study of the transfer of eyespot resistance from *Ae.*
ventricosa to wheat (Gale et al., 1983). Resistance to *Pseudocero-*
sporella herpotrichoides is present in some wheat varieties in
which it is primarily determined by chromosome 7A (Law et al., 1975).
Potential reinforcement of this resistance has been transferred from
Ae. ventricosa (Doussinault and Dosba, 1977) and has been incorporated in the variety VPM 1 where it is probably determined by chromosome 7D (Jahier et al., 1979). Gale et al. (1983) have sought a
biochemical marker for the 7D resistance to facilitate its manipulation in breeding programmes. The marker studied was α-amylase
determined by the α-Amy-2 loci which are carried on homoeologous
group 7 chromosomes in wheat. Using polyacrylamide isofocussing
gels with enzymes extracted from endosperm half-grains exposed to
gibberellic acid, 16 isozyme bands occur (Fig. 1). Three of these
bands (numbers 9, 13 and 16) are determined by the α-Amy-D2 allele
on chromosomes 7D. These are absent from VPM 1 and are replaced
by two new bands (numbers 8a and 12a) that are expressed in *Ae.*
ventricosa but not in wheat. Consequently, a segment of the *Ae.*
ventricosa chromosome homoeologous to chromosome 7D has been transferred in the selection for eyespot resistance and it also includes
the α-Amy-D2 allele of *Ae. ventricosa*. However, the *Ae. ventricosa*
chromosome segment must be large since in breeding experiments

α-Amy-D2 alleles and eyespot resistance segregate independently.
This also, unfortunately, implies that the α-Amy-2 phenotype cannot
be used as a marker for the eyespot resistance, consequently
another marker must be found.

CONCLUSION

     Chromosome manipulation stands at a pivotal position between
the improvement of the scientific understanding of plant genetics
and metabolism and the practical exploitation of this understanding
in the form of improved genotypes. The mechanistic adjustment of
the chromosome complements of plants has led to marked expansions
in our knowledge of plant cytogenetics and this will no doubt continue. It has also helped to provide plant breeders with the logical comprehension of their biological material that enables well
founded breeding strategies to be formulated; so contributing to
the conversion of breeding into a precise technology. In addition,
chromosome manipulation is increasingly being used to construct
genotypes, for use in crop improvement, and they are often of kinds
that would not otherwise have been attainable. The prospects for
the practical exploitation of this methodology increase as the
scientific foundations on which it rests are consolidated.

     However, more rapid advance in these ways would be possible if
plant science possessed a better background comprehension of the
intermediary metabolism of plants. To take a simple example from
the work that we have described earlier; it is necessary to seek a
marker gene closely linked to the eyespot resistance introduced into
wheat from *Ae. vetricosa* since the test for resistance is ambiguous
on a single plant basis. However, no information is available on
the biochemical basis of resistance which might enable the presence
of the gene to be detected directly. Analogous difficulties are
created in tracking other kinds of disease and pest resistance in
breeding programmes and the handling of many other economic attributes would be facilitated by some understanding of their metabolic
basis. Consequently major encouragement should be given to research
likely to contribute to the provision of knowledge necessary to
underpin plant breeding, chromosome manipulation and plant genetic
engineering.

     The plant genetic engineer, standing upstream of the chromosome
manipulator, is even more in need of an understanding of the intermediary metabolites that are the precursors of those ultimate genotype characteristics, modification of which is the endpoint of his
endeavours. This is a matter of concern to those working in
chromosome manipulation with whole plants since the production and
processing of plant breeding material in the future may be expected
often to start with genetic engineering but subsequently to need
further refinement by cytogenetic methods. After this the plant

material will be suitable to build into variety production programmes. As the exploitation of recombinant DNA technology is
rapidly becoming a realistic possibility so cytogeneticists and
breeders are anxious to develop its products further but this will
be delayed unless urgent deficiencies of knowledge about plant
processes are removed.

LITERATURE CITED

Axford, D. W. E., McDermott, E. E., and Redman, D. G., 1979, Note
    on the sodium dodecyl sulfate test of breadmaking quality:
    comparison with Pelshenke and Zeleny tests, Cereal Chem., 56:
    582-584.
Bedbrook, J. R., Jones, J., O'Dell, M., Thompson, R., and Flavell,
    R. B., 1980, A molecular description of telomeric hetero-
    chromation in Secale species, Cell, 19:545-560.
Breese, E. L., Stephens, D. E., and Thomas, A. C., 1975, Tetraploid
    hybrid ryegrass: Allopolyploidy in action, pp. 46-49, in:
    "Ploidy in fodder plants," Nuesch, B., ed., Report of the
    Eucarpia Fodder Crops Section, Zurich.
Doussinault, G., and Dosba, F., 1977, An investigation into increas-
    ing the variability for resistance to eyespot in wheat. Eyespot
    variability in the subtrite *Triticinae*, Zeitschrift fur Pflan-
    zensuchtung, 79:122-133.
Gale, M. D., Scott, P. R., Law, C. N., Ainsworth, C. C., Hollins,
    T. W., and Worland, A. J., 1983, A+a-amylase gene from *Aegilops
    vetricosa* transferred to bread wheat together with a factor for
    eyespot resistance, Hereidty, In press.
Gerlach, W. L., 1977, N-banded karyotypes of wheat species,
    Chromosoma, 62:59-56.
Gill, B. S., and Kimber, G., 1974, Giemsa C-banding and the evolution
    of wheat, Proc. Nat. Acad. Sci. U.S.A., 48:4086-4090.
Hart, G. E., 1982, Biochemical loci of hexaploid wheat (*Triticum
    aestivum*), Zn=42, Genomes AABBDD, in: "Genetic Maps," O'Brien,
    S. J., ed., Nat. Cancer Inst., Frederick, MD.
Hutchinson, J., and Lonsdale, D. M., 1982, The chromosomal distri-
    bution of cloned highly repetitive sequences from hexaploid
    wheat, Heredity, 48:371-376.
Iordansky, A. B., Zurabishvilli, T. B., and Badaev, N. S., 1978,
    Linear differentiation of cereal chromosomes. I. Common
    wheat and its supposed ancestors, Theor. Appl. Genet.,
    51:145-152.
Jahier, J., Doussinault, G., Dosba, F., and Bourgeosis, F., 1979,
    Monosomic analysis of resistance to eyespot in variety "Roazon."
    Proc. fifth Int. Wheat Genet. Symp., New Delhi, pp. 437-440.
Johnson, R., 1976, Breeding for durable disease resistance,
    Proceedings of the 14th NIAB Crop Conference, 1976:48-56.
Jones, J. D. G., and Flavel, R. B., 1982, The structure amount

and chromosomal localization and defined repeated DNA sequences in species of the genus *Secale*, Chromosoma, 86:613-641.

Khush, G. S., 1973, Cytogenetics of aneuploids, Academic Press, New York & London.

Law, C. N., Payne, P. I., Worland, A. J., Miller, T. E., Harris, P. A., Snape, J. W., and Reader, S. M., 1983, Studies on genetic variation affecting grain protein type and amount in wheat, Report of the International Atomic Energy Agency, In press.

Law, C. N., Scott, P. R., Worland, A. J., and Hollins, T. W., 1975, The inheritance of resistance to eyespot (*Cercosporella herpotrichoides*) in wheat, Genet. Res., 25:73-79.

Law, C. N., and Worland, A. J., 1973, Aneuploidy in wheat and its uses in genetic analysis, Annual Report 1972, Plant Breeding Institute, Cambridge: 25-65.

Olsson, G., 1963, Induced ployploids in *Brassica*, pp. 179-192, in: "Recent plant breeding research," Akerberg, E., Hagberg, E., Olsson, A., and Tedin, O., eds., Almquist & Wiksell, Stockholm, Goteborg and Uppsala.

O'Mara, J. G., 1940, Cytogenetic studies on Triticale. I. A method of determining the effect of individual *Secale* chromosomes on *Triticum*, Genetics, 25:401-408.

O'Mara, J. G., 1951, Cytogenetic studies on Triticale. II. The kinds of intergeneric chromosome additions, Cytologia, 16:225-232.

Pardue, M. L., and Gall, J. G., 1970, Chromosomal localization of mouse satellite DNA, Science, 168:1356-1358.

Payne, D. I., Holt, L. M., Jackson, E. A., and Law, C. N., 1983, Wheat storage proteins: their genetics and their potential for manipulation by plant breeding, Proceedings of the Royal Society, In press.

Riley, R., Chapman, V., and Johnson, R., 1968, The incorporation of alien disease resistance in wheat by genetic interference with the regulation of meiotic chromosome synapsis, Genet. Res., 12:199-219.

Sears, E. R., 1953, Nullisomic analysis in common wheat, Am. Nat., 87:245-252.

Sears, E. R., 1954, The Aneuploids of wheat, Res. Bull. Mo. Coll. Agr. Exp. Stn., 572:1-59.

Sears, E. R., 1956, The transfer of leaf-rust resistance from *Aegilops umbellulata* to wheat, Brookhaven Symposia in Biology, 9:1-22.

Sears, E. R., 1962, The use of telocentric chromosomes in linkage mapping, Genetics, 47:983.

Van Niekerk, M. A., and Pienaar, R. de V., 1983, Morphology and linear C-band differentiation of *Triticum aestivum* L. em. Thell T. *aestivum* cv. "Chinese Spring" chromosomes, Cereal Res. Communications, 11:115-122.

Zeller, F. J., 1973, 1B/1R wheat-rye chromosome substitutions and translocations, pp. 221, Proc. fourth Int. Wheat Gen. Symp., Sears, E. R., and Sears, L. M. S., eds., University of Missouri Columbia, Missouri.

IN VITRO APPROACHES TO INTERSPECIFIC HYBRIDIZATION

AND CHROMOSOME MANIPULATION IN CROP PLANTS

G. B. Collins, N. L. Taylor, and J. W. DeVerna[1]

Department of Agronomy
University of Kentucky
Lexington, KY 40546

INTRODUCTION

The earliest report of an interspecific hybrid in plants is credited to Thomas Fairchild who in 1717 crossed carnation (Dianthus caryophyllus) and sweet-william (D. barbatus) (Allard, 1960). Despite this early successful effort, the potential of wide hybridization in agronomic species was not appreciated until after the rediscovery of Mendel's Laws. Since that time, the use of exotic germplasm has been significant in the improvement of a number of crops including wheat (Triticum sp.), potato (Solanum tuberosum), barley (Hordeum vulgare), tomato (Lycopersicum esculentum), tobacco (Nicotiana tabacum), sugarcane (Saccharum officinarum), and sunflower (Helianthus annuus) (Sanchez-Monge and Garcia-Olmedo, 1977; Stalker, 1980; Sharma and Gill, 1983; Bowley and Taylor, 1984). Overall, wide hybridization has been important when intraspecific diversity is lacking and where suitable germplasm can be found in species which can be crossed easily.

An assortment of barriers are inherent in attempts to utilize exotic germplasm for crop improvement. Successful sexual hybridization involves a sequence of events including pollen germination, pollen tube growth, fertilization, embryo and endosperm development, and seed maturation. Stebbins (1950) has

_____

[1] Professors of Agronomy and Graduate Research Assistant, respectively.

323

divided these hybridization barriers into two broad groups, namely prefertilization and postfertilization barriers. The prefertilization category includes those mechanisms which prevent fertilization and includes geographic isolation, apomixis, and pollen-pistil incompatibilities. Postfertilization barriers are a greater hinderance to hybridization and can be a result of ploidy differences, chromosome alterations, chromosome elimination, incompatible cytoplasms, seed dormancy, and hybrid breakdown. Of particular importance is the need to obtain hybrids that are fertile. The reader is referred to recent reviews on the subject of conventional hybridization for additional details (Hadley and Openshaw, 1980; Stalker, 1980).

Some interspecific barriers have been overcome by conventional approaches to hybridization. Nonconventional approaches utilizing the techniques of somatic cell genetics now afford improved potential in hybridization and the transfer of genetic information. These approaches can make the process of hybridization and hybrid utilization less tedious and can broaden the scope of "available" germplasm. The primary in vitro methods which have aided in wide hybridization of plants are embryo rescue, ovule culture, in vitro pollination and fertilization, and manipulations with protoplasts. These new techniques can be used to (1) obtain novel hybrid combinations, (2) transfer entire cytoplasms or specific cytoplasmic traits, (3) transfer organelles and (4) to transfer isolated chromosomes and genes.

The recent successes with in vitro approaches do not preclude the continued successful use of wide hybridization utilizing conventional methods for many plant species. The objective of this review is to describe and discuss the use of in vitro approaches to wide hybridization in plants with emphasis on the techniques and their successful application where conventional methods fail.

EMBRYO CULTURE

Status and Applications

The technique of culturing isolated embryos is not a new idea as the first attempt dates back to 1754 when Charles Bonnett used embryo culture to evaluate the role of cotyledons in germination (Schopfer, 1943). This research involved removing cotyledons from mature embryos of Phaseolus multiflorus and germinating the naked embryos in soil. In 1904 Hannig aseptically cultured nearly mature embryos of species belonging to the Cochlearia and Raphanus genera on a mineral salts media supplied with 10% raw cane sugar. From this landmark research on plant embryogenesis, it was concluded that salts, nitrogen, and carbon were all essential for embryo culture.

Embryo culture was first utilized for the purpose of interspecific hybridization in 1925 by Laibach who reported the successful rescue of <u>Linum</u> <u>austriacum</u> X <u>L</u>. <u>perenne</u> embryos. The rescue of embryos has been utilized earlier and more often than any other <u>in</u> <u>vitro</u> technique for the purpose of wide hybridization. The enormous applicability of embryo rescue can be seen in the list of over 80 interspecific and intergeneric hybrids that have been obtained over the years using this method (Collins and Grosser, 1984).

<u>In</u> <u>vitro</u> culture of hybrid embryos is useful in cross combinations where fertilization occurs and the embryo begins to develop but degenerates prior to full maturity due to: (1) the inability of the endosperm to carry out its normal role in supplying nutrients to the developing embryo, or (2) the maternal tissue being hostile or antagonistic to the development of the embryo. In other words, embryo culture is useful when the embryo has the ability to mature but is prevented from doing so. A number of reviews detail the technical aspects of embryo dissection and media requirements (North, 1976; Raghavan, 1976, 1977, 1980; Yeung et al., 1981; Williams et al., 1982; Collins and Grosser, 1984).

Strategies for Utilizing Embryo Culture

Manipulation of Explant Source

Hybrid embryos are rescued by aseptically removing embryos prior to their abortion and culturing them directly onto an artificial media. An important aspect of embryo culture is an à priori knowledge of the <u>in</u> <u>vivo</u> system for the species of interest. It is beneficial to identify the existing barrier(s) to hybridization, to select genotypes in which the chances for hybridization are optimum, and to develop a suitable culture system for the species being hybridized.

Attempts to hybridize perennial species of clover (<u>Trifolium</u> sp.) with red clover (<u>Trifolium</u> <u>pratense</u>) have involved such an à priori sequence of experiments. Long-term field persistence is lacking in red clover and considerable efforts have been directed to improving this trait. Hybridization is possible among the perennial species of <u>Trifolium</u>, but none of these species or hybrids will cross with red clover. Apparently, postzygotic barriers are the primary cause involved in the reproductive isolation of red clover from related perennials (Taylor et al., 1980). Since some interspecific crosses involving red clover

produced shriveled seed, it seemed possible that embryo rescue
could aid in hybridization (Taylor et al., 1983).

A second step in this series of experiments was the
identification of genotypes within each species with the greatest
chance of hybridization. The premise was that those combinations
producing the largest seeds, albeit non-viable, would be most
likely to produce embryos which could be rescued. Utilizing
crossing experiments between the two species, genetic differences
for seed size were identified (Taylor et al., 1983). Three
classes of hybrid seed were produced ranging in weight from about
0.2 to 0.4 mg/seed. Phillips et al. (1982) used those
combinations producing the largest seeds as crosses from which to
rescue hybrid embryos.

The identification of superior combining genotypes has also
been utilized for the rescue of hybrid embryos in soybean
(Glycine max) (Broué et al., 1982) and in common bean (Phaseolus
vulgaris) (Mok et al., 1978). Reciprocal cross differences also
can affect hybridization success (Sharma et al., 1980). It
appears then that genotypic factors which facilitate in vitro
hybridization are similar to those utilized in conventional
hybridization approaches [See Hadley and Openshaw (1980) for a
review of the latter].

The success of embryo rescue is influenced by the age of the
embryo at the time of its culture. Two basic stages of embryo
growth exist with regard to nutritional independence (Raghavan,
1976). The heterotrophic stage of growth is the period during
which the embryo depends on the endosperm for its nutrition; it
extends from fertilization up to approximately the heart stage.
The autotrophic phase of growth begins at the late heart stage.
This stage is significant for in vitro culture as the embryo
becomes sufficiently independent of the endosperm for subsequent
growth. As the embryo matures, its growth requirements for
artificial culture become less complex. Older, healthy embryos
are, therefore, easier to culture than younger ones. The culture
of young embryos is also more difficult due to their size, damage
during dissection, and sensitivity to osmotic shock
(P. Maheshwari and Rangaswamy, 1965). The stage at which embryos
can be rescued depends on the time at which endosperm abortion
occurs and the time at which subsequent embryo degeneration
results. For example, Phillips et al. (1982) evaluated the
influence of embryo age (11 to 23 days) on the potential for
Trifolium interspecific embryos to be cultured. A compromise was
reached between the number of hybrid embryos produced and maximum
embryo development. Subsequently, embryos were rescued between
14 and 19 days, which corresponded to the heart stage of

development. Gill et al. (1981) noted that in crosses of
Aegilops squarossa X T. boeticum, endosperms aborted at about 10
days and embryos aborted by 14 days following pollination.
Hybrids were obtained by rescuing embryos 10 days after
pollination.

Media Manipulation

Another critical consideration in embryo culture is the
composition of the culture media (Raghavan, 1977; Yeung et al.,
1981). In addition to mineral salts, media for the culture of
embryos generally consist of a carbon source, one or more
nitrogen sources, vitamins, hormones, and miscellaneous additions
such as coconut milk and casein hydrolysate. Media requirements
have been covered in detail by Raghavan (1976, 1980) and only
general points will be considered here.

The addition of carbohydrates to the media has been shown to
greatly enhance embryo survival. In its natural environment the
young embryo is bathed by an osmotic solution the concentration
of which must be duplicated in vitro. In addition to preventing
osmotic shock to the cultured embryo, sugars also have a
nutritional role (P. Maheshwari and Rangsawamy, 1965) and aid in
preventing precocious germination (Raghavan, 1976). High
concentrations of carbohydrates (12-18%) favor the pregerminal
phase and lower concentrations (less than 2%) favor the
postgerminal phase of growth (P. Maheshwari and Rangaswamy,
1965). Since older embryos require a lower osmotic media for
growth, cultured embryos are often transferred to media with
lower sugar concentrations.

The type of sugar utilized may be important in embryo
culture. In general, sucrose and occasionally glucose have been
reported to be superior carbohydrate sources (Andronescu, 1919;
van Overbeek et al., 1944; Lee, 1952; Amemiya et al., 1956a,b).

The source of nitrogen which is available for cultured
embryos can have a profound influence on survival. Most culture
media include an inorganic source of nitrogen; however, the
addition of an organic form has been shown to have a stimulatory
effect on embryo growth and development in some species (Hannig,
1904; Paris et al., 1953; Rijven, 1955; Monnier, 1978). The form
of organic nitrogen used is also important. Although aspargine
is beneficial in some experiments (Hannig, 1904; Lea et al.,
1979), glutamine is generally superior for a wide range of genera
(Paris et al., 1953; Rijven, 1955, 1956; Mok et al., 1978).
Glutamine alone does not always give the best response and a
mixture of amino acids may be more effective in some instances

(Stafford and Davies, 1979). An undefined nitrogen source such
as casein hydrolysate also can be beneficial (Taira and Larter,
1978).

    The effect of hormones on cultured embryos is not as
apparent as the effect of the carbon and nitrogen sources.
Monnier (1978) felt that hormonal supplements were not necessary
for embryo culture as embryos possess an endogenous hormone
supply. Nevertheless, some investigators feel that hormones may
have a profound influence on embryo development (Raghavan, 1980).
Phillips et al. (1982) found successful culture of interspecific
hybrid embryos of Trifolium required the auxin picloram. Though
not necessary in all cases, hormones have been shown to be
beneficial in the culture of immature embryos of a number of
species (Raghavan, 1976, 1980; Norstog, 1979). In other cases
hormones have had either no effect or have proven to be
inhibitory (Veen, 1963; Mok et al., 1978; Dolezel et al., 1980).

    Since embryos are usually cultured in the autotrophic stage,
the need for vitamins can be met internally (Raghavan, 1980).
However, the addition of vitamins to the culture media may aid in
the survival and growth of less mature embryos. Vitamins which
are frequently added to the culture media include thiamine,
nicotinic acid, pyridoxine, biotin, ascorbic acid, and
pantothenic acid.

    Other media components also influence embryo survival. The
addition of natural plant extracts such as coconut milk has been
shown to be beneficial (van Overbeek et al., 1942; Ball, 1959;
Colonna et al., 1971; Anagnostakis, 1977). The pH of the media
is critical (Choudhury, 1955; Taira and Larter, 1978), and a
range between 5.5 and 6.5 is satisfactory. For the most part,
embryos are cultured on an agar-solidified media and a low agar
concentration seems best (Randolph and Khan, 1960; Stoltz, 1971).
Solid media are superior to liquid media (van Overbeek et al.,
1944; Mauney, 1961; Dolezel et al., 1980).

Other Culture Variables

    An intact suspensor has been shown to be critical for embryo
development (Monnier, 1978; Yeung and Sussex, 1979). The
requirement for the suspensor may vary depending on the cultural
conditions employed. For instance, gibberellin was shown to
replace the need of young Phaseolus embryos for the suspensor
(Cionni et al., 1976). However, other factors can also be
influential, as the culture of suspensor-less embryos of P.
vulgaris X P. lunatus, was possible even when gibberellin was not
present in the culture media (Mok et al., 1978).

The effect of temperature and light on embryo development has not been critically evaluated. Temperature requirements for cultured embryos generally have been shown to be optimal in the range of 25°C to 30°C (Narayanaswami and Norstog, 1964). In culturing mature embryos of onion (<u>Allium cepa</u>), a seven day dark period following embryo removal was shown to have a positive effect on embryo survival (Dolezel et al., 1980). <u>Capsella bursa-pastoris</u> embryos were found to have an age-dependent response to light; this effect could be offset with the addition of hormones (Raghavan and Torrey, 1964). This response implies that the effect of light on embryos may relate to hormone metabolism. In general, the requirement for light in young embryos seems to be less than that of mature embryos, with more mature embryos benefitting from higher light intensities (North, 1976).

An alternative to the direct <u>in vitro</u> culture of embryos is their transplantation into a 'nurse' endosperm. Research on the nurse endosperm method of embryo culture was first initiated by Brown and Morris (1890) who cultured embryos of barley by transplanting them to embryo-less endosperm of barley and wheat. An interesting aspect of the endosperm culture method is the stimulatory effect that the endosperm of one species can have on the growth of other unrelated species (Ziebur and Brink, 1951). Embryo transplantation has been successful in obtaining many interspecific hybrids (DeLautour et al., 1978; Williams, 1978; Williams and DeLautour, 1980, 1981; Williams and Verry, 1981; Broue et al., 1982; Williams and Williams, 1983) and some intergeneric hybrids (Hall, 1954, 1956; Kruse, 1973, 1974). Although this technique is useful, the double dissection method requiring one normal endosperm for each hybrid embryo is tedious (Williams et al., 1982). The nurse endosperm method is useful in cases where a suitable media for isolated embryo culture has not been developed.

OVULE CULTURE

Status and Applications

Often plant embryos can be rescued without dissection. This has been accomplished by culturing intact flowers, ovaries, placenta attached ovules, individual ovules or even half ovules (Withner, 1942, 1943; Nitsch, 1949, 1951; N. Maheshwari and Lal, 1958, 1961a,b; Harberd, 1969; Reed and Collins, 1978). This specialized type of organ culture has been referred to as ovule or ovary culture, fertilized ovule culture, and recently the more general term of "<u>in ovulo</u> embryo culture" has been applied (Stewart and Hsu, 1977). Several reviews exist on the subject of

ovule culture (P. Maheshwari and Rangaswamy, 1965; Beasley et
al., 1974; Raghavan, 1976; Beasley, 1977).

Withner (1942), working with orchids (Cattleya sp.), was the
first to successfully culture excised ovules. He did this to
reduce the time that it takes to produce orchid seedlings.
Nitsch (1949) cultured immature tomato flowers and produced
mature fruits which ripened in vitro but were seedless. Later,
Nitsch (1951) reported the production of viable seeds from
cultured ovaries of tomato and gherkin (Cucumis anguria). Their
successful in vitro development depended on the flowers being
left on the plant for a period of time following pollination.
N. Maheshwari (1958) cultured excised ovules of Papaver
somniferum (opium poppy). This research is significant since
ovules containing embryos as young as the two celled proembryo
stage were successfully cultured to maturity. Despite these
early successes with ovule culture, the method was not used for
the purpose of interspecific hybridization until 1968 by Gadwall
et al., who used it to obtain hybrids of Abelmoschus species.
Table I lists the hybrids which have been obtained to date using
the technique of in ovulo embryo culture.

As with embryo culture, in ovulo culture of embryos helps to
bypass such postzygotic barriers as endosperm abortion and
abnormal proliferation of the integument cells (Reed and Collins,
1980; Douglas et al., 1983). In ovulo embryo culture is not
likely to be useful in those instances where maternal tissue or
the endosperm exerts an inhibitory action on the development of
the embryo. In these cases the culture of isolated embryos would
be advantageous.

Strategies for Utilizing In Ovulo Embryo Culture

Manipulation of Explant Source

In research where in ovulo embryo culture is attempted, an
important consideration for successful culture is the type of
explant that is used. The explant material may range from
individual ovules to intact flowers (Raghavan, 1976), and
depending on the genotype, cultural conditions, and other
variables, the success rate will vary.

In some species, the culture of isolated ovules appears to
be the preferred method. With cotton (Gossypium hirsutum),
attempts to culture isolated fruits (ovaries) were not successful
whereas the culture of isolated ovules has been possible (Beasley
et al., 1974). Douglas et al. (1983) compared the culture of
placenta-attached ovules to the culture of individual ovules from

the crosses <u>N</u>. <u>rustica</u> X <u>N</u>. <u>tabacum</u> and <u>N</u>. <u>rustica</u> X <u>N</u>. <u>glutinosa</u>. In both crosses, total yield of seedlings was superior when free ovules were cultured.

The culture of ovaries instead of individual ovules aids in embryo development in some species. Ovaries of <u>Cooperia</u> <u>pedunculata</u> could be cultured earlier and resulted in more normal appearing seeds than seeds from ovules cultured individually (Sachar and Kapoor, 1958). Similar results were found with cultured ovules and ovaries of <u>Zephyranthes</u> (Sachar and Kapoor, 1959). Ovules of <u>Zephyranthes</u> which were excised two days after pollination, took twice as long to mature and produced smaller seeds than those which developed <u>in vivo</u>. Also, the seeds did not germinate. In contrast, ovaries excised at the same stage contained embryos which grew nearly as fast and produced seeds of the same approximate size as those formed <u>in vivo</u>. In addition, the seed produced viable offspring. Thus, it appears that with certain species the ovary wall or placental tissue can provide components to the developing embryo which are not provided otherwise.

Floral organs such as placentas, calyx lobes, and seed hulls can exert either a beneficial or inhibitory effect on embryo growth and development. In cases where a beneficial effect results, the accessory tissue may aid in the uptake of metabolites from the media and/or in providing additional growth factors to the developing embryo (Raghavan, 1976). Development of wheat ovaries cultured <u>in vitro</u> was found to be positively influenced when the rachillae and palea were attached (Redei and Redei, 1955a, 1955b). Ovaries cultured without maternal tissue rarely developed whereas ovaries cultured with the floral organs could be cultured as early as two days after anthesis. Similar results have also been found for barley (LaCroix et al., 1962; LaCroix and Canvin, 1963). In <u>Hyoscyamus</u> <u>niger</u>, normal fruit development occurred only when the ovaries were cultured with the calyx attached (Bajaj, 1966). In <u>Althaea</u> <u>rosea</u> the presence or absence of the calyx had little influence on fruit growth <u>in vivo</u>, but <u>in vitro</u> the calyx affected not only fruit growth and maturation, but also embryo and endosperm maturation (Chopra, 1962).

The effect of accessory organs is probably dependent on the age of the explant source and is also likely to be influenced by composition of the nutrient media. The <u>in vitro</u> development of <u>Iberis</u> <u>amara</u> fruits and embryos from cultured flowers one day after pollination occurred only when the calyx was left intact (N. Maheshwari and Lal, 1958, 1961a). The effect of the calyx could be partly overcome by adding 5% sucrose to the medium which suggests that unique media requirements may be necessary for each type of explant source. The importance of the interaction between age and explant was also implied by the fact that the calyx had no effect on fruit maturation when ovules were collected more than eight days after pollination. In onion, fruit developed containing viable seeds only when the perianth was present (Guha and Johri, 1966). The addition of an aqueous extract of perianths to the culture media enhanced the growth of perianth-less ovaries, but this was not equal to the results obtained with an intact perianth.

As in embryo culture, à priori investigations should be carried out when in ovulo embryo culture is utilized. Factors that should be considered are embryo maturity, the effects of crossing direction, the identification of responsive genotypes, ploidy levels, and media considerations.

A correlation exists between the time at which ovules are cultured and subsequent embryo development. The stage at which in ovulo embryos have been cultured varies with the species and with the cultural conditions utilized. N. Maheshwari (1958) was able to culture ovules of poppy containing a two-celled proembryo and a few endosperm nuclei into viable seed. Sachar and Kapoor (1959) were successful only with the culture of Zephyranthes ovules containing embryos at the globular or later stages. Tobacco ovules exhibited the best growth when inoculated at the globular stage (Siddiqui, 1964). Stewart and Hsu (1977) cultured cotton ovules to maturity which were removed only two days after pollination (zygote stage). Ovules of C. pedunculata containing embryos at the zygote stage would grow only to the globular stage (Sachar and Kapoor, 1958). However, ovules cultured at the globular stage could be reared to maturity. It is clear that when using ovules, embryos can be cultured at a much earlier stage than would otherwise be possible.

The method used to measure ovule maturity can be important. For instance, pod, ovary, or fruit maturation is not necessarily an indication of embryo maturation and thus should not be used. Poppy ovaries and ovules reach their final dimensions long before endosperm and embryo maturation occurs (N. Maheshwari and Lal, 1961b). Hybrid embryo size and pod development also are not associated in Glycine (Ladinsky et al., 1979). The best method for determining embryo maturity is by direct histological evaluation.

The time after pollination that cultures should be initiated is influenced by the genotypes used and the environment in which the plants are grown. Reed and Collins (1978) used fertilized ovule culture to rescue hybrids of tobacco with three wild tobacco species. Crosses were made from plants grown in the field and the greenhouse, and ovules were cultured from five to 22 days following pollination. Depending on the genotypes involved, a relationship existed between the number of days the ovules were left on the plant before culturing and the environment in which they were grown. In the N. repanda X N. tabacum cross, the only plants obtained from greenhouse grown materials were from ovules cultured six days after pollination. In contrast, the 11-16 day interval was the only productive period for field grown material. The influence of environment was clearly illustrated with the N. nesophila X N. tabacum cross. Hybrids from this cross were produced only when the parents were grown in the field.

Reed and Collins (1978) also observed differences in the time required for the in vivo maturation of the ovules before culture, which varied with the genotype of the hybrid embryos. The maturation times which Reed and Collins (1978) found differed markedly from those observed for the Nicotiana interspecific crosses produced by Douglas et al. (1983). Differences of this extent illustrate the importance of evaluating each new genotypic combination. The importance of an appropriate in vivo development period has also been demonstrated with other crops including cotton (Beasley, 1971; Eid et al., 1973), tomato (Thomas and Pratt, 1981), poppy (N. Maheshwari and Lal, 1958), white clover (T. repens) (Nakajima et al., 1969), and rape (Brassica napus) (Harberd, 1969).

Media Manipulations

The media requirements for naked embryos are different from embryos cultured in ovulo. For cotton ovule growth, it has been shown that a high salt medium is superior to low salt formulations (Beasley, 1971; Stewart and Hsu, 1977). Agar is not required for the culture of cotton ovules and is deleterious to ovule growth (Beasley et al., 1974). Magnesium and boron were essential for fiber elongation, whereas the presence of manganese had only a slight effect.

The form and concentration of available nitrogen can influence the culture of in ovulo embryos. A high concentration of inorganic nitrogen (Eid et al., 1973) in the forms of KNO_3 (Beasley and Ting, 1973) and NH_4^+ (Stewart and Hsu, 1977) positively influenced the growth of cultured cotton ovules. Addition of organic forms of nitrogen has been found to have a beneficial effect on A. cepa and Zephyranthes in ovulo cultured embryos (Kapoor, 1959; Guha and Johri, 1966).

The addition of hormones to the media can be beneficial for the culture of ovules and ovaries (see review by Raghavan, 1976). Many responses are found depending on the genotype cultured, its stage of development at culture initiation, and its interaction with other components in the media. The effect of hormones may vary during the culture period, and thus may necessitate a sequential treatment of media. One must exercise caution as the addition of growth promoters may be beneficial for growth of the ovule and yet have no apparent effect on the embryo (Bajaj, 1964).

The effect on ovule culture of vitamins and miscellaneous additions such as endosperm, yeast extracts, and casein hydrolysate, have generally been beneficial. Vitamins B_1, B_2, and E, although not indispensable, increased ovary growth and seed fertility of Dendrobium nobile (P. Maheshwari and Rangaswamy, 1965). Endosperm extracts have aided in ovary maturation of tomato (Nitsch, 1949) and interspecific hybrids of Brassica (Inomata, 1978a). Extracts of yeast or casein hydrolysate had a positive influence on the growth and differentiation of embryos obtained from ovules of poppy (N. Maheswari and Lal, 1961b) and half ovules of B. campestris (Harberd, 1969).

With in ovulo embryo culture one would expect the requirement of the enclosed embryo for a high osmotic concentration to be largely satisfied by its surrounding, maternal environment. This seems to be the case as most experiments involving ovary and ovule culture have involved lower sugar concentrations than those using embryo culture. However, few reports exist in which this factor has been critically tested. Douglas et al. (1983) evaluated a range of sucrose levels on the development of cultured interspecific ovules of Nicotiana. Genotypic differences existed as sucrose at a concentration of 4% was best for N. rustica X N. tabacum and 2% sucrose was optimal for N. rustica X N. glutinosa. The highest level of sucrose evaluated (8%) increased the frequency of seedlings with abnormal morphology. Beasley and Ting (1973) evaluated a number of carbohydrates (galactose, glucose, mannose, sucrose, and fructose) for their effect on fiber development of isolated cotton ovules. Each carbohydrate had a specific effect on fiber development. For instance, galactose inhibited both ovule and fiber growth and caused the tissue to brown. Sucrose caused the fibers to elongate but browning occurred after 2 weeks. Glucose gave superior results with excellent fiber development. For the culture of D. nobile ovaries it was found that dissacharides were better than monosaccharides and, also, that maltose and lactose were superior to sucrose (Ito, 1961 from Maheshwari and Rangaswamy, 1965).

In Ovulo Embryo Culture vs. Isolated Embryo Culture

There have been a number of studies where both embryo culture and in ovulo embryo culture were attempted. Although the two techniques can be compared, caution should be exercised in drawing conclusions as many subtle factors such as the optimal time for culture initiation, unique media requirements, and environments probably influence the relative success of these methods. A few examples will be given to illustrate the results that this comparison may yield.

Muzik (1956) cultured both ovules and isolated embryos of Hevea brasiliensis at various stages of development. Cultures of both immature ovules and embryos failed to develop despite the additions of sugar, IAA, or coconut milk. However, embryos could be cultured if portions of the cotyledons remained attached to the embryo. Sharma et al. (1980) cultured hybrid embryos and ovules of S. melongena X S. khasianum from fruits of the same age. Only the naked embryos produced plants. The fact that the ovules did not respond is an indication that the barrier to hybridization may have been due to an inhibitory effect of the ovule on embryo development. This is supported by the work of Haynes (1954) who compared the culture of embryos, ovules, and embryos removed from 10-day-old ovule cultures of S. tuberosum. The culture of immature ovules was not successful; however, embryos removed from the conditioned ovules could be successfully cultured three days earlier than embryos that were not conditioned. This work implies that the inhibitory action does not occur early, and that these two techniques can be used together to culture younger embryos than is possible using either method alone.

Both fertilized ovule culture and embryo culture have been used in attempts to hybridize perennial Glycine species with soybean (Ladinsky et al., 1979; Broue et al., 1982; Newell and Hymowitz, 1982). Ladinsky et al. (1979) were unsuccessful in obtaining hybrids of perennial Glycine species with soybean using both conventional hybridization and embryo culture. However, later attempts using ovule culture allowed the direct hybridization of G. max with G. tomentella (Newell and Hymowitz, 1982). Broue et al. (1982) rescued hybrid Glycine embryos by using both embryo culture and ovule culture. The method chosen was based on the size of the embryo contained within the pod. Embryos large enough to be dissected were removed and placed on nurse endosperm. Conversely, in those crosses where embryos were small, whole ovules were transferred intact to a culture media. Hybrid cultures were produced in the combination [G. tomentella X G. canescens] X G. max in about equal frequencies for each of these methods. For the three other cross combinations which

responded, ovule culture was more successful than embryo culture.
Though Broue et al. (1982) did not make direct comparisons
between these two methods, the implication is that ovule culture
in Glycine is as effective as embryo culture in rescuing hybrids.
Also, ovule culture enables the rescue of embryos which abort at
an early stage.

Examples in the literature also document cases where ovule
culture has been more successful than embryo culture. Cotton is
a species in which a substantial amount of ovule culture research
has been conducted and, as such, could be used as a model. One
reason for the emphasis on ovule culture for cotton was that
early efforts to culture young embryos met with only limited
success (Beasley et al., 1974). Isolated embryos cultured before
the heart stage did not develop at all. In contrast, utilizing
ovule culture, cotton embryos can be cultured from the zygote to
the mature embryo stage (Stewart and Hsu, 1977, 1978).

Arisumi (1980a) evaluated 346 Impatiens interspecific
incompatible crosses and of these, 57 were identified as having
potential for in vitro rescue based on the presence of aborted
embryos. However, early embryo abortion necessitated the
development of the ovule culture method. Experiments with embryo
and ovule culture were carried out, and four new interspecific
hybrids were obtained, three by ovule culture and one by embryo
culture (Arisumi, 1980b). Embryo culture was successful only in
culturing well-developed embryos one to two weeks beyond the late
heart stage. In contrast, rescue by ovule culture was possible
with much younger, globular stage embryos.

In three cross combinations involving species of Abelmoschus
where both embryo and ovule culture were attempted, one was
successful with only ovule culture and another with only embryo
culture (Gadwal et al., 1968). The failure of ovule culture to
rescue A. esculentus X A. ficulneus was attributed to the
inhibitory action of the maternal envelope. Therefore, success
with embryo culture was apparently the result of the removal of
the maternal tissue.

Thomas and Pratt (1981), using Lycopersicon interspecific
combinations, have compared the culture of isolated embryos with
a modified method of ovule culture. Rather than dissecting
embryos, they cultured intact ovules onto a callus inducing
media. This 'embryo callus' method increased the frequency of
rescued embryos from less than 0.25% for embryo rescue to 12% for
the ovule culture method.

The method of in ovulo embryo culture has a number of
advantages over culturing isolated embryos. (1) It is much

easier in comparison to embryo culture, as the tedious task of embryo dissection is avoided. (2) As the embryos are not removed from their natural environment, less damage is incurred to them upon culture. (3) Ovule culture is also useful in those situations where embryo abortion occurs so early in the development of the seed that embryo dissection is not practical. (4) In small seeded species such as tobacco, ovule culture may be easier than embryo culture. (5) Ovary culture also may be easier than embryo culture in those crosses where premature fruit abscission is a problem (Shabde-Moses and Murashige, 1979). (6) Embryos cultured in ovulo draw upon both an endogenous and exogenous supply of nutrients; therefore, embryos cultured in ovulo may be easier to culture and can be cultured at earlier stages than isolated embryos. Despite these advantages, the culture of fertilized ovules remains unexploited in comparison to embryo culture (Murashige, 1978; Reed, 1978). Ovule culture has been used only with a small number of genera (Table 1) and this reflects a lack of use of the technique.

IN VITRO POLLINATION AND FERTILIZATION

Status and Applications

The technique of in vitro pollination and fertilization is most useful in hybridization attempts where the pollen tube fails to fertilize the egg or where premature floral abscission occurs. Prefertilization barriers may make it impossible for the pollen tube to grow into the ovary through the micropyle of the ovule.

Conventional techniques have been used to successfully bypass such prezygotic barriers (P. Maheshwari and Rangaswamy, 1965). In cases where premature floral abscission is the barrier, the application of hormones may prolong the life of the flower. Substances can be applied to the stigma to increase pollen germination. These substances include boric acid, organic solvents, and extracts of compatible anthers. In instances where slow pollen tube growth prevents fertilization, modification of the temperature or application of growth regulators may aid pollen growth. Stylar incompatibility can also be overcome by stump pollination, using foreign 'compatible' styles, shortening the style, and manipulating the ploidy levels of the parents. The technique of intraovarian pollination has also been used to effect pollination by bypassing the style (Kanta, 1960; P. Maheshwari and Kanta, 1961). This technique, which involves injecting a suspension of pollen grains directly into the ovary, was used with P. somniferum to produce normal fruits in vivo which contained viable seed (Kanta et al., 1962).

Table 1.   Interspecific and intergeneric crosses in which plants
           have been obtained via in ovulo embryo culture.

Parents	Reference
Abelmoschus esculentus X A. ficulneus	Gadwal et al. (1968)
Brassica oleracea X B. campestris	Harberd (1969)
B. campestris X B. oleracea	Inomata (1977, 1978a,b); Matsuzawa (1978)
B. chinensis X B. pekinensis	Inomata (1968)
Chrysanthemum boreale X C. japonense	Watanbe (1977)
C. makinoi X C. japonense	Watanbe (1977)
C. makinoi X C. ornatum	Watanbe (1977)
Glycine max X G. tomentella	Newell and Hymowitz (1982)
[G. tomentella X G. max] X G. max	Brouè et al. (1982)
Gossypium arboreum X G. amourianum	Stewart (1981)
G. arboreum X G. anomalum	Stewart (1981)
G. arboreum X G. australe	Stewart and Hsu (1978); Stewart (1981)
G. arboreum X G. barbadense	Stewart and Hsu (1978); Stewart (1981)
G. arboreum X G. harknessii	Stewart (1981)
G. arboreum X G. hirsutum	Stewart and Hsu (1978); Stewart (1981)
G. arboreum X G. longicalyx	Stewart (1981)
G. arboreum X G. mustelium	Stewart (1981)
G. arboreum X G. raimondii	Stewart (1981)
G. arboreum X G. somalense	Stewart (1981)

Parents	Reference
G. arboreum X G. stocksii	Stewart (1981)
G. arboreum X G. sturtianum	Stewart (1981)
G. arboreum X G. trilobum	Stewart (1981)
G. barbadense X G. arboreum	Stewart and Hsu (1978); Stewart (1981)
G. barbadense X G. australe	Stewart and Hsu (1978); Stewart (1981)
G. barbadense X G. herbaceum	Stewart and Hsu (1978); Stewart (1981)
G. barbadense X G. hirsutum	Stewart and Hsu (1978)
G. herbaceum X G. amourianum	Stewart (1981)
G. herbaceum X G. australe	Stewart (1981)
G. herbaceum X G. barbadense	Stewart and Hsu (1978); Stewart (1981)
G. herbaceum X G. harknessii	Stewart (1981)
G. herbaceum X G. hirsutum	Stewart and Hsu (1978); Stewart (1981)
G. herbaceum X G. longicalyx	Stewart and Hsu (1978); Stewart (1981)
G. herbaceum X G. somalense	Stewart and Hsu (1978); Stewart (1981)
G. herbaceum X G. stocksii	Stewart and Hsu (1978); Stewart (1981)
G. hirsutum X G. arboreum	Stewart and Hsu (1978); Stewart (1981)
G. hirsutum X G. amourianum	Stewart (1981)

(continued)

Table 1. Continued

Parents	Reference
G. hirsutum X G. australe	Stewart and Hsu (1978); Stewart (1981)
G. hirsutum X G. barbadense	Stewart and Hsu (1978)
G. hirsutum X G. bickii	Stewart (1981)
G. hirsutum X G. herbaceum	Stewart and Hsu (1978); Stewart (1981)
G. hirsutum X G. stocksii	Stewart and Hsu (1978); Stewart (1981)
Impatiens flaccida X I. repens	Arisumi (1980)
I. uguensis X I. epiphytica	Arisumi (1980)
I. uguensis X I. flaccida	Arisumi (1980)
Lolium perenne X Festuca rubra	Nitzsche and Henning (1976)
Lycopersicon esculentum X L. peruvianu	Thomas and Pratt (1981)
Nicotiana nesophila X N. tabacum	Reed and Collins (1978)
N. rustica X N. tabacum	Shizukuda and Nakajima (1982); Douglas et al. (1983)
N. rustica X N. glutinosa	Douglas et al. (1983)
N. stocktonii X N. tabacum	Reed and Collins (1978)

A method with potential for bypassing prezygotic barriers is in vitro pollination and fertilization. The usefulness of this technique depends on the ability to artifically grow zygotes in culture from the time of fertilization to maturity. In this regard, in vitro fertilization is an extension of fertilized ovule culture. Table 2 lists the crosses in which interspecific and intergeneric hybridization have been accomplished using the technique of in vitro pollination and fertilization.

Scientists at the University of Delhi were the first to use this method (Kanta et al., 1962). Their success lies partly in the fact that N. Maheshwari and Lal (1961b) previously were able to culture fertilized ovules of P. somniferum from the zygote or 2-celled proembryo stage to maturity. The methods used for Papaver are similar to that used for other species and will be described briefly. Anthers were collected at about the time of their dehiscence and sterilized by exposure to ultraviolet light. The ovaries were gathered on the day of anthesis, dipped in alcohol, and then lightly flamed for sterilization. The ovules with portions of the placentas were excised, placed onto the culture media, and dusted with pollen. Pollen germination occurred within 15 min and within 2 h, the surface of the ovules was covered with a mat of pollen tubes. Embryo and endosperm development was normal, and within 22 days, fully developed embryos were present.

Strategies for Utilizing In Vitro Pollination and Fertilization

Manipulation of Explant Source

As in embryo culture and ovule culture, preliminary steps should be taken before the method of in vitro pollination is used. Most of the steps are similar to those for intraovarian pollination and basically involve: (1) pollinating ovules at the proper stage; and, (2) providing a nutrient media to support pollen germination, pollen tube growth, and development of seeds to maturity (Kanta and P. Maheshwari, 1963b; Rangaswamy, 1977; Zenkteler, 1980). However, the major concern of in vitro fertilization is the development of fertilized ovules since fertilization is not usually the problem (Maheshwari and Rangaswamy, 1965). To accomplish in vitro pollination and fertilization, a knowledge of the following aspects of the species floral biology is beneficial (Rangaswamy, 1977): (1) anthesis, (2) dehiscence of the anthers, (3) pollination, (4) pollen germination, (5) pollen tube growth and ovule penetration, (6) fertilization, and (7) seed maturation.

As with in ovulo embryo culture, the explant source used for in vitro pollination can have a substantial influence on the

Table 2.   Interspecific and intergeneric hybridization via _in_ _vitro_
           pollination and fertilization.

Parents	Reference
A. Crosses in which viable embryos or plants have been obtained.	
Gossypium hirsutum X G. arboreum	Stewart (1981)
G. arboreum X G. hirsutum	Stewart (1981)
Melandrium album X M. rubrum	Zenkteler (1967)
M. album X Silene schafta	Zenkteler (1967, 1969)
M. album X Viscaria vulgaris	Zenkteler (1969)
M. rubrum X M. album	Zenkteler (1969)
Nicotiana tabacum X N. amplexicaulis	Lar'kina (1980); DeVerna and Collins (unpublished)
N. tabacum X N. benthamiana	DeVerna and Collins (unpublished)
N. tabacum X N. debneyi	Ternovskii et al. (1976)
N. tabacum X N. repanda	DeVerna and Collins (unpublished)
N. tabacum X N. rosulata	Ternovskii et al. (1976); Lar'kina (1980)
N. tabacum X N. rustica	Marubashi and Nakajima (1982)
N. alata X N. debneyi	Zenkteler et al. (1981)
Petunia parodii X P. inflata	Sink et al. (1978)
Zea mays X Z. mexicana	Dhaliwal and King (1978)
B. Crosses in which fertilization occurred but embryos did not mature.	
Melandrium album X Campanula persicifolia	Zenkteler et al. (1975)

Parents	Reference
M. album X Cucubalus baccifer	Zenkteler (1969)
M. album X Datura stramonium	Zenkteler (1970)
M. album X Dianthus carthusianorum	Zenkteler (1967)
M. album X Dianthus serotinus	Zenkteler et al. (1975)
M. album X Mathiola indica	Zenkteler (1980)
M. album X Minuartia kitaibelli	Guzowska (1971)
M. album X Minuartia laricifolia	Zenkteler et al. (1975)
M. album X P. hybrida	Zenkteler and Melchers (1978)
M. album X Silene alpina	Zenkteler et al. (1975)
M. album X S. friwaldskyana	Zenkteler et al. (1975)
M. album X S. tatarica	Zenkteler (1967, 1969)
M. album X S. vulgaris	Guzowska (1971)
M. album X Vaccaria pyramidata	Zenkteler et al. (1975)
Nicotiana tabacum X Hyoscyamus niger	Zenkteler and Melchers (1978)
N. tabacum X Petunia hybrida	Zenkteler and Melchers (1978)
N. tabacum X Physosclaina praealta	Zenkteler et al. (1981)
N. tabacum X Withamia somnifera	Zenkteler et al. (1981)
Trifolium ambiguum X T. repens	Richards and Rupert (1980)
T. ambiguum X T. hybridum	Richards and Rupert (1980)

results. In vitro pollination has been accomplished using an array of explant sources ranging from whole flowers to individual ovules. Whole flowers are more amenable because they usually require a less refined medium, and also the injury to the enclosed ovules is greatly reduced. However, the culture of intact flowers negates some of the advantages of in vitro fertilization, especially in terms of bypassing stigma and stylar barriers.

The pollination and fertilization of whole pistils in vitro has been accomplished with a number of species (Rao, 1965; Shivanna, 1965; Usha, 1965; Dulieu, 1966; Rao and Rangaswamy, 1972; Balatkova and Tupy, 1973). As in fertilized ovule culture, the presence of accessory organs often has a beneficial effect on embryo development. As an example, pollinations involving Petunia hybrida were carried out by culturing pistils with the ovary wall removed and applying the pollen directly to the ovules (Wagner and Hess, 1973). Partial removal of the style did not affect fertilization but its complete removal had deleterious effects. Also, the presence of the calyx had a positive effect on N. rustica fruit growth (Rao and Rangaswamy, 1972) as did sepals on P. violacea (Shivanna, 1965).

Pistils of T. repens and T. ambiguum cultured without floral accessory structures and pollinated in vitro showed no evidence of fertilization (Richards and Rupert, 1980). When pistils were cultured with attached calyx lobes and pedicels, fertilization occurred in both species, and plants were produced in T. repens. The successful in vitro pollination of Trifolium is significant as it is the first report of its kind for legumes. It also points out the difficulty of applying this technique to species with only a few ovules per ovary.

In vitro pollinated placenta attached ovules are frequently used as the explant source. However, it may be more difficult to rear embryos to maturity using this technique since the explant source is more restricted than whole pistils. In addition, this method is more likely to injure ovules than is pistil culture and the beneficial nutritive effect of the ovary wall is removed. Whole ovaries are often cultured with a portion of the ovary wall removed to expose the ovules (Zenkteller, 1965, 1967; Rangaswamy and Shivanna, 1972; Watanbe, 1977), or whole or part of the placenta with attached ovules can be cultured (P. Maheshwari and Kanta, 1961; Kanta and P. Maheshwari, 1963b; Balatkova and Tupy, 1972).

The most restricted and vulnerable explant source has been individual ovules. The culture of individual ovules requires an additionally refined media and pollination technique. Stewart's

(1981) success in obtaining interspecific hybrid embryos of
Gossypium was attributed to the highly developed cotton
fertilized ovule culture methods that were already available for
cotton. Several other workers have reported the use of isolated
ovules pollinated in vitro (Kameya et al., 1966; Guzowska, 1971;
Rao and Rangaswamy, 1972).

The maturity of the maternal tissue at the time of culture
has had a profound influence on the success of in vitro
pollination and fertilization. This was noted in the original
report on in vitro pollination and fertilization by Kanta et al.
(1962). These investigators were able to increase the percentage
of ovules fertilized by collecting the flowers 24 h after
anthesis rather than at anthesis as reported in their earlier
work. Likewise, in corn (Zea mays) four days post silking was
best for most genotypes (Gengenbach, 1977b). Differences in the
time of day in which pollinations were made also seemed to be a
factor in successful fertilization. It appears that ovules
collected after the time of anthesis are much more responsive to
fertilization in vitro than those collected prior to anthesis.

The period between the time of inoculation of the ovules
onto the culture media and pollen application has been evaluated
with corn (Raman et al., 1980). Pollen was applied at intervals
following ovary culture. The highest percentage of ovules
fertilized were those pollinated at the time of culture;
thereafter, the response decreased and reached 0 after 72 hours.

The genotype of the male and female parent has been shown to
be important in conventional hybridization (Hadley and Openshaw,
1980). The effect of genotype on in vitro pollination has been
compared in a few species (Niimi, 1976; Gengenbach, 1977b;
Richards and Rupert, 1980). Niimi (1976) was able to obtain
mature seeds from P. hybrida clone W166H via placental
pollination but not from clone K146BH, even though pollen
germination was observed. Gengenbach (1977b) evaluated 11
singlecross and 20 inbred corn genotypes for fertilization
percentage. Singlecross females had an average fertilization
percent (42%) higher than that of inbred-line females (18%). In
all the genotypes, fertilization ranged from 0 to 84% and
significant effects were attributed to male, female, and male X
female interactions. Environment also had a significant effect.
Richards and Rupert (1980) used different ploidy levels of T.
repens and T. ambiguum for in vitro selfing and hybridization.
Viable selfed embryos of T. ambiguum were obtained using a
48-chromosome race, but not for the 16 or 32-chromosome races.
In the interspecific cross, T. ambiguum X T. repens, the 32 and
48 chromosome races as pistillate parents were more tolerant of
the hybrid embryos than the 16 chromosome race. Thus genotype

and ploidy are factors which should be considered.

Pollen Collection and Application

With some species it is easy to obtain viable pollen whereas with other species it is difficult and preliminary evaluations may be beneficial. Guzowska (1971) examined members of the Caryophyllaceae and Cruciferae families to determine their potential as male parents. Pollen grains were applied under nonaseptic conditions onto single ovules (Caryophyllaceae) and placenta attached ovules (Cruciferae) in order to identify genotypes with high pollen germination and, therefore, high potential as male parents for fertilization in vitro. Subsequently, experiments with the responding genotypes were carried out under aseptic conditions.

A critical aspect to pollen viability is the method used to obtain large amounts of aseptic pollen. A technique which is still commonly reported in the literature, and the one originally used by Kanta et al. (1962), is disinfection of pollen by exposure to ultraviolet light. With some species, surface disinfection of the anthers prior to dehiscence is possible; subsequently, the anthers are allowed to dehisce under sterile conditions and the pollen is then ready to use (Petru et al., 1964; Balatkova et al., 1977).   Pollen may be produced aseptically by allowing flower buds to bloom in a sterile container (Zenkteler, 1967). A unique situation is possible with corn in which it is possible to let the silks hang outside of the petri dish and, as a result, pollination is aseptic even though carried out under nonaseptic conditions (Gengenbach, 1977a).

Additional factors also influence pollen viability. Difficulties in obtaining viable, aseptic pollen of Antirrhinum majus made it necessary to develop a new procedure for pollen collection (Balatkova and Tupy, 1973). Initially, all attempts at obtaining good pollen from isolated anthers failed and germinable pollen could only be obtained from anthers that were left on the stem. Suitable pollen was finally obtained by collecting inflorescences one day prior to anthesis, disinfecting them, and after a two-day incubation period collecting the pollen. The importance of having a viable source of pollen and monitoring its viability cannot be overemphasized.

The media upon which the pollen is cultured can affect its viability. Sladky and Havel (1976) found the germination ability of corn pollen culture in a synthetic media containing sucrose and agar to be low (2%). The germination capacity reached 90%, however, when fresh egg yolk was added. A comparison of media containing a series of sugars showed that pollen germination and

pollen tube growth of A. majus was best when lactose was used
(Balatkova and Tupy, 1973). Work by others (Sen and Verma, 1958;
Golubinskii, 1974) has shown that the germination capacity of
pollen is influenced by the presence of calcium and potassium.
Boric acid has also been reported to positively affect pollen
germination (Zubkova and Sladky, 1975).

Temperature has an influence on pollen tube growth
(P. Maheshwari and Rangaswamy, 1965). Maximum growth usually
occurs in the range of 20°C to 30°C. However, with styles of
Narcissus pseudonarcissus pollinated in vitro, fertilization and
subsequent seed set were increased when the incubation
temperature was lowered from 25°C to 15°C (Balatkova et al.,
1977). This probably relates to the natural habitat of the plant
as it normally flowers in the spring when the temperature is
lower.

The influence of the age of the pollen on in vitro
fertilization has been investigated with N. tabacum to determine
whether pollen tubes in which the second pollen mitosis had
occurred were capable of fertilization (Balatkova and Tupy,
1968). Pollen was cultured for 8 h and 20 h and then applied to
the ovules. Fertilization occurred in both treatments; however,
the germinated pollen in which the second mitotic division had
occurred (20 h) consistently fertilized fewer ovules. It appears
that young pollen is superior.

If viable pollen is placed on ovules that are moist, pollen
viability can be drastically decreased (Balatkova and Tupy,
1968). Zenkteler (1980) found that when water was present on the
surface of potato and tomato ovules the pollen tubes frequently
failed to germinate, and when they did, they later burst.

The amount of pollen that is applied to ovules has also been
shown to be important. To determine the efficiency of different
pollination techniques on corn, Raman et al. (1980) placed either
germinated or ungerminated pollen, at various concentrations, on
the stigma. Fertilization was independent of pollen germination;
the amount of pollen applied was the important factor. If only a
single pollen grain was applied to sectors of 4-8 ovules, an
average of 4 percent of the ovules developed. The application of
10 pollen grains increased this frequency to almost 50 percent.
Dulieu (1966), who pollinated pistils of tobacco in vitro, also
noted the relation of amount of pollen applied to fertilization
frequency.

Balatkova and Tupy (1972) found that the broad application
of pollen to the surface of tobacco placenta-attached ovules had
a negative effect on the amount of seed set. On the other hand,

pollen placed at a localized site on the placenta was more
efficient than pollen applied to the surface of the agar in
proximity to the placenta. Also, when pollen was applied only to
a portion of the placenta, seed formation occurred on the
opposite side (Balatkova and Tupy, 1968). In cotton, pollen
placed near the micropylar end of isolated ovules resulted in a
higher fertilization rate than pollen placed casually on the
ovule (Stewart, 1981). Niimi (1976), working with P. hybrida
placenta-attached ovules, compared the placement of pollen grains
directly on the placenta to placement of pollen on stigmas of
shortened styles which were placed in the vicinity of the
placenta. A substantially higher fertilization frequency
occurred when pollen was applied to the stigma of detached
pistils. Thus, it appears that the site of pollen application in
vitro can have a marked influence on fertilization success.

Intact pistils, placenta-attached ovules, and isolated
ovules all may be pollinated in vitro. Fertilization in vitro of
individual ovules is the most difficult. Seeds of A. majus could
be obtained only by pollinating the stigmas of entire pistils
cultured in vitro (Balatkova and Tupy, 1973). Seeds did not
develop when pollen was applied to placenta-attached ovules.
This was attributed to the inability of the pollen tube to
penetrate the micropyle. Fertilization and embryo development in
N. rustica was accomplished when entire pistils were pollinated
in vitro, but not when individual ovules were pollinated (Rao and
Rangaswamy, 1972).

SOMATIC HYBRIDIZATION

Status and Applications

The use of protoplasts for wide hybridization has unique
advantages over sexual methods for transferring individual genes,
organelles, and other foreign particles, creating novel
cytoplasmic mixtures, and combining unrelated genomes.

Protoplast fusion to achieve wide hybridization is unique
from other hybridization techniques in that it is an asexual
process. Prezygotic barriers are automatically eliminated and
the hybrid cell is free from maternally induced inhibitions.
Parasexual hybridization is a multistage process and its success
depends on the ability of the investigator to select the fused
protoplasts and to regenerate these into healthy plants (Bottino,
1975). On the other hand, the regeneration of hybrids may be
hindered by the incompatibility of the genomes involved, and
hybrids, if obtained may be sterile.

Efforts at producing hybrids by asexual means were initiated by Winkler at the beginning of this century (Melchers and Labib, 1974). He attempted to produce somatic hybrids between L. esculentum and S. nigrum by grafting. Through grafting he hoped to induce spontaneous cellular fusion at the graft junction. Putative hybrids were identified but their hybrid nature was later refuted (Brabec, 1949, 1954, 1965).

Hybridization using protoplasts was first attempted at about the same time as Winkler's work by Kuster (1909, 1910, see Schieder and Vasil, 1980). These early attempts were limited by the inability to generate sufficient viable protoplasts and fusions. Cocking (1960) was the first to enzymatically isolate protoplasts from plant tissues. Subsequently, a variety of enzyme preparations have become commercially available to isolate large quantities of protoplasts from virtually any plant source (Vasil and Vasil, 1980).

Procedures to generate a high frequency of fusions have also been developed. Although protoplasts can undergo fusion spontaneously, the frequency is low and, thus, methods to induce fusion are always used. One of the first methods involved the use of sodium nitrate (Power et al., 1970). Other methods include treatments with calcium at high pH, elevated temperatures, dextran, polyvinyl alcohol, synthetic phospholipids, and electrical stimulation (Keller et al., 1982). The most popular method involves the use of high molecular weight polyethylene glycol eluted at high pH with calcium ions. Although a high fusion frequency is obviously desirable, extensive and rapid fusion may have a negative effect on the survival of the cells (Cocking, 1976). Many factors relating to the donor source and physical and chemical components of the system influence the rate and frequency of protoplast fusion and heterokaryon survival (Keller et al., 1982).

Carlson et al. (1972) were the first to regenerate hybrid plants from fused protoplasts. Their success in generating the hybrid between N. glauca and N. langsdorffii was dependent on the prior knowledge of the unique growth characteristics of the parents and sexually derived hybrid cell lines. Under the cultural conditions employed, protoplasts isolated from either parent would not form callus, whereas only about 0.01% of the sexually derived hybrid plants would produce callus. In addition, the sexual hybrid was hormone autotrophic which allowed a further selective step. Since that time somatic hybridization has been used frequently to produce hybrid cell lines and plants (Table 3).

Table 3.   Interspecific and intergeneric hybrid plants or dividing
cells produced via somatic hybridization.

Parents	Reference
A. Interspecific somatic hybrid plants.	
Brassica oleracea + B. campestris	Schenck and Robbelen (1982)
Daucus carota + D. capillifolius	Dudits et al. (1977); Kameya et al. (1981); Toshiaki et al. (1981)
Datura innoxia + D. candida	Schieder (1980)
D. innoxia + D. discolor	Schieder (1978a)
D. innoxia + D. sanguinea	Schieder (1980)
D. innoxia + D. stramonium	Schieder (1978a)
Medicago sativa + M. falcata	Teoule (1983)
Nicotiana glauca + N. langsdorffii	Carlson et al. (1972); Smith et al. (1976); Chen et al. (1977); Chupeau et al. (1978); Uchimaya et al. (1983)
N. knightiana + N. sylvestris	Maliga et al. (1977); Menczel et al. (1978)
N. tabacum + N. alata	Nagao (1979)
N. tabacum + N. debneyi	Gleba et al. (1979); Kosakovs'ka (1980)
N. tabacum + N. glauca	Evans et al. (1980); Uchimiya (1982)
N. tabacum + N. glutinosa	Nagao (1979); Uchimiya (1982)
N. tabacum + N. knightiana	Maliga et al. (1978); Menczel et al. (1981)

Parents	Reference
N. tabacum + N. nesophila	Evans et al. (1981); Evans et al. (1982)
N. tabacum + N. otophora	Evans et al. (1983)
N. tabacum + N. repanda	Nagao (1982)
N. tabacum + N. rustica	Nagao (1978); Iwai et al. (1980); Douglas et al. (1981); Peitian et al. (1981)
N. tabacum + N. stocktonii	Evans et al. (1981)
N. tabacum + N. sylvestris	Melchers (1977); Zelcer et al. (1978); Evans (1979); Aviv and Galun (1980); Medgyesy et al. (1980); Hein et al. (1983)
Petunia hybrida + P. axillaris	Izhar and Power (1979)
P. hybrida + P. parodii	Power et al. (1976, 1977); Cocking et al. (1977); Izhar et al. (1983)
P. parodii + P. inflata	Power et al. (1979)
P. parodii + P. parviflora	Power et al. (1980)
Solanum nigrum + S. tuberosum	Binding et al. (1982)
S. tuberosum + S. chacoense	Butenko and Kuchko (1979)

B.   Interspecific fusion products observed to undergo division but not plantlet regeneration.

Corchorus ditorius X C. capsularis	Kumar et al. (1983)
Vicia hajastana + V. villosa	Kao et al. (1974)
V. hajastana + V. narbonensis	Rennie et al. (1980)

(continued)

Table 3.   Continued

Parents	Reference
Nicotiana tabacum + N. paniculata	Hein et al. (1983)

C.   Intergeneric somatic hybrid plants.

Parents	Reference
Arabidopsis thaliana + Brassica campestris	Gleba and Hoffmann (1978, 1979, 1980); Hoffmann and Adachi (1981)
Datura innoxia + Atropa belladonna	Krumbiegel and Schieder (1979, 1981)
Daucus carota + Aegopodium podagraria	Dudits et al. (1979)
Nicotiana tabacum + Petunia hybrida	Binding (1976); Xianghui et al. (1982)
N. tabacum + Salpiglossis sinuata	Nagao (1982)
Solanum tuberosum + Lycopersicon esculentum	Melchers et al. (1978)

D.   Intergeneric fusion products observed to undergo division but not plantlet regeneration.

Parents	Reference
Atropa belladonna + Nicotiana chinensis	Gleba et al. (1982)
A. belladonna + N. tabacum	Gosch and Reinert (1978)
Daucus carota + N. tabacum	Gosch and Reinert (1978); Harms and Oertti (1982); Hauptmann and Widholm (1983)
D. carota + Hordeum vulgare	Dudits et al. (1976)
D. carota + Petunia hybrida	Reinert and Gosch (1976); Gosch and Reinert (1978)
Glycine max + Brassica napus	Kartha et al. (1974)
G. max + Caragana arborescens	Constabel et al. (1975b)
G. max + Colchicum autumnale	Constabel et al. (1976)

Parents	Reference
G. max + H. vulgare	Kao et al. (1974)
G. max + Melilotus sp.	Constabel et al. (1975b)
G. max + Medicago sativa	Constabel et al. (1975b)
G. max + N. glauca	Constabel et al. (1976); Kao (1977); Wetter and Kao (1980)
G. max + N. langsdorffii	Constabel et al. (1976)
G. max + N. rustica	Constabel et al. (1976)
G. max + N. tabacum	Constabel et al. (1976); Wetter (1977); Chien et al. (1982)
G. max + Pisum sativum	Kao et al. (1974); Constabel et al. (1975a, 1976)
G. max + Vicia hajastana	Kao et al. (1974); Constabel et al. (1977)
G. max + Zea mays	Kao et al. (1974)
N. tabacum + Hyoscyamus muticus	Potrykus et al. (1983)
Petunia hybrida + Atropa belladonna	Gosch and Reinert (1976, 1978)
P. hybrida + Parthenocissus tricuspidata	Power et al. (1975)
P. hybrida + Solanum nigrum	Nehls (1978)
P. hybrida + Vicia faba	Binding and Nehls (1978); Horst and Nehls (1978)
Sorghum bicolor + Zea mays	Brar et al. (1980)
Vicia hajastana + Pisum sativum	Kao et al. (1974)

Somatic hybridization is unique from sexual hybridization in that amphidiploids are generated directly and also in that novel cytoplasmic combinations result. An interesting feature of somatic hybridization is the fate of the cytoplasmic organelles following fusion. Belliard et al. (1978) fused a cytoplasmic male fertile line of N. tabacum with a male sterile line. Restriction endonuclease analysis of the chloroplast DNA from regenerated hybrids revealed that only one of the two parental chloroplasts DNAs was present; apparently the chloroplast populations did not mix or the chloroplasts sorted out. The mitochondria, which are responsible for male sterility, were also analyzed using restriction endonucleases (Belliard et al., 1979). Unique DNA patterns were observed indicating mitochondrial DNA recombination had occurred as shown by the array of generated floral types. Similar results have also been found for other members of the Solanaceae (Chen et al., 1977; Melchers et al., 1978; Galun et al., 1982; Boeshore et al., 1983). Apparently, mitochondrial recombination and chloroplast segregation are common phenomenon in somatic hybrids.

Strategies for Heterokaryon Selection

Providing the appropriate conditions for regenerating plants from protoplasts and the ability to select the desired fusion products are of fundamental significance. In fact, having an appropriate selection method has been identified as being the single most important factor in obtaining somatic hybrids (Vasil et al., 1979). There are a variety of selection procedures which have been utilized and these have been categorized by several authors (Cocking, 1978; Schieder and Vasil, 1980; Keller et al., 1982; Widholm, 1982).

The selection of hybrids may be carried out immediately following the fusion event or after whole plants have differentiated. Selection at the cellular level is usually more efficient and requires either the à priori knowledge of unique parental and hybrid growth characteristics, the use of mutants, or the physical separation of hybrids from parental cell types. A few examples of these will be given to convey the importance of a selection scheme and the types that may be utilized.

Carlson et al. (1972) were successful in selecting the somatic hybrid N. glauca + N. langsdorffii because of the unique growth characteristics of the cell lines involved. This method is limited to generating somatic hybrids of species where sexual hybridization is possible. As a result, it has little potential for obtaining new hybrid combinations.

Complementation selection is an effective method for isolating hybrids. Glimelius et al. (1978) used complementary nitrate reductase mutants to obtain a large number of N. tabacum hybrid cell lines. Each of these mutants were recessive and represented a unique aspect of nitrate metabolism. The hybrid cell lines then, were unique in their ability to grow when nitrate was the only nitrogen source. This sort of selection scheme is most useful when the genetic and biochemical basis for each mutation is understood.

In cases where only one of the parents carries such a mutation, complementation selection may be carried out when the other parent has a unique characteristic which allows it to be distinguished from the hybrid. For example, Cocking et al. (1977) fused an albino suspension line of P. hybrida with mesophyll protoplasts of P. parodii. The inability of P. parodii protoplasts to grow past the small colony stage allowed for the selection of hybrid colonies by their green phenotype. This type of selection scheme should be applicable to a large number of systems as albinos have been identified in many species. Other complementary mutants can also be used but are more difficult to obtain and require genetic characterization.

Cell lines carrying resistance to specific chemicals can also be useful. In this case, hybrid cell lines are selected by their ability to grow in the presence of the compound and therefore the resistant trait must be dominant. White and Vasil (1979) fused cell lines of N. sylvestris carrying resistance to the amino acid analogues S-2-aminoethyl-cysteine and 5-methyl-tryptophan. The hybrids were selected by their ability to grow in the presence of both compounds. N. tabacum cell lines resistant to hydroxy urea and able to utilize glycerol as the sole carbon source were fused and hybrids were selected by culture in hydroxy urea and glycerol (Evola et al., 1983).

In utilizing somatic hybrids for breeding purposes, most selectable mutants are not useful because of undesirable effects. Selection schemes which do not involve mutants are more suitable. One such system is based on an expression of vigor in the hybrid lines. A major problem with the use of vigor expression is that the growth response of novel hybrids cannot be predicted.

Nehls (1978) utilized metabolic inhibitors to isolate hybrids. S. nigrum and P. hybrida were treated with diethylcarbonate and iodoacetate, respectively, prior to fusion. Each of these inhibitors irreversibly affect cell metabolism but are complementary. Following fusion three multicellular regenerates were identified one of which was verified as being a hybrid.

Visual screening of hybrid cells has the most potential for selection purposes. The advantages to visual selection are that mutant lines are not necessary and the recovery of hybrids derived from one fusion product is possible. Heterokaryons derived from mesophyll tissue and suspension cultures can be visually identified by the presence of chloroplasts and dense cytoplasmic strands. A more refined and universal method is the use of dyes for the identification of hybrid cell types (Galbraith and Galbraith, 1979; Galbraith and Mauch, 1980). Kao (1977) used a combination of serial dilutions and microdroplet culture to isolate and culture individual fusion products of soybean with N. glauca. Gleba and Hoffmann (1978) used a micropipette to mechanically isolate and culture single hybrid cells of Arabidopsis thaliana and B. campestris. Disadvantages to these methods are the difficulty of culturing the cells at such low densities and the considerable patience and dexterity required for handling the micropipette. Menczel et al. (1978) also used a micropipette to isolate heterokaryons of N. knightiana and N. sylvestris but cultured the heterokaryons in an albino nurse culture rather than in microdroplets. Menczel et al. (1981) later isolated somatic hybrids of N. knightiana and N. tabacum using a micropipette, but placed 35 heterokaryons into a 9 microliter droplet of culture meida. More recently manual isolation has been accomplished using a micromanipulator (Patnaik et al., 1982; Hein et al., 1983) affording a considerable improvement in protoplast isolation. Heterokaryons can be automatically identified and separated using a sorter-flow cytometer (Galbraith and Harkins, 1982), but to date no hybrid plants have resulted from such heterokaryon isolations. The lack of success with this method has been primarily due to technical difficulties which affect cell viability.

Somatic Incompatibility vs. Sexual Incompatibility

Of fundamental importance to somatic hybridization is the relationship between zygote incompatibility and somatic incompatibility. If the same mechanisms which cause zygotic incompatibility cause somatic incompatibility, then it is likely that somatic hybridization will not extend the limits of hybridization much beyond that possible via embryo culture or in vitro pollination and fertilization. A possible exception is that novel cytoplasmic combinations which can arise in parasexual hybrids might result in hybrids which otherwise are not possible. The subject of somatic incompatibility has recently been reviewed by Harms (1982, 1983) and the insights provided should be a catalyst to future research in the area.

The exact level at which somatic incompatibility acts is difficult to determine but it is clear that the fusion process

itself is not inhibitory. The fusion process is a physical
phenomenon and not a physiological one (Harms, 1983). Because it
is physical, it is nonspecific as demonstrated by the ability to
fuse protoplasts from plant and animal kingdoms (Jones et al.,
1976; Davey et al., 1978).

Once cells are fused, mechanisms exist which determine the
ability of hybrid cells to survive and differentiate. Those
mechanisms leading to somatic incompatibility are similar to
those for zygotic incompatibility and may relate to: asynchronous
cell divisions, the arrangement of the spindle apparatus and
alignment of the parental chromosomes, genomic balance, factors
affecting the timing and execution of critical processes such as
differentiation and morphogenesis, and unfavorable genome-plasmon
interactions (Harms, 1983).

Somatic incompatibility barriers may be possible to
overcome. Methods similar to those used for conventional
hybridization such as the manipulations of ploidy levels, the use
of aneuploids, and the identification of genetic factors which
regulate chromosome elimination, pairing, and timing of critical
processes may all be useful in overcoming these barriers.
Backfusions may facilitate the transfer of single genes,
chromosomes, and genomes.

X-irradiation has recently been used to aid in the transfer
of nuclear genes between distantly related genera (Dudits et al.,
1980; Gupta et al., 1982). Nitrate reductase-deficient cells of
Nicotiana were transformed by fusing them with X-irradiated
inactive protoplasts of Physalis and also Datura (Gupta et al.,
1982). Dudits et al. (1980) fused a nuclear albino mutant of
carrot (Daucus carota) with X-irradiated protoplasts of parsley
(Petroselinum hortense) and restored the green carrot phenotype.
Similar results have been achieved with cytoplasmic traits
(Zelcer et al., 1978).

Parasexual hybridization between distantly related species
frequently results in chromosome elimination (Power et al., 1975;
Kao, 1977; Gleba and Hoffmann, 1978) and this may be advantageous
in developing chromosome addition or substitutions lines. In
most instances, hybridization has been utilized only as a means
to transfer one or a few genes. Therefore, generating symmetric
somatic hybrids (hybrids containing the complete genome of each
parent) may not be necessary. An asymmetric hybrid, produced via
chromosome elimination may be all that is required, providing
that the proper chromosome(s) are retained. Asymmetric hybrids
may have an advantage over symmetric hybrids in their ability to
undergo morphogenesis and to produce viable gametes. For
example, Hoffmann and Adachi (1981) maintained hybrid cell lines

derived from the fusion of <u>A</u>. <u>thaliana</u> and <u>B</u>. <u>campestris</u>
("Arabidobrassica") which had not previously shown chromosome
elimination or shoot regeneration. After a period of time on
regeneration media, chromosome elimination in one of these cell
lines began to occur and was followed by shoot formation. From
that point, the morphogenetic potential of the line increased
dramatically and a wide range of morphological types were
produced.

Where chromosome elimination from somatic hybrids is not
desired, genetic stability may be obtained by backfusions.
Somatic hybrids of <u>G</u>. <u>max</u> + <u>N</u>. <u>glauca</u> exhibit preferential loss
of <u>N</u>. <u>glauca</u> chromosomes and after six months, only a few of the
<u>N</u>. <u>glauca</u> chromosomes remain (Kao, 1977). Wetter and Kao (1980)
'backfused' hybrid cell lines twice, over a period of seven
months, to stabilize the somatic hybrid. After a period of six
months following the second backfusion, numerous <u>N</u>. <u>glauca</u>
chromosomes were still present. The ability to backfuse is a
unique application of protoplasts and can also be useful in
instances where progeny of wide crosses are abnormal and useless.
Dudits et al. (1979) fused protoplasts of carrot (2N=18) with
<u>Aegopodium</u> <u>podagraria</u> (2N=42) to produce abnormal hybrid plants
which had the identical chromosome number of carrot, but with
<u>Aegopodium</u> specific traits. An additional carrot genome was
later introduced by backfusion and gave rise to plants which grew
and developed normally but were sterile (Dudits, 1981). Thus,
the use of backfusions is a potentially valuable plant breeding
tool both for sexually and asexually derived hybrids.

It is clear then, that somatic hybridization may be
extremely beneficial in transferring individual genes,
chromosomes, and even in the reconstruction of chromosomes
(Hoffmann and Adachi, 1981). Both somatic hybridization and
other <u>in</u> <u>vitro</u> methods have been attempted to obtain cross
combinations not normally possible utilizing conventional
hybridization methods. These can be used as an indication of the
potential of parasexual hybridization in obtaining novel
symmetric hybrids. Indeed, reports have shown that somatic
hybridization can be useful in overcoming both prezygotic and
postzygotic barriers (Melchers et al., 1978; Schieder, 1978;
Power et al., 1979, 1980; Evans et al., 1982; Nagao, 1982).

Endosperm failure prevents sexual hybridization of <u>N</u>.
<u>nesophila</u>, <u>N</u>. <u>stocktonii</u> and <u>N</u>. <u>repanda</u> with <u>N</u>. <u>tabacum</u> (Reed and
Collins, 1980). Reed and Collins (1978) were able to obtain
these three hybrids using fertilized ovule culture; however,
seedling lethality prevented the maturation of <u>N</u>. <u>repanda</u> X <u>N</u>.
<u>tabacum</u>. Evans et al. (1981) later attempted to obtain these
three hybrids using protoplast fusion and successfully recovered

the N. nesophila and N. stocktonii combinations. Somatic hybrid
combinations involving N. repanda divided but no hybrid plants
were recovered. These results support those of Reed and Collins
(1978) and point toward a relationship between sexual and somatic
incompatibility. However, Nagao (1982) later was able to obtain
the N. tabacum X N. repanda hybrid. Thus it appears that the
inability to bypass somatic incompatibility barriers may
sometimes relate to cultural and selection techniques.

Hybridization barriers between D. innoxia and Atropa
belladonna are at least prezygotic as reciprocal pollinations are
unable to effect fertilization (Krumbiegel and Schieder, 1981).
Krumbiegel and Schieder (1981) also attempted to hybridize these
species by culturing placenta-attached ovules following
pollination in vivo, by in vitro pollination, and by somatic
hybridization. Via somatic hybridization, a hybrid cell line
containing a reduced chromosome number was able to produce
shoots. The failure of the other methods led the authors to
conclude that protoplast fusion is the only means for obtaining
hybrids. Though this may be true, critical evaluation of their
results does not necessarily imply that all means of sexual
hybridization are impossible. The existence of prezygotic
barriers prevents the culture of unfertilized ovules as a means
of obtaining hybrids. In addition, the in vitro fertilization
system was unsuccessful even in self fertilizing Atropa and
Datura. Thus it appeared that their inability to obtain sexual
hybrids was at least a consequence of the failure of
fertilization. This emphasizes the importance of the available
technical capabilities in limiting hybridization.

ADDITIONAL IN VITRO MANIPULATIONS USEFUL
FOR WIDE HYBRIDIZATION

Bypassing Lethality

In hybridization attempts where seedlings die soon after
germination, initiating callus from the seedlings followed by the
induction of morphogenesis may aid in hybrid survival. The cross
of N. suaveolens X N. tabacum results in viable seeds, but the
seedlings die after the first or second true leaves emerge
(Lloyd, 1975). Lloyd (1975) transferred segments of cotyledons
from some of the young seedlings to a tissue culture media from
which shoots were differentiated which later were grown to
maturity. Similar results have been obtained from the cross N.
tabacum X N. amplexicaulis (DeVerna and Collins, unpublished
data). Seedlings can be derived from this cross only by using in
vitro pollination and fertilization. Soon after germination,
though, the shoots of the seedlings become necrotic and die. By

culturing cotyledons of prenecrotic seedlings, hybrid plants
could be generated which grew to maturity.

Chromosome Doubling and Induction of Genetic Changes

When interspecific hybrids are sterile, fertility can
frquently be induced by doubling the chromosome number with
colchicine. Nakamura et al. (1981) used an _in vitro_ colchicine
treatment of hybrid callus to induce partial doubling of _T._
crassum X _H._ _vulgare_ hybrids. Kasperbauer and Collins (1972)
devised an _in vitro_ procedure for chromosome doubling of haploid
plants of tobacco which involves the culture of midrib or root
tissue (Kasperbauer and Wilson, 1979). The method takes
advantage of the tendency of older tissues to undergo
endomitosis. Midvein cultures were used successfully by Reed
(1978) to double the chromosome number of _Nicotiana_ interspecific
hybrids.

As mentioned earlier, callus culture may frequently induce
intrachromosomal and interchromosomal changes. These genetic
changes sometimes may be beneficial in the induction of
morphogenesis (Hoffman and Adachi, 1981) or in increasing
fertility. Thus, culturing sterile hybrids as callus may
eventually result in the production of fertile plants. The topic
of heritable variations in cell culture has recently been
reviewed (Meins, 1983).

Organelle and Chromosome Transfer and the Use of Cytoplasts

In addition to somatic hybridization, protoplasts also have
the potential to take up and incorporate whole chromosomes,
nuclei, mitochondria, and chloroplasts. Methods have recently
been developed for the isolation of whole mitotic chromosomes
from plant protoplasts (Malmberg and Griesbach, 1980; Griesbach
et al., 1982). Their incorporation into protoplasts has also
been demonstrated (Szabados et al., 1981; Griesbach et al., 1982;
Malmberg and Griesbach, 1983); however, transferred chromosomes
have not been functionally expressed in plant cells. Chromosome
uptake has already been successful in functional gene transfer
with animal systems (Athwal and McBride, 1980).

The possibility of transferring cytoplasmic organelles
directly into protoplasts also exists (Bonnett et al., 1980;
McDaniel, 1981). The expression of transplanted organelles
depends on their functionality being maintained during isolation
and their subsequent uptake. No reports on the successful
transplantation and expression of mitochondria in higher plants

have been made, though this has been reported with Neurospora and Saccharomyces (McDaniel, 1981). Chloroplasts have been incorporated into protoplasts of several plant species (Carlson, 1973; Bonnett and Eriksson, 1974; Vasil and Giles, 1975; Davey et al., 1976); however, whether these have resulted in chloroplast integration is not clear (McDaniel, 1981). A more recent effort by Uchimiya and Wildman (1979) showed that transferred N. gossei chloroplasts were not expressed in N. tabacum.

Protoplasts can be fractionated into subprotoplasts, which contain the nucleus with a small amount of cytoplasm, and into cytoplasts, which are enucleated cells (Wallin et al., 1978). The subprotoplasts may be useful in hybridization research as only a small population of organelles are transferred. Wallin et al. (1978) fused miniprotoplasts of two complementary nitrate reductase lines and isolated fusion products that grew on a media containing nitrate as the nitrogen source. The results indicate that subprotoplasts are efficient vehicles for nuclear transfer. The uptake of nuclei directly into protoplasts without the benefit of the surrounding plasma membrane has also been accomplished but no expression of the nuclear material was detected (Potrykus and Hoffman, 1973; Potrykus and Lorz, 1976; Cocking, 1977; Lorz and Potrykus, 1978).

The transfer of cytoplasmic traits has been accomplished by fusing normal protoplasts with irradiated protoplasts (Zelcer et al., 1978; Aviv and Galun, 1978, 1980; Aviv et al., 1980; Galun et al., 1982; Menczel et al., 1982, 1983). Zelcer et al. (1978) fused X-irradiated protoplasts of a cytoplasmic male-sterile cultivar of N. tabacum with protoplasts of N. sylvestris. Three types of plants were regenerated. Some were true cytoplasmic hybrids having the N. tabacum male sterility trait, and the chromosome number of N. sylvestris. However, other types were obtained in which the genomic constitution was either uncertain or which contained the N. tabacum genome.

The use of cytoplasts would seem to have more potential than the use of irradiated protoplasts since the deleterious effects of mutations is avoided and the chances of 'escapes' are reduced. Maliga et al. (1982) isolated cytoplasts from maternally-inherited streptomycin resistant N. tabacum protoplasts and fused them with streptomycin sensitive N. plumbaginifolia protoplasts. Resistant clones were isolated and plants regenerated which contained the N. tabacum plastids as shown by restriction endonuclease analysis. An earlier report was made by Bracha and Sher (1981) who fused enucleated protoplasts in A. cepa but did not attempt culturing.

SUMMARY

The results obtained to date involving the use of in vitro methods to facilitate wide hybridization in plants are voluminous and impressive. The techniques of embryo culture, ovule culture, and in vitro pollination and fertilization represent an extension of the normal sexual hybridization process. Successes recorded in obtaining hybrids stem largely from circumventing prezygotic or postzygotic hybridization barriers. Numerous recent successful hybridizations were possible because of the development of improved tissue and cell culture systems for crop plants and attention given to genotypes used in hybridization attempts.

Interspecific and intergeneric hybridization utilizing the process of protoplast fusion will bypass the limits set by all sexual methods. In addition to combining complete genomes from two different species through protoplast fusion, this system affords unique opportunities for creating novel cytoplasmic combinations, transfer of individual chromosomes, transfer of cytoplasmic organelles, manipulation of male sterility, and for single gene transfer. Some caution must be noted with regard to the extent of hybridization possible between distantly related species. Although practically no limit exists to the physical fusion of protoplasts from widely divergent species, the restrictions imposed by somatic incompatibility have not been adequately addressed.

Regeneration of plants from the protoplast or single heterokaryon level is still a major hurdle for many important crop species before somatic cell fusion can be exploited to produce interspecific and intergeneric hybrids. Identification and selection of hybrids is also a limitation to the efficient application of cell fusion methods.

Most efforts to date have been directed to obtaining somatic cell hybrids and simply combining two diverse genomes. The challenge and possibly most important applications of somatic cell fusion in the future will likely involve more restricted manipulations of genes, organelles, cytoplasm, and single chromosomes. The use of protoplasts of different ploidy levels, aneuploids, mutants for cell cycle traits, chromosome elimination, chromosome pairing, and back fusion procedures are certain to be useful strategies in the manipulation of genes, single chromosomes, and genomes.

This paper (83-3-255) is published with the approval of the Dean and Director of the Kentucky Agricultural Experiment Station.

REFERENCES

Allard, R. W., 1960, "Principles of Plant Breeding," John Wiley and Sons, Inc., New York.

Amemiya, A., Akemine, H., and Toriyama, K., 1956a, Cultural conditions and growth of immature embryo in rice plant (Studies in the embryo culture in rice plant, 1), Bull. Nat. Inst. Agric. Sci. Tokyo Ser. D. (Plant Physiol.), 6:1-40.

Amemiya, A., Akemine, H., and Toriyama, K., 1956b, The first germinative stage and varietal differences in growth response of cultured embryo of rice plant (Studies on the embryo culture in rice plant, 2), Bull. Nat. Inst. Agric. Sci. Tokyo Ser. D. (Plant Physiol.), 6:41-60.

Anagnostakis, S. L., 1977, In vitro culture of immature embryos of American Elm, Horti. Sci., 12:44.

Andronescu, D. I., 1919, Germination and further development of the embryo of Zea mays separated from the endosperm, Am. J. Bot., 6:443-452.

Arisumi, T., 1980a, Chromosome numbers and comparative breeding behaviour of certain Impatiens from Africa, India, and New Guinea, J. Amer. Soc. Hort. Sci., 105:99-102.

Arisumi, T., 1980b, In vitro culture of embryos and ovules of certain incompatible selfs and crosses among Impatiens sp., J. Am. Soc. Hort. Sci., 105:629-631.

Athwal, R. S. and McBride, O. W., 1980, Chromosome-mediated gene transfer and microcell hybridization, in: "Genetic Improvement of Crops: Emergent Techniques," I. Rubenstein, B. Gengenbach, R. L. Phillips, and C. E. Green, eds., pp. 153-181, Univ. of Minn. Press, Minneapolis.

Aviv, D., Fluhr, R., Edelman, M., and Galun, E., 1980, Progeny analysis of the interspecific somatic hybrids: Nicotiana tabacum (cytoplasmic male sterile) plus Nicotiana sylvestris with respect to nuclear and chloroplast markers, Theor. Appl. Genet., 56:145-150.

Aviv, D. and Galun, E., 1978, Interspecific transfer of cytoplasmic male sterility by fusion between protoplasts of normal Nicotiana sylvestris and X-ray irradiated protoplasts of male-sterile N. tabacum, Z. Pflanzenphysiol., 90:397-407.

Aviv, D. and Galun, E., 1980, Restoration of fertility in cytoplasmic male sterile (CMS) Nicotiana sylvestris by fusion with X-irradiated N. tabacum protoplasts, Theor. Appl. Genet., 58:121-127.

Bajaj, Y. P. S., 1964, Development of ovules of Abelmoschus esculentus L. var. Pusa Sawani in vitro, Proc. Nat. Inst. Sci. India, 30B:175-185.

Bajaj, Y. P. S., 1966, Growth of Hyoscyamus niger ovaries in culture, Phyton., 23:57-62.

Balatkova, V. and Tupy, J., 1968, Test-tube fertilization in
    Nicotiana tabacum by means of an artificial pollen tube
    culture, Biol. Plant., 10:266-270.

Balatkova, V. and Tupy, J., 1972, Some factors affecting the seed
    set after in vitro pollination of excised placentae of
    Nicotiana tabacum L., Biol. Plant., 14:82-88.

Balatkova, V. and Tupy, J., 1973, The significance of the methods
    of stigmatal and placentae pollination in vitro in
    Antirrhinum majus L., seed and callus formation on
    placentae, Biol. Plant., 15:102-106.

Balatkova, V., Tupy, J., and Hrabetova, E., 1977, Seed formation
    in Narcissus pseudonarcissus L. after placentae pollination
    in vitro, Plant Sci. Lett., 8:17-21.

Ball, E., 1959, Growth of the embryo of Ginkgo biloba under
    experimental conditions. II. Growth rates of root and shoot
    upon media absorbed through the cotyledons, Am. J. Bot.,
    46:130-139.

Beasley, C. A., 1971, In vitro culture of fertilized cotton
    ovules, BioScience, 21:906-907.

Beasley, C. A., 1977, Ovule culture: fundamental and pragmatic
    research for the cotton industry, pp. 160-178, in: "Plant
    Cell, Tissue, and Organ Culture", I. Reinert and Y. P. S.
    Bajaj, eds., Springer-Verlag, New York.

Beasley, C. A. and Ting, I. P., 1973, The effects of plant growth
    substances on in vitro fiber development from fertilized
    cotton ovules, Am. J. Bot., 60:130-139.

Beasley, C. A., Ting, I. P., Linkins, A. E., Binbaum, E. H., and
    Delmer, D. P., 1974, Cotton ovule culture: A review of
    progress and a preview of potential, in: "Tissue Culture and
    Plant Science," H. E. Street, ed., pp. 169-192, Academic
    Press, New York.

Belliard, G., Pelletier, G., Vedel, F., and Quetier, F., 1978,
    Morphological characteristics and chloroplast DNA
    distribution in different cytoplasmic parasexual hybrids of
    Nicotiana tabacum, Mol. Gen. Genet., 165:231-238.

Belliard, G., Vedel, F., and Pelletier, G., 1979, Mitochondrial
    recombination in cytoplasmic hybrids of Nicotiana tabacum by
    protoplast fusion, Nature, 281:401-403.

Binding, H., 1976, Somatic hybridization experiments in
    Solanaceous species, Mol. Gen. Genet., 144:171-175.

Binding, H. and Nehls, R., 1978, Somatic cell hybridization of
    Vicia faba and Petunia hybrida, Mol. Gen. Genet.,
    164:137-143.

Binding, H., Jain, S. M., Finger, J., Mordhorst, G., Nehls, R.,
    and Gressel, J., 1982, Somatic hybridization of atrazine
    resistant biotype of Solanum nigrum with Solanum tuberosum,
    Part I: Clonal variation in morphology and atrazine
    sensitivity, Theor. Appl. Genet., 63:273-277.

Boeshore, M. L., Lifshitz, I., Hanson, M. R., and Izhar, S., 1983, Novel composition of mitochondrial genomes in Petunia somatic hybrids derived from cytoplasmic male sterile and fertile plants, Mol. Gen. Genet., 190:459-467.

Bonnett, H. T. and Eriksson, T., 1974, Transfer of algae chloroplasts into protoplasts of higher plants, Planta, 120:71-79.

Bonnett, H., Wallin, A., and Glimelius, K., 1980, The analysis of organelle behavior and genetics following protoplast fusion or organelle incorporation, in: "Genetic Improvement of Crops: Emergent Techniques," I. Rubenstein, B. Gengenbach, R. L. Phillips, and C. E. Green, eds., pp. 137-152, Univ. Minn. Press, Minneapolis.

Bottino, P. J., 1975, The potential of genetic manipulation in plant cell culture for plant breeding, Radiat. Bot., 15:1-16.

Bowley, S. R. and Taylor, N. L., 1984. Introgressive hybridization, in: "Handbook of Plant Science," B. R. Christie, ed., CRC Press, Boca Raton.

Brabec, F., 1949, Zytologische untersuchungen an den burdonen Solanum nigrum-lycopersicum, Planta, 37:57-95.

Brabec, F., 1954, Untersuchungen uber die natur winklerschen burdonen auf grund neuen experimentellen materials, Planta, 44:562-606.

Brabec, F., 1965, Pfropfung und chimaren, unter besonderer besucksichtigung der entwicklungsphysiologischen problematik, in: "Handuch der Pflanzenphysiologie," W. Ruhland, ed., Vol. XV, pp. 388-494, Springer-Verlag, Berlin and New York.

Bracha, M. and Sher, N., 1981, Fusion of enucleated protoplasts with nucleated miniprotoplasts in onion (Allium cepa L.), Plant Sci. Lett., 23:95-101.

Brar, D. S., Rambold, S., Constabel, F., and Gamborg, O. L., 1980, Isolation, fusion and culture of Sorghum bicolor cultivar GPR-168 and corn (Zea mays cultivar Punjab Local) protoplasts, Z. Pflanzenphysiol., 96:269-276.

Broue, P., Douglass, J., Grace, J. P., and Marshall, D. R., 1982, Interspecific hybridization of soybeans and perennial Glycine species indigenous to Australia via embryo culture, Euphytica, 31:715-724.

Brown, H. T. and Morris, G. H., 1890, Researches on the germination of some of the Gramineae, J. Chem. Soc., 57:458-528.

Butenko, R. G. and Kuchko, A. A., 1979, Physiological aspects of procurement, cultivation, and hybridization of isolated potato protoplasts, Fiziol. Rast., 26:1110-1119.

Carlson, P. S., 1973, The use of protoplasts for genetic research, Proc. Nat. Acad. Sci., U.S.A., 70:598-602.

Carlson, P. S., Smith, H. H., and Dearing, R. O., 1972,
    Parasexual interspecific plant hybridization, Proc. Nat.
    Acad. Sci., U.S.A., 69:2292-2294.
Chen, K., Wildman, S. G., and Smith, H. H., 1977, Chloroplast DNA
    distribution in parasexual hybrids as shown by polypeptide
    composition of fraction I protein, Proc. Nat. Acad. Sci.,
    U.S.A., 74:5109-5112.
Chien, Y.-C., Kao, K. N., and Welter, L. R., 1982, Chromosomal
    and isozyme studies of Nicotiana tabacum - Glycine max
    hybrid cell lines, Theor. Appl. Genet., 62:301-304.
Chopra, R. N., 1962, Effect of some growth substances and calyx
    on fruit and seed development of Althaea rosea Cav., in:
    "Plant Embryology - A Symposium," pp. 170-181, Council of
    Scientific and Industrial Research, New Delhi.
Choudhury, B., 1955, Embryo culture technique. III. Growth of
    hybrid embryos (Lycopersicon esculentum x Lycopersicon
    peruvianum) in culture medium, Indian J. Hort., 12:155-156.
Chupeau, Y., Missonier, C., Hommel, M. C., and Goujaud, J., 1978,
    Somatic hybrids of plants by fusion of protoplasts:
    observation on the model system Nicotiana glauca - Nicotiana
    langsdorffii, Mol. Gen. Genet., 165:239-246.
Cionini, P. G., Bennici, A., Alpi, A., and D'Amato, F., 1976,
    Suspensor, gibberellin and in vitro development of Phaseolus
    coccineus embryos, Planta, 131:115-117.
Cocking, E. C., 1960, A method for the isolation of plant
    protoplasts and vacuoles, Nature, 187:962-963.
Cocking, E. C., 1976, Fusion and somatic hybridization of higher
    plant protoplasts, in: "Microbial and Plant Protoplasts," J.
    L. Peberdy, A. H. Rose, H. J. Rogers, and E. C. Cocking,
    eds., Academic Press, London, pp. 189-200.
Cocking, E. C., 1977, General properties of isolated protoplasts
    and uptake of foreign genetic material, in: "Nucleic Acids
    and Protein Synthesis in Plants," L. Bogorad and J. H. Weil,
    eds., pp. 321-327, Plenum Press, New York and London.
Cocking, E. C., 1978, Selection and somatic hybridization, in:
    "Frontiers of Plant Tissue Culture," T. A. Thorpe, ed., pp.
    151-158, Univ. Calgary Press, Calgary.
Cocking, E. C., 1980, Protoplasts: Past and present, in:
    "Advances in Protoplast Research: Proc. 5th Int. Protoplast
    Symposium 1979 Szeged Hungary," L. Ferenczy and G. L.
    Farkas, eds., pp. 3-15, Pergamon Press, Oxford.
Cocking, E. C., George, D., Price-Jones, M. J., and Power, J. B.,
    1977, Selection procedures for the production of
    interspecies somatic hybrids of Petunia hybrida and Petunia
    parodii. II. Albino complementation selection, Plant Sci.
    Lett., 10:7-12.

Collins, G. B., and Grosser, J. W., 1984, Embryo culture, in: "Cell Culture and Somatic Cell Genetics of Plants V. 1 Laboratory Techniques," I. K. Vasil, ed., Academic Press, New York.

Colonna, J. P., Cas, G., and H. Rabechault, H., 1971, Mise au point d'une methode de culture in vitro d'embryons de cafeiers, Application à deux varietes de cafeiers cultinés, C. R. Acad. Sci. (Paris), 272:60-63.

Constabel, F., Dudits, D., Gamborg, O. L., and Kao, K. N., 1975a, Nuclear fusion in intergeneric heterokaryons, Can. J. Bot., 53:2092-2095.

Constabel, F., Kirkpatrick, J. W., Kao, K. N., and Kartha, K. K., 1975b, The effect of canavanine on the growth of cells from suspension cultures and on intergeneric heterokaryons of canavanine sensitive and tolerant plants, Biochem. Physiol. Pflanzen., 168:319-325.

Constabel, F., Weber, G., Kirkpatrick, J. W., and Pahl, K., 1976, Cell division of intergeneric protoplast fusion products, Z. Pflanzenphysiol., 79:1-7.

Constabel, F., Weber, G., and Kirkpatrick, J. W., 1977, Chromosome compatability of intergeneric cell hybrids of Glycine max X Vicia hajastana, C. R. Acad. Sci. (Paris), 285:319-322.

Davey, M. R., Clothier, R. H., Balk, M., and Cocking, E. C., 1978, An ultrastructural study of the fusion of cultured amphibian cells with higher plant protoplasts, Protoplasma, 96:157-172.

Davey, M. R., Frearsson, E. M., and Power, J. B., 1976, Polyethylene glycol-induced transplantation of chloroplasts into protoplasts: An ultrastructural assessment, Plant Sci. Lett., 7:7-16.

DeLautour, G., Jones, W. T., and Ross, M. O., 1978, Production of interspecific hybrids in Lotus aided by endosperm transplants, N. Z. J. Bot. 16:61-68.

Dhaliwal, S. and King, P. J., 1978, Direct pollination of Zea mays ovules in vitro with Z. mays, Z. mexicana, and Sorghum bicolor pollen, Theor. Appl. Genet., 53:43-46.

Dolezel, J., Novak, F. J., and Luzny, J., 1980, Embryo development and in-vitro culture of Allium cepa and its interspecific hybrids, Z. Pflanzenzuecht., 85:177-184.

Douglas, G. C., Keller, W. A., and Setterfield, G., 1981, Somatic hybridization between Nicotiana rustica and Nicotiana tabacum: 2. Protoplast fusion and selection and regeneration of hybrid plants, Can. J. Bot., 59:220-227.

Douglas, G. C., Wetter, L. R., Keller, W. A., and Setterfield, G., 1983, Production of sexual hybrids of Nicotiana rustica-tabacum and N. rustica-glutinosa via in vitro culture of fertilized ovules, Z. Pflanzenzuecht., 90:116-129.

Dudits, D., 1981, Backfusion with somatic protoplasts as a method in genetic manipulation of plants, Acta. Biol. Acad. Sci. Hung., 32:215-218.

Dudits, D., Fejer, O., Hadlaczky, G., Koncz, C., Lazar, G. B., and Howath, G., 1980, Intergeneric gene transfer mediated from plant protoplast fusion, Mol. Gen. Genet., 179:283-288.

Dudits, D., Hadlaczky, G., Bajszar, G., Koncz, C., Lazar, G., and Horvath, G., 1979, Plant regeneration from intergeneric cell hybrids, Plant Sci. Lett., 15:101-112.

Dudits, D., Hadlaczky, G. Y., Levi, E., Fejer, O., Haydu, Z. S., and Lazar, G., 1977, Somatic hybridization of Daucus carota and Daucus capillifolus by protoplast fusion, Theor. Appl. Genet., 51:127-132.

Dudits, D., Kao, K. N., Constabel, F., and Gamborg, O. L., 1976, Fusion of carrot and barley protoplasts and division of heterokaryocytes, Can. J. Genet. Cytol., 18:263-269.

Dulieu, H. L., 1966, Pollination of excised ovaries and culture of ovules of Nicotiana tabacum L., Phytomorphology, 16:69-75.

Eid, A. A. H., Delange, E., and Waterkeyn, L., 1973, In vitro culture of fertilized cotton ovules. I. The growth of cotton embryos, Cellule, 69:361-371.

Evans, D. A., 1979, Somatic hybridization within the genus Nicotiana, Plant Physiol., 63(5, supplement):117.

Evans, D. A., Bravo, J. E., Kut, S. A., and Flick, C. E., 1983, Genetic behaviour of somatic hybrids in the genus Nicotiana: N. otophora + N. tabacum and N. sylvestris + N. tabacum, Theor. Appl. Genet., 65:93-101.

Evans, D. A., Flick, C. E., Kut, S. A., and Reed, S. M., 1982, Comparison of Nicotiana tabacum and Nicotiana nesophila hybrids produced by ovule culture and protoplast fusion, Theor. Appl. Genet., 62:193-198.

Evans, D. A., Flick, C. E., and Jensen, R. A., 1981, Disease resistance: Incorporation into sexually incompatible somatic hybrids of the genus Nicotiana, Science, 213:907-909.

Evans, D. A., Wetter, L. R., and Gamborg, O. L., 1980, Somtic hybrid plants of Nicotiana glauca and Nicotiana tabacum obtained by protoplast fusion, Physiol. Plant., 48:225-230.

Evola, S. V., Earle, E. D., and Chaleff, R. S., 1983, The use of genetic markers selected in vitro for the isolation and genetic verification of intraspecific somatic hybrids of Nicotiana tabacum, Mol. Gen. Genet., 189:441-446.

Gadwal, V. R., Joshi, A. B., and Iyer, R. D., 1968, Interspecific hybrids in Abelmoschus through ovule and embryo culture, Ind. J. Genet. Plt. Breed., 28:269-274.

Galbraith, D. W. and Galbraith, J. E. C., 1979, A method for the identification of fusion of plant protoplasts derived from tissue cultures, Z. Pflanzenphysiol., 93:149-158.

Galbraith, D. W. and Harkins, K. R., 1982, Cell sorting as a means for isolating somatic hybrids, in: "Proc. 5th Intl. Cong., Plant Tissue and Cell Culture," A. Fujiwara, ed., pp. 617-618, The Jpn. Assoc. for Plant Tissue Cult., Tokyo.

Galbraith, D. W. and Mauch, T. J., 1980, Identification of fusion of plant protoplasts, Z. Pflanzenphysiol., 98:129-140.

Galun, E., Arzee-Gonen, P., Fluhr, R., Edelman, M., and Aviv, D., 1982, Cytoplasmic hybridization in Nicotiana: Mitochondrial DNA analysis in progenies resulting from fusion between protoplasts having different organelle constitutions, Mol. Gen. Genet., 186:50-56.

Gengenbach, B. G., 1977a, Development of maize caryopses resulting from in vitro pollination, Planta, 134:91-93.

Gengenbach, B. G., 1977b, Genotypic influences on in vitro fertilization and kernel development in maize, Crop Sci., 17:489-492.

Gill, B. S., Waines, J. G., and Sharma, H. C., 1981, Endosperm abortion and the production of viable Aegilops squarrosa X Triticum boeoticum hybrids by embryo culture, Plant Sci. Lett., 23:181-187.

Gleba, Y. Y. and Hoffmann, F., 1978, Hybrid cell line Arabidopsis thaliana + Brassica campestris: No evidence for specific chromosome elimination, Mol. Gen. Genet., 257-264.

Gleba, Y. Y. and Hoffmann, F., 1979, Arabidobrassica: Plant genome engineering by protoplast fusion, Naturwissenschaften, 66:547-554.

Gleba, Y. Y. and Hoffman, F., 1980, Arabidobrassica: A novel plant obtained by protoplast fusion, Planta, 149:112-117.

Gleba, Y. Y., Momot, V. P., Cherep, N. N., and Skarzynskaya, M. V., 1982, Intertribol hybrid cell lines of Atropa belladonna (X) Nicotiana chinensis obtained by cloning individual protoplast fusion products, Theor. Appl. Genet., 62:75-79.

Gleba, Y. Y., Piven, N. M., Komarnitskii, I. K., and Sytnik, K. M., 1979, Parasexual cytoplasmic hybrids (cybrids) of Nicotiana tabacum + N. debneyi obtained by protoplast fusion, Doklady Akademii Nauk SSSR, 240:1223-1226.

Glimelius, K., Ericksson, T., Grafe, R., and Mullen, A. J., 1978, Somatic hybridization of nitrate reductase-deficient mutants of Nicotiana tabacum by protoplast fusion, Physiol. Plant., 44:273-277.

Golubinskii, I. H., 1974, ed., Biologiya Prorastaniya Pyltsy. [The Biology of Pollen Growth], Naukova dumka, Kiev.

Gosch, G. and Reinert, J., 1976, Nuclear fusion in intergeneric heterokaryocytes and subsequent mitosis of hybrid nuclei, Naturwissenschaften, 63:534.

Gosch, G., Reinert, J., 1978, Cytological identification of colony formation of intergeneric somatic hybrid cells, Protoplasma, 96:23-36.

Griesbach, R. J., Malmberg, R. L., and Carlson, P. S., 1982, An improved technique for the isolation of higher plant chromosomes, Plant Sci. Lett., 24:55-60.

Guha, S., and Johri, B. M., 1966, In vitro development of ovary and ovules of Allium cepa L., Phytomorphology, 16:353-364.

Gupta, P. P., Gupta, M., and Schieder, O., 1982, Correction of nitrate reductase defect in auxotrophic plant cells through protoplast-mediated intergeneric gene transfers, Mol. Gen. Genet., 188:378-383.

Guzowska, I., 1971, In vitro pollination of ovules and stigmas in several species, Genet. Pol., 12:261-265.

Hadley, H. H. and Openshaw, S. J., 1980, Interspecific and intergeneric hybridization, in: "Hybridization of Crop Plants," W. R. Fehr and H. H. Hadley, eds., pp. 261-272, Amer. Soc. Agron. and Crop Sci. Soc. Amer., Madison, Wisconsin.

Hall, O. L., 1954, Hybridization of wheat and rye after embryo transplantation, Hereditas, 40:453-458.

Hall, O. L., 1956, Further experiments in embryo transplantation, Hereditas, 42:261-262.

Hannig, E., 1904, Zur physiologie pflanzlicher embryonen. I. Ueber die cultur von cruciferen-embryonen ausserhalb des embrosacks, Bot. Zeit., 62:45-80.

Harberd, D. J., 1969, A simple effective embryo culture technique for Brassica, Euphytica, 18:425-429.

Harms, C. T., 1982, Somatic incompatibility - limits to remote somatic hybridization, Int. Assoc. Plant Tissue Cult., Newsletter, 38:6-12.

Harms, C. T., 1983, Somatic incompatibility in the development of higher plant somatic hybrids, Q. Rev. Biol., 58:325-353.

Harms, C. T. and Oertli, J. J., 1982, Complementation and expression of amino acid analog resistance studied by intraspecific and interfamily protoplast fusion, in: "Plant Tissue Culture," A. Fujiwara, ed., pp. 467-468, Jpn. Assoc. Plant Tissue Cult., Tokyo.

Hauptmann, R. M. and Widholm, J. M., 1983, Carrot and tobacco somatic cell hybrids selected by amino acid analog resistant complementation, TCA Report, 17:7-8.

Haynes, F. L., 1954, Potato embryo culture, Am. Potato J., 31:282-288.

Hein, T., Przewozny, T., and Schieder, O., 1983, Culture and selection of somatic hybrids using an auxotrophic cell line, Theor. Appl. Genet., 64:119-122.

Hoffman, F. and Adachi, T., 1981, Arabidobrassica: Chromosomal recombination and morphogenesis in asymmetric intergeneric hybrid cells, Planta, 153:586-593.

Horst, B. and Nehls, R., 1978, Somatic cell hybridization of Vicia faba + Petunia hybrida, Mol. Gen. Genet., 164:137-143.

Inomata, N., 1968, In vitro culture of ovaries of Brassica
    hybrids between 2X and 4X. I. Cultured medium, Jpn. J.
    Breed., 18:139-148.
Inomata, N., 1977, Production of interspecific hybrids between
    Brassica campestris and Brassica oleracea by culture in
    vitro of excised ovaries. I. Effects of yeast extract and
    casein hydrolysate of the development of excised ovaries,
    Jpn. J. Breed., 27:295-304.
Inomata, N., 1978a, Production of interspecific hybrids between
    Brassica campestris and Brassica oleracea by culture in
    vitro of excised ovaries. Part 2. Effects of coconut-milk
    and casein hydrolysate on the development of excised
    ovaries, Jpn. J. Genet., 53:1-12.
Inomata, N., 1978b, Production of interspecific hybrids in
    Brassica campestris X B. oleracea by culture in vitro of
    excised ovaries: Development of excised ovaries in the
    crosses of various cultivars, Jpn. J. Genet., 53:161-174.
Iwai, S., Nagao, T., Nakata, K., Kawashima, N., and Matsuyama,
    S., 1980, Expression of nuclear and chloroplastic genes
    coding for Fraction 1 protein in somatic hybrids of
    Nicotiana tabacum cultivar Bright Yellow and Nicotiana
    rustica var. rustica, Planta, 147:414-417.
Izhar, S. and Power, J. B., 1979, Somatic hybridization in
    Petunia: A male sterile cytoplasmic hybrid, Plant Sci.
    Lett., 14:49-56.
Izhar, S., Schlecter, M., and Swartzberg, D., 1983, Sorting out
    in somatic hybrids of Petunia and prevalence of the
    heteroplasmon through several meiotic cycles, Mol. Gen.
    Genet., 190:468-474.
Jones, C. W., Mastrangelo, D. A., Smith, H. H., Lin, H. Z., and
    Meck, R. A., 1976, Interkingdom fusion between human (HeLa)
    cells and tobacco hybrid (GGLL) protoplasts, Science,
    193:401-403.
Kameya, T., Hinata, K., and Mizushima, U., 1966, Fertilization in
    vitro of excised ovules treated with calcium chloride in
    Brassica oleracea L., Proc. Jpn. Acad., 42:165-167.
Kameya, T., Horn, M. E., and Widholm, J. M., 1981, Hybrid shoot
    formation from fused Daucus carota and D. capillifolius
    protoplasts, Z. Pflanzenphysiol., 104:459-466.
Kanta, K., 1960, Intraovarian pollination in Papaver rhoeas L.,
    Nature, 188:683-684.
Kanta, K. and Maheshwari, P., 1963a, Intraovarian pollination in
    some papaveraceae, Phytomorphology, 13:215-229.
Kanta, K. and Maheshwari, P., 1963b, Test-tube fertilization in
    some angiosperms, Phytomorphology, 13:230-237.
Kanta, K., Rangaswamy, N. S., and Maheshwari, P., 1962, Test-tube
    fertilization in a flowering plant, Nature, 194:1214-1217.
Kao, K. N., 1977, Chromosomal behaviour in somatic hybrids of
    soybean - Nicotiana glauca, Mol. Gen. Genet., 150:225-230.

Kao, K. N., Constabel, F., Michayluk, M. R., and Gamborg, O. L., 1974, Plant protoplast fusion and growth of intergeneric hybrid cells, Planta, 120:215-227.

Kapoor, M., 1959, Influence of growth substances on the ovules of Zephyranthes, Phytomorphology, 9:313-315.

Kartha, K. K., Gamborg, O. L., Constabel, F., and Kao, K. N., 1974, Fusion of rapeseed and soybean protoplasts and subsequent division of heterokaryocytes, Can. J. Bot., 52:2435-2436.

Kasperbauer, M. J. and Collins, G. B., 1972, Reconstitution of diploids from leaf tissue of anther-derived haploids in tobacco, Crop Sci., 12:98-101.

Kasperbauer, M. J. and Wilson, H. M., 1979, Haploid plant production and use, in: "Nicotiana: Procedures for Experimental Use," R. D. Durbin, ed., pp. 33-39, U.S. Dept. Agric., Tech. Bull. 1586.

Keller, W. A., Setterfield, G., Douglas, G., Gleddie, S., and Nakamura, C., 1982, Production, characterization, and utilization of somatic hybrids of higher plants, in: "Application of Plant Cell and Tissue Culture to Agriculture and Industry," D. T. Tomes, B. E. Ellis, P. M. Harney, K. J. Kasha, R. L. Peterson, eds., pp. 81-114, Univ. Guelph, Guelph.

Kosakovs'ka, I. V., 1980, Subunit structure of D-ribulose-1,5-diphosphate carboxylase of parasexual hybrids, Nicotiana tabacum and Nicotiana debneyi, UKR Bot. Z.H., 37:86-88.

Krumbiegel, G. and Schieder, O., 1979, Selection of somatic hybrids after fusion of protoplasts from Datura innoxia Mill. and Atropa belladonna L., Planta, 145:371-375.

Krumbiegel, G. and Schieder, O., 1981, Comparison of somatic and sexual incompatibility between Datura innoxia and Atropa belladonna, Planta, 153:466-470.

Kruse, A., 1973, Hordeum x Triticum hybrids, Hereditas, 73:157-161.

Kruse, A., 1974, An in vivo/vitro embryo culture technique, Hereditas, 77:219-224.

Kumar, P. M., Das, K., Sinha, R. R., Mukherjee, P., and Sen, S. K., 1983, Interspecific somatic protoplast fusion products in cultivated jute species, Basic Life Sci., 22:237-247.

LaCroix, L. J. and Canvin, D. T., 1963, The role of light and other factors in the growth and differentiation of barley embryos, Plant Physiol., 38(Suppl.):694.

LaCroix, L. J., Naylor, J., and Larter, E. N., 1962, Factors controlling embryo growth and development in barley (Hordeum vulgare L.), Can. J. Bot., 40:1515-1523.

Ladinsky, G., Newell, C. A., and Hymowitz, T., 1979, Wide crosses in soybeans: Prospects and limitations, Euphytica, 28:421-423.

Laibach, F., 1925, Das Taubwerden von Bastardsmen und die Kunstaiche Aufzucht fruh absterbender Bastardembryonen, Z. Bot., 17:417-459.

Lar'Kina, N. I., 1980, Overcoming incompatibility between Nicotiana species by means of in vitro pollination, Plant Breed. Abst., 50:690 (#8067).

Lea, P. J., Hughes, J. S., and Miflin, B. J., 1979, Glutamine and asparagine-dependent protein synthesis in maturing legume cotyledons cultured in vitro, J. Exp. Bot., 30:529-537.

Lee, A. E., 1952, The growth of excised immature sedge embryos in culture, Bull. Torrey Bot. Club., 79:59-62.

Lloyd, R., 1975, Tissue culture as a means of circumventing lethality in an interspecific Nicotiana hybrid, Tob. Sci., 19:4-6.

Lorz, H., Potrykus, I., 1978, Investigations on the transfer of isolated nuclei into plant protoplasts, Theor. Appl. Genet., 53:251-256.

Maheshwari, N., 1958, In vitro culture of excised ovules of Papaver somniferum, Science, 127:342.

Maheshwari, N. and Lal, M., 1958, In vitro culture of ovaries of Iberis amara L., Nature, 181:631-632.

Maheshwari, N. and Lal, M., 1961a, In vitro culture of ovaries of Iberis amara L., Phytomorphology, 11:17-23.

Maheshwari, N. and Lal, M., 1961b, In vitro culture of excised ovules of Papaver somniferum L., Phytomorphology, 11:307-314.

Maheshwari, P. and Kanta, K., 1961, Intraovarian pollination in Eschscholzia californica Cham., Argemone mexicana L. and A. ochroleuca Sweet, Nature, 191:304.

Maheshwari, P. and Rangaswamy, N. S., 1965, Embryology in relation to physiology and genetics, in: "Advances in Botanical Research," R. D. Preston, ed., Vol. 2, pp. 219-321, Academic Press, New York.

Maliga, P., Kiss, Z. R., Nagy, A. H., and Lazar, G., 1978, Genetic instability in somatic hybrids of Nicotiana tabacum and Nicotiana knightiana, Mol. Gen. Genet., 163:145-151.

Maliga, P., Lazar, G., Joo, F., Nagy, A. H., and Menczel, L., 1977, Restoration of morphogenetic potential in Nicotiana by somatic hybridization, Mol. Gen. Genet., 157:291-296.

Maliga, P., Lorz, H., Lazar, G., and Nagy, F., 1982, Cytoplast-protoplast fusion for interspecific chloroplast transfer in Nicotiana, Mol. Gen. Genet., 185:211-215.

Malmberg, R. L. and Griesbach, R. J., 1980, The isolation of mitotic and meiotic chromosomes from plant protoplasts, Plant Sci. Lett., 17:141-147.

Malmberg, R. L. and Griesbach, R. J., 1983, Chromosomes from protoplasts--isolation fractionation and uptake, in: "Genetic Engineering of Plants," T. Kosuge, C. P. Meridith, and A. Hollaender, eds., Basic Life Sci. V. 26, pp. 195-201, Plenum Press, New York.

Marubashi, W. and Nakajima, T., 1982, Interspecific hybridization between Nicotiana tabacum L. and N. rustica by test-tube pollination and ovule culture, in "Plant Tissue Culture," A. Fujiwara, ed., pp. 775-776, The Jpn. Assoc. Plant Tissue Cult., Tokyo.

Matsuzawa, Y., 1978, Studies on the interspecific hybridization in genus Brassica. Part 1. Effects of temperature on the development of hybrid embryos and the improvement of crossability by ovary culture in interspecific cross Brassica campestris x Brassica oleracea, Jpn. J. Breed., 28:186-196.

Mauney, J. R., 1961, The culture in vitro of immature cotton embryos, Bot. Gaz., 122:205-209.

McDaniel, R. G., 1981, Possibilities for organelle transfer, in: "Genetic Engineering for Crop Improvement," K. O. Rachie and J. M. Lyman, eds., pp. 185-207, Rockefeller Foundation, U.S.A.

Medgyesy, P., Menczel, L., and Maliga, P., 1980, The use of cytoplasmic streptomycin resistance: Chloroplast transfer from Nicotiana tabacum into Nicotiana sylvestris, and isolation of their somatic hybrids, Mol. Gen. Genet., 179:693-698.

Meins, Jr., F., 1983, Heritable variation in plant cell culture, Ann. Rev. Plant Physiol., 34:327-346.

Melchers, G., 1977, Microbial techniques in somatic hybridization by fusion of protoplasts, in: "International Cell Biology," B. R. Brinkley and K. R. Porter, eds., pp. 207-215, Rockefeller Univ. Press, New York.

Melchers, G. and Labib, G., 1974, Somatic hybridization of plants by fusion of protoplasts. I. Selection of light resistant hybrids of "haploid" light sensitive varieties of tobacco, Mol. Gen. Genet., 135:277-294.

Melchers, G., Sacristan, M. D., and Holder, A. A., 1978, Somatic hybrid plants of potato and tomato regenerated from fused protoplasts, Carlsberg Res. Commun., 43:203-218.

Menczel, L., Lazar, G., and Maliga, P., 1978, Isolation of somatic hybrids by cloning Nicotiana heterokaryons in nurse culture, Planta, 143:29-32.

Menczel, L., Galiba, G., Nagy, F., and Maliga, P., 1982, Effect of radiation dosage on efficiency of chloroplast transfer by protoplast fusion in Nicotiana, Genetics, 100:487-495.

Menczel, L., Nagy, F., Kiss, Z. R., and Maliga, P., 1981,
Streptomycin resistant and sensitive somatic hybrids of
Nicotiana tabacum plus Nicotiana knightiana: Correlation of
resistance to Nicotiana tabacum plastids, Theor. Appl.
Genet., 59:191-195.

Menczel, L., Nagy, F., Lazar, G., and Maliga, P., 1983, Transfer
of cytoplasmic male sterility by selection for streptomycin
resistance after protoplast fusion in Nicotiana, Mol. Gen.
Genet., 189:365.

Mok, D. W. S., Mok, M. C., and Rabak-Oarihanta, A., 1978,
Interspecific hybridization of Phaseolus vulgaris with P.
lunatus and P. acutifolius, Theor. Appl. Genet., 52:209-215.

Monnier, M., 1978, Culture of zygotic embryos, in: "Frontiers of
Plant Tissue Culture," T. A. Thorpe, ed., pp. 277-286, Univ.
Calgary Press, Calgary.

Murashige, T., 1978, The impact of plant tissue culture on
agriculture, in: "Frontiers of Plant Tissue Culture," T. A.
Thorpe, ed., pp. 15-26, Univ. Calgary Press, Calgary.

Muzik, T. J., 1956, Studies on the development of the embryo and
seed of Hevea brasiliensis in culture, Lloydia, 19:86-91.

Nagao, T., 1978, Breeding by somatic hybridization based on
protoplast fusion. 1. The combination Nicotiana tabacum and
Nicotiana rustica, Jpn. J. Crop Sci., 47:491-498.

Nagao, T., 1979, Somatic hybridization by fusion of protoplasts.
II. The combinations of Nicotiana tabacum and N. glutinosa
and N. tabacum and N. alata, Jpn. J. Crop Sci., 48:385-392.

Nagao, T., 1982, Somatic hybridization by fusion of protoplasts.
III. Somatic hybrids of sexually incompatible combinations
Nicotiana tabacum + Nicotiana repanda and Nicotiana tabacum
+ Salpiglossi sinuata, Jpn. J. Crop Sci., 51:35-42.

Nakajima, T., Doyama, Y., and Matsumoto, H., 1969, In vitro
culture of excised ovules of white clover, Trifolium repens
L., Jpn. J. Breed., 19:373-378.

Nakamura, C., Keller, W. A., Fedak, G., 1981, In vitro
propagation and chromosome doubling of a Triticum crassum x
Hordeum vulgare intergeneric hybrid, Theor. Appl. Genet.,
60:89-96.

Narayanswami, S. and Norstog, K., 1964, Plant embryo culture,
Bot. Rev., 30:587-628.

Nehls, R., 1978, The use of metabolic inhibitors for the
selection of fusion products of higher plant protoplasts,
Mol. Gen. Genet., 166:117-118.

Newell, C. A. and Hymowitz, T., 1982, Successful wide
hybridization between the soybean and a wild perennial
relative G. tomentella H., Crop Sci., 22:1062-1065.

Niimi, Y., 1976, Effect of "Stylar Pollination" on in vitro seed
setting of Petunia hybrida, Jpn. Hort. Sci., 45:168-172.

Nitsch, J. P., 1949, Culture of fruits in vitro, Science, 110:499.

Nitsch, J. P., 1951, Growth and development in vitro of excised ovaries, Am. J. Bot., 38:566-577.

Nitzsche, W. and Henning, L., 1976, Fruchtknotenkultur bei Grasern, Z. Pflanzenzuecht., 77:80-82.

Norstog, K., 1979, Embryo culture as a tool in the study of comparative and developmental morphology, in: "Plant Cell and Tissue Culture," W. R. Sharp, P. O. Larsen, E. F. Paddock, and V. Raghavan, eds., p. 179-202, Ohio State University Press, Columbus, Ohio.

North, C., 1976, In vitro culture of plant material as an aid to hybridization, Acta Hort., 63:67-74.

Paris, D., Rietsema, J., Santina, S., and Blakeslee, A. F., 1953, Effects of amino acids, especially aspartic and glutamic acid and their amides, on the growth of Datura stramonium embryos in vitro, Proc. Nat. Acad. Science, U.S.A., 39:1205-1212.

Patnaik, G., Cocking, E. C., Hamill, J., and Pental, D., 1982, A simple procedure for the manual isolation and identification of plant heterokaryons, Plant Sci. Lett., 24:105-110.

Peitian, W., Jiayu, C., Shimin, Z., Jinxiang, X., and Lianqing, W., 1981, Interspecific hybrid plants by protoplast fusion between Nicotiana tabacum and N. rustica, Ann. Report Inst. Genet. Academia Sinica, pp. 124-125.

Petru, E., Hrabetova, E., and Tupy, J., 1964, The technique of obtaining germinating pollen without microbial contamination, Biol. Plant., 6:68-69.

Phillips, G. C., Collins, G. B., and Taylor, N. L., 1982, Interspecific hybridization of red clover (Trifolium pratense L.) with T. sarosiense Hazsl. using in vitro embryo rescue, Theor. Appl. Genet., 62:17-24.

Potrykus, I. and Hoffmann, F., 1973, Transplantation of nuclei into protoplasts of higher plants, Z. Pflanzenphysiol., 69:287-289.

Potrykus, I. and Lorz, H., 1976, Organelle transfer into isolated protoplasts, in: "Cell Genetics in Higher Plants," D. Dudits, G. L. Farkas, and P. Maliga, eds., pp. 183-190, Akademiai Kiado, Budapest.

Potrykus, I., Shillito, R. D., Jia, J., and Lazar, G. B., 1983, Auxotroph complementation via protoplast fusion in Hyoscyamus muticus and Nicotiana tabacum, in: "Genetic Engineering in Eukaryotes," P. F. Lurquin and A. Kleinhofs, eds., p. 253, Plenum Press, New York.

Power, J. B., Frearson, E. M., Hayword, C., George, D., Evans, P. K., Berry, S. F., and Cocking, E. C., 1976, Somatic hybridization of Petunia hybrida and Petunia parodii, Nature, 263:500-502.

Power, J. B., Frearson, E. M., Hayword, C., and Cocking, E. C., 1975, Some consequences of the fusion and selection of Petunia and Parthenocissus protoplasts, Plant Sci. Lett., 5:197-207.

Power, J. B., Berry, S. F., Chapman, J. V., and Cocking, E. C., 1980, Somatic hybridization of sexually incompatible petunia: Petunia parodii, Petunia parviflora, Theor. Appl. Genet., 57:1-4.

Power, J. B., Berry, S. F., Frearson, E. M., and Cocking, E. C., 1977, Selection procedures for the production of interspecies somatic hybrids of Petunia hybrida and Petunia parodii: I. Nutrient media and drug sensitivity complementation selection, Plant Sci. Lett., 10:1-6.

Power, J. B., Berry, S. F., Chapman, J. V., Cocking, E. C., and Sink, K. C., 1979, Somatic hybrids between unilateral cross-incompatible Petunia species, Theor. Appl. Genet., 55:97-100.

Power, J. B., Cummins, S. E., and Cocking, E. C., 1970, Fusion of isolated plant protoplasts, Nature, 225:1016-1018.

Raghavan, V., 1976, "Experimental Embryogenesis in Vascular Plants," Academic Press, New York.

Raghavan, V., 1977, Applied aspects of embryo culture, in: "Plant Cell, Tissue and Organ Culture," J. Reinert and Y. P. S. Bajaj, eds., pp. 375-397, Springer-Verlag, Berlin.

Raghavan, V., 1980, Embryo culture, Int. Rev. Cytol. Suppl., 11B:209-240.

Raghavan, V. and Torrey, J. G., 1964, Effects of certain growth substances on the growth and morphogenesis of immature embryos of Capsella in culture, Plant Physiol., 39:691-699.

Raman, K., Walden, D. B., and Greyson, R. I., 1980, Fertilization in Zea mays by cultured gametophytes, J. Hered., 71:311-314.

Randolph, L. F. and Khan, R., 1960, Growth response of excised mature embryos of Iris and wheat to different culture media, Phytomorphology, 10:43-49.

Rangaswamy, N. S., 1977, Applications of in vitro pollination and in vitro fertilization, in: "Applied and Fundamental Aspects of Plant Cell, Tissue and Organ Culture," J. Reinert and Y. P. S. Bajaj, eds., pp. 412-425, Springer-Verlag, Berlin.

Rangaswamy, N. S. and Shivanna, K. R., 1972, Overcoming self-incompatibility in Petunia axillaris. III. Two-site pollinations in vitro, Phytomorphology, 21:284-289.

Rao, P. S., 1965, The in vitro fertilization and seed formation in Nicotiana rustica L., Phyton, 22:165-167.

Rao, P. S. and Rangaswamy, N. S., 1972, In vitro development of the pollinated pistils of Nicotiana rustica L., Bot. Gaz., 133:350-355.

Rèdei, G. and Rèdei, G., 1955a, Rearing wheats from ovaries cultured in vitro, Acta Bot. Acad. Sci. Hungar., 2:183-186.

Rèdei, G. and Rèdei, G., 1955b, Developing wheat embryos excised
    from ovaries cultured in vitro, Experientia, 11:387-388.
Reed, S. M., 1978, Interspecific hybridization in Nicotiana
    through in vitro culture of fertilized ovules, Ph.D.
    Dissertation, Univ. Ky., Lexington.
Reed, S. M. and Collins, G. B., 1978, Interspecific hybrids in
    Nicotiana through in vitro culture of fertilized ovules, J.
    Hered., 69:311-315.
Reed, S. M. and Collins, G. B., 1980, Histological evaluation of
    seed failure in three Nicotiana interspecific hybrids, Tob.
    Sci., 24:154-156.
Reinert, J. and Gosch, G., 1976, Continuous division of
    heterokaryons from Daucus carota and Petunia hybrida
    protoplasts, Naturwissenschaften, 63:534.
Rennie, P. J., Weber, G., Constabel, F., and Fowke, L. C., 1980,
    Dedifferentiation of chloroplasts in interspecific and
    homospecific protoplast fusion products, Protoplasma,
    103:253-262.
Richards, K. W. and Rupert, E. A., 1980, In vitro fertilization
    and seed development in Trifolium, In Vitro, 16:925-931.
Rijven, A. H. G. C., 1955, Effects of glutamine, asparagine and
    other related compounds on the in vitro growth of embryos of
    Capsella bursapastoris, Konkinkl. Nederl. Akad. Wetensch.
    Proc., C58:368-376.
Rijven, A. H. G. C., 1956, Glutamine and asparagine as nitrogen
    sources for the growth of plant embryos in vitro: A
    comparative study of 12 species, Austr. J. Biol. Sci.,
    9:511-517.
Sachar, R. C. and Kapoor, M., 1958, Influence of kinetin and
    gibberellic acid on the test tube seeds of Cooperia
    pedunculata Herb., Naturwissenschaften, 45:552-553.
Sachar, R. C. and Kapoor, M., 1959, In vitro culture of ovules of
    Zephyranthes, Phytomorphology, 9:147-156.
Sanchez-Monge, E. and Garcia-Olmedo, F., 1977, "Interspecific
    Hybridization in Plant Breeding," Proc. Eighth Con.
    Eucarpia, Escuela Tècuica Superior de Ingenieros Agrònomos,
    Ciudad Univ., Madrid, Spain.
Schenck, H. R. and Robbelen, G., 1982, Somatic hybrids by fusion
    of protoplasts from Brassica oleracea and B. campestris, Z.
    Pflanzenzuecht, 89:278-288.
Schieder, O., 1978a, Somatic hybrids of Datura innoxia Mill +
    Datura discolor Bernh. and of Datura innoxia Mill + Datura
    stramonium L. var. tatula L. I. Selection and
    characterization, Mol. Gen. Genet., 162:113-119.
Schieder, O., 1978, Production and uses of metabolic and
    chlorophyll deficient mutants, in: "Frontiers of Plant
    Tissue Culture," T. A. Thorpe, ed., pp. 393-401, Univ.
    Calgary Press, Calgary.

Schieder, O., 1980, Somatic hybrids between a herbaceous and two tree Datura species, Z. Pflanzenphysiol., 98:119-127.

Schieder, O. and Vasil, I. K., 1980, Protoplast fusion and somatic hybridization, in: "Int. Rev. Cytol. Suppl. 11B Perspectives in Plant Cell and Tissue Culture," I. K. Vasil, ed., pp. 21-46, Academic Press, New York.

Schopfer, W. H., 1943, "Plants and Vitamins," Chronica Botanica Co., Waltham.

Sen, B. and Verma, G., 1958, in: "Modern Developments in Plant Physiology," P. Maheshwari, ed., p. 118, New Delhi.

Shabde-Moses, M. and Murashige, T., 1979, Organ culture, in: " Nicotiana: Procedures for Experimental Use," R. D. Durbin, ed., pp. 40-51.

Sharma, D. R., Chowdhury, J. B., Ahuja, U., and Dhankhar, B. S., 1980, Interspecific hybridization in the genus Solanum: A cross between S. melongena and S. khasianum through embryo culture, Z. Pflanzenzuecht., 85:248-253.

Sharma, H. C. and Gill, B. S., 1983, Current status of wide hybridization in wheat, Euphytica, 32:17-31.

Shivanna, K. R., 1965, In vitro fertilization and seed formation in Petunia violacea Lindl., Phytomorphology, 15:183-185.

Shizukuda, N. and Nakajima, T., 1982, Production of interspecific hybrids between Nicotiana rustica L. and N. tabacum L. through ovule culture, Jpn. J. Breed., 32:371-377.

Siddiqui, S. A., 1964, In vitro culture of ovules of Nicotiana tabacum L. var. N.P. 31, Naturwissenschaften, 51:517.

Sink, K. C., Power, J. B., and Natarello, N. J., 1978, The interspecific hybrid Petunia parodii and its relevance to somatic hybridization, Theor. Appl. Genet., 53:205-208.

Sladky, Z. and Havel, I., 1976, The study of the conditions for the fertilization in vitro in maize, Biol. Plant., 18:469-472.

Smith, H. H., Kao, K. N., and Combatti, N. C., 1976, Interspecific hybridization by protoplast fusion in Nicotiana, confirmation and extension, J. Hered., 67:123-128.

Stafford, A. and Davies, D. R., 1979, The culture of immature pea embryos, Ann. Bot., 44:315-321.

Stalker, H. T., 1980, Utilization of wild species for crop improvement, Adv. Agron., 33:111-147.

Stebbins, G. L., Jr., 1950, Variation and evolution in plants, Columbia Univ. Press, N.Y.

Stewart, J. M., 1981, In vitro fertilization and embryo rescue, Environ. Exp. Bot., 21:301-315.

Stewart, J. M. and Hsu, C. L., 1977, In ovulo embryo culture and seedling development of cotton (Gossypium hirsutum L.), Planta, 137:113-117.

Stewart, J. M. and Hsu, C. L., 1978, Hybridization of diploid and
    tetraploid cottons (Gossypium sp.) through in ovulo embryo
    culture, J. Hered., 69:404-408.
Stoltz, L. P., 1971, Agar restriction of the growth of excised
    mature Iris embryos, J. Am. Soc. Hort. Sci., 96:681-684.
Szabados, L., Hadlaczky, G., and Dudits, D., 1981, Uptake of
    isolated plant chromosomes by plant protoplasts, Planta,
    151:141-145.
Taira, T. and Larter, E. N., 1978, Factors influencing
    development of wheat-rye hybrid embryos in vitro, Crop Sci.,
    18:348-350.
Taylor, N. L., Collins, G. B., Cornelius, P. L., and Pitcock, J.,
    1983, Differential interspecific compatibilities among
    genotypes of Trifolium sarosiense and T. pratense, in:
    "Prod. Fourteenth Int. Grassl. Cong.," A. Smith, ed.
Taylor, N. L., Quarles, R. F., and Anderson, M. K., 1980, Methods
    of overcoming interspecific barriers in Trifolium,
    Euphytica, 29:441-450.
Teoule, E., 1983, Somatic hybridization between Medicago sativa
    L. and Medicago falcata L., C. R. Acad. Sci. (Paris),
    297:13-16.
Ternovskii, M. F., Shinkareua, I. K., and Larkina, N. I., 1976,
    Production of interspecific tobacco hybrids by the
    pollination of ovules in vitro, Sov. Genet. (Eng. Transl.
    Genetika), 12:1209-1213.
Thomas, B. R. and Pratt, D., 1981, Efficient hybridization
    between Lycopersicon esculentum and L. peruvianum via embryo
    callus, Theor. Appl. Genet., 59:215.
Toshiaki, K., Horn, M. E., and Widholm, J. M., 1981, Hybrid shoot
    formation from fused Daucus carota and Daucus capillifolius
    protoplasts, Z. Pflanzenphysiol., 104:459-466.
Uchimiya, H., 1982, Somatic hybridization between male sterile
    Nicotiana tabacum and N. glutinosa through protoplast
    fusion, Theor. Appl. Genet., 61:69-72.
Uchimiya, H., Ohgawara, T., Kato, H., Akiyama, T., and Harads,
    H., 1983, Detection of two different nuclear genomes in
    parasexual hybrids by ribosomal RNA gene analysis, Theor.
    Appl. Genet., 64:117-118.
Uchimiya, H. and Wildman, S. G., 1979, Non-translation of foreign
    genetic information for fraction 1 protein under
    circumstances favorable for direct transfer of Nicotiana
    gossei isolated chloroplasts into N. tabacum protoplasts, In
    Vitro, 15:463-468.
Usha, S. V., 1965, In vitro pollination in Antirrhinum majus L.,
    Curr. Sci., 34:511-513.
van Overbeek, J., Conklin, M. E., and Blakeslee, A. F., 1942,
    Cultivation in vitro of small Datura embryos, Am. J. Bot.,
    29:472-477.

van Overbeek, J., Siu, R., and Haagen-Smit, A. J., 1944, Factors affecting the growth of Datura embryos in vitro, Am. J. Bot., 31:219-224.

Vasil, I. K., Ahuja, M. R., and Vasil, V., 1979, Plant tissue cultures in genetics and plant breeding, Adv. Genet., 20:127-215.

Vasil, I. K. and Giles, K. L., 1975, Induced transfer of higher plant chloroplasts into fungal protoplasts, Science, 190:680.

Vasil, I. K. and Vasil, V., 1980, Isolation and culture of protoplasts, in: "Int. Rev. Cytol. Suppl. IIB Perspectives in Plant Cell and Tissue Culture," I. K. Vasil, ed., pp. 1-19, Academic Press, New York.

Veen, H., 1963, The effect of various growth regulators on embryos of Capsella bursa-pastoris growing in vitro, Acta Bot. Neerl., 12:129-171.

Wagner, G. and Hess, D., 1973, In vitro - befruchtungen bei Petunia hybrida, Z. Pflanzenphysiol., 69:262-269.

Wallin, A., Glimelius, K., and Eriksson, T., 1978, Enucleation of plant protoplasts by cytochalasin B, Z. Pflanzenphysiol., 87:333-340.

Watanabe, K., 1977, Successful ovary culture and production of F_1 hybrids and androgenic haploids in Japanese Chrysanthemum species, J. Hered., 68:317-320.

Wetter, L. R., 1977, Isoenzyme patterns in soybean - Nicotiana somatic hybrid cell lines, Mol. Gen. Genet., 150:231-235.

Wetter, L. R. and Kao, K. N., 1980, Chromosome and isoenzyme studies on cells derived from protoplast fusion of Nicotiana glauca with Glycine max x Nicotiana glauca cell hybrids, Theor. Appl. Genet., 57:273-276.

White, D. W. R. and Vasil, I. K., 1979, Use of amino acid analogue-resistant cell lines for selection of Nicotiana sylvestris somatic cell hybrids, Theor. Appl. Genet., 55:107-112.

Widholm, J. M., 1982, Selection of protoplast fusion hybrids, in: "Plant Tissue Culture," A. Fujiwara, ed., pp. 609-612, The Jpn. Assoc. Plant Tissue Cult., Tokyo.

Williams, E., 1978, A hybrid between Trifolium repens and T. ambiguum obtained with the aid of embryo culture, N. Z. J. Bot., 16:499-506.

Williams, E. G. and DeLautour, G., 1980, The use of embryo culture with transplanted nurse endosperm for the production of interspecific hybrids in pasture legumes, Bot. Gaz., 141:252-257.

Williams, E. G. and DeLautour, G., 1981, Production of tetraploid hybrids between Ornithopus pinnatus and Ornithopus sativus using embryo culture, N. Z. J. Bot., 19:23-30.

Williams, E. G. and Verry, I. M., 1981, A partially fertile
    hybrid between Trifolium repens and T. ambigiuum, N. Z. J.
    Bot., 19:1.
Williams, E. G., Verry, I. M., and Williams, W. M., 1982, Use of
    embryo culture in interspecific hybridization, in: "Plant
    Improvement and Somatic Cell Genetics," I. K. Vasil, W. R.
    Scowcroft, and K. J. Frey, eds., pp. 119-128, Academic
    Press, New York.
Williams, W. M. and Williams, E. G., 1983, Use of embryo culture
    with nurse endosperm for interspecific hybridization in
    pasture legumes, in: "Proc. Fourteenth Int. Grassland
    Congress," A. Smith, ed., pp. 163-165.
Withner, C. L., 1942, Nutrition experiments with orchid
    seedlings, Am. Orchid Soc. Bull., 11:112-114.
Withner, C. L., 1943, Ovule culture: a new method for starting
    orchid seedlings, Am. Orchid Soc. Bull., 11:261-263.
Xianghui, L., Wenbin, L., and Meijuan, H., 1982, Somatic hybrid
    plants from intergeneric fusion between tobacco tumor B653
    and Petunia hybrida W43 and expression of LPDH, Scientia
    Sinica (B), 25:611-619.
Yeung, E. C. and Sussex, I. M., 1979, Embryogeny of Phaseolus
    coccineus: The suspensor and growth of the embryo-proper in
    vitro, Z. Pflanzenphysiol., 91:423-433.
Yeung, E. C., Thorpe, T. A., and Jensen, C. J., 1981, In vitro
    fertilization and embryo culture, in: "Plant Tissue Culture:
    Methods and Applications in Agriculture," T. A. Thorpe, ed.,
    pp. 253-271, Academic Press, New York.
Zelcer, A., Aviv, D., and Galun, E., 1978, Interspecific transfer
    of cytoplasmic male sterility by fusion between protoplasts
    of normal Nicotiana sylvestris and X-ray irradiated
    protoplasts of male-sterile N. tabacum, Z. Pflanzenphysiol.,
    90:397-408.
Zenkteler, M., 1965, Test-tube fertilization in Dianthus
    caryophyllus L., Naturwissenschaften, 52:645-646.
Zenkteler, M., 1967, Test-tube fertilization of ovules in
    Melandrium album Mill. with pollen grains of several species
    of the Caryophyllaceae family, Experientia, 23:775-776.
Zenkteler, M., 1969, From Zenkteler, M., 1980, PTPN Pr. Kom.
    Biol., 32:1-71.
Zenkteler, M., 1970, Test-tube fertilization of ovules in
    Melandrium allium Mill. with pollen grains of Datura
    stramonium L., Experientia, 26:661-662.
Zenkteler, M., 1980, Intraovarian and in vitro pollination, in:
    "Perspectives in Plant Cell and Tissue Culture," I. K.
    Vasil, ed., Intern. Rev. Cytol. Suppl. 11B, pp. 137-156,
    Academic Press, New York.

Zenkteler, M., Slusarkiewicz-Jarzina, A., and Woźna, J., 1981, Cytological investigations of hybrid plants of Nicotiana alata and Nicotiana debneyi: Obtained by in vitro pollination of ovules, Bull. Soc. Amis Sci. Lett. Poznań Ser. D, Sci. Biol., 21:79-84.

Zenkteler, M. and Melchers, G., 1978, In vitro hybridization by sexual methods and by fusion of somatic protoplasts, Theor. Appl. Genet., 52:81-90.

Zenkteler, M., Misiura, E., and Guzowska, I., 1975, Studies on obtaining hybrid embryos in test tubes, in: "Form, Structure and Function in Plants," B. M. Johri, ed., pp. 180-187, Sarita Prakashan, Meerut, India.

Ziebur, N. K. and Brink, R. A., 1951, The stimulative effect of Hordeum endosperms on the growth of immature plant embryos in vitro, Am. J. Bot., 38:253-256.

Zubkova, M. and Sladky, Z., 1975, The possibility of obtaining seeds following placental pollination in vitro, Biol. Plant., 17:276-280.

THE SIGNIFICANCE OF DOUBLED HAPLOID VARIATION

P. S. Baenziger, D. T. Kudirka, G. W. Schaeffer
and M. D. Lazar

Formerly, Research geneticist, Field Crops Laboratory,
Plant Genetics and Germplasm Institute, Beltsville
Agricultural Research Center, Beltsville, Maryland
20705 (now research manager, Crop Improvement, Monsanto
Agricultural Products Company), research specialist,
Wheat Cell Biology, Monsanto Agricultural Products
Company, St. Louis, MO 63167 and research plant
physiologist, Tissue Culture and Molecular Genetics
Laboratory, Plant Physiology Institute, Beltsville
Agricultural Research Center, Beltsville, Maryland
20705, and research scientist, Biology Group, Alberta
Research Council, Edmonton, Alberta, Canada T6G 2C2.

INTRODUCTION

Doubled haploid breeding has aroused considerable interest in
recent years as methods of creating doubled haploids have become
more efficient. This subject has been reviewed by a number of
authors (Reinert and Bajaj, 1977; Nitzsche and Wenzel, 1977; Sink
and Padmanabhan, 1977; Kihara, 1979; Maheshwari et al., 1980;
Hu and Shao, 1981; Kasha and Reinbergs, 1981; Collins and Genovesi,
1982; Maheshwari et al., 1982; Schaeffer and Baenziger, 1982;
Baenziger and Schaeffer, 1983). An extensive review of the earlier
literature was written by Kimber and Riley (1963). With that in
mind, the purpose of this report will not be to extensively review
the past and current literature, but rather will attempt to discuss
those aspects that plant breeders need to be aware of with respect
to doubled haploidy. The two main topics of this report will be 1)
the breeding theory that supports the use of doubled haploids and
2) how doubled haploids are currently being used in breeding
programs with particular emphasis on how theory differs from
practice and why. To be comprehensive, methods for creating
doubled haploids and their future uses will also be discussed.

Methods of Creating Doubled Haploids

Of the many methods of creating doubled haploids, the two most widely used are interspecific or intergeneric hybridization and anther culture. Interspecific hybridization may produce haploids by parthenogenesis (Rowe, 1974) or by chromosome elimination (Kasha and Kao, 1970; Kasha, 1974). Interspecific hybridization for the production of haploids by chromosome elimination is being widely used in barley (Hordeum vulgare L.) using the wild species H. bulbosum L. (2x) as a pollinator (Kasha and Reinbergs, 1981). 'Mingo" barley was released in 1979 and is a doubled haploid produced utilizing this method (Ho and Jones, 1980). H. bulbosum has also been shown to produce haploids with other Hordeum spp., and Triticum spp., (including some of the species formerly know as Aegilops) (Kasha, 1974; Barclay, 1976; Miller and Chapman, 1976; Chapman and Miller, 1977; Shigenobu and Sakamoto, 1977; Subrahmanyam, 1980). Interspecific hybridization is also effective in producing maternal haploids in tobacco (Nicotiana tabacum L.). In this procedure N. africana Mermx. is the pollinator (Burk et al., 1979).

In 1964, Guha and Maheshwari (1964) discovered that anthers from Datura innoxia Mill. could be cultured and haploid plants regenerated. This process later confirmed by Nitsch and Nitsch (1969) in tobacco has in succeeding years been extended to numerous plant species (Schaeffer et al., 1979; Collins and Genovesi, 1982; Maheshwari et al., 1982). In this procedure, immature anthers are placed on specialized media where the microspores will form either callus or embryoids, and after transferring to a regenerative medium haploid plants are regenerated. The main advantage of this method is that potentially every microspore within an anther can form a plant. Burk et al. (1979) reported that in tobacco an average of 3.7 plants were regenerated from each of over 2,000 anthers. The range of plants produced per anther was 0 to 60. While some of these plants may, and probably did, originate from the same microspore and hence would be clones, it is evident that in some genera a high production of haploids from anther culture is obtainable. Unfortunately, for most genera, particularly the grasses, the rate of success is much lower at present (see Picard et al., 1978; Miao et al., 1978; Brettell et al., 1981; Bullock et al., 1982; Foroughi-Wehr et al., 1982; Genovesi and Collins, 1982; Rines, 1983). However, progress is being made to improve the success rate in those genera where research is being actively pursued. In wheat, anther calli can routinely be produced in 15% of the anthers cultured utilizing selected cultivars.

Some haploids occur spontaneously. Haploidy can be induced by physical agents such as irradiation (Pandey and Phung, 1982) and chemical agents (see Lacadena, 1974). There are also genetic systems that enhance the production of haploids such as the haploid

initiator factor in barley (Hagberg and Hagberg, 1980; Hagberg and
Hagberg, 1981); and the indeterminant gametophyte gene in
(Kermicle, 1969) or stock 6 in corn (Zea mays) (Coe, 1959) (see
Sarkar, 1974). Genetic factors in the presence of alien cytoplasms
will also produce haploids (Kobayashi and Tsunewaki, 1980). Often
one embryo of a twin seedling is haploid (Kappert, 1933 in flax
(Linum usitatissimum L.; Namikawa and Kawakami, 1934 in wheat).
Polyembryonic seed of male sterile plants will also produce
haploids (Kenworthy et al., 1973). Semigamy, an abnormal
fertilization process in which an unreduced sperm nucleus enters an
unreduced egg cell but does not fuse with the egg cell, will also
produce doubled haploids (Turcotte and Feaster, 1974).

While there are a number of very different methods of creating
haploids and doubled haploids, there are a number of characteristics that are important to all of the methods. Of prime
significance to plant breeders is the importance of the genotype.
Genotypic differences within T. aestivum have been associated with
crossability with H. bulbosum (Snape et al., 1979; Falk and Kasha,
1983). Genotypic differences within H. bulbosum and H. vulgare
will also affect the number of haploids produced (Jensen, 1975;
Pickering and Hayes, 1976; Simpson et al., 1980). Genotypic
differences are extremely common in anther culture studies and some
authors have suggested that breeding for anther culturability would
be an efficient (perhaps the most efficient) way of increasing the
number of plants from anthers (Picard and DeBuyser, 1977; Wenzel et
al., 1977; Picard et al., 1978; Jacobsen and Sopory, 1978; Bullock
et al., 1982; Foroughi-Wehr et al., 1982; Raquin, 1982; Liang et
al., 1982; Rines, 1983; Lazar et al., 1983a, 1983b). Anther
culture itself has been suggested as a selection method for
improving anther culturability (Picard and DeBuyser, 1977; Rives
and Picard, 1977) though in other studies using different genotypes
anther culture did not appear to be effective in selecting for
improved anther culturability (Lazar et al., 1983a). It has been
recently estimated that heritabilities of both callus formation and
plantlet regeneration in a population of five spring wheat
cultivars were of the order of 0.6 to 0.7 (Lazar et al., 1983b),
suggesting that improving the responsiveness of currently
unresponsive types may not be very difficult. Also of critical
importance are the donor plant growth conditions (temperature and
light), donor plant health and growth stage, donor plant pre and
post treatment (often thermal shock is important), and the culture
medium (Sunderland, 1978 and 1979; Genovesi and Collins, 1982;
Maheshwari et al., 1982; Baenziger and Schaeffer, 1983).
Microspore development including vegetative and generative nuclei
development is affected by variations in plant growth conditions
(Bennett, 1976). Often one medium is used for callus induction, a
second medium for callus growth and maintenance, and a third for
plant regeneration. The initial medium may be "conditioned" with
anthers to enhance callus formation of the inoculated anthers (Xu

et al., 1981). All of the above factors will interact with each
other. For example, Dunwell (1976), Bernard (1977), Picard et al.
(1978), and Durr and Fleck (1980) all report seasonal response
variations for anther culturability for donor plants grown in
greenhouses. Lazar et al. (1983a) have further shown there is
significant genotype by donor plant environment interaction for
anther culturability not only due to changes in magnitude, but also
to reversals of order. Hence, it should not be surprising that
comparisons between studies and research groups may be difficult,
and at times seem contradictory.

    With the exception of techniques involving unreduced gametes
to directly produce doubled haploids, the techniques described
above will produce sterile haploid plants. Three principal
doubling techniques have been used to produce doubled haploids
which when left to grow to maturity should be fertile. They are
spontaneous doubling which often occurs during callus development
and embryo rescue, regeneration from pith callus as is done in
tobacco (Kasperbauer and Collins, 1972), and the use of colchicine
(Jensen, 1974). Colchicine has been the most widely used method.

DOUBLED HAPLOID BREEDING THEORY

    Much of the doubled haploid breeding theory has been reviewed
by Kasha and Reinbergs (1981), Collins and Genovesi (1982) and
Baenziger and Schaeffer (1983). Despite these recent reviews, the
topic will be discussed in detail as it greatly explains the
interest in doubled haploid research among plant breeders.

    The main advantage of doubled haploid breeding is that it can
very rapidly produce homozygous lines. Homozygous or "pure-
breeding" lines are marketed directly to growers in self-pollinated
crops. Homozygous or "inbred" lines are used to make hybrids which
are marketed to growers in many of the cross-pollinated crops. In
crops having long growing seasons, such as winter wheat, the time
savings can be substantial. Even more time can be saved in
biennial or multi-year crops that are marketed as pure-breeding
lines or that are marketed as hybrids from inbreds. Doubled
haploids have relatively less advantage in short season crops where
other methods that rapidly produce homozygous lines such as single
seed descent (hereafter abbreviated SSD; Grafius, 1965; Brim, 1966)
can be used. A comparison of a doubled haploid winter wheat
breeding program using anther culture versus a conventional winter
wheat breeding program is shown in Table 1 (excerpted from
Baenziger and Schaeffer, 1983). Both the conventional and doubled
haploid breeding programs begin with the introduction of variation
by sexual hybridization. It is in the F_1 generation that the two
methods diverge. In the doubled haploid breeding programs, anthers
from F_1 plants are excised and the microspores are induced to form
callus, then plants are regenerated which in turn are doubled to

Table 1.   Comparison of doubled haploid and conventional breeding
methods for winter wheat.

time[a]	Doubled haploid	Conventional
5-7/0	Select parents	
8-9/1	Vernalize seed, transplant to greenhouse	
11/1	Make crosses to produce F_1 seed	
12/1	Harvest seed, vernalize seed	
2-3/1	Transplant to greenhouse	Transplant to field
4/1	Excise anthers	
7/1	Calli forming	Harvest F_2 seed
9/2	Double and regenerate plants	
10/2	Transplant to field	Plant F_2 seed
7/2	Harvest doubled haploid	Harvest F_3 seed
7/3	Select and harvest doubled haploid rows	Harvest F_4 seed
7/4	Harvest preliminary doubled haploid yield trial	Harvest F_5 seed
7/5	Harvest advanced doubled haploid yield trial	Select and harvest F_6 head rows
7/6	Harvest advanced doubled haploid yield trial	Harvest preliminary yield trial

[a]Month/year. Year 1 starts with vernalizing parental seed for crossing.

P. S. Baenziger and G. W. Schaeffer, "Dihaploids via Anthers
Cultured in Vitro," from Lowell D. Owens, ed., Genetic Engineering:
Applications to Agriculture, volume 7 of the Beltsville Symposia
in Agricultural Research, Rowman & Allanheld, 1983, pp. 272.

restore fertility. The doubled haploids should represent the gametic array of the cross in a diploid homozygous form. While homozygosity is reached within two years of the cross, it should be remembered that additional years are needed to increase the seed prior to replicated selection trials. In the conventional breeding program, the F_1 plants can be transplanted to the field to produce F_2 seed. In general, the generations are advanced one generation per year. While SSD has been used in spring wheat programs, it has rarely been used in winter wheat programs due to the time needed to vernalize winter wheat. The years needed for generation advance are also used for seed increase. Five years after the cross, doubled haploids are being harvested from the advance yield trial, whereas conventionally derived lines are being selected for placement into the preliminary yield trial from the F_5 head row nursery. The F_5 heads were taken from F_4 plants which theoretically will still have 12% of their genes segregating. Hence as opposed to the doubled haploids, which should be entirely homozygous and need no further repurification, the conventionally derived material may need to be reselected for purity, requiring additional time and expense. Recognizing that every breeder organizes his program differently and that the time saving presented here may differ somewhat with different breeding methods, on average a minimum of two years should be saved by the use of doubled haploids. This time saving comes from that portion of a breeding program devoted to developing inbred or homozygous lines. Regardless of the breeding method, an intensive evaluation program, usually requiring a minimum of four years and many locations, will still be required.

Doubled haploids can also be used very effectively in a selection program. In Table 2 (excerpted from Baenziger and Schaeffer, 1983), a comparison between a doubled haploid population and the F_2 generation from a cross segregating for three dominant and for three recessive genes has been made. Considering the case of three recessive genes first, based on phenotypic selection 1/8 of the doubled haploid population versus 1/64 of the F_2 population would be selected. In both cases all of the selections will breed true for the genes. However, 1/8 fewer doubled haploids will have to be tested to have the same number of true breeding plants in a doubled haploid population, as would be selected from the F_2 population. Now considering the case of three dominant genes, 1/8 of the doubled haploid population would be phenotypically selected and every selected plant would breed true. No further evaluation for the trigenic trait(s) would be necessary. In the F_2 population, 27/64 of the population would be selected. However, of the F_2 population only 1/64 are true breeding. Therefore, 27/64 (42%) of the F2 population would be selected to insure the inclusion of the desired 1/64 (2%) true breeding lines. To separate the segregating material from homozygous material, F_3 families will need to be grown. Only 1/27 (4%) will be homozygous

Table 2. Comparison of doubled haploids and F_2 generation selection for a trigenic trait

Mode of inheritance	Doubled haploid	F_2 generation
3 Recessive Genes		
Proportion of population	$(1/2)^3$	$(1/4)^3$
Phenotypic selection	1/8	1/64
True breeding	1/8	1/64
3 Dominant Genes		
Proportion of population	$(1/2)^3$	$(3/4)^3$
Phenotypic selection	1/8	27/64
True breeding	1/8	1/64

P. S. Baenziger and G. W. Schaeffer, "Dihaploids via Anthers Cultured in Vitro," from Lowell D. Owens, ed., Genetic Engineering: Applications to Agriculture, volume 7 of the Beltsville Symposia in Agricultural Research, Rowman & Allanheld, 1983, p. 273.

and 26/27 (96%) will be segregating. If a selection is made in a segregating family, additional testing will be needed in the succeeding generations until the desired homozygous line is developed. A more generalized and thorough discussion of these considerations is presented by Nei (1963).

Of the numerous methods available to plant breeders, SSD (Grafius, 1965; Brim, 1966) is most similar to doubled haploid breeding in that both methods have as a goal the rapid development of homozygous lines followed by selection and evaluation. One of the major differences between these two methods is that in doubled haploid breeding there is only one opportunity for recombination if F_1 plants are used. In SSD recombination can occur in every generation, though the effects of recombination lessen with each cycle of inbreeding. Riggs and Snape (1977), using computer simulation, studied the effects of linkage in SSD and double haploid populations for a quantitative trait determined by 8 loci. As expected, the SSD population consistently had a higher frequency of recombination than did the doubled haploid population. When linkage was unimportant, there was no difference between the SSD and doubled haploid populations. However, when linkage was present, predominantly in the coupled phase, the doubled haploid population had higher means and variances than did the SSD population. When repulsion linkage was present, the SSD population had higher means and variances than did the doubled haploid population. Using a mixed coupling and repulsion model, both populations had similar means and variances. They suggested that since most breeders choose parents so that the critical genes are in the repulsion phase, or in the mixed coupled and repulsion phase, that SSD is the preferred breeding method of the two. The advantages of SSD over doubled haploidy would lessen in crops having large numbers of chromosomes, hence decreasing the effects of linkage. If repulsion phase linkages are important, Snape and Simpson (1981) suggest using F_2's rather than F_1's as the source of doubled haploids. Using F_2's has the advantages of allowing more recombination to occur and selecting among segregating types so as not to create a totally random sample of the population. Intermating F_2's or waiting until the F_3 generation gave insufficient improvements over using F_2's to warrant waiting an extra generation.

In previous discussions of the benefits of doubled haploidy, some plant breeders have commented that early generation selection is very useful. It allows breeders to select within a relatively small plant population for lines which contain all of the desired genes (Sneep, 1977; Nass, 1979). However, early generation selection often is done on individual plants and is hampered by dominance and epistasis effects, by unmeasured genotype x environment interactions, and by the lack of evaluation techniques sufficiently accurate to identify superior lines containing the desired genes (Knott, 1972; Knott and Kumar, 1975; Thakare and Qualset, 1978; Weber, 1982). Whan et al. (1982) found that the same level of improvement was made by selecting in early generations as in the late generations. They postulated that the loss of high yielding lines by waiting until later generations was counteracted by being better able to evaluate those lines in later generations because the lines were more homozygous and homogeneous.

Doubled haploids can also be very useful in recurrent selection programs for population improvement. The first important paper compared the efficiency changes in recurrent selection based on individual selection, clonal selection, general combining ability selection, and on reciprocal recurrent selection using conventionally derived diploid individuals to the same recurrent selection procedures using doubled haploids (Griffing, 1975). The phenotypic variances for diploid and doubled haploid populations are as follows:

Diploid: $6^2_P = 6^2_A + 6^2_D + 6^2_E$

Doubled Haploid: $6^2_P = 26^2_A + 6^2_A + 6^2_E$

where 6^2_P is the phenotypic variance, 6^2_A is the additive genetic variance, 6^2_D is the dominance genetic variance, and 6^2_E is the environmental variance. Using unrestricted population sizes and assuming recurrent selection cycles require the same amount of time, diploid recurrent selection based on clonal or general combining ability was superior to diploid recurrent selection based on individual selection. This superiority largely disappeared when population size was restricted. However, the very great advantages of doubled haploid recurrent selection (in one case almost 6 times more efficient if recurrent selection cycle lengths are the same) when compared to diploid recurrent selection technique did not disappear when population size was restricted as would be necessary in breeding programs. In these comparisons, relative cycle length is quite important, as is the assumption that equally large diploid and doubled haploid population can be developed. To exploit the superiority of doubled haploid recurrent selection methods, procedures for the rapid and efficient creation of doubled haploids must be developed.

An exceptionally useful extension of Griffing's (1975) models was done by Scowcroft (1978) who considered what happens when the diploid and doubled haploid population sizes are unequal. In many crops, doubled haploid techniques are not well developed and it would be extremely difficult to routinely create large numbers of doubled haploids. The diploid population was assumed to have 1000 individuals of which 50 (5%) would be selected. Considering the case when only 20 doubled haploids could be made and 5 (25%) are selected, the relative efficiency of doubled haploid recurrent selection compared to diploid recurrent selection (assuming the recurrent selection cycle lengths were the same) ranged from .84 (high heritability, $6^2_P = 6^2_{2A}$) to 3.55 (low heritability due large dominance variation, $6^2_D = 176^2_A$, $6^2_E = 0$). Scowcroft concluded that before using doubled haploid recurrent selection one should know: 1) the relative length of the recurrent selection cycle, 2) how many fertile doubled haploids can be produced, and 3) the type of gene action and the importance of the environment.

Choo and Kanenberg (1978), using a computer simulation, compared the efficiency of doubled haploid recurrent selection using mass selection to diploid recurrent selection using mass selection and S_1 selection for diploid cross fertilized species. Response to doubled haploid mass selection was 1.4 times faster than diploid mass selection and equal to S_1 selection. If replicated yield trials were used for the doubled haploids, the response would be faster than S_1 selection. The improved response by using doubled haploids was due to the absence of dominance variation and the doubling of additive variation. More desirable genes were lost in the mass selected doubled haploid population, which the authors speculated was due to the small sample sizes used in the simulation.

One final area of plant selection where doubled haploidy may have a major impact will be the selection of mutants from cell cultures. As described by Chaleff (1983), selection at the haploid level allows direct selection for both dominant and recessive traits. After chromosome doubling, both dominant and recessive genes will be expressed in the fertile plant. To induce variability in cell culture, mutagens are often used. Mutagenesis at the haploid level should also be more efficient than at the diploid level when recessive traits are involved, in that only a single mutation at a locus is needed for expression in the haploid, whereas the more unlikely double mutation at the same locus is required in the diploid cell. Another advantage of using haploid cells for selection that is often overlooked, particularly when mutagenesis is involved, is that every altered trait of significance should be expressed in the regenerated doubled haploid. Using diploid cells, many of the undesirable mutations will be hidden in the regenerated plants as they will be recessively inherited mutations with only one allele of the two at any one locus affected. Undesired recessive alleles are lost, even with intense selection, very slowly from populations (Falconer, 1960). Ideally, the plant breeder would be able to choose the most phenotypically normal plant from a number of doubled haploids carrying the desired genes. Backcrossing, if needed, also can be used to transfer the desired gene back into the parental line.

Besides the direct application of doubled haploids in breeding theory, doubled haploids offer great advantages to the field of statistical genetics. Diallel mating systems of conventionally developed lines which are often used to determine gene action are limited because there is 1) an inability to determine if the genes are independently distributed in the parents (critical for interpreting the results), and 2) no epistasis is assumed (Baker, 1978). Diallels using doubled haploids can be used to estimate additive and additive x additive genetic variance and for population improvement (Choo et al., 1979). Additionally, doubled haploids from diallels can be used in selective diallel mating systems (Jensen, 1970) and elite doubled haploids identified in the population improvement program can be released directly for commercial sale. The effect of gene association and linkage on the estimation of genetic variances can be detected in doubled haploid diallel experiments (Choo and Reinbergs, 1979). Additive x additive x additive variance can be estimated from doubled haploid diallel experiments (Choo, 1980). Doubled haploids can also be used to determine the mean recombinational value and its variance (Choo, 1981a) which is important as the work by Riggs and Snape (1977) and Snape and Simpson (1981) emphasized the detrimental effects on doubled haploid breeding due to linkage. Finally Choo

(1981b) showed that doubled haploids from F_2 plants also could be used to estimate additive and additive x additive genetic variances, as well as the number of segregating genes. The F_2 derived doubled haploid population may contain as much as 50% more of the best recombinant when linkage is present than does the F_1 derived doubled haploid population. The frequency of the best recombinant is the same in both the F_2 and F_1 when linkage is not present, again emphasizing the importance of linkage. Methods of determining mean recombinational values from the F_2 and BC_2 derived doubled haploid populations and the parents were also developed (Choo, 1981b). Methods for determining gene interaction in doubled haploid populations from F_1 plants have also been developed (Choo and Reinbergs, 1982a), as well as methods to determine the number of genes that will segregate in a cross (Choo and Reinbergs, 1982b).

From the above discussion, it should be apparent that doubled haploids theoretically have great potential both in breeding and genetics. Their current applications are discussed below.

DOUBLED HAPLOIDY IN PRACTICE

While practical doubled haploid breeding is becoming more common in many crops, this discussion will refer to research on barley (primarily using interspecific hybridization and some anther culture) and tobacco (primarily using anther culture and more recently interspecific hybridization) since they are examples of species that readily produce doubled haploids, though by different routes. Wheat (where haploids are recovered by anther culture) will be used as an example where the doubled haploid technologies are currently being developed.

Doubled haploidy in barley has been reviewed by Kasha and Reinbergs (1981). Probably the most dramatic demonstration of doubled haploid barley breeding has been the release of 'Mingo' (Ho and Jones, 1980) and 'Rodeo' barley by Ciba-Geigy Seeds Ltd., a company that breeds barley exclusively by doubled haploid breeding methods. The cross from which Mingo was derived was made in 1974. By 1979, after two years of regional performance testing, Mingo was licensed for sale and was the highest yielding barley in Ontario. Kasha and Reinbergs (1981) reported that 'Gwylan', another doubled haploid was to be marketed in New Zealand and that programs in at least 16 countries were using interspecific hybridization to produce barley doubled haploids.

Considerable field research on barley doubled haploids was done before the method was widely accepted. Barley doubled haploids were found to be no different from homozyogous lines developed by the pedigree and SSD methods for agronomic traits such as yield, heading date, and plant height. The mean, range, genetic variances, and frequency of desirable genotypes were similar for the three methods of producing homozygous lines (Park et al., 1976). Data from this study were re-analyzed by Choo et al. (1982) to show that the distribution of doubled haploid lines was very similar to the distribution of SSD lines. Hence, while the SSD lines had a greater chance for recombination, SSD produced no more recombinants than did the doubled haploids. When doubled haploid lines were compared to homozygous lines derived from bulk populations, the bulk derived lines had higher means and lower variation among lines than did the doubled haploid lines (Song et al., 1978). This result was explained by natural selection acting on the bulk populations which would tend to remove lower yielding lines, thus increasing yield while decreasing variation. The doubled haploid lines represented a random sampling of unselected materials. When the ten best doubled haploid derived lines were compared to the ten best bulk derived lines, no differences were found for any of the traits. Reinbergs et al. (1976) showed doubled haploids derived from barley bulks could be used to rank the bulk performance, thus the cross performance. Only twenty doubled haploids from the cross were needed to accurately determine the bulk mean, genetic variance, and frequency of desired types. Crosses with good breeding potential could be selected on the basis of their doubled haploids having high overall means and a high frequency of superior lines. Reinbergs et al. (1978) determined that doubled haploids were not different from check cultivars for yield stability using the methods of Eberhart and Russell (1966). Hence, the complete homozygosity of doubled haploids did not result in significantly less environmental buffering than would be found in conventionally developed cultivars which are heterogenous. From the above, it should be clear that doubled haploidy is a successful breeding method for developing high yielding, agronomically acceptable barley cultivars.

In tobacco, most of the agronomic research on doubled haploids has been done using anther-derived doubled haploids. Little published research is available on the agronomic performance of maternally derived doubled haploids from interspecific hybridization. Unfortunately, the results in tobacco are not as clear as in barley concerning the efficacy of doubled haploid breeding. Much of the tobacco research has studied the question of whether the techniques involved in creating doubled haploids induce variation, and, if variation is induced, what is its source? These questions have been actively pursued, unfortunately often with contradictory results which may be due in part to the use of different experimental materials and designs.

We have already discussed the importance of plant genotype on the successful creation of haploids and doubled haploids. The plant genotype also appears to have a marked effect on doubled haploid agronomic performance. Burk et al. (1972) reported reduced vigor in doubled haploids of flue cured tobacco, whereas Collins et al. (1974) found no reduction in vigor in burley tobacco doubled haploids compared to the parents. Burk and Matzinger (1976) compared a highly inbred (hence homozygous) line with its anther culture derived doubled haploids and found the doubled haploids, on average, yielded less than the inbred line. They attributed this to the process of anther culture preferentially inducing deleterious variation. However, not all doubled haploids yielded less than the inbred cultivar, indicating that not all of the doubled haploids were inferior. This deleterious variation, when present, may be caused by reduced residual heterozygosity in the inbred (doubled haploids would be more homozygous, hence may exhibit inbreeding depression) or by mutational changes. Aneuploidy has also been reported in haploid tobacco plants (Collins et al., 1972). In flue cured tobacco, doubled haploids of two conventionally derived inbreds, their F_1's, and their F_2's yielded less than conventionally derived inbreds, their F_1's, and their F_2's (Arcia et al., 1978). These results indicated that the loss of residual heterozygosity in the doubled haploids probably does not fully explain their observed reduced yields, since the doubled haploid F_1's should be highly heterozygous and equivalent in yield to the conventionally derived F_1's if the loss of residual heterozygosity were the main cause for reduced doubled haploid yield. In a further attempt to determine if residual heterozygosity can explain the doubled haploid yield reduction, Brown and Wernsman (1982) intermated, within inbred type, doubled haploids from two flue cured inbreds. If doubled haploidy partitioned residual heterozygosity, then it should be re-established by intermating the doubled haploids. The doubled haploids and their hybrids yielded less than the inbred source. Mutations were believed to be the cause of reduced vigor. Additional support for the hypothesis that anther culture induces variation was reported by Burk et al. (1979). They found unexpected genetic ratios in the haploids and doubled haploids derived from an F_1 plant that was heterozygous for three monogenically controlled traits indicating that the gametic array was not accurately expressed. The genetic ratios may have been affected by the presence of clones, however, that should have been a random event and should not have affected the overall ratios. One interesting aspect of this study was that the genetic ratio determined using haploids was identical to the genetic ratio determined using doubled haploids, as would be expected. However, it also means that genes for disease resistance are expressed at the hemizygous level. If chromosome doubling and subsequent plant maintenance is a difficult or resource intensive step, a plant breeder could screen his haploid population, discard haploids with undesirable traits in the seedling stage, and only double and grow to maturity those doubled haploids having the desired traits. Also, in flue cured tobacco (Brown et al., 1983) and in N. sylvestris (De Paepe et al., 1981) additional yield reductions were observed with additional cycles of doubled haploidy. If all of the residual heterozygosity was lost with the first cycle of doubled haploidy, no additional yield reduction would occur with additional cycles of doubled haploidy if that were the only cause of the yield reductions. De Paepe et al. (1982) found heritable quantitative and qualitative changes in nuclear DNA in plants from consecutive cycles of doubled haploidy. Total nuclear DNA and repeated sequences increased with additional doubled haploid cycles. Dhillon et al. (1983-) also found increased DNA and heterochromatin content in doubled haploids when compared to the source cultivar, which suggested an amplification of DNA sequences during doubled haploid formation. Hence, within flue cured tobacco and N. sylvestris, anther culture appears to induce heritable deleterious variation somewhat correlated to DNA amplification and possibly caused by mutagenesis. These results are contrary to the recent report of Deaton et al. (1982) who found in burley tobacco no consistent reduced vigor in doubled haploids when compared to their source cultivars, and that a diallel of doubled haploids had similar genetic performances to diallels previously reported for conventionally derived inbreds. Residual heterozygosity could explain the small differences when present among doubled haploids in this study. Alternatively, the mutagenic effects of the culture and/or doubling processes may be mutated in barley genetic backgrounds.

The breeding ramifications of using a system to create doubled haploids that may also induce variation has been discussed in a number of papers. Schnell et al. (1980) compared SSD and doubled haploid derived lines of flue cured tobacco. The SSD lines had a higher mean and lower variance than the doubled haploid lines. They recommended SSD breeding as preferable to doubled haploid breeding. However, Burk and Chaplin (1980) found some doubled haploids of flue cured tobacco F_1's were superior to either parent indicating that even in flue cured tobacco, doubled haploidy can create useful germplasm. Obviously much research remains to be done to truly understand the tobacco anther culture system.

Each anther culture system is different and will have its own triumphs and difficulties. Hence, while tobacco anther culture is the most advanced and studied anther culture system for developing doubled haploids, one should extrapolate carefully from its results. For example, tobacco yields which are measured by leaf weights similar to forages are greatly different from yields in grain crops. Reduced plant height can be very beneficial in many crops as it is a potentially new source of semi-dwarfing genes (Schaeffer, 1982).

Among grain crops, considerable progress in developing anther culture derived doubled haploids has been made in wheat. As opposed to tobacco where haploid plants are readily obtained from microspores, wheat, until recently, has been difficult to culture

(Research Group 301, 1976). Since a medium is not available upon which many genotypes can be readily cultured (Schaeffer et al., 1979), more of the wheat anther culture research has concentrated on other factors, particularly plant genotype, and to improve the efficiency of anther culture. As described in the section on breeding theory, F_1 or F_2 plants are the preferred explants sources for anther culture. Because wheat anther culture is a heritable trait (Bullock et al., 1982; Lazar et al., 1983b), F_1's having one parent that can be successfully anther cultured and one parent that does not anther culture will often anther culture successfully. Careful selection of parents or direct breeding for anther culturability can increase the number of genotypes available for doubled haploid breeding. While the population sizes were small, both Bullock et al. (1982) and Henry et al. (1980) found the full gametic array expressed in doubled haploids of F_1's. Bullock et al. (1982) also were able to screen for disease resistance at the hemizygous level which may increase the efficiency of doubled haploid breeding as previously discussed.

There are few studies on wheat doubled haploid agronomic performance. As early as 1973, Ouyang et al. reported obtaining superior doubled haploids to either parent from anther culturing an F_1 plant. De Buyser et al. (1981), in replicated experiments, obtained doubled haploids that were superior to standard check cultivars. In an attempt to determine if anther culture may induce variation in wheat as it does in some tobacco germplasm, Baenziger et al. (1983) compared doubled haploid lines derived from 'Kitt' to the cultivar Kitt and to two SSD lines of Kitt. The doubled haploids yielded significantly less (averaged 16%) than the cultivar or the SSD lines. However, the individual doubled haploid yield ranged from 64% to 100% of the cultivar yield and most of the doubled haploids were not significantly different from the cultivar Kitt. The results of this study are similar to reports in tobacco (Burk and Matzinger, 1976) in that deleterious variation was produced by anther culture. This study needs to be repeated with more doubled haploids and SSD lines and with different cultivars to determine the level of induced variation and the relationship of genotype to induced variation. In a practical sense, however, when the frequency of doubled haploids exhibiting deleterious variation is known, a plant breeder could increase the population of doubled haploids from an F_1 so that there are sufficient doubled haploids to accurately represent the gametic array in addition to those exhibiting deleterious variation. Although the technique may skew a population towards inferiority, excellent lines can be selected as occurred with De Buyser et al. (1981). Of course, the breeder will have to evaluate more lines than if the technique did not induce deleterious variation, but our experience is that some deleterious variation is visually obvious and requires little effort to discard. If the variation is subtle, the inferior lines would be discarded in the preliminary evaluation nursery.

Little is known about the cause of variation in phenotype of doubled haploids recovered by anther culture. Baenziger et al. (1983) were unable to determine the cause of the variation in the cultivar Kitt. Chromosomal changes are known to occur in cells during tissue culture (see Hu et al., 1980; De Buyser and Henry, 1980; Hao et al., 1981; Hu et al., 1981; Liang et al., 1982; Kudirka et al., 1983a; Kudirka et al., 1983b) but the lowered yield potential of some of the doubled haploids derived from the cultivar Kitt was not due to chromosome loss or gain. The progeny of these doubled haploids have been testcrossed back to the parental cultivar in order to determine whether the deleterious changes are correlated to chromosomal rearrangements. However, these crosses have not yet been analyzed. Presumably, point mutations are also induced during tissue culture by certain components of the tissue culture medium such as 2,4-dichlorophenoxy acid. In this case, the hypothesis is difficult to test since specific genes governing the complex agronomic trait of yield are largely unknown, most certainly many in number, and sequestered both in nuclear and organellar compartments. However, Baenziger et al. (1983) were able to determine that the lowered yields of some doubled haploids of the cultivar Kitt were not due to reduced seed weight or spike fertility, pollen inviability, or changes in seed storage proteins.

From the above discussion, it may seem that induced variation from anther culture is only detrimental to crop improvement and that tissue culture should be avoided if interspecific hybridization or other methods of creating doubled haploids are available. This is not the case. Anther culture has the potential for producing a plant from every microspore within an anther and will continue to be considered the potentially most efficient method of creating haploid plants. While tissue culture often induces variation, some of the induced variation (called somaclonal variation) is beneficial (Larkin and Scowcroft, 1981). For example, anther culture derived haploids in both rice (Oryza sativa) (Schaeffer and Sharp, 1981; Schaeffer, 1982) and rape (Sacristan, 1982) have exhibited beneficial variation compared to their source cultivars. Even major chromosomal changes such as chromosome loss provide opportunities for rapidly developing nullisomics and other materials of great cytogenetic value (Moore and Collins, 1982; Kudirka et al., 1983a). Anther culture induced chromosomal rearrangements could be used to break unwanted linkage groups as is needed in interspecific and intergeneric crosses (Feldman, pers. commun.). It should also be remembered that tissue culture does not always induce variation. For example, Chung-Mong et al. (1982) were able to use anther derived haploids and doubled haploids to determine the linkage relationships of two loci in rice. The haploid and doubled haploid population is equivalent to a testcross population in linkage analysis studies. This would have been impossible if somaclonal variation were large.

As most of the statistical genetics models involving doubled haploidy have only recently been described, little published information is available on their practical uses. Dunbier and Bingham (1975) used doubled haploids to test the theory of maximum heterozygosity in alfalfa (Medicago sativa).

Haploids have been successfully used in biochemical selection experiments. Disease resistance has been enhanced in tobacco (Carlson, 1973) and rape (Sacristan, 1982) using disease toxins or their analogues as the selection agent. Both rice (Schaeffer, 1981; Schaeffer and Sharpe, 1981, 1983), and alfalfa (Reish et al., 1981) have enhanced amino acid content using amino acid analogues as selection agents. The use of anther-derived callus from rice selected for resistance to an analog of lysine produced regenerated plants whose progeny had increased seed storage protein as well as increased lysine when expressed as percent of total protein amino acids in the seed (Schaeffer and Sharpe, 1981). It is the first evidence with cereals that in vitro techniques might be useful for the recovery of unique cell types utilizing inhibitors which provide intense selection pressure in the cereals. Few researchers have regenerated plants from haploid tissues or anther calli even though there are examples of plants regenerated from diploid callus (see Schaeffer, 1981). Bourgin (1978) used haploid tobacco protoplasts for the recovery of valine resistant cells and ultimately plants. Chlorate resistance was used to recover cells and plants lacking nitrate reductase (Muller and Grafe, 1978). Resistant diploid lines from plants that are used as a crop in tetraploid or polyploid form have been used to recover resistance to toxins of Phytophthora in potato (Solanum tuberosum) (Behnke, 1979) and ethionine in alfalfa (Reisch et al., 1981). Other examples of in vitro selection with the recovery of plants include: salt tolerance in tobacco (Nabors et al., 1980); methionine sulfoximine in tobacco (Carlson, 1973) 5-methyltryptophan in carrot (Daucas carota L.) (Widhol, 1978), and potato (Carlson and Widholm, 1978); lysine plus threonine (Gengenbach et al., 1978); the herbicide Isopropyl-N-phenyl carbamate in tobacco (Aviv and Galun, 1977); Picloram in toabacco (Chaleff and Parsons, 1978); streptomycin and BUdR in tobacco (Maliga et al., 1973; and Maliga et al., 1973); glycerol in tobacco (Chaleff and Parsons, 1978); and aminoethyl cysteine resistance in rice (Schaeffer, 1978).

In summary, as the methods of creating doubled haploids and breeding theory for their use have improved in the last ten to fifteen years, the critical field experiments to determine if the methods of creating doubled haploids induce variation and if the theory can be used in practice have begun. Doubled haploid breeding is already a reality for a number of crops and more crops will soon be bred using doubled haploids.

FUTURE PROSPECTS FOR DOUBLED HAPLOIDS

Doubled haploid breeding theory and practice are still
evolving. In many crops there are definite needs to increase the
efficiency of producing doubled haploids which will come only with
additional fundamental understanding of the mechanisms of haploid
induction. As described for anther culture by Maheshwari et al.
(1982), "...after two decades the mechanism of induction of
haploidy remains as mysterious as ever. Until we understand this,
the whole practice will remain as much an art as a science." Much
of the research will concentrate on non-genotype-specific methods
and these methods should be evaluated by agronomic performance. For
example, even in tobacco, the most efficient anther culture system,
procedures to create haploids should not be measured by how many
anthers produce calli, or how many green plants can be regenerated
from an anther, but by how those regenerated plants compare to
conventionally derived lines. Is induced variation ubiquitous or
is it specific to certain genotypes and culture methods? Also, if
the induction of variation is common, how can it be used or even
enhanced where needed? Is the variation different from existing
variation? Variation is critical for crop improvement. If
deleterious variation is not common or can be limited, the time and
cost effectiveness of doubled haploid production and breeding
methods must be assessed in addition to technical difficulties
encountered with doubled haploids. It will require extensive
research programs and long term commitments to evaluate haploidy
procedures, both in laboratories and in the field. In barley, the
most efficient interspecific hybridization system, procedures are
desired that would not require hand emasculation or environmental
chambers.

The ability to rapidly achieve homozygosity remains the main
attraction of doubled haploid breeding. As previously mentioned,
doubled haploid breeding will be most useful for crops having long
growing seasons such as winter wheat and barley or multi-year
crops. Even in short season crops, such as spring wheat or barley,
doubled haploidy may make more efficient use of time or resources.
For example, Mingo, described earlier, is a spring barley. Doubled
haploids will be effectively used in recurrent selection programs
and for traits greatly affected by the environment (such as drought
tolerance) or by gene dominance (as in pyramiding genes for disease
resistance). When too few doubled haploids can be made from a
cross for effective line selection, the doubled haploids can be
used to evaluate the cross which should allow breeders to concentrate their efforts on those crosses most likely to give high
yielding lines. While one of the drawbacks of doubled haploid
breeding is that inbreeding is achieved without natural or
artificial selection, inbreeding without selection can also be
advantageous. Intergenotypic competition in segregating bulks may
be detrimental to plant types that perform well in pure stands.

For example, semi-dwarf wheat plants are outcompeted in mixed stands with tall plants (Khalifa and Qualset, 1975). The release of Gwylan also demonstrates an advantage for inbreeding without selection at the inbreeding site. International crop development centers that supply germplasm throughout the world may find that inbreeding without selection will allow them to better create germplasm for regions that are different from their research centers though the evaluation nurseries would be larger. The international centers may also find inbreeding without selection most useful if the techniques for creating doubled haploids are not easily done at small research stations. The small research station could send material to a central research laboratory to have the doubled haploids developed. Because these doubled haploids are unselected, the genotypic array will be the same, regardless where the inbreeding was done. On a more local scale, the segregating populations and breeding nurseries are often grown at only one or two locations within an area. The evaluation nurseries are generally grown at more sites which tend to better represent the potential growing area of the cultivar. Artificial and natural selection at the breeding nursery site may be detrimental to the identification of the best cultivar for the evaluation sites.

A second future use of doubled haploidy will be to develop cytogentic and cytoplasmic stocks. The loss or gain of one or more chromosomes or parts of chromosomes can be readily fixed upon doubling to create nullisomic and tetrasomic stocks in commercial cultivars. The nullihaploids would be good recipients for transferred chromosomes. Upon doubling, chromosome substitution lines would be formed. It should also be remembered that many aneuploid stocks were derived from crosses between haploid and diploid plants (Sears, 1954). Similarly, chromosomal rearrangements can be used to develop homozygous translocation and inversion stocks or to potentially break-up linkage blocks needed in genetic exchange. Doubled haploid plants contain paternally inherited cytoplasmic organelles. As these paternal organelles are lost in normal sexual fertilization, microspore (paternal) cytoplasm may not be well conserved genetically and may provide novel cytoplasmic variation.

A third future use of doubled haploidy will be in statistical genetics and linkage analyses. In the case of statistical genetics, doubled haploids allow more of the standard model assumptions to be tested. Also, the Mendelian basis for quantitative variation can readily be determined by intermating doubled haploids having similar and different phenotypes from a cross exhibiting quantitative variation for the character in question. The genetic segregation ratios are simplified by the reduction in the number of segregating loci in the crosses. Individual genes having small effects should be identified (Qualset, 1978). As the doubled haploid population is the equivalent of a testcross, it is an extremely efficient way to determine linkage relationships.

A fourth future use will be to develop genetic stocks that cannot be made by conventional methods. For example, homozygosity cannot be obtained for dominant genes controlling male sterility. Plants heterozygous for the the dominant allele controlling male sterility are sterile and cannot be selfed or pollinated by another heterozygous (hence sterile) plant for the trait. If early microspore development is unaffected by the sterility gene, the plant could be anther cultured to obtain homozygous doubled haploid lines. If early microspore development is affected, homozygous lines could still be developed by doubling haploids from interspecific hybridization.

A fifth future use will be in physiological studies in which uniform standard plant materials are needed. Intracultivar variation adds to the unexplained experimental error and can make experiment-to-experiment comparisons less precise. Doubled haploids have no intracultivar variation and should improve the experimental precision and provide a uniform standard to monitor experiment-to-experiment variation.

Finally, haploidy and double haploidy will be widely used in biochemical selection because selection at the haploid level allows recessive and dominant genes to be selected. Once the desired haploid type is identified, doubling will fix the desired trait as well as any other trait that may have been altered. Chromosome doubling will also save a generation for selfing to obtain homozygosity for a heterozygous trait from diploid selection schemes.

SUMMARY

The prospects for doubled haploid breeding are bright. As with much of science, doubled haploid breeding theory has preceded the applications. The well developed theories will be a strong force in support of the technical developments needed to allow the empirical experiment to validate and mold the theory. Doubled haploidy will certainly not replace existing breeding methods, but for certain breeding objectives and where haploids can be created, it will be very useful. It will have important indirect impact on physiology and development. It will directly impact population and inbred development, statistical genetics, cytogenetics, biochemical selection and may become one of the most powerful plant breeding methods for crop improvement.

ACKNOWLEDGEMENT

The authors wish to acknowledge the help of Drs. L. G. Burk, G. B. Collins, K. M. Ho, K. J. Kasha, H. W. Rines, and E. A. Wernsman who provided additional information of their current research used in this paper; of Drs. G. R. Bauchan, L. W. Briggle, and D. H. Smith, Jr., for critical review of the manuscript; and of Mrs. D. Wray and Ms. P. Listman for typing the manuscript. The research was in part supported by BARD Project No. US 234-80.

REFERENCES

Arcia, M. A., Wernsman, E. A., and Burk, L. G., 1978, Performance of anther-derived dihaploids and their conventionally inbred parents as lines, in F_1 hybrids, and in F_2 generations, Crop Sci., 18:413-418.

Aviv, D. and Galun, E., 1977, Isolation of tobacco protoplasts in the presence of isopropyl-N-phenylcarbamate and their culture and regeneration into plants, Z. Pflanzenphysiol., 83, 267-273.

Baenziger, P. S., and Schaeffer, G. W., 1983, Dihaploid via anthers cultured in vitro, in: "Beltsville Symposia in Agricultural Research VII, Genetic Engineering: Applications to Agriculture," L. D. Owens, ed., Rowman & Allanheld, Totowa, New Jersey, p. 269-284.

Baenziger, P. S., Wesenberg, D. M., Schaeffer, G. W., Galun, E., and Feldman, M., 1983, Variation among anther culture derived doubled haploids of 'Kitt' wheat, in: "Proc. Sixth Inter. Wheat Genet. Symp.," S. Sakamoto, ed., Kyoto, Japan, In press.

Baker, R. J., 1978, Issues in diallel analysis, Crop Sci., 18:533-536.

Barclay, I. R., 1976, A study of the genetics and mechanism of genome and chromosome loss in cereals, 149pp., Ph.D thesis, Univ. of Cambridge.

Behnke, M., 1979, General resistance to late blight of Solanum tuberosum plants regenerated from callus, Theor. Appl. Genet. 56:151-152.

Bennett, M. D., 1976, The cell in sporogenesis and spore development, in "Cell Division in Higher Plants", M. M. Yoeman, ed., Academic Press, New York, p. 161-198.

Bernard, S., 1977, Study of some factors controlling in vitro androgenesis of hexaploid Triticale, Ann. Amelior. Plantes, 27(6):639-655.

Bourgin, J. P., 1978, Valine-resistant plants from in vitro selected tobacco cells, Mol. Gen. Genet., 161:225-230.

Brettell, R. I., Thomas, S. E., and Wernicke, W., 1981, Production of haploid maize plants by anther culture, Maydica, 26:101-111.

Brim, C. A., 1966, A modified pedigree method of selection in soybeans, Crop Sci., 6:220.

Brown, J. S., Wernsman, E. A., and Schnell, II, R. J., 1983, Effect of a second cycle of anther culture on flue cured tobacco lines of Nicotiana tabacum L., Crop Sci., 23: In press.

Brown, J. S. and Wernsman, E. A., 1982, Nature of reduced productivity of anther-derived dihaploid lines of flue-cured tobacco, Crop Sci., 22:1-5.

Bullock, W. P., Baenziger, P. S., Schaeffer, G. W., and Bottino, P.
    J., 1982, Anther culture of wheat F$_1$'s and their reciprocal
    crosses, Theor. Appl. Genet., 62:155-159.

Burk, L. G. and Chaplin, J. F., 1980, Variation among anther-
    derived haploids from a multiple disease-resistant tobacco
    hybrid, Crop Sci., 20:334-338.

Burk, L. G., Chaplin, J. F., Gooding, G. V., and Powell, N. T.,
    1979, Quantity production of anther-derived haploids from a
    multiple disease resistant tobacco hybrid. 1. Frequency of
    plants with resistance or susceptibility to tobacco mosaic
    virus (TMV), potato virus Y (PVY), and root knot (RK),
    Euphytica, 28:210-218.

Burk, L. G., Gerstel, D. U., and Wernsman, E. A., 1979, Maternal
    haploids of Nicotiana tabacum L. from seed, Science, 206:585.

Burk, L. G., Gwynn, G. R., and Chaplin, J. F., 1972, Diploidized
    haploids from aseptically cultured anthers of Nicotiana
    tobacum, J. Hered., 63:355-360.

Burk, L. G. and Matzinger, D. F., 1976, Variation among doubled
    haploid lines obtained from anthers of Nicotiana tobacum L.,
    J. Hered., 67:381-384.

Carlson, P. S., 1973, Methionine sulfoximine-resistant mutants of
    tobacco, Science, 180:1366-1368.

Carlson, J., and Widholm, J., 1978, Separation of two forms of
    anthranilate synthetase from S-methyltryptophan susceptible and
    resistant cultured Solanum tuberosum cells, Physiol. Plant.,
    44:251-255.

Chaleff, R. S., 1983, Isolation of agronomically useful mutants
    from plant cell cultures, Science, 219:676-682.

Chaleff, R. S., and Parsons, M. F., 1978, Direct selection in vitro
    for herbicide-resistant mutants of Nicotiana tabacum, Proc.
    Nat. Acad. Sci. U.S.A., 75:5104-5107.

Chapman, V., and Miller, T. E., 1977, Haploidy in the genus
    Aegilops, Wheat Info. Serv., 44:21-22.

Choo, T. M., 1980, Doubled haploids for estimating additive
    epistatic genetic variances in self-pollinated crops, Can. J.
    Genet. Cytol., 22:125-127.

Choo, T. M., 1981a, Doubled haploids for estimating mean and
    variance of recombination values, Genetics, 97:165-172.

Choo, T. M., 1981b, Doubled haploids for studying the inheritance
    of quantitative characters, Genetics, 99:525-540.

Choo, T. M., Christie, B. R., and Reinbergs, E., 1979, Doubled
    haploids for estimating genetic variances and a scheme for
    population improvement in self-pollinating crops, Theor. Appl.
    Genet., 54:267-271.

Choo, T. M., and Kannenberg, L. W., 1978, The efficiency of using
    doubled haploids in a recurrent selection program in a diploid,
    cross-fertilized species, Can. J. Genet. Cytol., 20:505-511.

Choo, T. M., and Reinbergs, E., 1979, Doubled haploids for estimat-
    ing genetic variance in the presence of linkage and gene
    association, Theor. Appl. Genet., 55:129-132.

Choo, T. M., and Reinbergs, E., 1982a, Analyses of skewness and kurtosis for detecting gene interaction in a doubled haploid population, Crop Sci., 22:231-235.

Choo, T. M., and Reinbergs, E., 1982b, Estimation of the number of genes in doubled haploid populations of barley (Hordeum vulgare), Can. J. Genet. Cytol., 24:337-341.

Choo, T. M., Reinbergs, E., and Park, S. J., 1982, Comparison of frequency distribution of doubled haploid and single seed descent lines in barley, Theor. Appl. Genet., 61:215-218.

Chung-Morg, C., Chi-Chang, C., and Ming-Hwa, L., 1982, Genetic analysis of anther-derived plants of rice, J. Hered., 73:49-52.

Coe, E. H., Jr., 1959, A line of maize with high haploid frequency, Amer. Nat., 93:381-382.

Collins, G. B., and Genovesi, A. D., 1982, Anther culture and its application to crop improvement, in: "Applications of Plant Cell and Tissue Culture to Agriculture and Industry," D. T. Tomes, B. E. Ellis, P. M. Harvey, K. J. Kasha and R. L. Peterson, eds., Univ. of Guelph, Guelph, Ontario, p. 1-24.

Collins, G. B., Legg, P. D., and Kasperbauer, M. J., 1972, Chromosome numbers in anther-derived haploids of two Nicotiana species, J. of Hered., 63:113-118.

Collins, G. B., Legg, P. D., and Litton, C. C., 1974, The use of anther-derived haploids in Nicotiana. II. Comparison of doubled haploid lines with lines obtained by conventional breeding methods, Tob. Sci., 18:40-42.

Deaton, W. R., Legg, P. D., and Collins, G. B., 1982, A comparison of barley tobacco doubled-haploid lines with their source inbred cultivars, Theor. Appl. Genet., 62:69-74.

De Buyser, J. Y. Henry, R. Laur, and Lonnet, P., 1981, Utilisation de l'androgenese in vitro dans des programmes de selection du ble tendres (Triticum aestivum L.), Z. Pflanzenzüchtg. 87:290-299.

De Buyser, J. and Henry, Y., 1980, Induction of haploid and diploid plants through in vitro anther culture of haploid wheat (n=3x=21), Theor. Appl. Genet., 57:57-58.

De Paepe, R., Bleton, D., and Grangbe, F., 1981, Basis and extent of genetic variability among doubled haploid plants obtained by pollen culture in Nicotiana sylvestris, Theor. Appl. Genet., 59: 177-184.

De Paepe, R., Prat, D., and Huguet, T., 1982, Heritable nuclear DNA changes in double haploid plants obtained by pollen culture of Nicotiana sylvestris, Pl. Sci. Lett., 28:11-28.

Dhillon, S. J., Wernsman, E. A., and Miksche, J. P., 1983, Evaluation of nuclear DNA content and heterochromatin changes in anther-derived dihaploids of tobacco (Nicotiana tabacum) cv. Coker 139, Can. J. Genet. Cytol., 25:169-173.

Dunbier, M. W., and Bingham, E. T., 1975, Maximum heterozygosity in alfalfa: Results using haploid-derived autotetraploids, Crop Sci., 15:527-531.

Dunwell, J. M., 1976, A comparative study of environmental and developmental factors which influence embryo induction and

growth in cultured anthers of <u>Nicotiana</u> <u>tabacum</u>, <u>Envir</u>. <u>and</u>
     <u>Exp</u>. <u>Bot</u>., 16:109-118.
Durr, A. and Fleck, J., 1980, Production of haploid plants of
     <u>Nicotiana</u> <u>langsdorffii</u>, <u>Plant</u> <u>Sci</u>. <u>Lett</u>., 18:75-79.
Eberhart, S. A., and Russell, W. R., 1966, Stability parameters for
     comparing varieties, <u>Crop</u> <u>Sci</u>., 6:36-40.
Falconer, D. S., 1960, Introduction to quantitative genetics, 365
     pp., The Ronald Press Company, New York.
Falk, D. E., and Kasha, K. J., 1983, Genetic studies of the
     crossability of hexaploid wheat with rye and <u>Hordeum</u> <u>bulbosum</u>,
     <u>Theor</u>. <u>Appl</u>. <u>Genet</u>., 64:303-307.
Foroughi-Wehr, B., Friedt, W., and Wenzel, G., 1982, On the genetic
     improvement of androgenetic haploid formation in <u>Hordeum</u>
     <u>vulgare</u> L., <u>Theor</u>. <u>Appl</u>. <u>Genet</u>., 62:233-239.
Gengenbach, B. G., Walter, T. J., Green, C. E., and Hibberd, K. A.,
     1978, Feedback regulation of lysine, threonine, and methionine
     biosynthetic ensymes in corn, <u>Crop</u> <u>Sci</u>., 18:472-476.
Genovesi, A. D. and Collins, G. B., 1982, <u>In</u> <u>vitro</u> production of
     haploid plants of corn via anther culture, <u>Crop</u> <u>Sci</u>.,
     22:1137-1144.
Grafius, J. E., 1965, Short cuts in plant breeding, <u>Crop</u> <u>Sci</u>.,
     5:377.
Griffing, B., 1975, Efficiency changes due to use of doubled
     haploids in recurrent selection methods, <u>Theor</u>. <u>Appl</u>. <u>Genet</u>.,
     46: 367-386.
Guha, S., and Maheshwari, S. C., 1964, <u>In</u> <u>vitro</u> production of
     embryos from anthers of <u>Datura</u>, <u>Nature</u> (Lond.), 204:497.
Hagberg, A., and Hagberg, G., 1980, High frequency of spontaneous
     haploids in the progeny of an induced mutation in barley,
     <u>Hereditas</u>, 93: 341-343.
Hagberg, G., and Hagberg, A., 1981, Haploid initiator gene barley,
     <u>in</u>: "Barley Genetics IV, Proc. of the 4th Int. Barley Genet.
     Symp.," M.J.C. Asher, ed., Edinburgh, U.K., p. 686-689.
Hao, S., He, M., Xu, Z., Zou, M., Hu, H., Xi, Z., and Ouyang, J.,
     1981, Analysis of meiosis of the anther-derived haploid plants
     in wheat (<u>Triticum</u> <u>aestivum</u>), <u>Sci</u>. <u>Sin</u>., 24:861-868.
Henry, Y., DeBuyser, J., and Le Brun, J., 1980, Androgenese sure
     des bles tendres (<u>Triticum</u> <u>aestivum</u>) en cours de selection. 3.
     Electrophorese de gliadines de quelque haploides doubles, <u>Z</u>.
     <u>Pflanzenzüchtg</u>., 85:322-327.
Ho, K. M., and Jones, G. E., 1980, Mingo Barley, <u>Can</u>. <u>J</u>. <u>Plant</u>
     <u>Sci</u>., 60:279-280.
Hu, H., and Shao, Z., 1981, Advances in plant cell and tissue
     culture in China, <u>Adv</u>. <u>Agron</u>., 34:1-13.
Hu Han, Z. Y. Xi, Ouyang, J. W., Hao Shui, M. Y. He, Xu, Z. Y., and
     Zou, M. Z., 1980, Chromosome variation of pollen mother cell of
     pollen-derived plants in wheat (<u>Triticum</u> <u>aestivum</u> L.), <u>Sci</u>.
     <u>Sinica</u>, 23 (7): 905-914.
Hu Han, Z. Y. Xi, Jing, J., and Wang, X., 1981, Production of
     pollen-derived wheat aneuploid plants through anther culture,

in: "Cell and Tissue Culture Techniques for Improvement of
Cereal Crops", Beijing, China, p. 767-778.

Jacobsen, E., and Sopory, S. K., 1978, The influence and possible
recombination of genotypes on the production of microspore
embryoids in anther cultures of Solanum tuberosum and dihaploid
hybrids, Theor. Appl. Genet., 52:119-123.

Jensen, C. J., 1974, Chromosome doubling techniques in haploids,
in: "Haploids in High Plants, Proc. of the 1st. Int. Symp.,"
June 10-14, Univ. Guelph Press, Ontario, p. 153-190.

Jensen, C. J., 1975, Barley monoploids and doubled monoploids:
techniques and experience, in: "Barley Genetics III," H. Gaul,
ed., Karl Thienig, Muchen, p. 316-345.

Jensen, N. F., 1970, A diallel selective mating system for cereal
breeding, Crop Sci., 10:629-635.

Kappert, H., 1933, Erbliche polyembryonic bil Linum usitatissimum,
Bio. Zentrabl., 53:276-307.

Kasha, K. J., 1974, Haploids from somatic cells, in: "Haploids and
higher plants, Proc. 1st. Intern. Symp.," K. J. Kasha, ed.,
Univ. of Guelph, Guelph, Canada, p. 67-87.

Kasha, K. J., and Kao, K. N., 1970, High frequency haploid
production in barley (Hordeum vulgare L.), Nature, 225:874-876.

Kasha, K. J., and Reinbergs, E., 1981, Recent developments in the
production and utilization of haploids in barley, in: "Barley
Genetics IV, Proc. of the 4th Int. Barley Genet. Symp.," M.J.C.
Asher, ed., Edinburgh, U. K., p. 655-665.

Kasperbauer, M. J., and Collins, G. B., 1972, Reconstitution of
diploids from leaf tissue of anther-derived haploid in tobacco,
Crop Sci., 12:98-101.

Kenworthy, W. J., Brim, C. A., and Wernsman, E. A., 1973,
Polyembryony in soybeans, Crop Sci., 13:637-639.

Kermicle, J. L., 1969, Androgenesis conditioned by mutation in
maize, Science, 166:1422-1424.

Khalifa, M. A., and Qualset, C. O., 1975, Intergenotypic
competition between tall and dwarf wheats. II. In hybrid
bulks, Crop Sci., 15:640-644.

Kihara, H., 1979, Artificially raised haploids and their uses in
plant breeding, Seiken Ziko, 27-28: 14-29.

Kimber, G., and Riley, R., 1963, Haploid angiosperms, Bot. Rev.,
29:490-531.

Knott, D. R., 1972, Effects of selection for F_2 plant yield on
subsequent generations in wheat, Can. J. Plant Sci.,
52:721-726.

Knott, D. R., and Kumar, J., 1975, Comparison of early generation
yield testing and a single seed descent procedure in wheat
breeding, Crop Sci., 15:295-299.

Kobayashi, M., and Tsunewaki, K., 1980, Haploid induction and its
genetic mechanism in alloplasmic common wheat, J. of Hered.,
71:9-14.

Kudirka, D. T., Schaeffer, G. W., and Baenziger, P. S., 1983a,
Induction of genetic variability in wheat through anther

culture, in: "In vitro techniques for agriculture," Y.P.S. Bajaj, ed., In press.

Kudirka, D. T., Schaeffer, G. W., and Baenziger, P. S., 1983b, Cytogenetic characteristics of wheat plants regenerated from anther calli of 'Centurk', Can. J. Genet. Cytol., 25:513-517.

Lacadena, J. R., 1974, Spontaneous and induced parthenogenesis and androgenesis, in: "Haploids and Higher Plants, Proc. of the 1st. Int. Symp.," K. J. Kasha, ed., University of Guelph, Guelph, Ontario, p. 13-32.

Larkin, P. J., and Scowcroft, W. R., 1981, Somaclonal variation - a novel source of variability from cell cultures for plant improvement, Theor. Appl. Genet., 60:197-214.

Lazar, M. D., Schaeffer, G. W., and Baenziger, P. S., 1983a, Cultivar and cultivar x environment effects on the development of callus and polyhaploid plants from anther cultures of wheat, Theor. Appl. Genet., 67:273-277.

Lazar, M. D., Schaeffer, G. W., and Baenziger, P. S., 1983b, Combining abilities and heretability of callus formation and plantlet regeneration in wheat (Triticum aestivum L.) anther cultures, Theor. Appl. Genet., In press.

Liang, G. H., Sangduen, N., Heyne, E. G., and Sears, R. G., 1982, Polyhaploid production through anther culture in common wheat, J. Hered., 73:360-364.

Maheshwari, S. C., Tyagi, A. K., and Malhotra, K., 1980, Induction of haploidy from pollen grains in Angiosperms - the current status, Theor. Appl. Genet., 58:193-206.

Maheshwari, S. C., Rashid, A., and Tyagi, A. K., 1982, Haploids from pollen grains - retrospect and prospect, Amer. J. Bot., 69:865-879.

Maliga, P., A., Breznovits, A. SZ., and Marton, L., 1973, Streptomycin-resistant plants from callus culture of haploid tobacco, Nature New Biology, 244:29-30.

Maliga, P., Marton, L., and Breznovits, A. SZ., 1973, 5-bromodeoxyuridine-resistant cell lines from haploid tobacco, Plant Sci. Lett., 1:119-121.

Miao, S. H., Kuo, C. S., Kwei, Y. L., Sun, A. T., Ku, S. Y., Lu, W. L., Wang, Y. Y., Chen, M. L., Wu, M. K., and Hang, L., 1978, Induction of pollen plants of maize and observation on their progeny, in: "Proc. of Symp. on Plant Tissue Culture," May 25-30, Science Press, Peking, p. 23-34.

Miller, T. E., and Chapman, V., 1976, Aneuhaploids in bread wheat, Genet. Res., 28:37-45.

Moore, G., and Collins, G. B., 1982, Isolation of nullihaploids from diverse genotypes of Nicotiana tobacum, J. Hered., 73:192-196.

Muller, A., and Grafe, R., 1978, Isolation and characterization of cell lines of Nicotiana tabacum lacking nitrate reductase, Mol. Gen. Genet., 161:67-76.

Nabors, M. W., Gibbs, S. E., Bernstein, C. S., and Meis, M. E., 1980, NaCl-tolerant tobacco plants from cultured cells, Z.

Pflanzenphysiol., 97:13-17.

Namikawa, S., and Kawakami, S., 1934, On the occurrence of haploid, triploid and tetraploid plants in twin seedlings of common wheat, Proc. Imp. Acad., Tokyo, 10:668-671.

Nass, H. G., 1979, Selecting superior spring wheat crosses in early generations, Euphytica, 28:161-167.

Nei, M., 1963, The efficiency of haploid method of plant breeding, Hered., 18:95-100.

Nitsch, J. P., and Nitsch, C., 1969, Haploid plants from pollen grains, Science, 163:85-87.

Nitzsche, W., and Wenzel, G., 1977, Haploids in Plant Breeding, Verlag Paul Parey, Berlin and Hamburg, 101p.

Ouyang, T. W., Hu, H., Chuang, C. C., and Tseng, C. C., 1973, Induction of pollen plants from anthers of Triticum aestivum L. cultured in vitro, Sci. Sin., 16:79-95.

Pandy, K. W., and Phung, M., 1982, 'Hertwig Effect' in plants: induced parthenogenesis through the use of irradiated pollen, Theor. Appl. Genet., 62:295-300.

Park, S. J., Walsh, E. J., Reinbergs, E., Song, L. S. P., and Kasha, K. J., 1976, Field performance of doubled haploid barley lines in comparison with lines developed by the pedigree and single seed descent methods, Can. J. Plant Sci., 56: 467-474.

Picard, E., and De Buyser, J., 1977, High production of embryoids in anther culture of pollen derived homozygous spring wheats, Ann. Ameloir. Plantes., 27:483-488.

Picard, E., De Buyser, J., and Henry, Y., 1978, Technique de production d'haploides de ble par culture d'antheres in vitro, Le Selectionneur Francais., 26:25-37.

Pickering, R. A., and Hayes, J. D., 1976, Partial incompatibility in crosses between Hordeum vulgare L. and Hordeum bulbosum L., Euphytica, 25:671-678.

Qualset, C. O., 1978, Mendelian genetics of quantitative characters with reference to adaptation and breeding in wheat, in: "Proc. Fifth Int. Wheat Genetics Symp.," S. Ramanajam, ed., New Delhi, India, p. 577-590.

Racquin, C., 1982, Genetic control of embryo production and embryo quality in anther culture of Petunia, Theor. Appl. Genet., 63:151-154.

Reinbergs, E., Park, S. J., and Song, L. S. P., 1976, Early identification of superior barley crosses by the doubled haploid technique, Z. Pflanzenzüchtg., 76:215-224.

Reinbergs, E., Song, L. S. P., Choo, T. M., and Kasha, K. J., 1978, Yield stability of doubled haploid lines of barley, Can. J. Plant Sci., 58:929-933.

Reinert, J., and Bajaj, Y. P. S., 1977, Anther culture: Haploid production and its significance, in: "Applied and fundamental aspects of plant cell, tissue, and organ culture," J. Reinert and Y. P. S. Bajaj, ed., Springer-Verlag, Berlin, p. 251-267.

Reish, B., Duke, S. H., and Bingham, E. T., 1981, Selection and characterization of ethionine-resistant alfalfa (Medicago sativa L.) cell lines, Theor. Appl. Genet., 59:89-94.

Research Group 301, 1976, A sharp increase of the frequency of
    pollen-plant induction in wheat with potato medium, Acta Genet.
    Sin., 3:30-31.

Riggs, T. J., and Snape, J. W., 1977, Effects of linkage and
    interaction in a comparison of theoretical populations derived
    by diploidized haploid and single seed descent methods, Theor.
    Appl. Genet., 49:111-115.

Rines, H. W., 1983, Oat anther culture: genotype effects on callus
    initiation and the production of a haploid plant, Crop Sci.,
    23:268-272.

Rives, M., and Picard, E., 1977, A case of genetic assimilation:
    selection through androgenesis or parthenogenesis of haploid
    producing systems (an hypothesis), Ann. Amel. Plantes.,
    27:489-491.

Rowe, P. R., 1974, Parthenogenesis following interspecific
    hybridization, in: "Haploids and higher plants, Proc. 1st
    Intern. Symp.," K. J. Kasha, ed., Univ. of Guelph, Guelph,
    Canada, p. 43-52.

Sacristan, M. D., 1982, Resistance responses to Phoma lingam of
    plant regenerated from selected cell and embryogenic cultures
    of haploid Brassica napus, Theor. Appl. Genet., 61:193-200.

Sarkar, K. R., 1974, Genetic selection techniques for the
    production of haploids in plants, in: "Haploids and Higher
    Plants, Proc. of the 1st. Int. Symp.," K. J. Kasha, ed.,
    University of Guelph, Guelph, Canada, p. 33-41.

Schaeffer, G. W., 1978, Seeds from rice cells resistant to
    S(2-aminoethyl)-L-cysteine, in: "Abstracts, 4th Int. Cong.
    Plant Tissue and Cell Culture," T. Thorpe, ed., University of
    Calgary, Calgary, Canada, p. 140.

Schaeffer, G. W., 1981, Mutations and cell selections: increased
    protein from regenerated rice tissue culture, Environmental and
    Experimental Botany, 21:333-345.

Schaeffer, G. W., 1982, Recovery of heritable variability in
    anther-derived doubled-haploid rice, Crop Sci., 22:1160-1164.

Schaeffer, G. W., and Baenziger, P. S., 1982, Anther culture and
    pollen plant regeneration in wheat (Triticum aestivum L. em
    Thell), Proc. 5th International Congress Plant Tissue and Cell
    Culture, Tokyo, Japan, 553-556.

Schaeffer, G. W., Baenziger, P. S., and Worley, J., 1979, Haploid
    plant development from anthers and in vitro embryo culture of
    wheat, Crop Sci., 19:697-702.

Schaeffer, G. W., and Sharpe, F. T., Jr., 1981, Lysine in seed
    protein from S-aminoethyl-L-cysteine resistant anther-derived
    tissue cultures of rice, In Vitro, 17:345-352.

Schaeffer, G. W., and Sharpe, F. T., Jr., 1983, Mutations and cell
    selections: Genetic variation for improved protein in rice,
    Chapter 17, in: "Proc. of Beltsville Symposia for Agricultural
    Research," L. D. Owens, ed., Rowman and Allanheld, Totowa, NJ.,
    p. 237-254.

Schnell, R. J., II, Wernsman, E. A., and Burk, L. G., 1980,
    Efficiency of single-seed-descent vs. anther-derived dihaploid
    breeding methods, Crop Sci., 20:619-622.
Scowcroft, W. R., 1978, Aspects of plant cell culture and their
    role in plant improvement, in: "Proc. of a Symposium on Plant
    Tissue Culture," May 25-30, 1978, Science Press, Peking, p.
    181-198.
Sears, E. R., 1954, The aneuploids of common wheat, Missouri Agr.
    Exp. Sta. Res. Bull., 572:59.
Shigenobu, T., and Sakamoto, S., 1977, Production of a polyhaploid
    plant of Aegilops corossa 6X pollinated by Hordeum bulbosum,
    Jap. J. Genet., 52:397-402.
Simpson, E., Snape, J. W., and Finch, R. A., 1980, Variation
    between Hordeum bulbosum genotypes in their ability to produce
    haploids of barley, Hordeum vulgare, Z. Pflanzenzüchtg,
    85:205-211.
Sink, K. C., Jr., and Padmanabhan, V., 1977, Anther and pollen
    culture to produce haploids: progress and application for the
    plant breeder, Hort. Sci., 12:143-148.
Snape, J. W., and Simpson, E., 1981, The genetical expectations of
    doubled haploid lines derived from different filial
    generations, Heredity, 42:291-298.
Snape, J. W., Chapman, V., Moss, J., Blanchard, E. E., and Miller,
    T. E., 1979, The crossabilities of wheat varieties with Hordeum
    bulbosum, Theor. Appl. Genet., 64:303-307.
Sneep, J., 1977, Selection for yield in early generations of
    self-fertilizing crops, Euphytica, 26:27-30.
Song, L. S. P., Park, S. J., Reinbergs, E., Choo, T. M., and Kasha,
    K. J., 1978, Doubled haploid vs. the Bulk Plot method for
    production of homozygous lines in barley, Z. Pflanzenzüchtg,
    81:271-280.
Subrahmanyam, N. C., 1980, Haploidy from Hordeum interspecific
    crosses, 3 trihaploids of Hordeum arizonicum and of Hordeum
    lechlei, Theor. Appl. Genet., 56:257-264.
Sunderland, N., 1978, Strategies in the improvement of yields in
    anther culture, in: "Proc. of Symp. on Plant Tissue Culture,"
    May 25-30, 1978, Science Press, Peking, p. 65-86.
Sunderland, N., 1979, Comparative studies of anther and pollen
    culture, in: "Plant Cell and Tissue Cultures: Principles and
    Applications," W. R. Sharp, P. O. Larsen, E. F. Paddock, and V.
    Raghavan, eds., Ohio State Univ. Press, Columbus, U.S.A., p.
    203-219.
Thakare, R. B., and Qualset, C. O., 1978, Empirical evaluation of
    single-plant and family selection strategies in wheat, Crop
    Sci., 18:115-118.
Turcotte, E. L., and Feaster, 1974, Semigametic production of
    cotton haploids, in: "Haploids and Higher Plants," Proc. of the
    1st. Int. Symp., K.J. Kasha, ed., Univ. of Guelph, Guelph,
    Ontario, p. 53-64.
Weber, W. E., 1982, Selection in segregating genetrations of

autogamous species. I. selection response for combined
selection, Euphytica, 31:493-502.

Wenzel, G., Hoffman, F., and Thomas, E., 1977, Increased induction
and chromosome doubling of androgenetic haploid rye, Theor.
Appl. Genet., 51:81-86.

Whan, B. R., Knight, R., and Rothjen, A. J., 1982, Response to
selection for grain yield and harvest index in F_2, F_3, and F_4
derived lines of two wheat crosses, Euphytica, 31:139-150.

Widholm, J. M., 1978, Selection and characterization of a Daucus
carota L. cell line resistant to four amino acid analogues, J.
Exp. Bot., 29:1111-1116.

Xu, Z. H., Huang, B., and Sunderland, N., 1981, Culture of barley
anthers in conditioned media, J. Exp. Botany, 32(129):767-778.

USE OF PROTOPLASTS: POTENTIALS AND PROGRESS

E. C. Cocking

Plant Genetic Manipulation Group
Department of Botany, University of Nottingham
Nottingham, NG7 2RD, U.K.

INTRODUCTION

In the Stadler Genetics Symposium on Gene Manipulation in
Plant Improvement it is particularly timely to review what has
been accomplished with the use of protoplasts since the comprehensive assessment made in 1981 at the Royal Society Discussion
Meeting on Plant Genetic Manipulations (Cocking, 1981). At that
Meeting the main attributes of protoplasts were highlighted,
namely that because protoplasts are single cells they provide
opportunities for genetic manipulations at the single cell level.
Because the cell wall is absent they can be fused with other
protoplasts, thereby providing opportunities for somatic hybridization. Moreover, the absence of the cell wall also facilitates
uptake of DNA enabling evaluations of transformation to be carried
out at a level comparable to those of microbial transformations.
The fact that, in an increasing number of species, regeneration
of plants is possible from such protoplasts (Davey, 1983) is
enabling such somatic cell genetics to be combined with sexual
procedures (Davey and Kumar, 1983).

It has become increasingly clear during the past five years
that the use of molecular and somatic genetics in plant breeding
should facilitate the selection and enhancement of the production
of plants with desirable characteristics. However, as recently
re-emphasized (Cocking et al., 1981) such techniques are still very
much at the experimental stage and several drawbacks must be overcome before they can be applied to agricultural practice, but the
effort should be well worthwhile. In any survey of the use of
protoplasts in these respects it is necessary to identify those

415

special attributes that protoplasts possess that make them useful
for molecular and somatic genetics.

CLONING OF PLANTS FROM PROTOPLASTS

Plant tissue culture per se appears to be an unexpectedly
rich source of genetic variation (Meins, 1983), and the ability
to dissect whole plants into single cells, through the use of
protoplasts, has now stimulated a considerable effort to find out
whether such genetic variation can be enhanced by protoplast
cloning. Using cultured leaf explants of tomato, it was shown by
Evans and Sharp (1983) that single gene mutations occurred in the
plants regenerated from tissue culture. This recovery of single
gene mutations provides good evidence that plant tissue culture
can be mutagenic. It was observed that in most cases the mutations
were similar to those that occur spontaneously, or after mutagenic
treatment of tomato seed. Seed treatment often produces mosaics,
since not all cells in the seed are equally affcted by the mutagen,
but interestingly Evans and Sharp observed no mosaics in their
tissue culture derived plants. This indicated that the shoots were
each derived from a single cell of the callus, suggesting perhaps
that regeneration from isolated single tomato (Lycopersicum
esculentum) cells, such as protoplasts, is not an essential prerequisite.

In studies using a diploid tobacce (Nicotiana sylvestris), it
was observed that protoplast culture induced a high frequency of
recessive mutations affecting qualitative and quantitative characters of plant morphology or plant growth; and Prat (1983) suggested
that mutations induced from the actual culture of protoplasts
might restrict its application in the multiplication of genotypes.
It would, however, be unwise to attempt to generalize in this
respect. Some of the difficulties encountered in the experimental
design of experiments aimed at studying genetic variation arising
from cultured somatic tissue have been highlighted by Orton (1983);
and it was suggested that protoplasts could be used as a means to
circumvent problems associated with the lack of definition of the
explant, but the necessary assumption of genetic equality of different protoplasts still remains. Several workers have recently
emphasized that variability generated by passage through tissue
culture will only be useful if it is not available from conventional means, recapitulating the conclusions of Shirvin (1978), in
reviewing natural and induced variation in tissue culture, that it
may have special use for the plant breeder who is working with
crops which are unique, highly-adapted, or in which the sexual
apparatus is disturbed - sugar cane (Saccharum officinarum) and
potato (Solanum tuberosum) immediately spring to mind. Attempts
have been made to assess this in detail using potato protoplasts
(Shepard, Bidney and Shahin, 1980). Evidence is required that
such variation produces dominant alleles and that the variants are

of the cause of this variation, if we are to understand and perhaps
control the system.   In this connection the investigations of
Brettel, Thomas and Ingram (1980) are particularly pertinent.
Tissue cultures were initiated from immature embryos of maize (Zea
mays) carrying Texas male-sterile (TMS) cytoplasm, and plants were
regenerated after selection to T-toxin.   Fertile, T-toxin resistant
plants were obtained from the unselected control cultures as well
as from the selected material.   One regenerant, from an unselected
culture, was fertile and T-toxin sensitive.   This indicated that
under the conditions of tissue culture there occurred in the TMS-
cytoplasm material a stable genetic change which was responsible
for the reversion to male-fertility and T-toxin resistance.   Clearly, mobile genetic elements (Sharpiro, 1983) may be involved; and
in vitro culture may accentuate these tendencies.   At present we
do not know whether protoplasts will provide an extra dimension in
this respect.   Often, however, it is beneficial to work with such
single cell systems because large numbers can be readily obtained
and the handling of the selection systems is made easier.   This
has been well illustrated from the extensive studies on the isolation of agronomically useful mutants from plant cell cultures
where protoplasts have been utilized (Chaleff, 1983).   Even if the
genetic variation is not completely novel, it may be advantageous
to clone plants from protoplasts to speed up the selection of
desired plant types.

SOMATIC HYBRIDIZATION BY PROTOPLAST FUSION

     The recent 6th International Protoplast Symposium provided a
forum, both in its excellent presentation of posters (Potrykus
et al., 1983a) and in its lectures (Potrykus et al., 1983b), for
much that is most up to date on somatic hybridization by protoplast
fusion.   Additionally much has also recently been written on genetic
transfer in plants through interspecific protoplast fusion (Shepard
et al., 1983) and on the genetic analysis of somatic hybrid plants
(Gleva and Evans, 1983); and it will not be useful to repeat what
has already been presented elsewhere at length.

     Generally agriculture has been very successful in utilizing
sexual hybridization and mutation breeding for crop improvement.
Ideally one would like to be able to transfer genes, specific for
the desired agronomic improvement.   Often many such genes are
involved, and only very rarely have the genes actually been
identified.   Where possible direct transfer of such genes into
plants using suitable vectors is desirable, and as we shall see
later, protoplasts offer additional opportunities in this respect.
Particularly relevant in this general context is recent work in
Nicotiana which has shown that irradiation of pollen provides a
valuable tool for plant breeders to be able to affect limited gene
transfer sexually between donor and recipient genotypes (Pandey,
1975; Jinks et al., 1981).   Irradiated pollen has been utilized to

facilitate limited gene transfer in wheat (Snape et al., 1983) and in barley (Powell et al., 1983). The exciting vista is that it may be possible to obtain limited gene transfer even between sex-aully isolated species by protoplast fusion, if one of the fusion partners is suitable irradiated (Cocking, 1983). With suitable selection procedures, this would be equivalent to have directly transferred such genes using suitable vectors; it would avoid the need to identify the genes required in plant improvement and be readily applicable to polygenic transfers. Concerted efforts are now required to evaluate the extent of limited gene transfer that can be obtained as a consequence of protoplast fusions. The plant breeder will often wish to use such fusions to transfer only a single or a small number of characters from one parent to another.

The cytoplasmic mix obtained from protoplast fusions is novel, with the opportunity for the production of hybrids coupled with the opportunity for the formation of mitochondrial recombinants (Cocking, 1981). Important advances have been made in recent years in our knowledge of the segregation of cytoplasmic traits in higher plant somatic fusion hybrids (Fluhr, 1983).

With both of these two vistas arising from the use of proto-plasts in mind, it will be most useful to itemize some of the breeding objectives in which various facets of somatic hybridization are being utilized here at Nottingham (Cocking, 1983). It has been suggested that protoplast fusion may provide a means of introducing characteristics for leaf tannin production from Sainfoin (Onobrychis viciifolia), and Birdsfoot Trefoil (Lotus corniculatus),which act as anti-bloat factors, into the main forage legumes (white clover) (Trifolium repens) and alfalfa (Medicago sativa) grown extensively in Europe and the USA. Sexual incompatibility prevents the intro-duction of such desirable characters by conventional means. Irra-diation of Sainfoin and Birdsfoot Trefoil protoplasts is being assessed to achieve the limited gene transfer required for the selection of white clover and alfalfa partial somatic hybrids. Regeneration of plants from protoplasts of the parental forage legumes has already been achieved (Davey, 1983).

The utilization of protoplast fusions for the improvement of Solanum species has been stimulated by advances in their protoplast cultural capability. Kowalczyk et al. (1983) have demonstrated that plants can be regenerated from S. viarum and S. dulcamara leaf mesophyll protoplasts. S. viarum has been suggested as the main alternative source of steroid drug precursors to the medicinal yams. It is an annually grown prickly undershrub which has a wide distribution throughout the Indian subcontinent. The steroidal glycoalkaloid, solasodine, is contained in its mature fruits. Improvements required are a spineless strain to facilitate harvest-ing, an increase in the solasodine content of the berries and resistance to vascular wilt disease caused by the fungus Fusarium

oxysporium. Protoplast technology offers the opportunity to improve
the agronomic characters of S. viarum through 'protoplast cloning',
and somatic fusions with other alkaloid-producing species such as
S. dulcamara. The combination of superior characters, through somatic hybridization, between these two species could result in an
increase in total plant solasodine yield, a reduction in spines and
a greater environmental tolerance. Irradiation of the S. dulcamara
protoplasts could facilitate these limited gene transfers.

     N. tabacum is the most widely grown commercial non-food plant
in the world and it is susceptible to bacterial wild fire disease
and fungal black shank. N. rustica carries resistance to both
types of disease, but sexual crosses with N. tabacum lack fertility.
It is also possible to produce amphiploid somatic hybrids between
these two species by protoplast fusions. Leaf protoplasts of wild
type N. tabacum can be fused with wild type suspension culture protoplasts and heterokaryons isolated directly (Patnaik et al., 1982)
and cultured to regenerate whole fertile somatic hybrids (Patnaik
et al., 1984). Fertile somatic hybrids can also be produced
(Pental et al., 1984) by fusion of wild type N. rustica cell suspension protoplasts with leaf protoplasts of the double mutant of
N. tabacum which is both NR^- and SR^+ (Hamill et al., 1983). In
this somatic hybrid the presence of the black ovary wall gene from
N. rustica can be readily demonstrated; and irradiation is being
currently assessed to determine its influence on the degree of
gene transfer between these two species, and to determine the
readiness with which disease resistance genes could be transferred
from N. rustica without the necessity for backcrossing.

     Flax (Linum usitatissimum) improvement using protoplast fusions
is at an earlier stage of development. Flax (2n=30) is an important source of natural fibres, and its seeds contain oil and significant amounts of protein. Several of the wild species (2n=18)
of Linum possess many agronomically valuable genes for disease
resistance, for instance to flax rust, and drought resistance.
Attempts at sexual crosses between these wild species and flax
have failed to produce seeds. Here again protoplast fusions, including irradiation of those of the wild species, could be used to
bring about the desired gene flow. This work can now proceed since
whole flax plants can be regenerated from flax root and cotyledon
protoplasts (Barakat and Cocking, 1983). At a somewhat comparable
stage of development is the use of protoplast fusions for lettuce
improvements. Here the main breeding objective is the introduction
of more durable resistance to the fungus (Bremia lactucae) which
causes downy mildew. Horizontal resistance to Bremia lactucae is
available in the wild lettuce species, Lactuca perennis. As yet
it has not been possible to cross lettuce (L. sativa) with L.
perennis sexually, so somatic hybridization, coupled with irradiation procedures, is a logical avenue to explore. Encouragingly
whole plant regeneration has been obtained from protoplasts of

lettuce roots, cotyledons and in vitro grown shoots (Berry et al.,
1982) and leaf protoplasts (Engler and Crogan, 1982). Studies on
rice (Oryza sativa) improvement by fusion of protoplasts are at an
earlier stage of development, basically because it is not yet possible to regenerate whole plants from rice protoplasts. Breeding
objectives have been discussed by Swaminathan (1982), and include
transfer of cytoplasmic male sterility factors by protoplast fusions,
and fusions with protoplasts of wild rice species to enhance disease
resistance properties and transfer of salt resistance characteristics.

As discussed by Swaminathan (1982) the possibility of using
the water fern Azolla, together with its associated blue-green alga,
as a nitrogen fixing green manure crop suitable for rice culture
has been recognized by many researchers, agricultural administrators
and peasants. Sexual crossing of Azolla species has not been successful; and attempts are being made to cross Azolla species by
protoplast fusions to improve the strains, particularly in relation
to enhanced biomass.

Very significant improvements in the Brassicas have resulted
from protoplast fusions. Schenck and Robbelen (1982) have shown
that re-synthesis of Brassica napus from its ancestral diploids
(B. oleracea and B. campestris) is possible via somatic protoplast
fusion.

Recently Pelletier et al., (1983) have extended these studies
to a more comprehensive investigation of intergeneric cytoplasmic
hybridization in the Cruciferae by protoplast fusion. The agronomic importance of this work is considerable because the plants
produced will probably be useful for Brassica hybrid seed production. B. napus plants were regenerated after fusion between
protoplasts bearing cytoplasms of different genera. One type of
hybrid produced had B. napus chloroplasts and cytoplasmic male
sterility (CMS) trait from Raphanus sativus, and another type had
chloroplasts of a triazine resistant B. campestris and CMS trait
from R. sativus, with the nucleus of B. napus.

Rapeseed oil contains appreciable quantities of fatty acids
with chain lengths greater than the usual 18 carbon atoms. Significant amounts of poly unsaturated acids are also present including
significant amounts of linoleic and α-linolenic acids, but not
γ-linolenic acid (Stumpf and Pollard, 1983); and it may be possible
to modify the nature of these fatty acids using protoplast fusions.
One breeding objective, which is also of course applicable to a
range of other species, including rapeseed, is to convert linolenic
acid to γ-linolenic acid by fusion of protoplasts of B. napus with
those of Oenothera biennis (Evining Primrose). The Evening Primrose is a rich source of poly unsaturated fatty acid γ-linolenic
acid, and is unique in plants in possessing the gene for the pro-

duction of the enzyme delta-6-desaturase, which catalyses the conversion of linoleic acid to γ-linolenic acid. By somatic hybridization, including perhaps the irradiation of the Evening Primrose
protoplasts to fragment their DNA so as to facilitate gene flow,
it might thereby be possible to obtain rapeseed plants with some
γ-linolenic acid synthetic capability. Moreover, this breeding
objective, involving a single gene transfer, would be a very suitable challenge for the assessment of transformation for crop
improvement. Increased supplies of γ-linolenic acid for human
consumption are required because it seems likely that a functional
deficiency of γ-linolenic acid can cause degeneration of arteries.

Hybrid vigour is well known in sexual hybridization, and it
has been suggested (Bingham, 1983) that somatic hybridization may
produce an even greater vigour in hybrids - critical evaluation of
this suggestion is required, since it could result in enhanced
yields in many crops, including, for instance, alfalfa. An attractive feature for the improvement of certain species is that somatic
hybridization enables the whole genomes to be combined. Meiotic
segregations are thereby avoided. Moreover, additional genetic
variation could arise from the actual culture of the heterokaryons.

DIRECT GENE TRANSFER USING PROTOPLASTS

As previously pointed out (Cocking, 1981) in many of the
experiments of plant transformation it is not necessary, at least
initially, to use protoplasts. This is often the case when using
the normal infective process of Agrobacterium to introduce Ti
plasmids, modified by in vivo recombination, into host plants.
Tumour cells produced by such Agrobacteria can be subsequently
suitably screened (Cocking et al., 1981). At present it has
proved possible to develop most of the basic knowledge required,
particularly that relating to the transfer of foreign genes into
plants without the use of protoplasts. (Caplan et al., 1983;
Murai et al., 1983). However, some of the limitations of the use
of Agrobacterium as a delivery system are becoming apparent.
Agrobacterium as such cannot, at present, be utilized for DNA
transfer into monocotyledenous plants which include the important
cereal food crops. One solution may be the use of Agrobacteria
with non-oncogenic Ti plasmid vectors in combination with new
selectable marker genes (Caplan et al., 1983). Another attractive
alternative is to eliminate the use of Agrobacteria, as such, and
to utilize isolated plasmids interacting with protoplasts. This
alternative has been available since the first stable transformation of higher plants by isolated DNA was achieved, involving
the incubation of Petunia ssp. suspension cell protoplasts with Ti
plasmids in the presence of poly-L-ornithine. Cell colonies were
selected for the acquired tumorous characteristics of phytohormone
independence, opine synthesis and multiplication when grafted to
healthy host stem explants (Davey et al., 1980; Davey et al., 1979).

Currently there has been a resurgence of interest in protoplast systems in this respect and a number of different delivery
systems are being evaluated (Steinbiss and Broughton, 1983).
Quantitative comparisons of DNA uptake have indicated that much
vector DNA is apparently 'wasted': perhaps direct methods such as
microinjection and electrically-mediated transfer will prove more
efficient (Freeman et al., 1984). The increasing success in getting
crop protoplasts to undergo sustained divisions is an additional
stimulus (Davey, 1983). The abiding challenge, if transformation
of plants is to advance significantly further, is not only to be
able to extend direct transfer of genes to monocotyledenous species,
but to eliminate the need for the use of Agrobacteria as the
delivery system. Encouragingly it has recently been reported that
this may be possible (Ledoux et al., 1983), without even having to
use protoplasts.

GENERAL OUTLOOK

In 1981 it was concluded (Cocking et al., 1981) that, although
it seemed likely that plant breeding using well established conventional methods would continue to make the key contribution to meet
the demands of increased food production, multiple gene transfer
mediated by somatic hybridization would assume increasing significance in the improvement of forage legumes, vegetables and fruit
crops; and that in the short term protoplast cloning would be used
to enhance and release somatic variability. It was also suggested
that it would be advantageous to attempt to incorporate into the
recipient crop species some limited, perhaps single, genetic attribute of a donor species using a transformation procedure, thereby
eliminating the need for backcrossing and recurrent selection. In
1984 one now sees this transformation procedure as either involving
direct DNA transfer or, perhaps more realistically for plant
improvement, the use of limited gene transfer by protoplast fusions.

REFERENCES

Barakat, M. N. and Cocking, E. C., 1983, Plant regeneration from
        protoplast-derived tissues of Linum usitatissimum L. (Flax),
        Plant Cell Reports, 2:314-317.
Berry, S. F., Lu, D. Y., Pental, D., and Cocking, E. C., 1982,
        Regeneration of plants from protoplasts of Lactuca sativa L.,
        Z. Pflanzenphysiol., 108:31-38.
Bingham, E. T., 1983, in: "Ciba Foundation Symposium 97, Better
        crops for food," p. 130-143.
Brettel, R. I. S., Thomas, E., and Ingram, D. S., 1980, Reversion
        of Texas male-sterile cytoplasm maize in culture to give
        fertile, T-toxin resistant plants, Theor. Appl. Genet., 58:
        55-58.
Caplan, A., Herrar-Estrella, L., Inze, D., Van Haute, E., Van
        Montagu, M. Schell, J., Zambryski, P., 1983, Introduction of

genetic material into plant cells, Science, 222:815-821.

Chaleff, R. S., 1983, Isolation of agronomically useful mutants from plant cell cultures, Science, 219:676-682.

Cocking, E. C., Davey, M. R., Pental, D., and Power, J. B., 1981, Aspects of plant genetic manipulation, Nature, 193:256-270.

Cocking, E. C., 1983, Applications of protoplast technology to agriculture, in: "Protoplasts 1983, Lecture Proceedings 6th Intern. Protoplast Symposium," Experientia Supplementum, Vol., 46:123-126.

Cocking, E. C., 1981, Opportunities from the use of protoplasts, Phil. Trans. R. Soc. Lond. B., 292:447-568.

Davey, M. R., Cocking, E. C., Freeman, J., Pearce, N., and Tudor, I., 1980, Transformation of Petunia protoplasts by isolated Agrobacterium plasmids, Plant Science Lett., 18:307-313.

Davey, M. R., Cocking, E. C., Freeman, J., Draper, J., Pearce, N., Tudor, I., Hernalsteens, J. P., Beuckeleek, M. De, Van Montague, M., and Schell, J., 1979, The use of plant proto- plasts for transformation by Agrobacterium and isolated plasmids, in: "Advances in Protoplast Research, Proceedings 5th Intern. Protoplast Symposium," Szeged, pp. 425-

Davey, M. R., 1983, Recent developments in the culture and regener- ation of plant protoplasts, 1983, in: "Lecture Proceedings 6th Intern. Protoplast Symposium," Potrykus, I., ed., Experientia Supplementum, Vol., 46:19-29.

Davey, M. R., and Kumar, A., 1983, Higher plant protoplasts - retrospect and prospect, in: "Plant Protoplasts," K. L. Giles, ed., Intern. Rev. Cytology, 16:219-299.

Engler, D. E., and Grogan, R. G., 1982, Isolation, culture and regeneration of lettuce leaf mesophyll protoplasts, Plant Science Lett., 28:223-229.

Evans, D. A., and Sharp, W. R., 1983, Single gene mutations in tomato plants regenerated from tissue culture, Science, 221: 949-951.

Fluhr, R., 1983, The segregation of organelles and cytoplasmic traits in higher plant somatic fusion hybrids in Protoplasts 1983, in: "Lecture Proceedings 6th Intern. Protoplast Sympos- ium," Potrykus, I., ed., Experientia Supplementum Vol., 46: 85-92.

Freeman, J. P., Draper, J., Davey, M. R., Cocking, E. C., Gartland, K. M. A., Harding, K., and Pental, D., 1984, A comparison of methods for plasmid delivery into plant protoplasts, Plant and Cell Physiology, In press.

Gleba, Y. Y., and Evans, D. A., 1983, Genetic analysis of somatic hybrid plants, in: "Handbook of Plant Cell Culture, Vol. 1," D. A. Evans, and W. Sharp, eds., Macmillan Press. pp. 322-357.

Hamill, J. D., Pental, D. Cocking, E. C., and Muller, A. J., 1983a, Production of a nitrate reductase deficient streptomycin resistant mutant of Nictoiana tabacum for somatic hybridization studies, Heredity, 50:197-200.

Jinks, J. L., Caligari, P. D. S., Ingram, N. R., 1981, Gene transfer in *Nicotiana rustica* using irradiated pollen, *Nature*, 291: 586-588.

Kowalczyk, T. P., Mackenzie, I. A., and Cocking, E. C., 1983, Plant regeneration from organ explants and protoplasts of the medicinal plant *Solanum khasianum* C. B. Clarke var., *chatterjeeanum* sengupta (Syn. *Solanum viarum* Dunal), *Z. Pflanzenphysiol.*, 111:55-68.

Ledoux, L., Charles, P. Diels, L., Mergeay, M., Loppes, R., Frere, J. M., Matusznska, Y., Merckaert, C., Hooghe, R., Castiau, C., Ryngaert, A. M., Remys, J., Hooyberghs, L., Gilles, J., 1983, Transformation of plants with pBR plasmids applied to germinating seeds, *Proceedings Intern. Congress Genetics*, New Delhi 1983, Abstract No. 724. pp

Meins, F., 1983, Heritable variation in plant cell culture, *Ann. Rev. Plant Physiol.*, 34:327-346.

Murai, M., Sutton, D. W., Murray, M. G., Slightom, J. L., Merlo, D. J., Reichect, N. A., Sengupta-Gopalan, C., Stock, C. A., Barker, R. F., Kemp, J. D., and Hall, T. C., 1983, Phaseolin gene from bean is expressed after transfer to sunflower via tumor-inducing plasmid vectors, *Science*, 222:476-482.

Orton, T. J., 1983, Experimental approaches to the study of soma clonal variation, *Plant Molecular Biology Reporter*, 1(2):67-76.

Pandey, K. K., 1975, Sexual transfer of specific genes without gametic fusion, *Nature*, 256:310-313.

Patnaik, G., Cocking, E. C., Hamill, J. D., and Pental, D., 1982, A simple procedure for the manual isolation and identification of plant heterokaryons, *Plant Science Lett.*, 24:105-110.

Patnaik, G., Hamill, J. D., Pental, D., and Cocking, E. C., 1984, Somatic hybridization by the direct isolation of heterokaryons, Golden Jubilee Number, *Proceedings Indian Academy of Science (Plant Science)*, In press.

Pelletier, G., Primard, C., Vedel, F., Chetrit, P., Remy, R., Rousselle, A., and Renard, M., 1983, Intergenetic cytoplasmic hybridization in Cruciferae by protoplast fusion, *Mol. Gen. Genet.*, 191:244-250.

Pental, D., Hamill, J. D., and Cocking, E. C., 1984, Somatic hybridization using a double mutant of *Nicotiana tabacum*, *Heredity*, In press.

Potrykus, I., Harms, C. T., Hinnen, A., Hutter, R., King, P. J., and Shillito, R. D., 1983a, Protoplasts 1983, Proceedings 6th Intern. Protoplast Symposium, *Experientia Supplementum, Vol.*, 45, pp. 366.

Potrykus, I., Harms, C. T., Hinnen, A., Hutter, R., King, P. J., and Shillito, R. D., 1983b, Protoplasts 1983, Lecture Proceedings 6th Intern. Protoplast Symposium, *Experientia Supplementum, Vol. 46*, pp. 269.

Powell, W., Caligari, P. D. S., and Hayter, A. M., 1983, The use of pollen irradiation in barley breeding, *Theor. Appl. Genet.*, 65:73-76.

Prat, D., 1983, Genetic variability induced in *Nicotiana sylvestris* by protoplast culture, Theor. Appl. Genet., 64:223-230.

Riley, R., 1983, Discussion in Ciba Foundation Symposium 97, Better crops for food, pp. 188-189.

Schenck, H. R., and Robbelen, G., 1982, Somatic hybrids by fusion of protoplasts from Brassica oleracea and B. campestris, Z. Pflanzenzuchtg, 89:278-288.

Shapiro, J. A., Mobile genetic elements, Academic Press, New York, 1983.

Shepard, J. F., Bidney, D., Barsby, T., and Kemble, R., 1983, Genetic transfer in plants through interspecific protoplast fusion, Science, 219:683-688.

Shepard, J. F., Bidney, D., and Shahin, E., 1980, Potato proto-plasts in crop improvement, Science, 208:17-24.

Shirvin, R. M., 1978, Natural and induced variation in tissue culture, Euphytica, 27:241-266.

Snape, J. W., Parker, B. B., Simpson, E., Ainsworth, C. C., Payne, P. I., and Law, C. N., 1983, The use of irradiated pollen for differential gene transfer in wheat, Triticum aestivum, Theor. Appl. Genet., 65:103-111.

Steinbiss, H. H., and Broughton, W. J., 1983, Methods and mechan-isms of gene uptake in protoplasts, Intern. Rev. Cytol. Suppl., 16:191-208.

Stumpf, P. K., and Pollard, M. R., 1983, in: "High and low crucic acid rapeseed oils," Krassel, J. K. G., ed., Academic Press. pp. 131-141.

Swaminathan, M. S., 1982, Biotechnology research and third world agriculture, Science, 218:967-972.

SOMACLONAL VARIATION: THEORETICAL

AND PRACTICAL CONSIDERATIONS

T. J. Orton

Agrigenetics Corporation
Applied Genetics Laboratory
3375 Mitchell Lane
Boulder, Colorado   80301

INTRODUCTION

The traditional view of somatic mutations in higher eukaryotes is that they are rare - roughly 10^{-6} in frequency - and of little consequence in population dynamics. Such mutations should only have an evolutionary impact when they occur in the germline and enhance viability. Recent observations are challenging these views and suggesting new ways in which organisms can alter themselves or be altered genetically without the intervention of sex. For example, evidence from a large and growing number of reports has shown that mutations are present at much higher than expected frequencies when higher plant cells or tissues are induced to proliferate in vitro for even a brief period of time (see the following section). In other studies, mechanisms have been described which cause high somatic mutation frequencies in otherwise normal tissues of higher plants and animals in vivo (Kidwell et al., 1977; McClintock, 1978). Further evidence suggests that mutations are indeed present at detectable frequencies in the differentiated somatic tissues of both seed and vegetatively propagated plants.

In a review, Larkin and Scowcroft (1981) introduced the useful term 'somaclonal variation' to denote variation arising within or from populations of cultured cells. The phenomenon of somaclonal variation has been reviewed extensively and from several stand points: cytology (D'Amato, 1975, 1977; Sunderland, 1977; Bayliss, 1981; Constantin, 1981), cytoplasmic (Pring et al., 1981), morphology/physiology of regenerated plants (Skirvin, 1978; Chaleff, 1981; and Larkin and Scowcroft, 1981, 1983),

genetic/epigenetic forms (Meins, 1983), biological significance
and speculations as to causes (Chaleff, 1981; Larkin and
Scowcroft, 1983; Meins, 1983; Orton, 1983b) and experimental
approaches (Orton, 1983a). In the present review, I shall
attempt to build upon the foundations established collectively by
these earlier efforts with emphasis on underlying mechanisms and
applications in agriculture.

NOMENCLATURE

    It is obvious that common terminology is a desirable feature
of any language. Hence, it is unfortunate that a rift has
emerged regarding the nomenclature of plants regenerated from
cultures and ensuing self-pollinated generations. Chaleff (1981)
recognized the need for a unified nomenclature and introduced the
R, R_1 R_2, etc. (respectively regenerate, first generation selfed
progeny, second generation selfed progeny, etc.) system, which is
now in wide use. Variations on this system include the use of
'P' instead of 'R' (Sibi, 1976) and the addition of a zero
subscript to denote the population of primary regenerated plants
(Edallo et al., 1981).

    Larkin et al. (1983) have recently proposed a new system:
SC1, SC2, SC3, etc., equivalent respectively to R, R1, R2, etc.
The SC refers to 'somaclone', and was introduced primarily to
avoid confusion with the established 'R' system. The authors
argue that the passage of cells or tissues through culture often,
if not usually, results in the introduction of mutations. If
cells and corresponding regenerated plants are diploid, mutations
should be heterozygous. Hence, the 'R' designation is numeri-
cally inconsistent with the established 'filial' nomenclature
denoting hybridity and successive selfed generations (i.e., F_1,
F_2, F_3, etc.) and that where variation has been introduced mutagenesis (i.e., M_1, M_2, M_3, etc.)

    Supporters of the 'R' system counter with the following
points: The "1" subscript denoting primary regenerated plant is
potentially misleading because it presumes the existence of
heterozygous mutations in plants regenerated from cell and tissue
cultures, thus introducing a bias with respect to the presence or
absence of genetic variation. Those who propose to use cell and
tissue culture as a tool for efficient, high fidelity plant propagation would certainly protest the presumption of variation.
Perhaps the strongest arguments in favor of the 'R' system is
that it has historical precedent (Sibi, 1976) and is already in
wide use and broadly accepted.

    The disagreement is rooted in the context in which the
culturing process is regarded: as cloning vs. mutagenic.

Debates on the subject conducted during the summer of 1983 have
indicated that agreement on a unified system is unlikely in the
near future. It is hoped that a broadly subscribed organization,
such as the International Association of Plant Tissue Culture,
will consider the issue and provide an official sanction on
unified nomenclature.

THEORETICAL CONSIDERATIONS

Gross Chromsomal Mutations In Vitro

     Chromosomal variation has been known to exist in genetic and
crown gall tumors since the 1930's (summarized in D'Amato, 1952).
Mitra and Steward (1961) first demonstrated the existence of
chromosomal anomalies in untransformed cultured cells. They
reported on the existence of polyploidy, aneuploidy, and abnormal
cell divisions in their carrot (Daucus carota L.) suspension
cultures. Progress has been slow in this area due to difficulties in controlling experimental materials (e.g., explant
source, source genotype, developmental status of cultured cells)
and the lack of diagnostic genetic markers in morphologically
uniform somatic tissues. It has been anticipated that the possibilities provided by new molecular tools will provide genetic
markers to study variation using informational probes (Rivin et
al., 1983; Cullis, 1983; Orton, 1983a,b).

     Cultured cells cannot be considered equivalent in all spe-
cies, culture regimes, and developmental fluxes. Therefore, it
is also impossible to extract general principles from the body of
descriptive literature on chromosomal variation. It can be
concluded that gross chromsomal rearrangements have been observed
frequently enough that they should be suspected in all cultures.
Factors which appear to contribute to the types and degree of
chromosomal variation in cultured tissues include: preexisting
cytological mutations in the explant (D'Amato, 1975); developmental status of the explant (Banks-Izen and Polito, 1980; Meins,
1983); genotype (Sacristan, 1971; Balzan, 1978; Ogura, 1978;
McCoy et al., 1982; Browers and Orton, 1982a; Meins, 1983);
apparent developmental status of the cultured cells (Orton,
1980a); and culture regime (Torrey, 1965; Demoise and Partanen,
1969; Bayliss, 1973, 1975; Libbenga and Torrey, 1973). It is
difficult to interpret patterns of the appearance, accumulation,
and loss of chromosomal variation over time in culture, as actual
mutation rates, selective fitnesses, and interactions of factors
are often completely unknown. The predomination of certain
cultures by particular karyologically abnormal cells clearly
indicates that they have a selective advantage over others in the
culture (Ashmore and Gould, 1981; Ogihara, 1982).

What are the earliest detectable events with regard to the de novo generation of gross chromosomal changes in populations of cultured cells? Cytophotometric comparisons between explant and corresponding adventitious callus cells shortly after initiation shows a general tendency toward polyploidization (summarized in D'Amato, 1975). However, in one study, only cells of lower ploidies (as determined by direct chromosome counts) were observed in mitosis as compared with higher ploidies (determined cytophotometrically) among interphase cells (Bennici et al., 1968). Polyploid cells and aneuploid derivatives frequently persist and even predominate in older, established cultures (D'Amato, 1975; Bayliss, 1980) but it is not known whether they are descendants of early or recent polyploidization events. The relationship of polyploidy with other forms of genetic variation, e.g., aneuploidy, deletions, translocations, inversions, transpositions, and point mutations, is also unknown.

Cytological observations of cultured cells and regenerated plants have led Sacristan (1971) and McCoy et al. (1982) to speculate that asynchrony between mitosis and the synthesis of late-replicating heterochromatic DNA might be involved in partial or whole chromosome loss. In a recent study employing more direct molecular approaches, Durante et al. (1983) showed that A+T and G+C rich satellite fractions are amplified within hours after tobacco (Nicotiana tabacum) pith explants were placed on callus induction medium. These fractions are localized in constitutive heterochromation. The authors discuss this finding in light of mechanisms underlying developmental fluxes, but do not consider that such amplification may also constitute an early event in the generation of chromosomal mutations by introducing asynchrony of DNA synthesis and mitosis.

Murata and Orton (1982) observed frequent end-to-end chromosome fusions in young (2-3 subcultures) celery (Apium graveolens) cultures. Fusions among acrocentric chromosomes were not random with respect to arm length: short arm-short arm were significantly more frequent than expected. Chromatin bridges were observed frequently in this culture, presumably arising as a consequence of opposite separation of the active centromeres of dicentric chromosomes resulting from fusion. It should be pointed out that the frequency of chromatin bridges does not necessarily correspond to that predicted solely by the frequency of cells containing multicentric chromosomes. Murata and Orton (1983) observed that 95% of the cells in an 18-month old celery culture contained at least one multicentric chromosome. Assuming independent assortment of centromeres, at least 47.5% of all anaphase cells would have been expected to exhibit bridges. Only 21.0% actually showed bridges, suggesting either that some of the constrictions were silent, or that assortment of centromeres was not random with respect to chromatid.

Duplications and deletions will result depending upon the
precise location where the bridge ruptures. It is unknown
whether the broken ends subsequently heal or fuse to initiate a
breakage-fusion-bridge (BFB) cycle (McClintock, 1938, 1942).
McClintock (1978) has accumulated convincing evidence that arti-
fical initiation of BFB cycles in maize (Zea mays) embryos culmi-
nates in the rapid and extensive reorganization of the nuclear
genome not involved in the primary BFB lesion. The evidence
further implicates independent two-element systems, one (the
controlling element) which codes for a restriction enzyme or
transposase, and another element which is the actual transposing
sequence. The controlling element also appears to cause chromosomal rearrangements in addition to transposition. McClintock
speculates that such systems constitute a normal response to
severe stress. Similar mechanisms may be responsible for the
extensive genome rearrangements observed in cultured cells. They
may be invoked by events such as fusions resulting in dicentric
chromosomes (discussed above) or by premature condensation and
disjunction of underreplicated chromatin as speculated by McCoy
et al. (1982). It should also be recognized that plant callus
cultures are initiated analogous to a wound response, which is a
condition that could activate such stress responses.

A large and growing number of reports have demonstrated that
cells containing gross chromosomal changes are sometimes able to
participate in organogenic or embryogenic processes and persist
in the differentiated tissues of regenerated plants. The general
observation is of attenuation of variation from culture to regenerate (Sacristan and Melchers, 1969; Orton, 1980; Ogihara, 1981;
Browers and Orton, 1982b). In some instances, specific chromosomal anomalies have been clearly associated with regenerative or
developmental incompetence (Mouras and Lutz, 1983; Mahfouz et
al., 1983; Orton, 1983c). Finally, while cytological anomalies
frequently appear to be fixed within single regenerated plants
(as expected if the plant were regenerated from a small number of
clonally related cell initials and the ensuing mitoses were
highly conservative), numerous independent observations of cytological chimeras have also been reported (Sacristan and Melchers,
1969; Novak, 1980; Orton, 1980a,b; Bajaj et al., 1981; Ogihara,
1981; McCoy et al., 1982; Browers and Orton, 1982b). These
observations point to alternative possibilities: 1) regeneration from a larger, genetically diverse population of cell
initials; or 2) mitotic instability after regeneration.

Mendelian Variation In Vitro

Mendelian variation is defined as that which does not result
from gross chromosomal changes (described in the previous
section), is heritable, and segregates in crosses with individuals exhibiting a disparate phenotype. This class includes, but

is not limited to, point or insertional mutations of major genes
having a qualitative effect or minor genes having more subtle,
quantitative effects. What are the types and degrees of
Mendelian variation in populations of cultured plant cells? What
are the mechanisms which underlie their appearance? How do they
behave over time? Can general principles be elucidated which
apply to a broad spectrum of plant species and culture regimes?
Unlike gross chromosomal variation, where the mutation can be
deduced directly, Mendelian variation must rely on a corresponding structural or functional change for detection. At
our present state, we have insufficient knowledge of the fine
structural organization of the nuclear genome and the primary
function of gene products to permit such experiments to be conducted directly.

Several reports have appeared recently which provide a genetic analysis of morphological and physiological variation in
regenerated plants (see next section). These reports constitute
the best evidence that mutations affecting functions which are
manifest in whole plants can occur in somatic tissues. However,
it is abundantly clear that gross chromosomal, and perhaps
Mendelian, mutations impair the regenerative competency of
cultured plant cells (Torrey, 1967; Mahfouz et al., 1983; Mouras
and Lutz, 1983). Hence, in at least some instances the variation
manifested by regenerated plants misrepresents the total genetic
variation in the original culture by an unknown degree.

Oono (1978, 1981) reported on the existence of heritable
variation among populations of plants regenerated from rice
(Oryza sativa) calli. The donor plant, a dihaploid derived from
the pure line 'Norin 8', was presumably homozygous not-
withstanding the subsequent observation of variation. Only 28.1%
of the 1121 primary regenerates (referred to as D_1) were considered normal with respect to five selected characters, while
43.9% were variant in one character and 28.0% variant in two or
more characters. Examination of selfed progeny of regenerates,
referred to as D_2 and D_3, showed that mutations affecting plant
height, heading date, grain number, grain weight, panicle length
and number, chlorophyll pigmentation, and salt tolerance were
simply inherited. Segregation in the 'D_2' generation and rapid
fixation by the 'D_3' generation, indicative of simple inheritance, was observed for many characters tested, including several
of agronomic importance. One dwarf mutation did not segregate in
the 'D_2' generation, and was shown not to be maternally
inherited, suggesting the existence of a double mutation.

Edallo et al. (1981) regenerated 110 plants ('R_0') from scu-
tellar calli of the maize inbreds W64A and S65 after up to eight
subcultures. One of these plants was tetraploid and another
apparently trisomic, while the remaining 108 were presumptive

diploids. Of the presumed diploids, 31 exhibited partial to
complete loss of fertility. The remaining 77 'R$_0$' plants were
selfed twice and analyzed for variation. Simple inheritance was
demonstrated for 17 morplological characters, and the average
frequency of simple mutations was estimated at 1.0 per diploid R$_0$
plant.

Prat (1983) reported on the generation of mutations during
protoplast isolation, culture, and regeneration from inbred and
derivative dihaploid Nicotiana sylvestris. The two source plants
gave rise to primary regenerates which exhibited a similar
spectrum of ploidy changes: slightly more than half were
polyploid, slightly less than half diploid, and no aneuploids.
The diploid primary regenerates were not detectably different
from the protoplast donor plants. Many of the selfed progenies
of primary regenerates (8 of 13 from the inbred and of 8 from the
dihaploid) contained new recessive mutations not ever observed in
the source plants or their progenies. Two of the plants had
double mutations. Quantitative depressive size variation was
observed among progenies not exhibiting striking mutant phenotypes.

Plants exhibiting mutations were regenerated from leaf
callus cultures of the processing tomato (Lycopersicon esculentum)
pure line 'UC82B' (Evans and Sharp, 1983). The authors state that
chromosomal variation (particularly autotetraploidy) was observed
frequently among primary regenerates, but did not provide any
further data. Selfed seed was obtained from 230 fertile diploid
'R' plants, and large replicated field plots were examined over
two years. Variants were carefully tested to determine their
mutational origin and potential allelisms to known markers.
Thirteen independent mutations were apparently present in the
primary regenerates, none of which were observed in seed-
propagated UC82B. Eight of these behaved as recessives while one
(virescence) did not fit the expected 3:1 ratio (42:4). The
authors conclude that this mutation might still be recessive, the
segregation being confounded by sublethality, as is often
observed with chlorophyll deficiencies. Three of the mutations
behaved as dominant, while one was homozygous recessive. A
number of the mutations had similar phenotypes (4 male steriles,
2 jointless pedicel, 2 indeterminates), and the data to
demonstrate the independence of these mutants was not presented.
One mutation conditioning orange fruit was found to be allelic to
a known marker (t) while another causing jointless pedicels was
found not to be allelic to j, although allelism tests with the
other known jointless marker, j-2, were not reported.

Larkin et al. (1983) have conducted a similar study in
wheat (Triticum aestivum). Extensive genetic studies were conducted on 142 primary regenerated plants (denoted SC$_1$) from
immature embryo calli of cv. 'Yaqui 50E' after 2-3 subcultures.

'SC_1' plants were self pollinated, and 'SC_2' and 'SC_3' generations evaluated for the appearance of new variation. An average
of 16.2% of all 'SC_2' plants exhibited partial to complete loss
of fertility. All of 50 'SC_2' plants examined had apparently
normal chromosomal complements. Among fertile 'SC_2' families, a
broad range of morphological and biochemical variants were
observed. Pooled data showed that "awned", "white glume", "white
kernel" and "reduced wax" were all new recessive mutations. Most
height variants appeared to be inherited quantitatively, but at
least one case of a major gene and simple inheritance was
observed. Heading date was also quantitatively inherited, and
both earlier and later (than Yaqui 50E) types were noted.
Heritable changes in seed gliadin patterns were also observed at
an average of 3.0 band changes per 'SC_2' individual. Finally,
heritable insensitivity to GA_3 – induced amylase synthesis and
ABA repression of GA_3 – induced amylase synthesis were reported.
Thus, mutations ranged from altered electrophoretic mobility to
those affecting physiological processes and simple and complex
morphological changes. No such variation was observed among seed
progenies of Yaqui 50E. Although it is impossible to reconstruct
the various individual mutation frequencies among individual
'SC_1' plants, it would appear that observable mutations occur
within an order of magnitude of 1.0 per plant when considering
only awns, glume and grain color, and waxiness. It is reasonable
to assume that the actual frequency is much higher than this.

Plants regenerated from mesophyll protoplasts of the lettuce
(Lactuca sativa) inbred cv. 'Climax' were also found to exhibit a
high degree of heritable variation (Engler and Grogan, 1983).
The most frequent form of variation observed among primary regenerates ('P_0') was reduced seed set ($\leq$ 100 seeds as compared to
$>$ 5000 seeds per control plant). Twenty of 119 'P_0' plants
exhibited poor seed set, and 12 of these were found to be
tetraploid. Fifteen 'P_0' plants with normal fertility were all
diploid. The reasons for poor seed set among the 8 apparent
diploids were not investigated. Only 'P_0' plants producing
$\geq$ 100 seeds were used in subsequent genetic studies. Segregation
of striking qualitative variation was observed among 13 of the 99
'P_1' families examined. Prevalent among characters affected were
altered growth rate or size, altered leaf or cotyledon morphology, and altered green pigmentation. Normal:variant ratios were
not significantly different from 3:1 in 11 of the 13 families,
consistent with the existence of single recessive mutations
masked by wild type dominant alleles in the 'P_0' plants. The
remaining two did not segregate, indicative of a homozygous or
cytoplasmic mutation in 'P_0' plants. Since complementation tests
were not conducted, it was impossible to tell which, if any, of
the morphologically similar mutations were independent.
Moreover, additional quantitative variation was observed among
'P_1' families, but was not analyzed due to difficulties in

interpretation. The average head size and associated variance of
'P_1' progeny were, respectively, significantly less and greater
than the control (cv. 'Climax'), suggesting possible mutations
affecting vegetative growth and development as well. Hence,
while the frequency of definitive mutations in 'P_0' plants was on
the order of 0.1, this would appear to be a rather gross
underestimate.

An interesting study of the relative phylogeny of mutations
occurring in a plant cell culture was conducted by Fukui (1983).
Callus cultures were initiated from seeds of the rice cv.
'Nipponbare'. Only one of over 100 two week old calli thus initiated could be induced to regenerate – a genotype effect? An
interesting array of single recessive mutations was observed
among regenerated plants: four independent, unlinked mutations
were present in patterns which suggested the sequence of their
occurrence in the callus: early heading date, albino, and short
culm or sterility. The numbers of plants having mutations 0, 1,
2, 3, and 4, were respectively 3, 4, 1, 0, and 4. Thus, the
average number of mutations per regenerated plant was estimated
at 1.83, although clonally related plants obviously shared the
same mutation.

A number of studies have demonstrated Mendelian variation in
androgenic plants. Such plants are usually derived directly from
microspores without an intervening free culture phase. In addition, microspores are haploid as compared to the presumed diploid
or polyploid status of somatic explants. For these reasons,
variation arising via androgenesis may not be strictly comparable
to cultures initiated from somatic cells. Recognizing these
problems in drawing comparisons with cultured diploid tissues,
variation among androgenic plants of tobacco and oilseed rape
(Brassica napus) is analyzed below.

Burk and Matzinger (1976) observed numerous phenotypic differences among selfed families of colchicine doubled dihaploids
from the highly inbred tobacco cv. 'Coker 139'. In general,
dihaploid families exhibited reduced vigor and growth, but
increased alkaloid and reducing sugar content as compared with
the donor cultivar. No further experiments were conducted to
elucidate the genetic behavior of the responsible mutations.

DePaepe et al. (1981) reported the findings of similar but
more revealing experiments. They observed an increase in heritable variation following each of several recurrent androgenic
cycles from homozygous N. sylvestris. This evidence would appear
to be sufficient to rule out residual heterozygosity in the original donor plant(s) as a potential source of variation. Many of
the mutations were homozygous, as would be expected if the mutations were present in haploid cells prior to regeneration. Some

of the mutations segregated 1:3 (and hence were recessive), indicating either, as the authors conjecture, that diploid androgenic plants arose from fusions of genetically divergent generative and vegetative gametophytic nuclei, or that mutations arose after the chromosome doubling event. Some forms of variation were observed among doubled haploid ('D.H.') plants at frequencies much higher than expected if caused by random mutations. Examples included a general loss of vigor and 'crumpled' appearance of the leaves, which were observed in all D.H. regenerates. Hypothesizing that genetic changes which were more extensive than simple random point mutations were responsible for these heritable changes, they undertook a molecular study contrasting variant and normal lines (De Paepe et al., 1983). Using a number of independent approaches, they were able to show that the nuclear DNA content of D.H. plants increased by an average of 10, and up to over 20%, over that of the original plant. Buoyant density separations, reassociation kinetics, and intensity of electrophoretically separated Eco RI fragments all pointed to amplification of both A-T and G-C rich highly repeated sequences and the number of inverted repeats. This may represent a general phenomenon, at least in the Solanaceae, because similar independent observations of increased DNA content have been reported among androgenic doubled haploids of N. tabacum cv. 'Coker 139' (Dhillon et al., 1983) and Datura innoxia (Sangwan-Norreel, 1983).

Of 45 dihaploid plants regenerated from anthers of inbred oilseed rape lines 24/72 and cvs. 'Tower' and 'Egra', 28 gave rise to uniform progeny exhibiting one variant morphological character, 8 produced uniform progeny, but with reduced vigor and sterility, and 9 gave rise to progeny which were segregating for one or more characters (Hoffman et al., 1982). While the original donor lines were characteristically and stably low in glucosinolates, there was a clear tendency toward a marked increase in glucosinolate content among androgenic dihaploids. When androgenic embryos were maintained in culture, they produced adventitious diploid embryos from which selfed progeny showed segregation for a large number of new mutations.

Another approach to the study of Mendelian variation arising in plant cell cultures is to use phenotypes selectable both in vitro and in vivo. Chaleff and Parsons (1978) exposed non-mutagenized N. tabacum cell cultures to highly toxic concentrations of the herbicide pichloram. Plants were regenerated from seven surviving colonies and tested to determine the stability, tissue specificity, and inheritance of resistance. Two of the seven colonies were unstable, and one could not be regenerated. Picloram resistance (PmR) was expressed in regenerated plants from the remaining four colonies, and proved to be inherited in each as a single dominant mutation. At least three of these four mutations were found to assort independently (summarized in

Chaleff, 1982). The story was amended somewhat by Chaleff (1981)
as follows: five PmR resistant colonies were caused by separate
mutations, three of which were dominant and two semidominant.
Two of the dominant mutations are linked and the others assort
independently. The estimated frequency of the original mutations
in the original cell culture was 2.7×10^{-6}. However, more than
half of the cell lines were also resistant to hydroxyurea
(Chaleff and Keil, 1981). Analysis of the progeny of regenerates
from three such cell lines showed the character to be conditioned
by a single dominant mutation that segregated independently of
PmR in two cases, and non-independently (indicating linkage) in
the third.

Still another approach to the study of Mendelian variation
in cultured cells is to begin with explant material from multiply
heterozygous plants and to determine the effect(s) of cell
culture on the stability of those heterozygous phenotypes. It is
thus theoretically possible to detect a number of types of mutations affecting the genotype at known loci. One such study, protoplasts were isolated from cotyledons of A_1Yg/a_1yg (yellow
green) tobacco plants and regenerated plants from subsequent
callus colonies (Barbier and Dulieu, 1980). Approximately 3.5%
of the regenerated plants had green leaves, caused either by
reversions or deletions (of the mutant recessive allele). The
frequency of wild type plants did not increase over time in
culture, prompting the authors to conclude that mutations were
either present in the original cotyledons, or occurred only
during the early stages of callus growth. Interestingly, manifestations of increased variation over time in culture were
observed for numerous other characters.

Lorz and Scowcroft (1983) isolated leaf mesophyll protoplasts of N. tabacum plants heterozygous at the Su (aurea or
sulfur) locus (Su/Su). The known alleles at this locus are also
codominant: Su/Su - green; Su/Su - yellow green; Su/Su - albino.
They isolated more than 2,100 morphogenic calli and analyzed more
than 8,000 shoots. Twenty of these colonies (0.92%) produced
only green (Su/Su) or white (Su/Su) shoots, interpreted as
variation pre-existing in the cotyledon. Fifty-nine (2.75%) were
heterogeneous, giving rise to shoots of chimeral pigmentation,
presumed to have arisen as de novo variation during the period of
cell growth in vitro. Alternatively, the mutations could have
occurred during early shoot development after regeneration. The
frequency of variant colonies increased over time or if colonies
were exposed to the mutagen N-methyl-N'-nitro-N-nitrosoguanidine.
Callus growth under stress or altered nitrogen regime did not
affect the frequency of mutations. Interestingly, yellow green
plants regenerated from older cultures tended to exhibit significantly higher frequencies of twin - spotting than those from
younger cultures, suggesting that additional genetic changes

which predisposed plants to more frequent somatic cross-overs also accumulated over time in culture.

Celery plants were synthesized which were heterozygous at four known loci, three of which were linked, and experiments were conducted to determine the effects of culture on the phenotypic stability at these loci (Orton, 1983c). In addition, two of the linked loci encoded allozymes of phosphoglucomutase (PGM) and shikimic acid dehydrogenase (SDH) which could be visualized directly in cultured tissues (technique reviewed by Lassner and Orton, 1983). Also, heterozygous plants were constructed in two different cytoplasmic backgrounds. Callus was initiated from immature petiole tissues and cultured via bulk serial transfer for approximately one year at which time clones were isolated from each culture (i.e. the multiply heterozygous lines in the two different cytoplasmic backgrounds termed 'reciprocals'). Clones of one reciprocal always exhibited the heterozygous pheno-types at both the Pgm-2 and Sdh-1 locus, but 25.8% of clones of the other exhibited loss of the fast-migrating electromorph at the Pgm-2 while remaining 100% heterozygous at the Sdh-1 locus. No reversion from Pgm-2S to Pgm-2FS was observed, and Pgm-2S cells gradually came to predominate until, after approximately 2.5 years, most cultures contained no detectable Pgm-2FS cells. No consistent gross karyological differences could be discerned between Pgm-2S and Pgm-2FS cells, and the relative stability at Sdh-1, closely linked to Pgm-2, was inconsistent with the hypothesis that simple chromosome loss or random deletion were responsible for loss of Pgm-2F. The variant Pgm-2S phenotype was transmitted stably into regenerated plantlets, although they were karyologically abnormal and failed to develop normally and even-tually died; hence, inheritance studies were not possible. The unusual manifestations of the phenomenon could be explained by postulating the existence of a mobile element which inserted at high frequency in or near the Pgm-2F allele resulting in its inactivation.

CYTOPLASMICALLY INHERITED MUTATIONS

The structural and functional organization of higher plant chloroplast (ct) genomes is now fairly well understand (Edelman et al., 1983). Mitochondrial (mt) DNA has been more difficult to isolate, is generally larger than ctDNA, and occurs in multipar-tite configurations (Quetier and Vedel, 1977; Spruill et al., 1980; J. D. Palmer, pers. comm.). Hence, detailed maps of mtDNA have been slower in emerging. The wealth of knowledge regarding the organization and sequences of higher plant ct and mt genomes will make it possible to gain tremendous insights on the role of cytoplasmic genomes in existing and potential genetic diversity.

Although it is reasonable to assume that most of the genetic
information is nuclear, it is clear that important energy trans-
ducing functions are encoded in ct and mt DNA's. However, at
least some of the cytoplasmic information either encodes
directly, or secondarily, or interacts with nuclear gene products
to control such characters as morphology/and vigor, (R. Frankel,
pers. comm.) and floral development (Pring and Levings, 1978 and
others too numerous to list). In the case of the Southern corn
leaf blight disease, a major component of the pathogen's attack
system appears to be a toxin which destabilizes the inner
mitochondrial membrane of susceptible, but not resistant lines
(Mathews et al., 1979). Tentoxin, a substance secreted by
Alternaria tenuis, inhibits energy coupling in chloroplasts of
susceptible plants by binding with the CF_1-ATPase complex, which
is encoded in the chloroplast (Steele et al., 1976). The
triazine herbicides exert their phytotoxic effects by binding
with the 32 KD photosystem II protein which is also encoded in
the chloroplast (Arntzen et al., 1982). Finally, because the
transcriptional and translational apparati of both chloroplasts
and mitochondria are similar to those of prokaryotes, and
distinct from those utilized by nuclear genes, a number of inhibitors are capable of distinguishing resistance mutations of
cytoplasmic or nuclear origin. Hence, although it is presently
impossible to quantify the contribution of cytoplasmic genes to
the total phenotype, analysis of a small cross section of mutations has revealed a substantial number which are of cytoplasmic
orgin and have diverse phenotypic effects.

It is clear from the previous sections that the frequency of
nuclear-encoded variation mutations is increased when somatic
cells or tissues are passaged through in vitro culture as compared to conventional seed or vegetative propagation. To what
extent is cytoplasmically-encoded variation altered by in vitro
culture? Sibi (1976) observed morphological variation in plants
regenerated from lettuce cell cultures which was maternally
inherited. Aside from this, few reports have appeared which
demonstrate unequivocally that mutations incurred in vitro may be
cytoplasmic. An example which has been extensively studied is in
connection with T-cms cytoplasm in Zea mays. An excellent summary has been provided by Pring et al. (1981). This form of CMS
in maize is extremely stable under sexual propagation, but a high
frequency to reversion to fertile is observed among plants regenerated from scutellum-derived callus cultures. In three
separate studies, the proportions of male fertile: male sterile
regenerates from T-cms derived calli were 0/54, 8/143, and 31/60
(Pring et al., 1981). It was speculated that differences in
variant frequency might have been a consequence of different
nuclear background or culture conditions. In almost all cases,
reversion to male fertility occurred concomitantly with reversion from susceptibility to resistance to Helminthosporim maydis

race T. This does not appear to be a case of simple pleiotropy, since at least one susceptible and male fertile plant has been recovered (Brettell et al., 1982). Direct inspection of mitochondrial DNAs from different male fertile revertant regenerates digested with Xho I and electrophoresed on agarose showed a variable number of missing fragments, one of which was common to all lines tested. The mtDNA's of revertants were still extremely similar to T-cms mtDNA, and strikingly different from mtDNA of N cytoplasm. Concomitant with the loss of the Xho I fragment(s), there was a loss or reduction in the levels of a 13 KD polypeptide from mitochondrial extracts (Dixon et al., 1982).

While there is little evidence to suggest that striking ctDNA or mtDNA variants arise at an accellerated rate as compared to somatic tissues in vivo, a number of unusual observations suggest novel ways that cytoplasmic genomes can be altered in nature. Prominent among these is the growing body of evidence in the Solanaceae that recombination occurs among heterogeneous mitochondria juxtaposed by protoplast fusion (Belliard et al., 1979; Boeshore et al., 1983). Plants regenerated from unfused parental protoplasts exhibited no detectable variation in restriction patterns. Hence, if recombination does occur in somatic cells in vitro and in vivo, it would appear to do so faithfully. In a second set of observations, mt - associated plasmids exhibit a distribution that suggests trans-compartmental movement (Palmer et al., 1983; Kemble et al., 1983). Homologous sequences have been discovered in the nucleus, chloroplast and mitochondrion of spinach (Timmis and Scott, 1983), the mitochondrial and nuclear genome of yeast (Farrelly and Butow, 1983), and chloroplast and mitochondrial genome of maize (Stern and Lonsdale, 1982), further suggesting the possibility of natural transcompartmental transpositions. Such phenomena have not yet been observed directly, and involvement in the short term generation of genetic diversity is presently unknown.

SYNTHESIS

Can any common thread be found among these complex observations which allows unified hypotheses to be constructed regarding the mechanisms underlying somaclonal variation? Numerous recent independent reports provide good data to substantiate the simple inheritance of many variant characters which emerge from plant cultures (summarized in Table 1). The frequencies of such mutations are inordinately high, from approximately 1.0 to 100% per selected locus per regenerated plant, much too high to suppose the involvement of rare, random hit processes. In all of the studies cited regarding Mendelian variation observed from cultured cells, a limited spectrum of mutations having qualitative effects was observed. This could have been

Table 1. Summary of experiments in which unselected simple mendelian variation has been observed
and verified in plants regenerated from plant somatic cell cultures.

Species	Culture Type	Culture Age at Time of Regeneration	Mutation	Inheritance	(%) Estimated Frequency in R Plants*
Oryza sativa [a] dihaploid of cv. 'Norin 8'	seed-derived callus	?	chlorophyll deficiency	?	17.1
			reduced fertility	?	63.1
			plant height	?	18.9
			altered heading date	?	5.2
			altered morpholgy		1.7
Zea mays [b] inbreds W64A, S65	scutellar callus	1 to 8 months	Endosperm: defective, opaque, etched, white cap, germless; Seedling: lethal, abnormal, yellow- green, pale green, albino, fine stripe, yellow stripe, virescent, reduced, dwarf, viviparous	all dominant or recessive (individual mutations not specified)	NR [b]

(continued)

Table 1. Summary of experiments in which unselected simple mendelian variation has been observed and verified in plants regenerated from plant somatic cell cultures. (Continued)

Species	Culture Type	Culture Age at Time of Regeneration	Mutation	Inheritance	(%) Estimated Frequency in R Plants*
Nicotiana[c] sylvestris inbred and derivative dihaploid	mesophyll protoplast derived calli	approx. one week	male sterile	double recessive ?	9.5**
			dwarf	recessive	4.8
			late	recessive	4.8
			albino	recessive	4.8
			rosette necrosis/ variegation	complex	4.8
			lethal cotyledon necrosis	dominant ?	4.8
			wide leaves (+ others)	complex	4.8
			male sterile/dwarf	?	4.8
Triticum[d] aestivum cv. Yaqui 50E	embryo callus	2-3 subcultures	awned	recessive	45.1
			white glume	recessive	12.7
			white kernel	recessive	5.9
			reduced wax	recessive	2.1
Lactuca[e] sativa cv. climax (a pure line)	mesophyll protoplast derived calli	approx. 1 month	slightly dwarfed; dark cotyledons	recessive	5.6

Species	Culture Type	Culture Age at Time of Regeneration	Mutation	Inheritance	(%) Estimated Frequency in R Plants*
Lactuca sativa (cont'd)			light green; crinkled cotyledons	recessive	1.4
			yellow cotyledons	recessive	1.4
			severe dwarf; heart cotyledons	recessive	1.4
			large seedling	?	1.4
			severe dwarf	recessive	1.4
			albino	recessive	1.4
			small seedling	?	1.4
			severe dwarf; dark cotyledons	recessive	1.4
Oryza[f] sativa cv 'Nipponbare'	seed callus	2 weeks	early albino short culture sterile	recessive recessive recessive recessive	75.0 42.0 33.0 33.0
Apium[g] graveolens PI 169001 x PI 279225	petiole	6 months	frilly leaf	recessive	1.8

(continued)

Table 1. Summary of experiments in which unselected simple mendelian variation has been observed and verified in plants regenerated from plant somatic cell cultures. (Continued)

Species	Culture Type	Culture Age at Time of Regeneration	Mutation	Inheritance	(%) Estimated Frequency in R Plants*
Lycopersicon esculentum[h] cv. 'UC82B'	leaf callus	3 to 4 weeks	male sterile	recessive	1.7
			jointless-1	recessive	0.4
			jointless-2	double recessive	0.4
			tangerine fruit	recessive	0.4
			green base	dominant	0.4
			albino	recessive	0.4
			virescent	probably recessive	0.4
			indeterminant	dominant	0.4
			mottled	recessive	0.4

* Per cent of all original regenerated plants bearing the mutation in the germ line. Like mutations were grouped (i.e., assumed to be identical) if no data to prove independence was provided.

** Results of two separate experiments, one on a pure line and another on a dihaploid derivative, were bulked.

a Oono, 1981
b Edallo et al., 1981. Mutation frequencies reported as 0.4 to 2.3 mutations per regenerated plant but not reported according to individual mutation.
c Prat, 1983
d Larkin et al., 1983
e Engler and Grogan, 1983
f Fukui, 1983
g Orton, (Submitted).
h Evans and Sharp, 1983

due to sampling errors or non-randomness of the mutation process in vitro. An earlier review (Orton, 1983b), proposed that the actual 'mistake rate' which generates diversity might be the same in plant tissues in vivo and in vitro, and that observed differences in manifestations of variation were a consequence of selective fitness. If this is true, it implies that cells carrying these recessive mutations enjoy a tremendous selective advantage over normal cells in a given culture. Alternatively, if such mutations do not impart a selective advantage, mutation rates on the order of 10^{-2} to 10^{1} must be postulated. For example, in the study of rice reported by Fukui (1983), 9 of 12 (75%) regenerated plants exhibited the early heading mutation. Assuming that these plants represented a random sample of genotypes in the culture, a point mutation on the first replication cycle should have yielded 75% normal and 25% mutant cells and corresponding clones and regenerated plants. Alternatively, perhaps mutant clones arose from pre-existing mutations in the embryo. It is clear from the studies of Barbier and Dulieu (1980) and Lorz and Scowcroft (1983) that, while a small fraction of the observed mutations were pre-existing, most arose during the culture phase. If these observations can be generalized across a broad spectrum of species, explants, and culture techniques, the most logical conclusion is that mutations occur at an extremely high rate some time during growth in vitro.

Most of the mutations analyzed thusfar behave recessively, implying loss of function (Table 1). Such mutations could conceivably result from point mutations, partial or complete deletions, or insertions within the coding or control regions of the gene. It is impossible to distinguish among these possibilities based on information available at present. Further, while it is interesting to note parallels between the possible activities of transposable elements in generating karyological (discussed earlier) and Mendelian variation, all of the studies reporting on Mendelian variation show that regenerated mutant plants are karyologically normal. Thus, it is unlikely that these recessive mutations could have been caused by insertions of transposable elements stimulated by a breakage-fusion-bridge cycle. Again, perhaps the culture state itself calls up the stress response which stimulates the movement of such elements. A mutation affecting flower color in alfalfa (Medicago sativa) has been recovered from cell cultures which is highly unstable both in vitro and in vivo, suggesting in involvement of a transposable element (E. T. Bingham, pers. comm.). Cullis (1983) has reported the instability of a transposable-like element in callus cultures of a 'plastic' flax (Linum usitatissimum) cv. 'Stormont Cirrus' variety concomitantly with an increase of certain repeated DNA fractions. While it is tempting to elevate these results to generality, it must be recognized that this particular

flax variety has been shown to be genetically unstable when
exposed to certain environmental stimuli, and hence may not be
strictly representative.

Orton (1983a) analyzed experimental approaches which could
hopefully yield more fundamental information regarding the mechanisms underlying somaclonal variation. There exists a clear need
to develop molecular probes as markers for the direct visualiza-
tion of mutations, such as those advanced by Rivin et al. (1983).
With molecular probes of sequences having known function (e.g.,
Adh and Ds of maize; Peacock, 1983), experiments to elucidate the
basis and significance of somaclonal variation will progress by
orders of magnitude.

Practical Applications

The potential of somaclonal variation for crop improvement
has been suggested frequently (Heinz and Mee, 1971; Nickell and
Heinz, 1973; Liu and Chen, 1976; Skirvin, 1978; Shepard et al.,
1980; Larkin and Scowcroft, 1981). From the standpoint of utility, knowledge of fundamental underlying mechanisms and genetic
vs. epigenetic origins of somaclonal variation are not
necessarily critical. Hence, emphasis in this section will be
placed on sources of variation emerging directly from somatic
cells or tissues, that have impacted or may impact economically
on the crop. Prominent among these are vegetatively propagated
crops which often defy classical breeding and genetic approaches.
Crops in which significant or notable attempts have been made to
generate and utilize somaclonal variation for crop improvement
will be considered separately.

Potato

Few world crops rival the potato, Solanum tuberosum, as a
target for improvement via somaclonal variation. First, the crop
is normally propagated and established vegetatively from tuber
pieces. Second, some of the best cultivars presently in use are
sexually impotent, and breeding of such lines proceeds entirely
by clonal selection. Where classical breeding approaches are
used, the enormous levels of heterozygosity encountered and
inbreeding depression make the selection of superior lines a difficult task as compared to other crops. The potato is also considered to be a autotetraploid, which further complicates
inheritance studies and breeding.

Independent reports of successful regeneration of whole
potato plants from leaf mesophyll protoplast-derived calli first
appeared in the mid 1970's (Behnke, 1976; Shepard and Totten,
1977). Since that time, Shepard and his colleagues have invested

a major effort into the manipulation of protoplasts for crop improvement. Early results (Matern et al., 1978) showed that variation existed among callus clones derived from protoplasts, and could be transmitted into corresponding regenerated plants. Large scale field studies were initiated for purposes of determining the types and degrees of variation observed among regenerated potato cv. 'Russet Burbank' 'protoclones' and to select and evaluate any of anticipated economic significance (summarized in Shepard et al., 1980; Secor and Shepard, 1981; Bidney and Shepard, 1981; Shepard, 1982). Furthermore, the efficiency of the system has been improved such that approximately 50% of protoplast-derived calli can now be successfully regenerated. In experiments started in 1977, 1,700 protoclones were evaluated in a Montana field and 396 were selected for further evaluation. This original population exhibited exceedingly high levels of phenotypic variation as compared to a 'Russet Burbank' population established from tuber pieces, and only the most promising or strikingly altered lines, were selected. In 1978, the population was narrowed further to 63 protoclones in Idaho, and were subsequently evaluated in North Dakota and Colorado. Using 26 tuber, foliar, and floral characteristics, all 63 clones were determined to differ significantly from 'Russet Burbank', and all but three clones varied in more than one character; one clone differed significantly in seventeen characters (Secor and Shepard, 1981). Clones are generally stable through vegetative propagation, and 10 lines appear to perform consistently better in replicated trials than the parent variety. One of these will be advanced to regional trials in 1984 for potential release. This clone is significantly earlier, more blocky, higher setting, and more uniform than 'Russet Burbank' (Secor, pers. comm.). The potential of this approach has spawned a number of independent efforts worldwide which are pursuing somaclonal variation in potato in regional breeding programs.

One conspicuous component of the spectrum of somaclonal variants seems to be resistance or enhanced tolerance to pathogens as compared to the derivative cultivar. For example, 'Russet Burbank' is susceptible to Alternaria solani and several races of Phytophthora infestans. However, 1.0% of calli derived from 'Russet Burbank' protoplasts were resistant to the A. solani pathotoxin, and all but one of these clones also exhibited resistance to the pathogen as whole plants. Of 800 protoclones tested, 2.5% were resistant to P. infestans race 0, and showed some tolerance to other races as well. All of these resistances appeared stable through normal vegetative propagation. Similar variation has been observed using other potato cultivars (Thomas et al., 1982), and indications of enhanced disease resistance have also been observed among androgenic plants (Behnke, 1980; Wenzel, and Uhrig, 1981).

Van Harten et al. (1981) noted a significant degree of variation emerging in populations of plants regenerated via adventitious meristem culture techniques. Hence, it would appear that variation is a common feature of cultured potato tissues across a broad range of developmental contexts and genotypes. However, the observation by Wenzel et al. (1979) that plants regenerated from protoplasts of dihaploid plants failed to exhibit striking variation (using techniques similar to others) is circumstantial evidence that mutations may occur and accumulate over time in tuber-propagated clones (Brettell and Ingram, 1979; Orton, 1983b). However, variation was observed as the age of these callus cultures was increased. Thus, the culture and regeneration of potato cells and tissues offers an extremely powerful breeding tool, and the economic impact of the technique in the foreseeable future is likely to be significant. The long-term impact of recent efforts to develop sexual breeding techniques which integrate haploidization, diploidization, selection for combining ability, and release and sale of controlled hybrid seed or propagules generated from fused dihaploid protoplasts is likely to be even greater.

SUGARCANE

Like the potato, sugarcane (Saccharium officinarum) is a vegetatively propagated polyploid crop in which breeding has proceeded largely by clonal selection and genetic analysis is virtually impossible. Despite early observations of the general recalcitrance of monocots to culture and regeneration, sugarcane was among the first crop species to be successfully regenerated (Heinz et al., 1969; Heinz and Mee, 1971). The most striking feature of those reports was indeed the tremendous degree of variation observed among regenerated plants, much of which was found to be stable through tiller propagation. Characters affected were general morphology, yield components, and disease resistance (Heinz et al., 1969; Heinz and Mee, 1971; Liu and Chen, 1976). These have been substantially reviewed up to quite recently (Nickell and Heinz, 1973; Heinz et al., 1977; Liu, 1981; Larkin and Scowcroft, 1981, 1983). Propagants from variant somaclones exhibiting enhanced yield and disease resistance are already being used for commercial sugar production, and the approach of generating and capturing somaclonal variation for sugarcane improvement will probably be the method of choice for a long time to come.

WHEAT

It is argued by the breeders of many major sexually-propagated crops that sufficient genetic variation exists in present germplasm collections to permit significant breeding gains well into the future. What is needed are techniques by which existing variation can be transmitted and/or selected more efficiently. Enormous stores of wheat germplasm are maintained worldwide. Why turn to the generation of presumably random mutations generated in culture as an alternative to already preexisting variation? First, there is an immediate advantage in the addition of single, beneficial mutations to existing superior varieties thus bypassing the need for lengthy backcrossing of existing mutations. Second, it appears that at least part of the spectrum of stable variation produced is inherited in a complex fashion. If beneficial true-breeding variants for characters such as yield, uniformity, and quality can be obtained in superior varieties directly, an alternative to the pedigree method may emerge. Finally, there exists a distinct possibility that benefical mutations that do not exist in the world germplasm collection will surface. Until basic research can illuminate the distinctions between somaclonal variants and those generated in mutation breeding approaches and perhaps determine ways in which its manifestations can be controlled, it would appear worthwhile to conduct empirical experiments on somaclonal variation for crop improvement in species with well established, effective breeding techniques such as wheat.

Ahloowalia (1982) first reported the existence of striking morphological and reproductive variation in hexaploid wheat, but did not determine the inheritance of variation. Unlike vegetatively propagated crops, it is essential to demonstrate a stable genetic basis for variation in seed-propagated crops. Extensive inheritance studies of somaclonal variation from wheat callus cultures were discussed in an earlier section (Larkin et al., 1983). Among the array of heritable variants obtained were several of potential economic significance. For example, white grain color mutations, such as those observed by Larkin et al. (1983), would be useful in situations where a red variety was desirable in all aspects except that the market called for a white wheat. Tiller number is a conspicuous component of yield, and it would be interesting to see how other components are altered among somaclonal variants for this character. Reduced height is desirable inasmuch as lodging resistance may be missing from an otherwise promising line. The observation of qualitative variation in gliadin protein composition suggests the potential for altering the baking and nutritional characteristics of wheat.

The finding that some of the heritable variation observed in wheat regenerates is potentially beneficial is certainly promising. However, an extensive program will be necessary generate and test somaclonal variants before its ultimate impact in wheat improvement can be judged.

RICE

Rice, a diploid monocot of tremendous world importance, is relatively easy to regenerate from cell cultures as compared to it's Gramineae relatives. The early experiments of Oono (1978, 1981) which provided a genetic characterization of variants arising from cell culture, have been described in an earlier section. Among the potentially useful mutations in those studies were enhanced yield components (grain number and weight, panicle number), plant height, heading date, and salt tolerance.

Following similar procedures and using a broader base of germplasm, Suenaga et al. (1982) also succeeded in obtaining a significant array of mutations among plants regenerated from rice callus cultures. Approximately 54% of regenerated plants were deduced to have contained one or more new mutations, most of which were undesirable (polyploidy, chlorophyll deficiencies, reduced fertility, and necrotic leaf spotting). Certain mutations, such as new dwarfisms and altered heading dates, may ultimately be useful, but further tests will be necessary.

Likewise, Zhao et al. (1982) regenerated plants from callus cultures of 23 rice varieties and found high levels of variation among their selfed progenies. Mutations were obtained which affected foliar morphology, floral morphology, grain shape, heading date, and fertility. Most characters were found to be fixed following selection among selfed progeny of regenerated plants, and remained stable through two more cycles of selfing. A number of lines with good characters were noted, and these are apparently being integrated into ongoing rice pedigree and backcross breeding programs. The further interesting observation was made that variation tended to be greater in upland than in lowland types.

The practical contributions of Fukui's (1983) study, described earlier, were: 1) desirable mutations (e.g., early heading and short culm) can be obtained after a very short time in vitro, and 2) lengthening the period of time may permit the accumulation of mutations in the same clone.

Anther culture has been pursued extensively as a means to obtain homozygous from heterozygous (e.g., F_1 hybrid) rice plants, thus reducing the period from original hybridization to final testing and release (Zhen-Hua, 1982). As of 1980, it was estimated that roughly 100,000 ha in China were used to grow rice varieties developed via anther culture. (Hu Han, pers. comm.). It is generally presumed that all variation observed among dihaploids obtained from microspores is a reflection of the assortment of alleles contributed by the original parents of hybrid. However, striking new variants are sometimes observed –

the most conspicuous example being albinism. Up to 90% of
plantlets regenerated from rice microspores may be albino
(Chih-Ching, 1982). Whether this is due to new mutations or
incompetancy of proplastids to proliferate in microspores is not
yet known. Schaeffer (1982) has obtained evidence of other mutations other than albinism among androgenic rice plants. Although
it is impossible to determine the precise frequency of mutational
events, several independent cases of dwarf mutations were noted.
The dwarf mutations were associated in all cases with a reduction
in leaf width. Other inherited characters observed among dwarf
plants were reduced number of grains per plant, reduced awn
length, reduced seed weight, increased tillering, and the tendency for open hulls. These characters were observed among
androgenic plants from a recognized pure line, cv. 'Calrose 76'.
Similar mutations arising among dihaploids regenerated from
microspores of multiple heterozygotes would certainly have been
lost in the sea of segregants. Hence, it appears that new mutations arising during the culture process may contribute to the
pool of genetic variation, both desirable and undesirable,
observed among androgenic rice plants produced for breeding
purposes.

The dihaploid approach would appear to have the advantages
of both generating pure lines and new mutations for direct selection very rapidly. Unfortunately, the yields of haploids from
rice are characteristically low (10^{-4}) and genotype-specific.
Yet haploid derivatives are already reported to be widely used in
China. Improvements in efficiency and generality of protocols
will make haploid breeding approaches an extremely attactive
alternative to the commonly used pedigree methods.

CORN

The procedure of choice for culturing corn somatic cells is
to derive organogenic (and now embryogenic) callus cultures from
immature embryo scutellum tissue. Reports have suggested that
this method yields genotypic fidelity (Green et al., 1977; McCoy
and Phillips, 1983). However, recent studies clearly demonstrate
the existence of new mutations among maize plants regenerated
using similar procedures. While it can be speculated that the
reasons for these opposing findings lie in subtle differences in
procedure or genotype, the paradox does provide some hope that
variation, at least in maize, might be controllable.

The Edallo et al. (1981) study, discussed earlier established that mutation frequencies among regenerated plants from
scutellum-derived callus cultures of inbred W64A and S65 were
remarkably high (approximately 1.0 unique simple mutation per
regenerated plant). However, among 110 regenerated plants, no

obviously desirable mutations were recovered. More recently, a
report by Beckert et al. (1982) has confirmed in maize what has
been suspected of plants regenerated from cultured cells in
general: that quantitatively inherited as well as simply
inherited characters are affected. Later, the same group
(Beckert et al., 1983) conducted an analysis of quantitative and
qualitative variants arising among plants regenerated from
scutellum-derived callus cultures of several maize inbreds.
Prominent among simply inherited mutations were dwarfs and
chlorophyll deficiencies. No variants were considered to be
obviously desirable and the effective use of existing genetic
variability was judged to be an approach preferable over the
blind recovery of variation from cell cultures.

The peculiar observation of T-cms/ H. maydis race T susceptibility instability in cultured tissues has provided a model
example of the potential of tissue culture in altering maternally
inherited characters (discussed in a previous section). The
extent to which this represents a general phenomenon is unknown.

ALFALFA

Reisch and Bingham (1981) conducted a phenotypic analysis of
plants regenerated from alfalfa callus cultures with and without
exposure to the mutagen ethyl methane sulfonate (EMS) and the
selective agent ethionine (a methionine analogue which inhibits
feedback-controlled methionine biosynthesis). A number of
morphological characters, including some affecting yield, were
obtained. Curiously, the frequency of variants was clearly
enhanced by ethionine selection, an observation which prompted
the authors to suggest that ethionine is mutagenic.
Alternatively, the compound could be acting as a selective agent
for variant cells in general, as suggested by Larkin and
Scowcroft (1983). However, a number of variants were obtained,
including the highest yielding clone, in the absence of EMS and
ethionine. More recent inheritance studies have revealed that
most of these, including a leafy/branched mutant which was speculated to enhance effective forage yield, have a simple genetic
basis (E. T. Bingham, pers. comm.). In an independent project,
non-mutagenized Regen-S callus cultures were exposed to toxic
culture filtrates of pathogenic Fusarium oxysporum suspensions.
Plants regenerated from resistant colonies were found to be also
resistant to the pathogen, and resistance segregates as if a
simple dominant mutation in crosses to susceptible alfalfa lines
(C. Hartman, pers. comm.).

It appears that potentially useful mutations occur spontaneously in cultured alfalfa tissues. Genetics and beeding of
alfalfa are stymied by self-incompatibility, inbreeding

depression, and polyploidy. Hence, the recovery of desirable
mutations in otherwise popular cultivars may present an attrative
alternative to improvement via recurrent backcrossing or hybridization and recurrent mass selection.

TOBACCO

Tobacco (and various Solanaceous relatives) remains as the
undisputed model system of choice in plant somatic cell manipulation. It is also one of the major species in which high levels
of somaclonal variation have been reported (reviewed by Larkin
and Scowcroft, 1981, 1983). Variation has been observed among
plants regenerated from a broad range of callus tissues (from
microspores, pith, leaf mesophyll). Characters for which variant
phenotypes have been observed include flowering date, leaf shape,
height, total vegetative yield, alkaloid content, sugar content,
chlorophyll content, CO_2, adsorption rate, disease-resistance,
and male sterility (Mousseau, 1970; Devreux and Laneri, 1974;
Oinuma and Yoshida, 1974; Burk and Matzinger, 1976; Popchristov
and Zaganska, 1977; Chaleff and Parsons, 1978; Burk et al., 1979;
Collins and Legg, 1980; Chaleff and Keil, 1981; De Paepe et al.,
1981; Deaton et al., 1982; Prat, 1983). Recently, a line developed from an androgenic tobacco plant containing a new mutation
conferring tobacco hornworm resistance has been released (Miles
et al., 1981). In vitro techniques are now routinely used in
direct association with established tobacco breeding programs,
and it seems reasonable to speculate that somaclonal variation
will progressively gain importance as a source of useful mutations in adapted, commercial backgrounds.

LETTUCE

Lettuce was among the first crop species in which somaclonal
variation was observed and characterized. Sibi (1976) noted
three qualitative (flat, glaucous leaves; yellow leaves; pale
midveins) and numerous quantitative variants among a population
of plants regenerated from cotyledon-derived callus cultures of
the cv. "Val d'Orge". The 'flat, glaucous leaves' variant was
shown to be maternally inherited, while the other two qualitative
characters diminished in intensity over selfed generations, and
could be environmentally corrected. Interestingly, it was
observed that hybrids between sexual descendants of regenerates
and seed-propagated "Val d'Orge", or among sexual descendants of
regenerates, exhibited hybrid vigor in the form of increased leaf
area as compared to self-pollinated progenies. This heterotic
effect was observed irregardless of which plant was used as
female parent. Sibi hypothesized that this hybrid vigor was a
consequence of the mixture of wild type and culture-induced

mutant organelles during gamete fusion, resulting in zygotes having a new, stable heterogeneous organelle genotypes. However, the genetic studies to verify this hypothesis were not reported. Subsequently, Y. Demarly and M. Sibi filed a patent application pertaining to the use of plant tissue culture as a tool for obtaining useful germplasm (No. 4003156 issued 18 Jan 1977).

Among nine variant characters observed among selfed progeny of lettuce plants regenerated from mesophyll protoplast-derived calli of the cv. "Climax", seven were conditioned by nuclear recessive mutations (Engler and Grogan, 1983). All of these variants were considered as horticulturally undesirable. The remaining two variants, increased and decreased seedling size, did not segregate upon selfing, and could therefore have been caused by homozygous nuclear or cytoplasmic mutations. The increased seedling size mutation was considered as potentially desirable from the standpoint of rapid stand establishment, although the effect was diminished during later developmental stages. No crosses were conducted to determine whether regenerated plants contained new mutations which might contribute to hybrid vigor, as described by Sibi (1976).

CONCLUDING REMARKS - PRACTICAL CONSIDERATIONS

Table 2 summarizes the examples of useful spontaneous variation from cell cultures discussed above and other additional cases. This list is not exhaustive, and may not reflect the action of comparable processes since a multiplicity of unrelated species, tissue sources, and culture methods were used. The first hints that useful variation arises spontaneously among cultured plant cells emerged in the late 1960's in the pioneering work of Heinz et al. (1969) with sugarcane. Until the early 1980's, such variation was considered by many to be artifactual and probably unstable. Since then, evidence has quickly mounted that this source of genetic variation, now commonly known as somaclonal (Larkin and Scowcroft, 1981), is stable, has high potential utility, and probably reflects natural mechanisms which maintain and generate variation in somatic tissues.

Tissue Cloning

One of the most obvious immediate applications of cell and tissue culture capabilities is to amplify single plants into large identical populations. Since culture techniques provide the possibility of working with entities as small as single cells, they are potentially much more efficient than conventional methods of asexual propagation (Murashige, 1974; Lawrence, 1981). In breeding and seed production, this technique theoretically permits the production of unlimited numbers of individuals which

Table 2. Crop species in which research has indicated the possible use of somaclonal variation for crop improvement

Crop	Normal mode of propagation	Ploidy level	Potentially useful somaclonal variants/mutations	References
Potato (Solanum tuberosum)	Vegetative (tuber)	4x (autotetraploid)	Yield, quality, uniformity, disease resistance	Matern et al., 1978 Secor and Shepard, 1981 Shepard et al., 1980 Shepard, 1982
Sugarcane (Saccharum officinarum)	Vegetative (tiller)	approx. 16x	Yield, sugar content, disease resistance	Heinz et al., 1969 Heinz and Mee, 1971 Liu and Chen, 1976 Larkin and Scowcroft, 1981, 1983
Wheat (Triticum aestivum)	Seed	6x (allohexaploid)	Grain color, height, tiller number, seed storage proteins	Ahloowalia, 1982 Larkin et al., 1983
Rice (Oryza sativa)	Seed	2x	Yield components, height, heading date, salt tolerance	Oono, 1978 Suenaga et al., 1982 Zhao et al., 1982 Schaeffer, 1982
Maize (Zea mays)	Seed	2x	Height, disease resistance, reversion to male fertility	Pring et al., 1981 Beckert et al., 1983
Lettuce (Lactuca sativa)	Seed	2x	Seedling vigor, heterosis	Engler and Grogan, 1983 Sibi, 1979

(continued)

Table 2. Crop species in which research has indicated the possible use of somaclonal variation for crop improvement (Continued)

Crop	Normal mode of propagation	Ploidy level	Potentially useful somaclonal variants/mutations	References
Alfalfa (Medicago sativa)	Seed	4x (autotetraploid)	Yield, disease resistance	Reisch and Bingham, 1981 C. Hartman (personal communication)
Tobacco (Nicotiana tabacum) and relatives	Seed	4x (allotetraploid) or 2x (diploid relatives)	Flowering rate, height, yield, alkaloid content, disease resistance, male sterility, insect resistance, herbicide resistance	Mousseau, 1970 Devreux and Lancri, 1974 Oinuma and Yoshida, 1974 Burk and Matzinger, 1976 Popchristov and Zaganska, 1977 Burk et al., 1979 Collins and Legg, 1980 Chaleff and Parsons, 1978 DePaepe et al., 1981 Deaton et al., 1982 Prat, 1983
Oats (Avena sativa)	Seed	4x (allotetraploid)	Height, heading date	Cummings et al., 1976
Rape (Brassica napus)	Seed	4x (allotetraploid)	Glucosinolate content	Hoffman, 1978 Wenzel, 1980
Onion (Allium cepa)	Seed	2x	Bulb size, shape	Novak, 1980

Crop	Normal mode of propagation	Ploidy level	Potentially useful somaclonal variants/mutations	References
Barley (Hordeum vulgare)	Seed	2x	Haploidy, introgression	Orton, 1980b
Geranium (Pelargonium hortorum)	Vegetative (cutting)	2x	Morphological characteristic (floral, vegetative), essential oil composition, anthocyanin pigmentation	Abo El-Nil and Hildebrandt, 1973 Skirvin and Janick, 1976
Celery (Apium graveolens)	Seed	2x	Disease resistance	Rappaport et al., 1982 Murakishi, 1982
Tomato (Lycopersicon esculentum)	Seed	2x	Jointless pedicals, male sterility	Evans and Sharp, 1983
Orange (Citrus sp.)	Cutting	2x	Salt tolerance	Kochba et al., 1982

are normally unstable through seed propagation (e.g., heterozygotes). Where adequate efficiencies of multiplication can be
obtained, it becomes possible to use these clonal micropropagation techniques for the production of the desired end products,
for example potted ornamentals or woody perennial or vegetable
transplants for direct field establishment. In long-term clonal
perennials, the technique has the added potential benefit of
purging tissues of resident pathogenic virus' if used in conjunction with meristem culture and heat therapy (Hollings, 1965).
Yet the observations described in this and numerous other reviews
would suggest that enhanced frequency of mutations is an inherent
feature of cell and tissue culture. Indeed, stable variants
arise at a detectable frequency even during the course of conventional vegetative propagation (e.g., 'sports' in woody perennials
Skirvin, 1978 and clonal variants in potato, J. Pavek, pers.
comm.), and these can, perhaps, also be included in the broad
definition of somaclonal variation. The clear and general trend
is that fidelity of genotype preservation decreases proportion-
ately with the complexity of the basic unit of propagation
(e.g., cutting ➤ axilary meristerm ➤ adventitious meristem ➤
unorganized tissues and cells; Lawrence, 1981). This could be a
consequence of the simple assortment of variant cells from
complex chimeral mixtures or selection for 'normal' genotypes in
differentiated vs. unorganized tissues (Orton, 1983b).

Spontaneously occurring variation clearly limits the use of
in vitro micropropagation techniques in agriculture. Despite
this, numerous micropropagation programs are operational in
United States ornamental and seed companies. It is hoped that a
greater understanding of the mechanisms responsible for somaclonal variation made possible by the advent of new molecular
technologies will suggest methods by which variation can be
controlled or eliminated during in vitro micropropagation. In
the meantime, those applying the technique will continue to use
empirically-derived methods to avoid variation, such as the use
of organized propagules and exotic culture/transfer regimes.

CONCLUDING REMARKS

The prevailing view of cell and tissue culture has been,
from the early 1970's until recently, that multicellular organisms could, given sufficient system development, be made to
behave at least transiently like microorganisms. This belief
fostered visions of in vitro mutation induction, selection, and
characterization that would usher in a new era of understanding
of the structure, function and developmental coordination of
higher plant genes. The weight of circumstantial evidence has
pointed to the naivete of this simplistic view. For example,

higher plants have at least two, and frequently more copies of homologous genes. Nuclear (and possibly cytoplasmic) genes are under specific controls such that their expression is not constitutive but fluctuates under different external environments and intrinsic developmental states. Populations of higher plant cells frequently exhibit properties which are not strictly the additive expections derived from constituent cells: e.g., Latent differentiation, critical plating densities, and cross-feeding. Perhaps the most striking departure from microbial culture principles is the observation of high frequencies of mutations in populations of cultured cells from higher plants. The environment of the test tube was developed to optimize the proliferation of microbial cultures. However, this environment is not natural to higher plant cells, and either permits the persistence of naturally-occurring variant cells, which would normally not survive, or actually causes (directly or indirectly) the variation to appear. New molecular techniques will hopefully illuminate the basic mechanisms responsible for the high frequency of mutations in cultured cells of higher plants, suggest methods by which it can be controlled for the benefit of mankind, and provide further evidence on the role of spontaneous somatic mutations in the generation of biological diversity.

ACKNOWLEDGEMENTS

I wish to express my appreciation to Drs. R. H. Lawrence and D. E. Engler for their critical review of the manuscript, E. T. Bingham, C. Hartman, D. E. Engler, G. A. Secor, and P. J. Larkin for information regarding research in progress or not yet published, Lynne Foote for excellent information services, and M. Cumpton and J. Kirk for excellent typing.

LITERATURE CITED

Abo El-Nil, M. M., and Hildebrandt, A. C., 1972, Morphological changes in geranium plants differentiated from anther cultures, In Vitro, 7:258.

Ahloowalia, B. S., 1982, Plant regeneration from callus culture in wheat, Crop Sci., 22:405-410.

Arntzen, C. J., Pfister, K., and Steinback, K. E., 1982, The mechanism of chloroplast triazine resistance: alterations in the site of herbicide action, in: "Herbicide Resistance in Plants," H. M. LeBaron and J. Gressel, ed., John Wiley & Sons, New York, pp. 185-214.

Ashmore, S. E., and Gould, A. R., 1981, Karyotype evolution in a tumour-derived plant tissue culture analyzed by Giemsa C-banding, Protoplasma, 106:297-308.

Bajaj, Y. P. S., Ram, H. K., Labana, K. S., and Singh, H., 1981, Regeneration of genetically variable plants from the anther derived callus of Arachis hypogaea and Arachis villosa, Plant. Sci. L., 23:35-40.

Balzan, R., 1978, Karyotype instability in tissue cultures derived from the mesocotyl of Zea mays seedling, Caryologia, 31:75-87.

Banks-Izen, M. S., and Polito, V. S., 1980, Change in ploidy level in calluses derived from two growth phases of Hedera helix L., the English ivy, Plant Sci. L., 18:161-167.

Barbier, M., and Dulieu, H. L., 1980, Effets genetiques observes sur des plantes de tabac regenerees a partir de cotyledons par culture in vitro, Ann. Amelior. Plantes, 30:321-344.

Bayliss, M. W., 1973, Origin of chromosome number variation in cultured plant cells, Nature, 246:529-530.

Bayliss, M. W., 1975, The effects of growth in vitro on the chromosome complement of Daucus carota L. suspension cultures, Chromosoma, 51:401-411.

Bayliss, M. W., 1980, Chromosomal variation in plant tissues in culture, Int. Rev. Cytol. Suppl., 11A:113-144.

Beckert, M., Pollacsek, M., and Cao, M. Q., 1982, Analysis of genetic variability in regenerated inbred lines, Maize Genet. Coop. Newsl., 56:38-39.

Beckert, M., Pollacsek, M., and Caenen, M., 1983, Etude de la variabilite genetique obtenue chez le mais apres callogenes et regeneration de plantes in vitro, Agronomie, 3:9-17.

Behnke, M., 1976, Kulturen isolierter von einigen dehaploiden Solanum-tuberosum-Klonen und ihre regeneration, Z. Pflanzenphysiol, 78:177-182.

Behnke, M., 1980, General resistance to late blight of Solanum tuberosum plants regenerated from callus resistant to culture filtrates of Phytophthora infestans, Theor. Appl. Genet., 56:151-152.

Belliard, G., Vedel, F., and Pelletier, G., 1979, Mitochondrial recombination in cytoplasmic hybrids of Nicotiana tabacum by protoplast fusion, Nature, 281:401-403.

Bennici, A., Buiatti, M., and D'Amato, F., 1968, Nuclear conditions in haploid Pelargonium in vivo and in vitro, Chromosoma, 24:194-201.

Bidney, D. L., and Shepard, J. F., 1981, Phenotypic variation in plants regenerated from protoplasts: the potato system, Biotechnol. and Bioeng., 23:2691-2701.

Boeshore, M. L., Lifshitz, I., Hanson, M. R., and Izhar, S., 1983, Novel composition of mitochondrial genomes in Petunia somatic hybrids derived from cytoplasmic male sterile and fertile plants, Molec. Gen. Genet., 190:459-467.

Brettell, R. I. S., and Ingram, D. S., 1979, Tissue culture in the production of novel disease-resistant crop plants, Biol. Rev., 54:329-345.

Brettell, R. I. S., Conde, M. F., and Pring, D. R., 1982, Analysis
    of mitochondrial DNA from four fertile maize lines obtained
    from a tissue culture carrying Texas cytoplasm, Maize Genet.
    Coop. Newsl., 56:13-14.
Browers, M. A., and Orton, T. J., 1982a, A factorial study of
    chromosomal variability in callus cultures of celery (Apium
    graveolens), Plant Sci. L., 26:65-73.
Browers, M. A., and Orton, T. J., 1982b, Transmission of gloss
    chromosomal variability from suspension cultures into regener-
    ated celery plants, J. Hered., 73:159-162.
Burk, L. G., and Matzinger, D. F., 1976, Variation among anther-
    derived doubled haploids from an inbred line of tobacco, J.
    Hered, 67:381-384.
Burk, L. G., Chaplin, J. F., Gooding, G. V., and Powell, N. T.,
    1979, Quantity production of anther-derived haploids from a
    multiple disease resistant tobacco hybrid I. Frequency of
    plants with resistance or susceptibility to tobacco mosaic
    virus (TMV), potato virus (PVY), and root knot (RK),
    Euphytica, 28:201-208.
Chaleff, R. S., 1981, "Genetics of High Plants - Applications
    of Cell Culture," Cambridge Univ. Press, Cambridge, pp. 24-40.
Chaleff, R. S., 1983, Isolation of agronomically useful mutants
    from plant cell cultures, Science, 214:676-682.
Chaleff, R. S., and Keil, R. L., 1981, Genetic and physiological
    variability among cultured cells and regenerated plants of
    Nicociana tabacum, Molec. Gen. Genet., 181:254-258.
Chaleff, R. S., and Parsons, M. F., 1978, Direct selection in
    vitro for herbicide resistant mutants of Nicotiana tabacum,
    Proc. Natl. Acad. Sci. U.S.A., 75:5104-5107.
Chih-Ching, C., 1982, Anther culture of rice and its signifi-
    cance in distant hybridization, in: "Rice Tissue Culture
    Planning Conference," Intl. Rice Res. Inst., Los Banos,
    Laguna, Phillipines, pp. 47-54.
Collins, G. B., and Legg, P. D., 1980, Recent advances in the
    genetic applications of haploidy in Nicotiana, in: "The
    Plant Genome," D. R. Davies, and D. A. Hopwood, ed., John
    Innes Charity, Norwich, England, pp. 197-213.
Constantin, M. J., 1981, Chromosome instability in cell and
    tissue cultures and regenerated plants, Env. Exptl. Bot.,
    21:359-368.
Cullis, C. A., 1983, Environmentally induced DNA changes in
    plants, CRC Critical Rev. in Plant Sci., 1:117-131.
Cummings, D. P., Green, C. E., and Stuthman D. D., 1976, Callus
    induction and plant regeneration in oats, Crop Sci., 16:465-470.
D'Amato, F., 1952, Polyploidy in the differentiation and func-
    tion of tissues and cells in plants, A critical examination
    of the literature, Caryologia, 4:311-357.

D'Amato, F., 1975, The problem of genetic stability in plant tissue and cell cultures, in: "Crop Genetic Resources for Today and Tomorrow," O. Frankel and J. G. Hawkes, ed., Cambridge Univ. Press, Cambridge, pp. 333-348.

D'Amato, F., 1977, Cytogenetics of differentiation in tissue and cell cultures, in: "Plant Cell, Tissue, and Organ Culture," J. Reinert and Y. P. S. Bajaj, ed., Springer-Verlag, Berlin, pp. 343-357.

Deaton, W. R., Legg, P. D., and Collins, G. B., 1982, A comparison of burley tobacco doubled-haploid lines with their source inbred cultivars, Theor. Appl. Genet., 62:69-74.

Demoise, C. F., and Partanen, C. R., 1969, Effects of subculturing and physical condition of medium on the nuclear behavior of a plant tissue culture, Am. J. Bot., 56:147-152.

DePaepe, R., Bleton, E., and Gnangbe, F., 1981, Basis and extent of genetic variability among doubled haploid plants obtained by pollen culture in Nicotiana sylvestris, Theor. Appl. Genet., 59:177-184.

DePaepe, R., Prat, D., and Huguet, T., 1983, Heritable nuclear DNA changes in doubled haploid plants obtained by pollen culture of Nicotiana sylvestris, Plant Sci. L., 28:11-28.

Devreux, M., and Laneri, V., 1974, Anther culture, haploid plants, isogemic line, and breeding research in Nicotiana tabacum L., in: "Polyploidy and Induced Mutations in Plant Breeding," IAEA, Vienna, pp. 101-107.

Dhillon, S. S., Wernsman, E. A., and Miksche, J. P., 1983, Evaluation of nuclear DNA content and heterochromation changes in anther-derived dihaploids of tobacco (Nicotiana tabacum) cv. 'Coker 139', Can. J. Genet. Cytol., 25:169-173.

Dixon, L. K., Leaver, C. J., Brettell, R. I. S., and Gengenbach, B. G., 1982, Mitochondrial sensitivity to Dreschlera maydis T-toxin and the synthesis of a variant mitochondrial polypeptide in plants derived from maize tissue cultures with Texas male-sterile cytoplasm, Theor. Appl. Genet., 63:75-80.

Durante, M., Geri, C., Grisvard, J., Guille, E., Parenti, R., and Buiatti, M., 1983, Variation in DNA complexity in Nicotiana glauca tissue cultures, I. Pith tissue dedifferentiation in vitro, Protoplasma, 114:114-118.

Edallo, S., Zucchinali, C., Perenzin, M., and Salamini, F., 1981, Chromosomal variation and frequency of spontaneous mutation associated with in vitro culture and plant regeneration in maize, Maydica, 26:39-56.

Edelman, M., Hallick, R. B., and Chua, N. H., 1982, "Methods in Chloroplast Molecular Biology," Elsevier, Amsterdam, pp. 1140.

Engler, D. E., and Grogan, R. G., 1983, Variation in lettuce plants regenerated from protoplasts, J. Hered, (Submitted).

Evans, D. A., and Sharp, W. R., 1983, Single gene mutations in tomato plants regenerated from tissue culture, Science, 221:949-951.

Farrelly, F., and Butow, R. A., 1983, Rearranged mitochondrial
    genes in the yeast nuclear genome, Nature, 301:296-301.

Fukui, K., 1983, Sequential occurrence of mutations in a growing
    rice callus, Theor. Appl. Genet., 65:225-230.

Green, C. E., Phillips, R. L., and Wang, A. S., 1977, Cytological
    analysis of plants regenerated from maize tissue cultures,
    Maize Genet. Coop. Newsl., 51:53-54.

Heinz, D. J., Mee, G. W. P., and Nickell, L. G., 1969, Chromosome
    numbers of some Saccharum species hybrids and their cell
    suspension cultures, Am. J. Bot., 56:450-456.

Heinz, D. J., and Mee, G. W. P., 1971, Morphologic, cytogenetic,
    and enzymatic variation in Saccharum species hybrid clones
    derived from callus tissue, Am. J. Bot., 58:257-262.

Heinz, D. J., Krishnamurthi, M., Nickell, L. G., and Maretzki, A.,
    1977, Cell tissue, and organ culture in sugarcane improvement,
    in: "Plant Cell, Tissue, and Organ Culture," J. Reinert and
    Y. P. S. Bajaj, ed., Springer-Verlag, Berlin, pp. 3-17.

Hoffmann, F., 1978, Mutation and selection of haploid cell
    culture system of rape and rye, in: "Production of Natural
    Compounds by Cell Culture Methods," A. W. Alfermann and E.
    Reinhard, ed., Gesellschaft fur Strahlen and Umweltforschung,
    Munich, pp. 319-329.

Hoffmann, F., Thomas, E., and Wenzel, G., 1982, Anther culture
    as a breeding tool in rape II. Progeny analysis of andro-
    genetic lines and induced mutants from haploid cultures,
    Theor. Appl. Genet., 61:225-232.

Hollings, M., 1965, Disease control through virus-free stock,
    Ann. Rev. Phytopathol., 3:367-396.

Kidwell, M. G., Kidwell, J. F., and Sved, J. A., 1977, Hybrid
    dysgenesis in Drosophila nelanogaster: a syndrome of aberrant
    traits including mutation, sterility, and male recombination,
    Genetics, 86:813-833.

Kemble, R. J., Mans, R. J., Gabay-Laughnan, S., and Laughnan, J.
    R., 1983, Sequences homologous to episomal mitochondrial DNAs
    in the maize nuclear genome, Nature, 304:744-747.

Kochba, J., Ben-Hayyim, G., Spiegel-Roy, P., Saad, S., and
    Neumann, H., 1982, Selection of stable salt-tolerant callus
    cell lines and embryos in Citrus sinensis and C. aurantium,
    Z. Pflanzenphysiol., 106:111-118.

Larkin, P. J., and Scowcroft, W. R., 1981, Somaclonal variation -
    a novel source of variability from cell cultures for plant
    improvement, Theor. Appl. Genet., 60:197-214.

Larkin, P. J., and Scowcroft, W. R., 1983, Somaclonal variation
    and crop improvement, in: "Genetic Engineering of Plants," T.
    Kosuge, C. P. Meredith, and A. Hollaender, ed., Plenum, New
    York, pp. 289-314.

Larkin, P. J., Ryan, S. A., Brettell, R. I. S., and Scowcroft, W.
    R., 1983, Heritable somaclonal variation in wheat, Theor.
    Appl. Genet., (In Press).

Lassner, M. W., and Orton, T. J., 1983, Detection of somatic
    variation, in: "Isozymes in Plant Genetics and Breeding; Part
    A," S. D. Tanksley and T. J. Orton, ed., Elsevier, Amsterdam,
    pp. 209-218.
Lawrence, R. H., 1981, In vitro plant cloning systems, Env.
    Exptl. Bot., 21:289-300.
Libbenga, K. R., and Torrey, 1973, Hormone-induced endoreduplica-
    tion prior to mitosis in cultured pea root cortex cells,
    Am. J. Bot., 60:293-299.
Liu, M.-C., and Chen, W.-H., 1976, Tissue and cell culture as
    aids to sugarcane breeding. I. Creation of genetic variation
    through callus culture, Euphytica, 25:393-403.
Liu, M.-C., 1981, In vitro methods applied to sugarcane improve-
    ment, in: "Plant Tissue Culture," T. A. Thorpe, ed., Academic
    Press, New York, pp. 299-323.
Lorz, H., and Scowcroft, W. R., 1983, Variability among plants
    and their progeny regenerated from protoplasts of Su/su
    heterozygotes of Nicotiana tabacum, Theor. Appl. Genet.,
    66:67-75.
Mahfouz, M. N., de Boucaud, M.-T., and Gaultier, J.-M., 1983,
    Caryological analysis of single cell clones of tobacco,
    Relation between the ploidy and the intensity of the
    caulognesis, Z. Pflanzenphysiol, 109:251-257.
Matern, V., Strobel, G., and Shepard, J., 1978, Reaction to phyto-
    toxins in a potato population derived from mesophyll protoplasts,
    Proc. Natl. Acad. Sci. U.S.A., 75:4935-4939.
Mathews, D. E., Gregory, P., and Gracen, V. E., 1979,
    Helminthosporium maydis race T-toxin induces leakage of
    NAD+ from T cytoplasm corn mitochondria, Plant Physiol.,
    63:1149-1153.
McClintock, B., 1938, The fusion of broken ends of sister half
    chromatids following chromatid breakage at meiotic anaphases,
    Missouri Agr. Exp. Res. Bull., 290:1-48.
McClintock, B., 1942, The fusion of broken ends of chromosomes
    following nuclear fusion, Proc. Natl. Acad. Sci. U.S.A.,
    28:458-463.
McClintock, B., 1978, Mechanisms that rapidly reorganize the
    genome, in: "Stadler Genet. Symp.," G. P. Redei, ed., 10:25-47.
McCoy, T. J., Phillips, R. L., and Rines, H. W., 1982, Cytogenetic
    analysis of plants regenerated from oat (Avena sativa) tissue
    cultures: high frequency of partial chromosome loss, Can. J.
    Genet. Cytol., 24:37-50.
McCoy, T. J., and Phillips, R. L., 1983, Chromosome stability in
    maize (Zea mays L.) tissue cultures and sectoring among
    regenerated plants, Can. J. Genet. Cytol., (In Press).
Meins, F., 1983, Heritable variation in plant cell culture,
    Ann. Rev. Plant Physiol., 34:327-346.
Miles, J. D., Chaplin, J. F., Burk, L. G., and Baumhover, A. H.,
    1981, Registration of I-35 tobacco germplasm, Crop Sci.,
    21:802.

Mitra, J., and Steward, F. C., 1961, Growth induction in cultures of Haplopappus gracilis II, The behavior of the nucleus, Am. J. Bot., 48:358-368.

Mouras, A., and Lutz, A., 1983, Plant tumor reversal associated with the loss of marker chromosomes in tobacco cells, Theor. Appl. Genet., 65:283-288.

Mousseau, J., 1970, Fluctuations induites par la neoformation de bourgeons in vitro, in: "Cultures des Tissues de Plantes," Stasbourg Coll. Intern. CNRS, 293:234-239.

Murakishi, H. H., 1982, Interactions of virus and cells, CSRS Report Project, MICL00919.

Murashige, T., 1974, Plant propagation through tissue cultures, Ann. Rev. Plant Physiol., 25:135-166.

Murata, M., and Orton, T. J., 1982, Analysis of karyotypic changes in suspension cultures of celery, in: "Proceedings of the Fifth International Congress of Plant Cell and Tissue Culture," A. Fujiwara, ed., Japanese Assoc. of Plant Tissue Culture, Tokyo, pp. 435-436.

Murata, M., and Orton, T. J., 1983, Chromosome structural changes in cultured celery cells, In Vitro, 19:83-89.

Nickell, L. G., and Heinz, D. J., 1973, Potential of cell and tissue culture techniques as aids in economic plant improvement, in: "Genes, Enzymes, and Populations," A. M. Srb, ed., Plenum, New York, pp. 109-128.

Novak, F. J., 1980, Phenotype and cytological status of plants regenerated from callus cultures of Allium sativum L., Z. Pflanzenzuchtg., 84:250-260.

Ogihara, Y., 1981, Tissue culture in Haworthia IV. Genetic characterization of plants regenerated from callus, Theor. Appl. Genet., 60:353-363.

Ogihara, Y., 1982, Tissue culture in Haworthia V. Character-ization of chromosomal changes in cultured callus cells, Jpn. J. Genet., 57:499-511.

Ogura, H., 1978, Genetic control of chromosomal chimerism found is a regenerate from tobacco callus, Jpn. J. Genet., 53:77-90.

Oinuma, T., and Yoshida, T., 1974, Genetic variation among doubled haploid lines of burley tobacco varieties, Jnp. J. Breed, 24:211-216.

Oono, K., 1978, Test tube breeding of rice by tissue culture, Trop. Agric. Res. Ser., 11:109-123.

Oono, K., 1981, In vitro methods applied to rice, in: "Plant Tissue Culture," T. A. Thorpe, Academic Press, New York, pp. 273-298.

Orton, T. J., 1980a, Chromosomal variability in tissue cultures and regenerated plants of Hordeum, Theor. Appl. Genet., 56:101-112.

Orton, T. J. 1980b, Haploid barley regenerated from callus cultures of Hordeum vulgare x H. jubatum, J. Hered, 71:280-282.

Orton, T. J., 1983a, Experimental approaches to the study of
    somaclonal variation, Plant Molec. Biol. Rep., 1:67-76.
Orton, T. J., 1983b, Genetic variation in somatic tissues,
    Method or madness? Adv. Plant Pathol., 2:153-189.
Orton, T. J., 1983c, Spontaneous electrophoretic and chromosomal
    variability in callus cultures and regenerated plants of
    celery, Theor. Appl. Genet., 65 (In Press).
Palmer, J. D., Shields, C. R., Cohen, D. B., and Orton, T. J.,
    1983, An unusual mitochondrial DNA plasmid in the genus
    Brassica, Nature, 301:725-728.
Palmer, J. D., and Shields, C. R., 1983, The tripartite structure
    of the Brassica campestris mitochondrial genome, Nature,
    (Submitted).
Peacock, W. J., 1983, Gene transfer in agricultural plants, in:
    "Advances in Gene Technology: Molecular Genetics of Plants
    and Animals," W. J. Whelan and J. Schultz, ed., Academic
    Press, New York, (In Press).
Popchristov, V. D., and Zaganska, N. A., 1977, Study of the seed
    progeny of regenerated plants obtained by the tissue
    culture of tobacco, in: "Uses of Plant Tissue Culture in
    Breeding," Ustav Experimentalni Botaniky CSAV, Prague,
    pp. 209-221.
Prat, D., 1983, Genetic variability induced in Nicotiana
    sylvestris by protoplast culture, Theor. Appl. Genet.,
    64:223-230.
Pring, D. R. and Levings, C. S. III, 1978, Heterogeneity of
    maize cytoplasmic genomes among male-sterile cytoplasms,
    Genetics, 89:121-136.
Pring, D. R., Conde, M. F., and Gengenbach, B. G., 1981,
    Cytoplasmic genome variability in tissue culture-derived
    plants, Env. Exptl. Bot., 21:369-377.
Pullman, G. S., and Rappaport, L., 1983, Tissue culture-induced
    variation for Fusarium yellows, Phytopathol., 73:818.
Quetier, F., and Vedel, F., 1977, Heterogeneous population of
    mitochondrial DNA molecules in higher plants, Nature,
    268:365-368.
Reisch, B., and Bingham, E. T., 1981, Plants from ethionine
    resistant alfalfa tissue cultures: variation in growth and
    morphological characteristics, Crop. Sci., 21:783-788.
Rivin, C. J., Zimmer, E. A., Cullis, C. A., Walbot, V., Huynh, T.,
    and Davis, R. W., 1983, Evaluation of genomic variability
    at the nucleic acid level, Plant Molec. Biol. Rep., 1:9-16.
Sacristan, M. D., 1971, Karyotypic changes in callus cultures
    from haploid and diploid plants of Crepis capillaris (L.),
    Wallr, Chromosoma, 33:273-283.
Sacristan, M. D., and Melchers, G., 1969, The caryological analysis
    of plants regenerated from tumorous and other callus cultures
    of tobacco, Molec. Gen. Genet., 105:317-333.

Sangwan - Norreel, B. S., 1983, Male gametophyte nuclear DNA content evolution during androgenic induction in Datura innoxia Mill., Z. Pflanzenphysiol., 111:47-54.

Schaeffer, G. W., 1982, Recovery of heritable variability in anther-derived doubled-haploid rice, Crop Sci., 22:1160-1164.

Secor, G., and Shepard, J. F., 1981, Variability of protoplast-derived potato clones, Crop Sci., 21:102-105.

Shepard, J. F., 1981, Protoplasts as sources of disease resistance in plants, Ann. Rev. Phytopathol., 19:145-166.

Shepard, J. F., and Totten, R. E., 1977, Mesophyll cell protoplasts of potato, Plant Physiol., 60:313-316.

Shepard, J. F., Bidney, D., and Shahin, E., 1980, Potato protoplasts in crop improvement, Science, 28:17-24.

Sibi, M., 1976, La notion de programme genetique chez les vegetaux superieurs, II. Aspect experimental, Obtention de variants par culture de tissues in vitro sur Lactuca sativa L. apparition de vigueur chez les chroisements, Ann. Amelior. Plant, 26:523-547.

Sibi, M., 1979, Expression of cryptic genetic factors in vivo and in vitro, in: "Broadening the Genetic Base of Crops," A. M. Van Harten and C. C. Zeven, ed., Pudoc, Wageningen, Netherlands, pp. 339-340.

Skirvin, R. M., 1978, Natural and induced variation in tissue culture, Euphytica, 27:241-266.

Skirvin, R. M., and Janick, J., 1976, Tissue culture-induced variation in scented Pelargonium spp., J. Amer. Soc. Hort. Sci., 101:281-290.

Spruill, W. M. Jr., Levings, C. S. III, and Sederoff, R. R., 1980, Recombinant DNA analysis indicates that the multiple chromosomes of maize mitochondria contain different sequences, Dev. Genet., 1:362-378.

Steele, J. A., Uchytil, T. F., Durbin, R. D., Bhatnagar, P., and Rich, D. H., 1976, Chloroplast coupling factor I:A species-specific receptor for tentoxin, Proc. Natl. Acad. Sci. U.S.A., 73:2245-2248.

Stern, D. B., and Lonsdale, D. M., 1982., Mitochondrial and chloroplast genomes of maize have a 12-kilobase DNA sequence in common, Nature, 299:698-702.

Suenaga, K., Abrigo, E. M., and Yoshida, S., 1982, Seed-derived callus culture for selecting salt-tolerant rices Part 1. Callus induction, plant regeneration, and variations in visible plant traits, I.R.R.I. Res. Paper Ser., 79:1-11.

Thomas, E., Bright, S. W. J., Franklin, J., Lancaster, V. A., Miflin, B. J., and Gibson, R., 1982, Variation amongst protoplast derived potato plants (Solanum tuberosum cv. "Maris Bard"), Theor. Appl. Genet., 62:65-68.

Timmis, J. N., and Scott, N. S., 1983, Sequence homology between spinach nuclear and chloroplast genomes, Nature, 305:65-67.

Torrey, J. G., 1965, Cytological evidence of cell selection by plant tissue culture media, in: "Plant Tissue Culture," P. R. White and A. R. Grove, ed., McCutchan Publ., Berkeley, pp. 473-483.

Torrey, J. G., 1967, Morphogenesis in relation to chromosomal constitution in long-term plant tissue cultures, Physiol. Plant, 20:265-275.

Van Harten, A. M., Bouter, H., and Broertjes, C., 1981, In vitro adventitious bud techniques for vegetative propagation and mutation breeding of potato (Solanum tuberosum L.), II. Significance for mutation breeding, Euphytica, 30:1-8.

Wenzel, G., 1980, Recent progress in microspore culture of crop plants, in: "The Plant Genome," D. R. Davies and D. A. Hopwood, ed., The John Innes Charity, Norwich, U.K., pp. 185-196.

Wenzel, G., Schieder, V., Prezewozny, T., Sopory, J. K., and Melchers, G., 1979, Comparison of single cell culture derived Solanum tuberosum L. plants and a model for their application in breeding programs, Theor. Appl. Genet., 55:49-55.

Wenzel, G., and Uhrig, H., 1981, Breeding for nematode and virus resistance in potato via anther culture, Theor. Appl. Genet., 59:333-340.

Zhao, C. Z., Zheng, K. L., Qi, X. F., Sun, Z. X., and Fu, Y. P., 1982, Characteristics of rice plants derived from somatic-tissue and their progenies, Acta Genetica Sinica, 9:320-324.

Zhen-Hua, Z., 1982, Application of anther culture techniques to rice breeding, in: "Rice Tissue Culture Planning Conference," Intl. Rice Res. Inst., Los Banos, Laguna, Philippines, pp. 55-62.

NUCLEAR ARCHITECTURE AND ITS MANIPULATION

Michael D. Bennett

Plant Breeding Institute
Maris Lane, Trumpington
Cambridge CB2 2LQ, U.K.

1. DIMENSIONS IN NUCLEAR ARCHITECTURE

Nuclear architecture has many dimensions and is perceived and portrayed by scientists in many different ways. These dimensions include anatomical, cytological, genetical, nucleotypical, molecular, temporal, spatial, developmental and evolutionary aspects. Moreover, in that genetical information is encoded in the anatomy of the DNA molecule, and genome size is correlated with the rate and duration of cell and plant development (Bennett, 1972 and 1977), then clearly these different dimensions are not independent of each other.

A symposium like this brings together scientists who study the nucleus at vastly different levels, and who study its form and function in quite different ways. Hopefully we are encouraged to see the nucleus through one another's eyes, and to develop a more integrated approach in our future work. Certainly, our ability to manipulate the nucleus for practical and beneficial ends in plant breeding and biotechnology will depend largely on an increased understanding of the nucleus as a self-replicating and self-regulating structure, library, computer, factory, communications system and time capsule.

Limitation of space precludes a detailed review here of all the dimensions in nuclear architecture listed above, or of how they interrelate. Besides, such a treatment is unnecessary as many of these aspects are discussed elsewhere in the present volume. Instead, this paper will concentrate on the subject of nuclear anatomy. In particular I shall discuss whether nuclear architecture

469

is important, and if so, whether it may become amenable to useful manipulation in crop plants.

2. NUCLEAR ANATOMY – EXTERNAL FORM AND FUNCTION

The nucleus was first seen as an anatomical feature (e.g. Brown, 1833). Since then it has become increasingly clear that nuclear size and shape are remarkably variable within an organism. For example, nuclear volume in a bread wheat variety varied more than 600-fold (Bennett et al., 1973), from 240 μm^3 in a haploid sperm nucleus, to over 160,000 μm^3 in an endopolyploid cell nucleus. The latter nucleus can be as large as a mature pollen grain (Fig. 1a and b), and like a pollen grain is visible to, and can be picked up on a needle using, the unaided eye. Variation in nuclear shape in the same variety includes the elongate mature sperm cell nucleus which is about 30 μm long and about 4 μm in diameter (Fig. 1a) with a smooth profile, as well as the sub-spherical antipodal cell nucleus which is up to 70 μm in diameter but often with a highly convoluted nuclear membrane (Fig. 1b and d).

Such anatomical variation in the size and shape of nuclei is not haphazard. The small size and streamlined shape of the transcriptionally inactive mature sperm nucleus is adapted for rapid transport of a haploid genome, over 5 mm in less than 30 minutes, from the pollen grain to the embryo sac, along a narrow pollen tube with an internal diameter of about 8 μm (Fig. 1c). Similarly, the peculiar anatomy of the antipodal cell nucleus is also of functional significance, reflecting the secretory role of the cell (Heslop-Harrison, 1972). The large nuclear size reflects the large number of genome and gene copies which it contains, while the convoluted nuclear surface, which greatly increases nuclear surface area, is presumably an adapatation which facilitates transport in and out of the nucleus.

It has been possible to manipulate ploidy level using colchicine for almost 50 years, and although the genetic code was elucidated only about 25 years ago, already genetic engineering offers the prospect of sophisticated manipulations of single genes in crop plants. Unfortunately our ability to manipulate the structure of the genome inside the nucleus is not matched by an equal ability to manipulate external features of nuclear architecture. Although size and shape were among the first dimensions of nuclear architecture to be seen, it is only in the last decade that the structural elements responsible, i.e. the cytoskeleton and nuclear matrix (see Introduction of Maraldi et al., 1983) were discovered. Consequently, should we wish to manipulate nuclear size and shape, for example to allow the large sperm nucleus of one species to pass along the narrower pollen tube or intracellular spaces of another species, we could not do it.

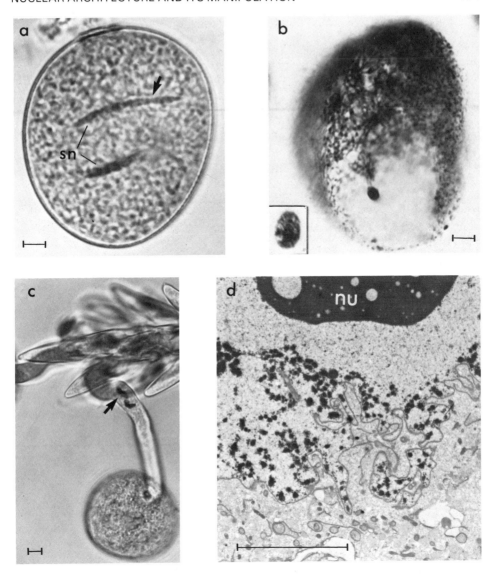

Fig. 1. (a) A <u>Triticum</u> <u>aestivum</u> pollen grain showing elongated
sperm nuclei (sn), one (arrowed) all in focus; and, at the
same magnification (b) an antipodal cell nucleus (about
512C) of <u>Hordeum</u> <u>vulgare</u> with (inset) a 4C ovary wall
nucleus; (c) the two sperm nuclei (arrowed) of <u>Hordeum</u>
<u>bulbosum</u> soon after pollination of a <u>H.</u> <u>vulgare</u> stigma
(top), and (d) the highly convoluted nuclear membrane
(arrowed) of an antipodal cell nucleus in <u>T.</u> <u>aestivum</u>
showing a nucleus (nu) deep in the nucleus and chromatin
clustered near the nuclear membrane. Bar = 5 μm.

3. NUCLEAR ANATOMY - INTERNAL FORM AND FUNCTION?

The examples described in the previous section clearly show
that the biological dictum that "function modifies form" holds true
for external features of nuclear architecture such as size and
shape. However, my purpose here is to look more deeply into this
relationship and to ask if function is also related to form with
respect to features of the internal architecture of nuclei,
especially the intranuclear arrangement of genomes and chromosomes.

a. DNA molecular form and function

Function and form are certainly causally related at the most
basic level of nuclear architecture, that of the base sequence of
DNA. The discovery of the structure of DNA, and the elucidation of
the genetic code showed that genes contain and express their
information in an anatomical language of chemical form. More
recently it has been shown that DNA can exist in vivo in more than
one form, i.e. as a right-handed double helix (the B form) or as a
left-handed double helix (the Z form) (Wang et al., 1979).
Moreover, a change in molecular conformation from right-handed turns
in B-DNA to left-handed turns in Z-DNA has the potential to act as a
switch, or a recognition signal affecting DNA function for many
kilobases (Pardue et al., 1984). Thus at these molecular levels
form may determine function, rather than the reverse.

At a somewhat higher level of genome architecture, namely the
chromosome of a phage or a bacterium, there is evidence of a spatial
ordering into clusters of genes whose products share related
temporal or biochemical functions (Lacadena et al., 1983). For
example, the genome of the phage T7 of Escherichia coli is divided
into three consecutive segments which correspond to three
transcriptional classes of genes (early, intermediate, and late)
involved, in functions related to the initiation of infection, phage
DNA replication, and phage morphogenesis and host lysis,
respectively. Thus, as Lacadena et al. (1983) noted, "in the
organisation of the phage T7 chromosome there is an apparent
relationship between structure (genome organisation) and function
(infective cycle)". These authors also cite examples in bacteria of
the significant clustering of structural genes coding for enzymes
involved in a determined metabolic pathway, either within an operon,
or within many operons themselves clustered in one small region of
the genome.

Thus there appears to be some relationship between the spatial
order of parts of the genome and their functions in primitive
organisms where the genome is a single chromosome. However, this
paper is concerned with the nuclei of higher plants, both species
and hybrids between them. Here the basic genome contains a number
(x) of different chromosome types, i.e. heterologues, and the

nucleus often contains more than one basic genome. Moreover, in the sporophyte of a diploid species, the two basic gametic haploid sets can originate from different parents (P1 and P2). At this more complex and gross level of internal nuclear architecture, does function modify form as it does with the external nuclear form such as shape, and/or, does form affect function as it does with the order of DNA base sequences in genes which the chromosomes contain?

Before any meaningful answer to these questions is possible, it is necessary to answer a simpler question, namely, "Is there an intranuclear spatial architecture?". In other words, "Are haploid genomes and their constituent chromosomes and genes spatially ordered in the nuclei of higher plants?". Clearly, if there is no architectural form regulating the spatial arrangement of genomes and chromosomes within the nucleus, then there can be no relationship between form and function at these levels.

4. STUDIES OF INTRANUCLEAR GENOME AND CHROMOSOME ARRANGEMENT IN
    GRASSES

a. Materials and methods

Work at the Plant Breeding Institute, Cambridge, has addressed these basic questions of internal nuclear architecture using the electron microscopic (EM), three dimensional (3-D), serial thin-section reconstruction method to study the placement of genomes, chromosomes and chromosome segments, e.g. centromeres, telomeres and nucleolus organisers. Full details of the techniques for serial sectioning and electron microscopy, reconstructing nuclei, identifying chromosomes, and obtaining accurate three dimensional positions of identified structures have been published previously (Bennett et al., 1979 and 1981; Finch et al., 1981; Bennett et al., 1982; and, Heslop-Harrison and Bennett, 1983b), as has a discussion of the major advantages of the EM, 3-D reconstruction method over most other methods for investigating chromosome disposition (Bennett, 1984a).

Complete reconstructions of more than 300 mitotic and premeiotic nuclei, mainly at metaphase, have been made. The great majority of these were from root tips or anthers of about twenty grass species or hybrids, mostly diploids, including important cereal grain crops such as Hordeum vulgare (barley) and Secale cereale (rye) and their wild relatives such as Hordeum bulbosum and Secale africanum. Between 60 and 100 sections 0.1 μm thick were typically required to reconstruct individual nuclei of these materials.

Using simple morphological characters, such as (1) relative chromosome volume, (2) arm volume ratio, and (3) the characteristic appearance of an expressed nucleolar organiser region (NOR), it was

usually possible to identify unequivocally every chromosome within a
single reconstructed nucleus in most materials studied (e.g.
H. vulgare cv. Tuleen 346, 2n = 2x = 14; Bennett et al., 1982).
Similarly, in many hybrids it was also possible to identify the
parental origin of each chromosome within a single reconstructed
cell (e.g. in the diploid hybrid H. vulgare cv. Sultan x
S. africanum; Finch et al., 1981).

From an examination of a reconstructed nucleus it was also
possible to obtain very accurate estimates ($\pm$ 0.1 μm) for the three
dimensional positions of any identified chromosome segment, for
example, the centromere position, for each chromosome. Using a
microcomputer it was then easy to generate (1) the distances
separating all pairs of centromeres (Heslop-Harrison and Bennett,
1983a), and (2) to define the mean position about which all the
centromeres within a nucleus were distributed ($\bar{x}$, $\bar{y}$, $\bar{z}$, the mean
centromere point (MCP), Fig. 2). In nuclei where all the
chromosomes were identified as to their parental origin, it was also
simple to estimate the mean distance ($\bar{d}$) from the MCP of centromeres
on chromosomes from either parent, and to express this as a ratio,
$\bar{d}$.P1:$\bar{d}$.P2 (e.g. see Fig. 2). This ratio has proved to be important
in studies of haploid genome position in diploid hybrids.

Using the microcomputer, a rotation programme has been
developed which gives a two dimensional projection of nuclear
features, for example, of the distribution of all the centromeres
within a nucleus, seen from any required angle (Heslop-Harrison and
Bennett, 1983c). Analysis showed that centromere alignment was
almost perfect on a plane in 15 non-cold-treated mitotic metaphases
of diploid cereals where 2n = 14; only 2% of the total variance in
centromere position was out of the plane of the metaphase plate.
Consequently, almost all of the variation in centromere position can
still be seen in a polar view of such a metaphase plate obtained by
a rotation.

In addition to the above mentioned techniques for identifying
and displaying the positions and distribution of centromeres two
analyses have been developed for use on pooled results for replicate
reconstructed nuclei. One, a 'Procrustes' analysis mentioned in
Heslop-Harrison (1983), shows what is the mean 3-D arrangement of
heterologues. The other, described in full in Heslop-Harrison and
Bennett (1983a), shows whether a particular predicted order of
heterologues is significantly expressed.

b. Conclusions about genome, chromosome and gene distribution based
    on reconstructed nuclei

Experiments using the above mentioned materials and methods
have allowed several firm conclusions to be reached regarding the
spatial distribution of genomes and chromosomes studied. These

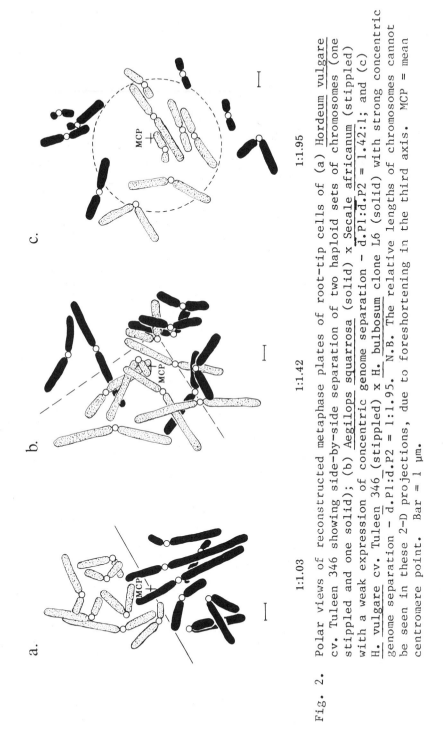

1:1.03   1:1.42   1:1.95

Fig. 2. Polar views of reconstructed metaphase plates of root-tip cells of (a) Hordeum vulgare cv. Tuleen 346 showing side-by-side separation of two haploid sets of chromosomes (one stippled and one solid); (b) Aegilops squarrosa (solid) x Secale africanum (stippled) with a weak expression of concentric genome separation – d.Pl:d.P2 = 1.42:1; and (c) H. vulgare cv. Tuleen 346 (stippled) x H. bulbosum clone L6 (solid) with strong concentric genome separation – d.Pl:d.P2 = 1:1.95. N.B. The relative lengths of chromosomes cannot be seen in these 2-D projections, due to foreshortening in the third axis. MCP = mean centromere point. Bar = 1 μm.

conclusions, many of which have already been published (Bennett, 1982, 1983, 1984a, b and c; Heslop-Harrison and Bennett, 1983a and b), are briefly summarised below.

First, studies of mitotic metaphase in root tips of several species included Aegilops umbellulata, H. vulgare and Zea mays provide no evidence of somatic association of homologues. Indeed, pooled results for each species and treatment showed that, as a class, centromeres of homologues were always on average further apart than centromeres of heterologues, and when results for the different species were pooled this difference became significant. The same result was obtained when cells at the last premeiotic mitosis in anthers of H. vulgare were studied (Bennett, 1984c).

Second, there is a highly significant tendency for centromeres of two haploid genomes to be spatially separated. Moreover, in interspecific and intergeneric hybrids, such separation inevitably separates centromeres of parental genomes. In general, two types of haploid genome separation have been commonly observed, namely, (1) a side-by-side arrangement which appears to be the predominant form in species, and (2) a concentric arrangement, which appears to be the predominant form in interspecific and intergeneric hybrids (Fig. 2).

Concentric separation of haploid parental sets was clearly under genotypic control as in different intergeneric hybrids, the same parental set comprising all large chromosomes was central in one hybrid, i.e. H. vulgare cv. Sultan x S. africanum (Finch et al., 1981; and Bennett, 1982), but peripheral in another, i.e. H. chilense x S. africanum (Schwarzacher et al., 1984). More evidence supporting this conclusion comes from the observation that in a hybrid between diploid species with chromosomes of similar sizes (H. marinum ssp. marinum x H. vulgare), each parental set consistently occupied the central or peripheral site in the embryo, but the reverse site in endosperm nuclei, of the same seeds (Finch and Bennett, 1983; and Finch, 1983). As the basic haploid genome tends strongly to occupy a substantially separate spatial domain within the deploid nucleus it was concluded that it is a real structural unit in nuclear architecture.

Third, the spatial distribution of heterologues is non-random within separate haploid sets. Indeed, the mean placement of centromeres of all the heterologues in a basic haploid set predicted by a model which orders biarmed chromosomes mainly so that pairs of chromosome arms of most similar volume and morphology from heterologues are adjacent (see Bennett, 1982 and 1983) has been shown using the analysis developed by Heslop-Harrison and Bennett (1983a). The model has predicted a significantly expressed mean order of 7 heterologous centromeres in the haploid genomes in all 4 grass species tested, using samples of only 10 or fewer reconstructed cells. Moreover, the predicted order of centromeres was significantly expressed in both somatic and germ line mitotic metaphase cells in H. vulgare cv. Tuleen 346 (Bennett, 1984c).

As different heterologues tend strongly to be arranged in a particular, and predictable, order, it was suggested that this order represents "the natural karyotype" (Bennett, 1984b) for the organism. Moreover, it was noted that if the chromosomes within a haploid set tend to occupy mean fixed relative positions with respect to one another, so also do all the genes which they carry. Thus, just as the linkage map shows the order of, and genetic distances between, genes within linkage groups, the natural karyotype may determine aspects of the real physical arrangement of genes both within and between chromosomes.

Fourth, the distribution of DNA present as low copy number coding sequences and as high copy number non-genic sequences seems to be non-random between heterologues of the spatially ordered karyotype. For example, there is a tendency for C-bands of equal size, to occur on adjacent arms of heterologues at similar distances from their centromeres (Bennett, 1982). Similarly, there appears to be a significant tendency for paralogous genes to cluster in domains formed of different but spatially adjacent linkage groups in Z. mays (Bennett, 1983).

The above results for reconstructed grass nuclei represent a considerable advance in our knowledge of internal nuclear architecture. Moreover, all that we have learnt tends to contradict the view widely held hitherto, which regarded the nucleus as "a loose assortment of molecules" (see Lewin, 1981), and believed that "the chromosomes just floated about like so many noodles in a soup" (see Walgate, 1983).

Taken together the results and conclusions for reconstructed cereal nuclei summarised above provide an unequivocally affirmative answer to the question, "Are haploid genomes and their constituent chromosomes and genes spatially ordered in the nuclei of higher plants?". Clearly, there is a strongly expressed internal nuclear architecture of spatially separated haploid genomes, containing spatially ordered heterologues. Moreover, related genes probably tend to be clustered in spatial domains extending over several chromosome arms on adjacent heterologues in the natural karyotype. It is important to note that as the order of heterologues within the natural karyotype is predictable given the relative chromosome arm sizes, then clearly the mean spatial order of chromosomes and of the genes which they carry reflects chromosome form.

Having established the existence of an internal nuclear architecture at the level of the chromosome and the basic haploid genome, it is now meaningful to ask whether this form has any functional significance.

If the internal architecture of the nucleus were of no significance, then all variation in the relative spatial locations of genomes, chromosomes and genes, provided they remained undamaged and within the nucleus, should have no effect on their mechanical behaviour, transcription, and phenotypic expression. However, there is a growing body of evidence both in the literature, and from our unpublished work on reconstructed nuclei, showing that intranuclear spatial position is important. These observations highlight the effects on chromosome behaviour and/or function of, (1) the position of a gene on the chromosome, (2) the position of a chromosome in the nucleus, and (3) the position of a genome within the nucleus. Together these observations, which are described below, contradict the view, mentioned above, which sees the contents of the nucleus as a genetic minestroni which can be stirred without altering or impairing its functional qualities.

5. THE IMPORTANCE OF INTRANUCLEAR POSITION – EXAMPLES FROM THE LITERATURE

An example showing very clearly that a chromosome's behaviour is not independent of its situation within the nucleus concerns neocentric activity in maize. The presence of the K10 knob in maize has a spectacular effect on meiosis. When it is present all knobs act as precocious centromeres (neocentromeres) at the two meiotic divisions. However, acentric fragments containing a knob, newly produced by crossing-over during first prophase, do not show such neocentric activity in the presence of K10 (Rhoades, 1952). "However we interpret this situation in detail ... we must conclude that the true centromere plays a decisive role in the manifestation of T-end [i.e. neocentric] activity and that its influence, unlike that of K10, is chromosome limited" (Lewis and John, 1963). It is not enough for both the centromere and the knob to be within the same cell and nucleus. Neocentric activity depends on their being attached, and perhaps on the flow of some material along the length of a chromosome arm (Darlington, 1957).

Several examples are known of particular chromosomes which regularly occupy a constant position in the nucleus. For example, the central position on the first metaphase plate of meiosis is regularly occupied by the sex chromosome(s) in several species such as the bug Cimex rotundus (see Plate XXVI in Darlington and La Cour, 1976), and the nematode Caenorhabditis elegans (Albertson and Thomas, 1982; and Albertson, personal communication). It is reasonable to ask if this arrangement is of functional significance for meiosis, especially in view of the results for reconstructed nuclei described below. However, the reason, if any, for this behaviour is unknown, although it may be significant that only a chromosome in this central position is equidistant from all other chromosomes and hence ideally placed to produce or control a radial gradient in the nucleus affecting the behaviour of all other

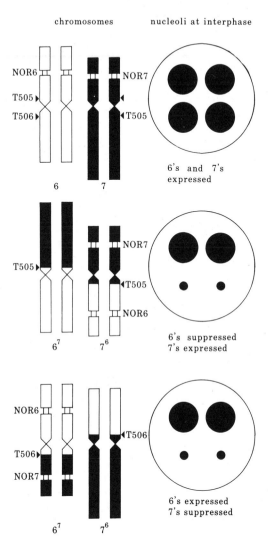

Fig. 3. A diagrammatic representation of the structure of the
         nucleolar organizing chromosomes (left) and of nuclei at
         interphase (right) in normal barley (top) and in lines with
         reciprocal translocations T505 (middle) and T506 (bottom),
         respectively, showing suppression of NORs translocated to
         the non-NOR bearing arm of chromosomes 6 or 7 (redrawn from
         Nicoloff et al., 1979).

chromosomes. What is certain is that the sex and other chromosomes which occupied this position in the ant-lions, Myrmeleontidae, behaved differently from others located elsewhere, segregating precociously in the central unit of the spindle (Hughes-Schrader, 1983).

In their recent seminal paper Agard and Sedat (1983) wrote:- "That chromosome organisation is important for gene expression is suggested by genetic studies. Perhaps the best example is the interaction of the zeste and white loci in Drosophila. Jack and Judd [1979] ... reported that repression only occurs if the two white alleles are physically adjacent or paired. The strength of repression was found to diminish with increasing separation. This suggests that there may be a close and defined association between loci on separate chromosomes, and that communication of regulatory signals may be spatially restricted."

Another example of a dramatic position effect has been carefully studied by Nicoloff et al. (1979). The standard karyotype of barley (H. vulgare) consists of 7 pairs of chromosomes, two pairs of which are nucleolar SAT chromosomes (pairs 6 and 7). Transcription of rDNA located in the nucleolar organisers (NORs) of these four chromosomes results in the formation of four primary nucleoli at late telophase, and up to four nucleoli per nucleus are commonly seen at mitotic interphase. However, diploid homo- and heterokaryotypes of barley translocation lines with only one satellite chromosome pair containing two NORs in opposite arms were found to show suppressed nucleolus formation by the transposed NOR (Fig. 3). The same was true for translocation lines with both NORs tandemly arranged into the same chromosome arm. When NORs were transposed to chromosomes without a NOR in the standard karyotype, the normal pattern of nucleolus formation remained unaffected. These observations suggest that suppression of transcription on transposed NORs represents a rather specific intra-chromosomal position effect in which 'correctly positioned' NORs influence the activity of 'incorrectly positioned' NORs transposed to the same chromosome (Nicoloff et al., 1979). Clearly in this case, the position of the rRNA genes in the complement affects their genetic expression.

6. THE IMPORTANCE OF INTRANUCLEAR POSITION - EXAMPLES FROM
    RECONSTRUCTED GRASS NUCLEI

a. Position and relative chromosome volume

Studies of reconstructed mitotic metaphase nuclei of S. cereale (rye, 2n = 2x = 14) show that chromosome behaviour can be strongly influenced by relative position in the nucleus (Bennett et al., 1982). For example, studies of two chromosome types with unmistakeable morphologies, namely, 1R which is the nucleolus organising chromosome, and 5R which has the highest arm ratio,

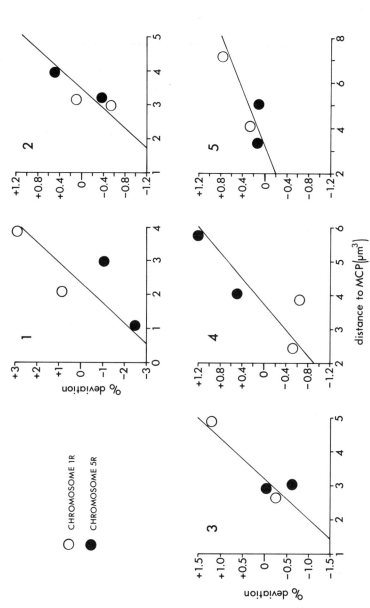

Fig. 4. The significant tendency for peripherally located chromosomes to be less condensed than their centrally located homologues in 5 root-tip metaphase cells of Secale cereale cv. King II. The deviation of the observed relative volumes of homologues of 1R and 5R within each cell from their mean relative volumes in all 5 cells is plotted against the distance of each centromere from the mean centromere position (MCP).

showed that these chromosomes were up to 23% larger than average when located peripherally, and up to 19% smaller than average when located centrally on the plate. Moreover, the deviation between the relative volumes of these chromosomes in single cells, and their mean relative volumes, was significantly correlated (r ranged from 0.80 to 0.93) with the distance of their centromeres from the MCP in each of five reconstructed nuclei (Fig. 4). Thus, chromosomes near the MCP tended to be more condensed than their homologues remote from it, showing that there are important gradients running across the radius of the metaphase plate which affect chromosome condensation. The cause of such gradients is unknown.

If, as seems likely, the arrangement of centromeres and chromosomes on the metaphase plate substantially reflects their interphase arrangement (Cremer et al., 1984), then it is reasonable to ask if the gradients affecting chromosome condensation at mitosis might also affect genetic activity during interphase, especially as nuclear volume at interphase and chromosome volume at metaphase are known to be positively correlated with metabolic activity in S. cereale (Bennett and Rees, 1969) and Vicia faba (Bennett, 1970). Indeed as a general rule, chromosome density is inversely related to genetic activity; decondensation or 'puffing' indicates activity, and vice versa. Thus, it is not unreasonable to suppose that peripheral chromosomes, which were less condensed at metaphase, had engaged in greater or more prolonged transcriptional activity in interphase than had the more condensed centrally located ones. This hypothesis would agree with evidence noted by Agard and Sedat (1983) "... that transcriptionally active genes are closely associated with the nuclear membrane." This possibility, that intranuclear position may affect transcriptional activity, is highly relevant to the results presented in section 6c, on the phenotype of hybrids with concentric genome separation.

b.   Intranuclear position and uniparental chromosome elimination

Many, though not all, hybrid combinations between a Hordeum species and some other grass species are mitotically unstable and exhibit the loss of chromosomes from just one parent. Often a complete haploid genome is eliminated during early embryo development, so that embryo culture regularly recovers plants with a complete haploid genome of one parent (Subrahmanyam and Kasha, 1973; Bothmer et al., 1983; and Finch, 1983). The degree of such instability varies greatly within a given combination of species, depending on the genotypes involved (Simpson et al., 1980). For example, crosses between H. vulgare cv. Sultan and diploid H. bulbosum clone L6 gave 100% haploid and no hybrid progeny, but H. vulgare cv. Tuleen 346 x H. bulbosum clone L6 gave about 50% of haploids and of diploid hybrids in its progeny.

Nuclei of several Hordeum hybrid combinations known to produce

haploids with chromosomes of only one parent, including H. vulgare
cv. Tuleen 346 x H. bulbosum clone L6, have been reconstructed at
Cambridge, and all showed a significant tendency for concentric
parental genome separation in cells with 13 or 14 chromosomes,
i.e. with at least six chromosomes from each of both parents.
Nuclei were also reconstructed for another Hordeum hybrid, H.
chilense x S. africanum which is not known to produce haploid
progeny, but is known to eliminate a chromosome of H. chilense
(R.A. Finch - pers. comm.). These cells also showed a significant
expression of concentric parental genome separation. Another
diploid hybrid, namely, H. vulgare x Psathyrostachys fragilis has
been reported to display both concentric parental genome separation
(Linde-Laursen and Bothmer, 1984; and Bothmer et al., 1984) and
uniparental chromosome elimination, based on light microscope
studies. Table 1 gives the direction of concentric genome
separation, i.e. which parental genome is central and which is
peripheral, and also lists the direction of uniparental chromosome
or genome elimination in all six diploid Hordeum hybrids studied.

Comparing the polarity of concentric genome separation with the
direction of uniparental chromosome elimination reveals a clear
relationship between intranuclear genome position and chromosome
loss. Thus, in all six Hordeum hybrids listed (Table 1) only
chromosomes of the genome which tends to be peripheral are
eliminated in the diploid tissues studied (P < 0.02). As most of
these combinations produce at least some hybrids which are
relatively stable, often retaining the complete hybrid genomes of
both parents in somatic and meiotic cells, it is concluded that
occupation of the peripheral position by one parental genome appears
to predispose to, but does not inevitably cause, mitotic elimination
of its chromosomes.

Another telling observation showing that the direction of
uniparental genome elimination is correlated with the peripheral
intranuclear genome position concerns the hybrid combination H.
vulgare x H. marinum ssp.. marinum where the polarity of concentric
parental genome separation and the direction of uniparental genome
elimination were both tissue specific in young seeds (Finch and
Bennett, 1983; and Finch, 1983). Thus, in the diploid embryo the
marinum genome was peripheral and was lost to leave haploid vulgare
nuclei, while in the triploid endosperm the situation was the
reverse, with the vulgare genome peripheral before elimination to
leave diploid marinum nuclei.

This observation, that the polarity of concentric genome
separation can be tissue specific, certainly supports the conclusion
that concentric parental genome separation is determined not as a
kind of packing phenomenon, but by genotypic control. If so, it
will be important to identify the gene(s) responsible for
controlling different intranuclear genome positions in such hybrids.

Table 1. The polarity of concentric parental genome separation compared with the direction of (1) uniparental genome or chromosome elimination, (2) dominance of parental characters in the phenotype, [continued on next page]

Hybrid (P1 x P2)	Peripheral genome (centromeres)	Relative sizes of peripheral & central genomes	Ratio of mean distance to MCP of centromeres $\overline{d}.P1:\overline{d}.P2$
COMBINATIONS WITH STRONG CONCENTRIC GENOME SEPARATION			
1. Hordeum vulgare cv. HP 40 x Psathyrostachys fragilis	[a]Psathyrostachys fragilis		no data
2. H. vulgare cv. Sultan x Secale africanum	Secale africanum	big around small	1:1.74
3. H. vulgare x Secale cereale (20 x R)	Secale cereale		1:1.71
4. H. marinum ssp. marinum x H. vulgare cv. Tuleen 346	Hordeum marinum		1.68:1
5. H. vulgare cv. Tuleen 346 x H. bulbosum clone L6	Hordeum bulbosum	equal sizes	1:1.56
COMBINATIONS WITH WEAK CONCENTRIC GENOME SEPARATION			
6. H. chilense x S. africanum	Hordeum chilense	small	1.32:1
7. Aegilops squarrosa PBI line x A x S. cereale cv. Petkus Spring	Aegilops squarrosa	around big	1.15:1

a. Results for P. fragilis x H. vulgare are for light microscopy alone from Bothmer et al., 1984; Linde-Laursen and Bothmer, 1984; or Linde-Laursen (pers. comm.).

Table 1 [continued] .. (3) nucleolar dominance in seven interspecific
    or intergeneric grass F$_1$ hybrids with strong or weak
    concentric genome separation.

Eliminated genome or chromosomes	Parent whose characters dominate the F$_1$ hybrid phenotype	Parental genome with suppressed NOR activity	Relative degree of suppression of NOR activity	Cell type of tissue studied
COMBINATIONS WITH STRONG CONCENTRIC GENOME SEPARATION				
1. [a]Psathyro-stachys fragilis	[a]Psathyro-stachys fragilis	[a]Psathyro-stachys fragilis	[a]Strong	[a]Root tip
2. Secale africanum	Secale africanum	Secale africanum	Strong	Root tip
3. Secale cereale	Secale cereale	Secale cereale	Strong	Root tip
4. Hordeum marinum	[b]—	Hordeum marinum	Strong	Young embryo
5. Hordeum bulbosum	Hordeum bulbosum	Hordeum bulbosum	Strong	Root tip & anther
COMBINATIONS WITH WEAK CONCENTRIC GENOME SEPARATION				
6. Hordeum chilense	Secale africanum	Secale africanum	Weak	Root tip
7. no elimination seen	Aegilops squarrosa	Secale cereale	Weak	Root tip

b. As uniparental genome elimination occurred in early embryo
    development no hybrid plant was obtained and this comparison was
    impossible.

Indeed, such genes may have been discovered already, but their real function not understood. I refer to the genes located on both arms of chromosome 2, and on one arm of chromosome 3, in H. vulgare with major effects on the frequency of uniparental genome elimination in H. vulgare x H. bulbosum (Ho and Kasha, 1975). An effect seen only in interspecific hybrids is unlikely to be the prime function of these genes. It is suggested, therefore, that the prime function of these genes may be to control aspects of the intranuclear spatial arrangement of chromosomes or haploid genomes at some stage(s) of development in H. vulgare itself. Perhaps such a function is expressed in wide-crosses, either to an abnormal degree, or at an abnormal time during development, thereby controlling concentric parental genome separation which is sometimes so pronounced as to cause the elimination at mitosis of some, or all, of the chromosomes in the peripheral genome. Be that as it may, the above results agree in showing that the occupation of different intranuclear positions can have profound mechanical and genetical consequences.

c. Intranuclear genome position and phenotypic dominance

The results above show clear effects of intranuclear position on chromosome behaviour. It was knowledge of these effects which led me to consider seriously whether the intranuclear position of a chromosome might also affect its genetic activity and hence its phenotypic expression. It seemed premature to try to answer that question for just a single chromosome. However, the fact that whole parental genomes tend strongly to occupy substantially different regions of the nucleus in some grass hybrids seemed to provide a powerful tool for testing whether the intranuclear position of a genome, i.e. whether it is peripheral or not, can affect its phenotypic expression.

Section 6a concluded by noting that peripheral chromosomes may have engaged in greater or more prolonged transcriptional activity than others during interphase, as indicated by their significantly increased volume and closer proximity to the nuclear membrane. If so, then one might expect that the phenotype of a hybrid with strongly expressed concentric parental genome separation would tend to be dominated by characters controlled by the peripheral genome. Preliminary studies strongly support this expectation.

Living plants, dried specimens, or photographs of the two parental species and their F_1 hybrid at vegetative and flowering stages, were placed side-by-side and compared for gross, mainly morphological, phenotypic characters, such as are normally used by plant breeders. This was done for two hybrids with strong concentric parental genome separation, H. vulgare cv. Sultan x S. africanum and H. vulgare cv. Tuleen 346 x H. bulbosum clone L6, and for H. chilense x S. africanum which has weak concentric genome separation. Opinions regarding the gross appearance of three other hybrids, compared with their parents, were obtained by consulting scientists who worked closely with these materials, namely: H. vulgare x S. cereale (20 x R) (B. Forster, pers. comm.), Aegilops squarrosa x S. cereale (T.E. Miller, pers. comm.); and H. vulgare x Psathyrostachys fragilis (I. Linde-Laursen, pers. comm.). Although the hybrids were all intermediate between the parents in gross phenotype, nevertheless, in all six instances the F_1 hybrid clearly tended to resemble one parent much more than the other at vegetative and/or flowering stages. Indeed, in several hybrids e.g. H. vulgare cv. Sultan x S. africanum and H. vulgare cv. Tuleen 346 x H. bulbosum clone L6, this tendency was so pronounced at flowering that the F_1 hybrid might even be confused with one of its parents by a casual observer.

Comparing the direction of this tendency with the direction and strength of genome separation (Table 1) showed, first, a clear tendency for the peripheral parental genome to dominate the gross phenotype in hybrids with strong concentric parental genome separation; and second, a suggestion that the direction of uniparental dominance may be affected by either the strength of concentric parental genome separation, and/or the relative sizes of the parental genomes and their chromosomes. Thus, in all four F_1 hybrid plants with strong genome separation, and in one of two hybrids with weak genome separation, the hybrid plants strongly resembled the parent whose genome occupied the peripheral intranuclear domain. For example, Table 2 presents information for 11 gross phenotypic characters for the F_1 hybrid H. vulgare cv. Sultan x S. africanum and its parents. These were all the characters which the parents differed for, and which could be scored on the day when the comparison was made. For all eleven characters, the F_1 hybrid resembled H. bulbosum, the parent whose haploid genome tended to be peripheral, more than H. vulgare whose genome was centrally located. This uniparental dominance is highly significant (P < 0.001) for this hybrid.

In one hybrid, H. chilense x S. africanum, the Secale centromeres tend to be central, yet Secale characters tend to dominate the hybrid phenotype. However, any deviation from the overall tendency for the peripheral genome to dominate the F_1 hybrid phenotype may be more apparent than real. First, the central genome is more deeply and more completely enveloped by the peripheral genome when the latter has a larger DNA content and chromatin volume than vice versa. Thus, when the 7 chromosomes of S. cereale (2C DNA amount = 16.6 pg) surround those of H. vulgare (2C = 11.1 pg) the peripheral genome must be on average farther from the nuclear membrane than when the haploid genome of H. chilense (2C = 11.0 pg) surrounds that of S. africanum (2C = 14.9 pg). Second, and perhaps more importantly, the ratio for genome separation (d.P1:d.P2) in Table 1 is for centromeres alone. The chromosomes of S. cereale and S. africanum are longer than those of H. vulgare and H. chilense.

Table 2.   A comparison of some gross, mainly morphological
characters of Hordeum vulgare cv. Sultan, Secale africanum
and their F_1 hybrid showing that in the hybrid there is a
significant tendency for characters of S. africanum (the
peripheral genome in the hybrid – see Table 1) to be
dominant over those of H. vulgare cv. Sultan (the central
genome). N.B. The arrow indicates which parent the F_1
resembles more closely for each character scored.

Character	Parent 1 – contributing the central genome – H. vulgare cv. Sultan	F_1 hybrid H. vulgare x S. africanum		Parent 2 – contributing the peripheral genome – S. africanum
1. Plant habit	annual	perennial	→	perennial
2. Plant glaucosity	+	++	→	++
3. Leaf hairs	glabrous	very hairy	→	hairy
4. Leaf sheath veins	green	red	→	red
5. Length of ligule (mm)	2.13	0.48	→	1.09
6. Length of auricle claw (mm)	2.69	1.13	→	0.64
7. Culm neck hairs	glabrous	hairy	→	hairy
8. Spikelets per node	3	1	→	1
9. Spikelet orientation to the rachis	radial	tangential	→	tangential
10. Florets per spike	1	2-3	→	2-3
11. Length of awns (mm)	140.8	37.6	→	17.6

Thus, when centromeres of the S. cereale genome are strongly
peripheral to those of H. vulgare, it is difficult for vulgare
chromosomes to occupy much of the peripheral intranuclear domain.
However, when the centromeres of H. chilense are weakly peripheral
to those of S. africanum, many segments of the africanum
chromosomes can easily occupy large parts of the peripheral
intranuclear domain. Thus, in hybrids like H. chilense x S.
africanum where concentric genome separation is only weakly
expressed for centromeres, and the centromeres of the physically
smaller genome are peripheral, chromosomes of the latter may not
occupy enough of the intranuclear peripheral domain at interphase to
dominate the hybrid phenotype. As noted above, the peripheral
genome dominated the phenotype as expected in all four hybrids where

genome separation was strong and the peripheral genome was much larger than, or of equal size to, the central genome (Table 1). It may be significant that the one possible exception to expectation was a hybrid with only weak genome separation, where the centromeres of small chromosomes were peripheral to those of large chromosomes, but where the relative locations at interphase of other segments, and of the genomes as a whole, was uncertain.

The above results are preliminary, and further examples of hybrids with strong concentric parental genome separation must be examined before any firm conclusions regarding genome position and phenotypic expression can be reached. Nevertheless, the available evidence strongly suggests that there may be a major effect of intranuclear genome position on the phenotype of hybrids, especially when concentric genome separation is pronounced, and the peripheral genome contains more DNA and larger chromosomes than the central genome.

It is important to note that the observations described above involved only gross, mainly morphological, phenotypic characters, such as might be scored by a plant breeder. It will be interesting to see whether or not biochemical characters, e.g. isozyme patterns, also show enhanced genetic activity and/or phenotypic dominance by the peripheral genome.

If genome position is causally related with the relative expression of parental characters in the phenotypes of hybrids, this will be of considerable interest for geneticists and practical significance for plant breeders. For example, it would interest geneticists if dominance could sometimes have a structural component, and reflect different intranuclear positions of different homoeoalleles from the two parents. It has often been noted in wide crosses that desirable characters of a given species are not expressed as strongly as expected in the phenotypes of its hybrids with another species, e.g. rye characters in triticale (X Triticosecale Wittmack). Perhaps this phenomenon is a consequence of concentric genome separation. If so, it may not be possible to obtain regularly the optimal phenotypic expression of both parental genotypes, until we understand: (1) how different forms of intranuclear genome separation are determined, and (2) how to manipulate and control this aspect of nuclear architecture.

d. Intranuclear genome position and nucleolar dominance

Nucleolar dominance in hybrids, the phenomenon where nucleolus organizers of one parental species are suppressed in the presence of nucleolus organisers from the other parental species, has been recognised since Navashin (1934). It occurs in hybrids between Hordeum species, between Hordeum and Secale, and between Aegilops or Triticum and Secale (see Kasha and Sadasivaiah, 1971). It also occurs regularly in each of the seven hybrids listed in Table 1.

Comparing the direction of nucleolar dominance with the
polarity of concentric parental genome separation in these hybrids
(Table 1) shows a strong tendency for nucleolar organisers (NORs) of
the central and peripheral genomes to be expressed and suppressed,
respectively. Indeed, dominance of central genome NORs over
peripheral genome NORs occurred regularly in diploid cells of all
five hybrids with strong concentric parental genome separation (P <
0.05). Moreover, this dominance was very strong insofar as no
evidence of activity in the suppressed organiser was seen, even at
the ultrastructural level. In contrast, nucleolar dominance was
much weaker in H. chilense x S. africanum and Ae. squarrosa x S.
cereale. Here evidence of occasional but limited activity at Secale
NORs was seen at the ultrastructural level, and in the former, at
the light microscopical level. Thus, for the reasons outlined in
section 6c, it is probably significant that: first, nucleolar
dominance was strong in all five hybrids where concentric genome
separation was strongly expressed, but weak in both hybrids where it
was weakly expressed; and second, that the genome whose centromeres
tended to be centrally located always expressed complete nucleolar
dominance except when the centromeres of small chromosomes tended
weakly to surround those of larger chromosomes, so that the relative
locations of NORs at interphase was unknown.

If the activity of genes affecting the somatic phenotype of
these hybrids is higher in the peripheral genome, as suggested
above, then one might expect NORs of their peripheral genomes to be
dominant. However, the opposite is normally true. Is there a
reason why rRNA genes behave differently from most other genes?
Ultrastructural studies, including reconstructions of whole nuclei
show that active NORs are almost invariably located in the central
domain of somatic interphase nuclei of all the hybrids listed in
Table 1, and their parental species. (N.B. The inner domain of a
spherical nucleus is defined as that half of the total nuclear
volume contained by approximately 80% of its inner radius.)
Moreover, their nucleoli tend strongly to be similarly located, and
rarely contact the nuclear envelope. Clearly, therefore, the
intranuclear distribution of active NORs in somatic cells is not
random. If it were, more than 50% of nucleoli would be in contact
with the nuclear envelope (Fig. 5a).

The NOR is unique in transcribing products which form so large
an associated intranuclear structure. Indeed, nucleolar material
occupied up to 8% of the total nuclear volume in cells of the
hybrids listed in Table 1, and of their parents. Consequently,
unlike any other active gene or chromosome segment, an active NOR
would seriously disrupt the normal form of somatic nuclear
architecture, were it to remain located near the nuclear surface
during interphase.

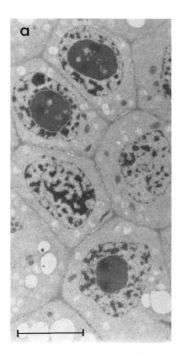
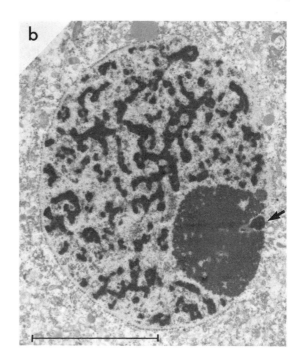

Fig. 5.   Sections of (a) interphase nuclei in a root-tip meristem of
           Hordeum chilense x Secale africanum, and, (b) a nucleus
           near the start of first meiotic prophase in Triticum
           aestivum cv. Chinese Spring, showing the typical positions
           of nucleoli in the central and peripheral nuclear domains,
           respectively, at these different developmental stages.
           Note that a nucleolus organizing chromosome (arrowed) is
           very close to the nuclear membrane in (b). Bar = 5 μm.

     The location of active NORs in somatic nuclei is clearly not
random. Neither is their central location the result of a packing
phenomenon. This can be concluded from the observation that near to
the start of meiosis the nucleolus regularly moves from a central to
a peripheral position, usually in contact with the nuclear membrane,
in all of the parental species of the hybrids listed in Table 1
(Fig. 5b). This transfer of active NORs from a central to a
peripheral domain is probably of functional significance. As noted
in section 5 the central position is regularly occupied by the same
chromosome(s) at meiosis in many species, and perhaps in grasses too
(Heslop-Harrison and Bennett, 1984). Be that as it may, it is clear
that the intranuclear location of nucleoli is regularly tissue
specific and, hence, must be under genotypic control. Moreover, if

the transcription of most genes is usually, or more efficiently, accomplished near the nuclear surface, then locating an active NOR and its large nucleolus in the central intranuclear domain may be of functional significance. Locating the NOR in the central domain is the only way to ensure that both (1) the nucleolus does not occupy any part of the peripheral domain, and (2) the average distance from the nuclear membrane of non-NOR chromatin in the nucleus, is near the minimum possible.

It seems fair to conclude, that first the behaviour of NORs in the present hybrids, i.e. to be active in a central, but suppressed in a peripheral intranuclear position, is no different from that seen in corresponding tissues of their parental species; and second, this behaviour which is probably the opposite to that for most other genes, may have the same functional significance in both parents and hybrids. Nicoloff et al. (1979) argued that the mechanism of nucleolar dominance for barley translocation stocks and interspecific hybrids is the same. Thus, it is suggested here that NOR activity is in general affected by intranuclear position, such that only NORs located in a domain where expression is normal for that cell type, usually form large nucleoli. In particular, it is suggested that nucleolar dominance in the hybrids listed in Table 1, exemplifies this phenomenon. No absolute relationship between intranuclear position and NOR activity is suggested, since NORs may be active in central or peripheral intranuclear domains in different tissues in these organisms. However, it is suggested that within the genetically controlled polarity normal for their somatic interphase nuclei, some site at, or near, the NOR registers its intranuclear location and activates rDNA transcription if it is central, but suppresses rDNA transcription if the NOR if peripheral.

e. The death of mapped genes on NOR bearing chromosome arms

As noted above, active NORs are regularly located in the central domain of mitotic interphase nuclei (Fig. 5a). Moreover, unpublished preliminary studies of reconstructed prophase and interphase nuclei in Hordeum hybrids show that much of the active NOR bearing arm is similarly located, while the non-NOR bearing arm of the same chromosome tends to occupy a more peripheral position, as shown diagrammatically in Figure 6. This regularly different intranuclear arrangement of the two arms of a single chromosome, known only for NOR bearing chromosomes, suggests a test of the hypothesis that a deep intranuclear position tends to inhibit the effectiveness of transcription for most genes. If this idea is true, then the NOR bearing arm of a biarmed nucleolus organising chromosome, which is invariably attached to a centrally located nucleolus, should be relatively poor in mapped genes. On the other hand, the other arm which lacks a NOR, and is free to occupy a peripheral position near to the nuclear membrane at interphase, should have a higher than normal frequency of mapped genes. These

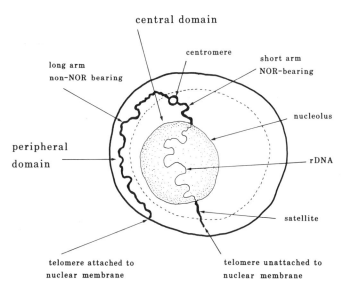

Fig. 6.   A diagrammatic representation of the typical intranuclear
          locations at interphase of the segments of a satellited
          chromosome with an active NOR in Hordeum vulgare and its
          diploid hybrids.

expectations have been tested in the only three crop plant species
with biarmed NOR chromosomes, and more than 100 genes mapped on
their genetic linkage maps, namely, H. vulgare, Lycopersicon
esculentum and Z. mays. Species with fewer mapped genes are unable
to provide meaningful tests. For the purposes of the test, the
expected number of genes per chromosome arm was calculated on the
assumption that the distribution of mapped genes should be directly
proportional to chromosome arm length or volume, and hence to arm
DNA content (Bennett et al., 1982). All three species tested had a
dearth of mapped genes, compared with expectation, in the NOR
bearing arm of the test chromosome. This dearth approached
significance in H. vulgare and Z. mays, was significant for L.
esculentum and very highly so for the pooled results for all three

Table 3.   The significant dearth of mapped genes on the NOR bearing arms of biarmed nucleolar organising chromosomes of three crop species

Species	Number of mapped genes or loci on genetic linkage map[a]	Linkage group with NOR	% of haploid complement[c] — In NOR bearing arm	% of haploid complement[c] — In non-NOR bearing arm	Number of mapped genes — In NOR bearing arm — Expected		Obtained	P of (Ob.) > Ob.	Number of mapped genes — In non-NOR bearing arm — Expected		Obtained	P of < Ob.
1. Hordeum vulgare L.	111	6[b]	6.66	7.14	7.39	>	3	0.057	7.93	<	9	0.603
2. Lycopersicon esculentum Mill.	222	2	1.34	8.76	2.97	>	0	0.050	19.45	<	34	0.999
3. Zea mays L.	171	6	3.00	5.83	5.13	>	2	0.110 (pooled) <0.001	9.97	<	13	0.800 (pooled) <0.01

a.   The numbers of mapped genes or loci on genetic linkage maps are taken from Figure 1 in Tsuchiya (1983), for H. vulgare; from O'Brien (1982), for L. esculentum; and from Neuffer et al. (1968), for Z. mays.   The NOR and all mapped centromeres are excluded from the count.

b.   H. vulgare has two chromosomes with NORs.   However, the centromere has not been mapped on chromosome 7, and so the above test can be applied only to linkage group 6, where the numbers of genes mapped to both arms are known.

c.   The percentage of the haploid complement in the NOR, and non-NOR bearing arms were calculated from chromosome lengths in L. esculentum obtained in light microscope studies of pachytene bivalents (Barton, 1950), and from EM reconstruction results for relative chromosome arm volume in root tip cells at metaphase for H. vulgare (Heslop-Harrison and Bennett, 1983a) and Z. mays (Bennett and Smith, unpublished).

species (Table 3). Moreover, the number of mapped genes on the
non-NOR bearing arms of the test chromosomes was higher than
expectation for all three species, significantly so for L.
esculentum and the pooled data.

The above test cannot distinguish between the effect of a
normally active major NOR on gene distribution on adjacent segments,
and the effect of intranuclear position itself, but either way the
results indicate some sort of position effect on gene distribution
in the karyotype. It should be noted that the expectations tested
above apply to organisms with only a single NOR in their haploid
complement which must be expressed to form the nucleolus (e.g. L.
esculentum and Z. mays), and to chromosomes whose NOR is invariably
expressed in organisms with more than one satellited chromosome
(e.g. chromosome 6 in H. vulgare). They may not apply to
NOR bearing chromosomes in organisms with several NOR bearing
homoeologues or heterologues per haploid genome (e.g. breadwheat and
man with, respectively, 4 and 5 NOR bearing chromosomes per haploid
genome) where NOR activity may be suppressed on different
heterologues or homoeologues in different cell types.

7. THE IMPORTANT ROLE OF INTRANUCLEAR POSITION

Evidence from reconstructed nuclei suggests that there may be
a relationship in grass species and hybrids between what at first
sight appears to be four unrelated characters: namely, chromosome
condensation, uniparental chromosome elimination, nucleolar
suppression, and a tendency for characters of one parent to dominate
the gross hybrid phenotype. The constant factor, which appears to
link these otherwise disparate characters into a unified pattern of
behaviour, is intranuclear chromosome position. A fifth character,
the distribution of genes on NOR bearing chromosomes, might be
similarly determined.

Figure 7 summarises this pattern diagrammatically showing that
occupation of a peripheral location predisposes for less condensed
chromatin at mitotic metaphase. In interspecific hybrids occupation
of the peripheral domain by one parental genome predisposes for its
elimination at mitosis. However, unless or until that occurs, it
predisposes for dominance of genes which determine the gross plant
phenotype, but for suppression of rDNA gene activity. Conversely,
occupation of a central nuclear position determines increased
relative chromatin condensation at metaphase in species, and in
hybrids it predisposes for uniparental genome retention to form
haploids, but weak phenotypic expression and nucleolar dominance.
It is not suggested that gene activity is controlled solely or even
mainly by variation in intranuclear position, but that this
character may be one important factor contributing to such control.
It is suggested that: (1) the activity or the effectiveness of all
genes may be affected by their intranuclear positions, (2) most

PERIPHERAL DOMAIN   CENTRAL DOMAIN

1. Chromosomes here tend to   1. Chromosomes here tend to
    condense less at mitotic metaphase   condense more at mitotic metaphase

2. Predisposes for mitotic   2. Predisposes for non-elimination
    chromosome elimination   of chromosomes at mitosis

3. Genes here tend to dominate   3. Genes here tend not to be expressed
    the plant phenotype   in the plant phenotype

4. NOR genes here   4. NOR genes here
    tend to be repressed   tend to be expressed

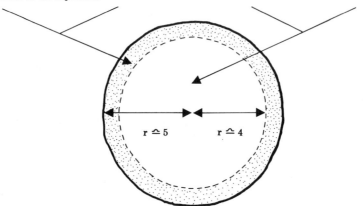

Fig. 7. A diagrammatic summary of the suggested effects on
         chromosome behaviour and plant phenotype which appear to
         depend on whether a chromosome or genome is in the central
         or the peripheral intranuclear domain in somatic cells.

genes are normally transcribed more actively or effectively near to
the nuclear surface, while rDNA genes, and perhaps some others, are
normally transcribed deeper in the nucleus. Moreover, it seems fair
to conclude that the internal nuclear architecture may indeed play
an important role in controlling aspects of chromosome behaviour and
gene activity and, thereby, the plant phenotype. In other words, at
this level too, function is related to form.

8.   THE NUCLEOTYPE AND INTRANUCLEAR POSITION EFFECTS

     The present work has emphasised the possible importance of

distance from the nuclear membrane for transcription in grass
hybrids. However, it is too soon to be certain which, if any,
aspects of nuclear architecture are causally responsible for
phenotypic effects. Nuclear gradients for polarity, latitude,
longitude, and depth may all be important, as may the position of a
gene on the chromosome, and of the chromosome within the natural
karyotype (Bennett, 1984b). Moreover, future studies should pay
attention to the possibility of nucleotypic effects given that the
determination of the order of heterologues within the natural
karyotype may have both a nucleotypic and a genotypic basis
(Bennett, 1982). Intranuclear position effects may depend more on
physical absolutes than on relative locations, so that the
expectations may differ greatly for different species with,
respectively, very small and very large genomes. For example,
transcription may be more affected by the absolute distance of a
gene from the nuclear surface and/or the nuclear surface area to
volume ratio, than by whether genes are relatively near to, or far
from, the nuclear envelope (Fig. 8). If so, intranuclear position
effects on gene activity and phenotypic expression might be weak or
absent in 1C to 4C nuclei of diploid species with very small
genomes, but increase in strength with increasing species DNA
C-value, nuclear diameter, and, decreasing nuclear surface area to
volume ratio.

     If intranuclear position effects do have such a nucleotypic
basis then this would have clear consequences for the internal
architecture of highly polytene or polyploid nuclei, including those
of species with low DNA C-values. In such large nuclei most
chromosome segments which are actively transcribed would be expected
to lie near the nuclear envelope, and this structure might be
expected to be convoluted to increase its area. Thus, it may be
significant that in the large polytene Drosophila salivary gland
nuclei studied by Mathog et al. (1984), "a large percentage of each
chromosome ... runs along the envelope's inner surface". Moreover,
the anatomy of the very large antipodal cell nuclei in grasses
accords with these expectations (Fig. 1d).

9. INTRANUCLEAR GENOME ARCHTECTURE AND ITS ROLE IN DEVELOPMENT

     The various observations described above lead one to ask
whether, within a species, features of the internal nuclear
architecture recur in nuclei of the same tissue but vary in a
regular manner between nuclei from different tissues or stages of
development; and, if so, whether such variation reflects or directs
developmental change in some higher organisms. Certainly there are
regular, stage-specific variations in the internal architecture of
interphase nuclei (e.g. see Fig. 5). Moreover, recent work on the
polytene chromosomes of Drosophila salivary glands (Agard and Sedat,
1983; Mathog et al., 1984) showed that each polytene chromosome arm
folds up in a characteristic way, and contacts the nuclear surface

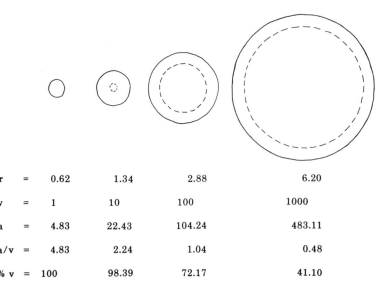

r	=	0.62	1.34	2.88	6.20
v	=	1	10	100	1000
a	=	4.83	22.43	104.24	483.11
a/v	=	4.83	2.24	1.04	0.48
% v	=	100	98.39	72.17	41.10

Fig. 8. A diagrammatic illustration of how nucleotypic characters
may affect the strength of intranuclear position effects if
gene action is affected by absolute distance from the
nuclear surface. Note that the nuclear volume (v)
increases as the cube of the radius (r) while nuclear
surface area (a) increases as its square, so that as v
increases the surface area to volume ratio (a/v) decreases,
as does the percentage of v within 1 micron of the nuclear
surface (% v) for nuclei where r > 1. N.B. Values of r, a
and v are um, um^2 and um^3, respectively, and the dotted
line is 1 um from the nuclear surface. Bar = 1 μm.

at specific sites. Thus, it is not difficult to envisage that some
major stages of development could each have a unique intranuclear
genome architecture with different chromosome arms or segments
regularly near to, or far from, the nuclear membrane and/or each
other, and that together these different structural features could
help to modulate the patterns of gene activity characteristic of
each developmental stage.

The question of whether or not the internal architecture of
nuclei is causally related to aspects of development is now of major
theoretical importance in biology, and may be of great practical
significance. Unfortunately, our knowledge of intranuclear
architecture is still woefully inadequate, especially for mitotic
interphase. Thus, it will not be possible to answer this question

without a broad and sustained research effort, and perhaps, without first developing new techniques which can concurrently resolve the locations and activities of many genes within a single nucleus.

REFERENCES

Agard, D.A., and Sedat, J.W., 1983, Three-dimensional architecture of a polytene nucleus, Nature, 302: 676-681.

Albertson, D.G., and Thomas, J.N., 1982, The kinetochores of Caenorhabditis elegans, Chromosoma, 86: 409-428.

Barton, D.W., 1950, Pachytene morphology of the tomato chromosome complement, Am. J. Bot., 37: 639-643.

Bennett, M.D., 1970, Natural variation in nuclear characters of meristems in Vicia faba, Chromosoma, 29: 317-335.

Bennett, M.D., 1972, Nuclear DNA content and minimum generation time in herbaceous plants, Proc. Roy. Soc. Lond. B, 181: 109-135.

Bennett, M.D., 1977, The time and duration of meiosis, pp 201-226 in: "A discussion on the meiotic process," organised by R. Riley, M.D. Bennett, and R.B. Flavell, Phil. Trans. Roy. Soc. Lond. B, 277: 183-376.

Bennett, M.D., 1982, Nucleotypic basis of the spatial ordering of chromosomes in eukaryotes and the implications of order for genome evolution and phenotypic variation, in: "Genome Evolution," G.A. Dover and R.B. Flavell, eds., Academic Press, London, pp 239-261.

Bennett, M.D., 1983, The spatial distribution of chromosomes, in: "Kew Chromosome Conference II," P.E. Brandham and M.D. Bennett, eds., George Allen & Unwin, London, pp 71-79.

Bennett, M.D., 1984a, Towards a general model for spatial law and order in nuclear and karyotypic architecture, in: "Chromosomes Today vol. 8," M.D. Bennett, A. Gropp and U. Wolf, eds., George Allen & Unwin, London, pp 191-202.

Bennett, M.D., 1984b, The genome, the natural karyotype and biosystematics, in: "Plant Biosystematics,: W.F. Grant, ed., Academic Press, Canada, pp 41-66.

Bennett, M.D., 1984c, Premeiotic events and meiotic chromosome pairing, in: "Controlling events in meiosis, 38th Symposium of the Society for Experimental Biology," C.W. Evans and H.G. Dickenson, eds. Company of Biologists, Cambridge, in press.

Bennett, M.D., and Rees, H., 1969, Induced and developmental variation in chromosomes of meristematic cells, Chromosoma, 27: 226-244.

Bennett, M.D., Rao, M.K., Smith, J.B., and Bayliss, M.W., 1973, Cell development in the anther, the ovule, and the young seed of Triticum aestivum L. cv. Chinese Spring, Phil. Trans. Roy. Soc. Lond. B, 266: 38-81.

Bennett, M.D., Smith, J.B., Simpson, S., and Wells, B., 1979, Intranuclear fibrillar material in cereal pollen mother cells, Chromosoma, 71: 289-332.

Bennett, M.D., Smith, J.B., and Ward, J.P., 1982, Ultrastructural studies of cereal chromosomes, Annual Report, Plant Breeding Institute, pp 75-78.

Bennett, M.D., Smith, J.B., Ward, J., and Jenkins, G., 1981, The relationship between nuclear DNA content and centromere volume in higher plants, J. Cell Sci., 47: 91-115.

Bennett, M.D., Smith, J.B., Ward, J.P., and Finch, R.A., 1982, The relationship between chromosome volume and DNA content in unsquashed metaphase cells of barley (Hordeum vulgare L. cv. Tuleen 346), J. Cell Sci., 56: 101-111.

Bothmer, R. von, Flink, J., Jacobsen, N., Kotimake, M., and Landstrom, T., 1983, Interspecific hybridisation with cultivated barley (Hordeum vulgare L.), Hereditas, 99: 219-244.

Bothmer, R. von, Jacobsen; N., Jorgensen, R.B., and Linde-Laursen, I., 1984, Haploid barley from the intergeneric cross Hordeum vulgare x Psathyrostachys fragilis, Euphytica, in press.

Brown, R., 1833, On the organs and mode of fecundation in Orchideae and Asclepiadeae, Trans. Linn. Soc., 16: 685-745. (Also privately printed, 1831.)

Cremer, T., Baumann, H., Nakanishi, K., and Cremer, C., 1984, Correlation between interphase and metaphase chromosome arrangements as studied by laser-UV-microbeam experiments, in: "Chromosomes Today, vol. 8," M.D. Bennett, A. Gropp and U. Wolf, eds., George Allen & Unwin, London, pp 203-212.

Darlington, C.D., 1957, Messages and movements in the cell, in: "Conference on Chromosomes," Willink, Zwolle, pp 1-33.

Darlington, C.D., and La Cour, L.F., 1976, "The Handling of Chromosomes," George Allen & Unwin, London.

Finch, R.A., 1983, Tissue-specific elimination of alternative whole parental genomes in one barley hybrid, Chromosoma, 88: 386-393.

Finch, R.A., and Bennett, M.D., 1983, The mechanism of somatic chromosome elimination in Hordeum, in: "Kew Chromosome Conference II," P.E. Brandham and M.D. Bennett, eds., George Allen & Unwin, London, pp 147-154.

Finch, R.A., Smith, J.B., and Bennett, M.D., 1981, Hordeum and Secale mitotic genomes lie apart in a hybrid, J. Cell Sci., 52: 391-403.

Heslop-Harrison, J., 1972, Sexuality in Angiosperms, in: "Plant Physiology A treatise," VIC, F.C. Steward, ed., New York and London, Academic Press, pp 133-289.

Heslop-Harrison, J.S., 1983, Chromosome disposition in Aegilops umbellulata, in: "Kew Chromosome Conference II," P.E. Brandham and M.D. Bennett, eds., George Allen & Unwin, London, pp 63-70.

Heslop-Harrison, J.S., and Bennett, M.D., 1983a, Prediction and analysis of spatial order in haploid chromosome complements, Proc. Roy. Soc. Lond. B, 218: 211-223.

Heslop-Harrison, J.S., and Bennett, M.D., 1983b, The spatial order of chromosomes in root-tip metaphases of Aegilops umbellulata,

Proc. Roy. Soc. Lond. B, 218: 225-239.

Heslop-Harrison, J.S., and Bennett, M.D., 1983c, The positions of centromeres on the somatic metaphase plate of grasses, J. Cell Sci., 64: 163-177.

Heslop-Harrison, J.S., and Bennett, M.D., 1984. Chromosome order - possible implications for development, in: "10th Symposium of the British Society for Developmental Biology," H. Macgregor and A.P. Swann, eds., supplement to J. Embryol. exp. Morphol., in press.

Ho, K.M., and Kasha, K.J., 1975, Genetic control of chromosome elimination during haploid formation in barley, Genetics, 81: 263-275.

Hughes-Schrader, S., 1983, Chromosomal segregation mechanisms in ant-lions (Myrmeleontidae, Neuroptera), Chromosoma, 88: 256-264.

Kasha, K.J., and Sadasivaiah, R.S., 1971, Genome relationships between Hordeum vulgare L. and H. bulbosum L., Chromosoma, 35: 264-287.

Lacadena, J.R., Jodar, B., and Ferrer, E., 1983, Suprachromosomal organisation: cytological rationale and experimental evidence, in: "Kew Chromosome Conference II," P.E. Brandham and M.D. Bennett, eds., George Allen & Unwin, London, pp 81-90.

Lewin, R., 1981, Do chromosomes cross-talk?, Science, 214: 1334-1335.

Lewis, K.R., and John, B., 1963, "Chromosome Marker", Churchill, London, pp 224-225.

Linde-Laursen, I., and Bothmer, R. von, 1984, Somatic cell cytology of the chromosome eliminating, intergeneric hybrid Hordeum vulgare x Psathyrostachys fragilis, Can. J. Genet. Cytol., in press.

Maraldi, N.M., Marinelli, F., Amati, S., Cocco, L., Santi, P., Papa, S., Galanzi, A., Caramelli, E., Antonucci, A., and Mazzotti, G., 1983, Association between centriole and nuclear matrix in human lymphocytes, J. Submicrosc. Cytol., 15: 883-888.

Mathod, D., Hochstrasser, M., Gruenbaum, Y., Saumweber, H., and Sedat, J., 1984, Characteristic folding pattern of polytene chromosomes in Drosophila salivary gland nuclei, Nature, 308: 414-421.

Navashin, M., 1934, Chromosome alterations caused by hybridisation and their bearing upon certain general genetic problems, Cytologia, 5: 169-203.

Neuffer, M.G., Jones, L., and Zuber, M.S., 1968, "The mutants of maize," Crop Science Society of America, Madison.

Nicoloff, H., Anastassova-Kristeva, M., Rieger, R., and Kunzel, G., 1979, 'Nucleolar dominance' as observed in barley translocation lines with specifically reconstructed SAT chromosomes, Theor. Appl. Genet., 55: 247-251.

O'Brien, J., 1982, "Genetic maps 2," National Cancer Institute, Frederick, Maryland, 21701, USA.

Pardue, M.L., Nordheim, A., Moller, A., Weiner, L.M., Stoller, B.D.,

and Rich, A., 1984, Z-DNA and chromosome structure, in:
"Chromosomes Today, vol. 8," M.D. Bennett, A. Gropp and U.
Wolf, eds., George Allen & Unwin, London, 34-35.

Rhoades, M.M., 1952, Preferential segregation in maize, in: "Corn
and Corn Improvement," Academic Press, New York, pp 123-219.

Schwarzacher, T., Finch, R.A., and Bennett, M.D., 1984, Genome
elimination in Hordeum hybrids, in: "Chromosomes Today, vol.
8," M.D. Bennett, A. Gropp and U. Wolf, eds., George Allen &
Unwin, London, p 339.

Subrahmanyam, N.C., and Kasha, K.J., 1973, Selective chromosomal
elimination during haploid formation in barley following
interspecific hybridisation, Chromosoma, 42: 111-125.

Simpson, E., Snape, J.W., and Finch, R.A., 1980, Variation between
Hordeum bulbosum genotypes in their ability to produce
haploids of barley, Hordeum vulgare, Z. Pflanzenzüchtg.,
85: 205-211.

Tsuchiya, T., 1983, Current linkage maps of barley, Barley Genetics
Newsletter, 13: 101-106.

Waldgate, R., 1983, Pattern of life, The Guardian, (May 12th), p 19.

Wang, A.H.-J., Quigley, F.J., Kolpak, J.L., Crawford, J.H., van
Boom, G., van der Marel, G., and Rich, A., 1979, Molecular
structure of a left-handed double helical DNA fragment at
atomic resolution, Nature, 282: 680-686.

SELECTING BETTER CROPS FROM CULTURED CELLS

Carole P. Meredith

Department of Viticulture and Enology
University of California
Davis, California 95616

INTRODUCTION

It is now well-established that populations of cultured plant cells are genetically variable and that new plant genotypes can be obtained from such cells, both by the selection of phenotypic variants with subsequent plant regeneration and by screening populations of plants regenerated from cultured cells. A critical question remains, however. Can this variability really contribute to the genetic improvement of crop plants?

Effective identification and exploitation of the genetic variation in cultured plant cells requires a clear understanding of both 1)the genetic basis of selection and 2)the phenotype being sought. The first requirement includes the recognition of the difference between a variant and a mutant, the use of appropriate selection methodology, and an understanding of the relationship (or lack thereof) between cellular and whole-plant phenotypes. The second requirement adds several other considerations that are related to the phenotype being sought--the design of an effective selection strategy, knowledge of the cellular manifestation of the phenotype, and an appreciation of the full complexity of the phenotype. A sound selection system must take into account all of these factors for to overlook even one will greatly reduce the probability of success.

VARIANTS VS. MUTANTS

While the term "mutant" is frequently, and casually, used in reports of selection in plant cell cultures, "variant" is usually a

more accurate word. The selected cells may represent novel phenotypes, but they are often insufficiently characterized to justify
their designation as novel genotypes. The exact definition of
"mutant" is certainly less clear today than it was a few years ago
but, in general, its use should be reserved for only those cases
where genetic stability and sexual transmission of a trait have been
demonstrated. Where this cannot be shown because of regeneration or
fertility difficulties, convincing molecular or biochemical evidence
may also be acceptable (e.g. altered nucleotide sequence or altered
gene product). There are certainly many bases for an altered
phenotype, both genetic and otherwise. It is becoming increasingly
clear that genetic changes are frequently the result of phenomena
other than the classical point mutation, e.g. gene amplification,
transposition events, chromosomal rearrangements. Where reasonably
stable, phenotypic variants resulting from these genetic phenomena
may fall within the working definition of "mutant".

To elucidate the genetic basis of a selected phenotype, one can
ask several questions. Is the phenotype stable in the absence of
selection? Is it expressed in regenerated plants or in cell cultures
derived therefrom? Is it transmitted to the progeny of regenerated
plants? Affirmative answers to all these questions provide
convincing evidence that a selected phenotype does indeed represent
an altered genotype. Unfortunately, these questions have not been
asked of the vast majority of selected cell lines. In a large number
of cases investigations have been so limited by the inability to
recover plants from the selected cells. For a small, but significant, minority of the variants reported, however, the selected
phenotype has been thoroughly examined for stability in the absence
of selection, expression in regenerated plants or cultures derived
therefrom, and sexual transmission. There is no question that _in
vitro_ selection can result in new plant genotypes.

SELECTION VS. SCREENING

Specific variants may be identified and isolated by the application of some selection pressure that permits the preferential
survival and/or growth of a desired phenotype, but selection is not
essential. A great deal of genetic variation (usually called somaclonal variation) may be expressed in plants regenerated from cell
cultures to which no selection has been applied. Some phenotypes may
not be expressed in cultured cells, and for others no effective
selection strategy may yet be conceivable. Those phenotypes are
better isolated by simply screening regenerated plants. Non-
selective screening can also be applied directly to cultured cells.
For example, pigment variants can be easily identified by visually
scanning large numbers of cultured cells (Schieder, 1976). But in
those cases in which the phenotype _is_ expressed in culture and a
selection strategy _can_ be conceived, the effective use of selection

pressure is likely to greatly increase the efficiency of recovery of new phenotypes. The particular selection method employed can greatly influence the nature of the variants obtained and so should be chosen with great care.

SELECTION METHODS

Negative selection, in which growing cells are preferentially killed and non-growing cells survive, is appropriate for obtaining auxotrophic mutants, those lacking a metabolic function such that their growth depends on a nutritional supplement. While negative selection methods have been employed with plant cells (Carlson, 1970; Polacco, 1979), they have proven difficult and most auxotrophs have been obtained by non-selective screening (e.g. King et al., 1980; Gebhardt et al., 1981). Positive selection, in which only the desired cell type grows, favors the isolation of resistant mutants. Most mutants obtained in plant cell cultures to date are the results of positive selection (Maliga, 1980).

Whether negative or positive selection methods are used, the culture method employed in selection experiments is also of critical importance. Selection can be imposed on virtually all kinds of cultures, but the results can vary markedly. Callus selection, in which the selection pressure is imposed on pieces of callus (e.g. Gengenbach et al., 1975) can be very inefficient. Cells within a callus piece are not uniformly exposed to the selective agent since not all are in direct contact with the medium. Callus cells are in close contact with each other, thus permitting cross-feeding and promoting escape from the selection. Those rare cells of the desired phenotype that may exist in a callus piece can easily go undetected if they are physically surrounded by non-growing cells all around them. So not only are the desired cells less likely to be detected with callus selection, but the surviving sectors are likely to consist of a mixed population of cells that includes many that simply escaped the selection pressure.

Selection in suspension cultures (e.g. Widholm, 1976) is also complicated by a number of factors. The selection pressure is applied to an entire flask of cells, and eventual recovery of growth in the flask is taken as an indication that the flask population is now enriched for the desired variant cell type. However, with this method it is not possible to discriminate between the rapid growth of a rare variant cell type and the slow adaptation to the selective agent of the entire cell population. In addition, it is not possible with this method to separate distinct variants arising from independent events. If the original cell population contained several variants of the desired phenotype, the recovered cell population will be a mixture and further characterization will certainly be confused.

Selection via <u>plated cells</u> (e.g. Chaleff and Parsons, 1978) overcomes the problems associated with selection in callus and suspension cultures. The cells can be plated directly on agar or modifications that permit the use of liquid medium and/or feeder cells can be incorporated to increase the efficiency of variant recovery (Weber and Lark, 1979; Horsch and Jones, 1980; Conner and Meredith, 1984). Selection with plated cells offers the advantage that all cells are uniformly exposed to the selective agent. The cells are not in contact so cross-feeding is minimized. Rare variants can easily be distinguished from general adaptation of the cell population. Rather than being mixed, distinct variants can be recognized as separate colonies on a plate, although some may certainly be chimeras since the plating units are often not single cells (Portnoy and Murphy, 1983). Additionally, selected colonies are not necessarily genetically distinct. They may be clonally derived from the same mutation event. Selection with plated cells is somewhat more laborious than either callus selection or selection in suspensions.

Some have chosen to employ <u>plated protoplasts</u> for selection experiments to minimize the selection of chimerical colonies. Protoplasts are, by definition, single cells and would be expected to give rise to a genetically homogeneous colony. The possibility of a chimerical colony still exists, however, since a colony may arise from two or more abutting protoplasts. An advantageous feature of employing protoplasts is that they can be isolated from differen- tiated tissues, such as leaf mesophyll (Nagata and Takebe, 1972) or root tissue (Gronwald and Leonard, 1982), thereby suggesting the possibility of imposing selection pressures on differentiated functions that may not be present in established cell cultures. Protoplast selection is by far the most laborious of the techniques, however. Its major advantage, that of insuring the single cell origin of selected variants, is not so absolute to eliminate the necessity of cloning variants to completely eliminate the possibility of chimeras.

A selection pressure may be applied suddenly (e.g. Chaleff and Parsons, 1978) or in a step-wise fashion (e.g. Gengenbach et al., 1977). Step-wise selection may favor novel genotypes resulting from two particular mechanisms of genetic change, mutations in organelle genomes and gene amplification, and both by the same principle. A cell carrying a single mutated mitochondrion or a single copy of a nuclear mutation might not express the mutant phenotype strongly enough to survive sudden rigorous selection, but might preferentially survive weak selection. Gradually increasing selection pressure may favor cells possessing an increased proportion of mutated mito- chondria in the organelle population or cells in which the mutated nuclear gene is sufficiently amplified to confer a stronger phenotype.

The ploidy level of the cells will also influence the recovery of mutants. Haploid cells should permit the increased recovery of recessive mutations while diploid cells favor dominant mutations. But haploid cells are difficult to obtain in most crop plants, and once obtained are difficult to maintain in a haploid condition.

EXPRESSION AND TRANSMISSION IN REGENERATED PLANTS

It is not always essential that a trait selected in cultured cells be expressed in a regenerated plant. For example, if the objective of the selection is to obtain a marker for use with cultured cells, it may only be necessary that the trait be expressed in culture. There are also cases in which expression in the regenerated plant may be required but sexual transmission of the trait is unimportant, for example in the case of a vegetatively propagated crop. In such a crop, stability of expression is essential but sexual transmission may be irrelevant. However, for the large majority of agricultural applications, expression and sexual transmission of the selected trait in regenerated plants are essential.

There are a number of reasons why a trait selected at the cellular level may not be expressed in regenerated plants, and they do not reflect the weakness of the approach, but rather the developmental complexity of higher plants. These reasons have been covered well by Chaleff (1983) and will be mentioned only briefly here. Obviously, many variants selected in vitro are not the result of genetic change, but rather a change in gene expression or biochemical activity. It is not surprising that such changes might not persist through the developmental turmoil of regeneration, especialy when many epigenetic events are not transmitted through zygotic embryogenesis in the normal life cycle of plants (Meins and Binns, 1979). Other variants may be the result of genetic changes of an unstable nature, e.g. gene amplification, and likewise may not persist. But even "real" mutations in the classical sense may not always be expressed in plants. For example, if the gene of interest (or an auxiliary gene whose function is required for expression of the first gene) is one that is developmentally regulated, it may be silent in the relevant cells in the plant. The phenotypic segregation of such mutations may be monitored in cultures derived from the progeny of regenerated plants. Additional possible explanations related to the metabolic complexity of higher plants have been presented by Chaleff. Suffice it to say that the absence of expression of a selected trait in a regenerated plant is not necessarily an indication that the trait was not the result of a stable genetic change.

It is to be expected, then, that even a soundly conceived

selection program may often not succeed in producing a new plant
phenotype. The many possibilities for non-expression of a selected
trait in a regenerated plant argue for large-scale selection
programs. If all the hopes of a selection program rest on a single
selected variant, the chance of failure is considerable. It may be
necessary to initially select a large number of variants to obtain
one stable mutant that is expressed in the appropriate tissue and at
the right time in the regenerated plant. Most reports of in vitro
selection are based on only one or very few selected variants. If
the variants prove unstable or do not persist through regeneration,
the experiment is often considered unsuccessful and further attempts
to obtain that particular phenotype are abandoned. However, had the
selection been pursued on a larger scale it might have produced the
desired result.

SELECTION STRATEGIES

For some phenotypes, e.g. drug resistance, the selection strategy is obvious--the cells are simply challenged with an inhibitory
drug concentration in the culture medium. However, most agriculturally significant traits are much more complex and it is critical that
the development of a selection strategy be given very careful
consideration.

Selection imposed on cells in culture acts on cellular
phenotypes, necessitating an understanding of the cellular manifestation of the desired whole plant modification. For many agriculturally significant traits, unfortunately, we do not yet know what
modifications must occur in cells to produce the desired change in
the plant. For other important traits, the significant foundation of
knowledge that does exist is easily overlooked in the course of
designing a selection program. In cases where a selection strategy
may initially seem obvious, the important complexities of the
phenotype may not have been considered.

PHENOTYPIC COMPONENTS

On the other hand, the apparent complexity of a trait may deter
investigators from attempting to isolate variants. But some plant
phenotypes that at first may seem inaccessible because of their
complexity may, upon further analysis, be reduced to their essential
components, some of which may be more amenable to selection in cell
cultures. A particular phenotype of agricultural interest may be the
sum of biochemical, physiological, morphological, and developmental
components which may reside at the cellular, tissue, or organ
levels of organization. While tissue and organ characteristics may
be inaccessible, cellular components may lend themselves to
selection.

This concept may be illustrated by the example of drought resistance. A plant may resist water stress by a number of mechanisms, e.g. deep roots, thick cuticle, stomatal closure, early leaf drop, osmoregulation (Parsons, 1982). It is difficult to conceive of an in vitro selection strategy for obtaining deeper roots or a thicker cuticle, both organ characteristics, but it might be possible to select cells with altered osmoregulatory properties (A. Handa et al., 1983). A plant regenerated from such cells may exhibit increased drought resistance. Thus, the modification of one component of a phenotype could contribute to an overall change in plant phenotype.

The concept of phenotypic components may also explain why genotypic differences observed in culture in response to a stress may not always correspond to those observed in whole plants. When genotypes are compared as cultured cells, only the cellular components are being compared. A variety with significant organ or tissue phenotypic components, but a negligible cellular component, would perform relatively well as a whole plant but relatively poorly in cell culture. In contrast, a variety with significant cellular components, but minor tissue and organ components would be expected to perform well both as a whole plant and in cell culture. Thus the lack of correlation of in vitro and whole plant data for a particular trait is not sufficient reason for abandoning the in vitro approach without further investigation.

PHENOTYPES AMENABLE TO SELECTION

It is not sufficient that the desired cellular phenotype be known; to be selectable it must also be expressed in culture. Selection pressure can obviously only act on a function being expressed. It is quite conceivable, however, that the trait of interest may not be expressed in cultured cells, and for a number of reasons. The gene(s) governing the trait of interest (or auxiliary genes) may be subject to regulation such that it is not expressed in cultured cells. Or expression may be in an unrecognized form because we do not yet adequately understand the cellular manifestation of the phenotype. The phenotype may also be one that requires structures or specialized processes that do not exist in cultured cells. Obviously such a trait is not accessible to selection at the cellular level, but variants might well be isolated by screening regenerated plants.

With these limitations in mind, then, certain phenotypes can be identified as less amenable to genetic manipulation via cell culture selection at the present time. Examples of these include modifications of plant architecture or flavor components. The architecture of a plant is the result of the integration of organ systems, a degree of organization that cannot be found in cell cultures. Flavor components are generally the products of secondary metabolic path-

ways, not all of which generally operate in cultured cells, and
certainly not with the subtle orchestration involved in the production of flavor. Some phenotypes, however, particularly those based
on fundamental cellular processes, seem more accessible. These
include (but are not limited to) resistance to salinity, to mineral
stresses, and, in some cases, to disease--all phenotypes of
tremendous agricultural significance. At the present time, it seems
likely that _in vitro_ selection may make agriculturally valuable
contributions in all three areas.

RESISTANCE TO SALINITY

     While it is now well-accepted that differential salt resistance
in plants is under genetic control (Epstein et al., 1980), the
genetic basis is not clear. Salt resistance is generally regarded as
a polygenic trait (e.g. Venables and Wilkins, 1978; Humphreys, 1982),
although monogenic control has been reported (Abel, 1969). The
effects of salinity on plants are complex. Not only does salinity
expose plants to osmotic stress, thus restricting water uptake, but
it can also produce specific ion toxicities (e.g. Na^+, SO_4^{+2}, Cl^-,
CO_3^{-2}). Some crops are more sensitive to specific ion effects than
others (Epstein, 1972). Efforts to generate salt resistant crops
should take both these effects into account. It is wise to consider
both the specific agricultural location for which salt resistance is
sought as well as any specific ion sensitivities in the crop under
study.

     Most _in vitro_ selection programs have used NaCl as the selective
agent (e.g. Croughan et al., 1978; Mathur et al., 1980; Nabors et
al., 1980; Tyagi et al., 1981; Kochba et al., 1982; Rangan and Vasil,
1983; Wong et al., 1983). While NaCl is clearly a very important
factor in salt-affected soils, there may well be other toxic ions
that play an important role in certain agricultural situations. NaCl
selection is not likely to produce genotypes with resistance to toxic
ions other than Na^+ or Cl^-. While it may result in genotypes with
resistance to osmotic stress, NaCl selection may also produce a
halophytic type of genotype that is highly tolerant to the ionic
stresses, and not so much to osmotic stress (since it might take up
the Na^+ and Cl^- freely and thus never encounter the full osmotic
stress). Such a genotype would not be expected to perform well in a
saline situation in which other ions were contributing to salinity if
the species were particularly sensitive to any of these ions. NaCl
selection may thus not be an appropriate selective agent in all
cases. Several investigators have employed salt mixtures (Yano et
al., 1982; Nyman et al., 1983) and might thereby be more closely
simulating the agricultural situation.

     Differential ion toxicities have been observed in salinity
studies with cultured cells. Carrot (_Daucus carota_) callus has been

shown to be more sensitive to inhibition by a salt mixture than by a mannitol solution of equal osmotic potential. These cells were also much more sensitive to K^+ than other ions (Goldner et al., 1977). Citrus cells are also differentially sensitive to ions, Cl^- being particularly toxic. The resistance of a selected citrus line for Cl^- was greatly influenced by the cation, with resistance being expressed to NaCl but lost when Cl^- was provided as KCl (Kochba et al., 1982).

Salinity is clearly inhibitory to cultured cells and thus can be studied in culture, but are salt resistance mechanisms also of a cellular nature? Can salt resistance be selected at the cellular level? Plants can resist salinity both by excluding the toxic ions and compensating osmotically via the synthesis of compatible solutes, and/or by freely taking up the ions so that osmotic stress is avoided (Flowers et al., 1977; Greenway and Munns, 1980). The latter mechanism requires that ion toxicities be overcome. Both mechanisms of resistance could conceivably be expressed by cells in culture.

In intergeneric, interspecific, and intraspecific comparisons, genetic differences in salt resistance have been shown to be expressed in cultured cells, thus at least some resistance mechanisms are cellular in nature. Callus of _Beta vulgaris_, a salt-resistant species, is more resistant than callus of _Phaseolus vulgaris_, a salt-sensitive species. However, the salt resistance of two halophytes (one with salt glands and the other with succulence) is lost in culture, demonstrating that their resistance is not cellular, but dependent upon higher levels of organization, whereas the resistance of _B. vulgaris_ is cellular (Smith and McComb, 1981a). Another halophyte, _Distichlis spicata_, also is more resistant in culture than two glycophytic species, _Nicotiana sylvestris_ and _Zea mays_ (Warren and Gould, 1982). Two barley species, _Hordeum jubatum_ and _H. vulgare_, were compared both as callus and as plants; _H. jubatum_ was more resistant in both states (Orton, 1980). Likewise, callus cultures of the salt-resistant wild tomato species _Lycopersicon peruvianum_ and _L. pennellii_ are more resistant than callus of the salt-sensitive cultivated tomato _L. esculentum_ (Tal et al., 1978). Finally, the salt resistance of four alfalfa (_Medicago sativa_) genotypes was compared both in callus and in plants. Resistance was found in only one of the genotypes and it was expressed both in the plant and the callus (Smith and McComb, 1981b).

The facts that salinity affects cultured cells and that salt resistance mechanisms do operate in cultured cells strongly suggest that salt-resistant plants can be obtained by _in vitro_ selection. It is somewhat surprising, therefore, that to date only one case of sexual transmission of salt resistance so obtained has been reported (Nabors et al., 1980). By repeated selection for improved growth in the presence of NaCl, a salt-resistant cell line of _N. tabacum_ was obtained from which plants were regenerated. The plants exhibited greater NaCl resistance than plants regenerated from unselected cells and an elevated level of salt resistance was also expressed in two subsequent generations of seed progeny from these plants. The inheritance of the increased resistance did not fit a recognizable pattern and the genetic basis for this trait has not yet been elucidated.

There are a number of other reports in which selected salt resistance is retained through the plant regeneration process. Salt resistance selected in citrus cell culture is expressed in somatic embryos regenerated from the cells (Kochba et al., 1982). Some degree of resistance is expressed in rice (Oryza sativa) plants regenerated from salt-resistant callus (Yano et al., 1982). In other cases, resistance has been shown to be retained in callus derived from plants regenerated from salt-resistant callus (Dix, 1979; Tyagi et al., 1981).

It may be possible to obtain salt-resistant plants by selecting for resistance to only the osmotic stress component of salinity. Tomato cells resistant to elevated concentrations of polyethylene glycol (PEG) also show enhanced resistance to NaCl (Bressan et al., 1981). Such cells presumably are resistant by virtue of their ability to accumulate osmoregulatory compounds (S. Handa et al., 1983).

A problem that may limit selection for PEG resistance, and perhaps also NaCl resistance, is the remarkable capacity of cultured cells to adapt to osmotic stress. While cell populations probably do contain genetic variants with enhanced resistance, it may be difficult to distiguish them from wild-type cells that have adapted to the stress (A. Handa et al., 1983). It may be necessary to screen a large number of resistant selections in order to identify a few stable variants.

The resistance mechanisms of selected cell lines have been elucidated in some cases. Salt-resistant cell lines selected in alfalfa, N. tabacum, N. sylvestris, and N. tabacum/gossii have been shown to not exclude salt ions and are thus able to maintain the osmotic gradient necessary for continued function (Croughan et al., 1978; Heyser and Nabors, 1981; Dix and Pearce, 1981; Watad et al., 1983). In alfalfa, the resistant cell line actually exhibits a halophytic mode of resistance, in that it accumulates higher ion concentrations than the unselected line and actually seems to require elevated salt for normal growth. With N_+ tabacum/gossii, both resistant and sensitive cells take up Na^+ from a saline medium but the resistant cells, as in the case of alfalfa, are able to maintain high K^+ concentrations, a characteristic of many true halophytes (Epstein, 1972). In these four cases, the selected cell lines are truly salt tolerant in that they can withstand high cellular salt concentrations. The situation is different in a salt-resistant citrus cell line in which resistance seems to be due to the exclusion of Na^+ and and Cl^- (Ben-Hayyim and Kochba, 1983). In this case salt stress is actually avoided. It is likely that the cells osmoregulate via the synthesis of some compatible solute to maintain their water status. That such osmoregulation can occur in cultured cells is strongly suggested by the studies of Bressan et al. (1982) with cells adapted to conditions of very negative water potential created by PEG. In this case the medium contained insufficient solutes to account for the degree of osmoregulation necessary for the observed continued growth of these cells, but their adaptation could be accounted for by osmoregulation via the synthesis of organic solutes.

As with intact plants, proline has been implicated in salt resistance in selected cell lines; but as is also the case with intact plants, there is no consensus on whether or not its role is adaptive. In N. sylvestris, a salt-resistant line did not accumulate as much proline as the unselected line, arguing for proline synthesis as a stress response rather than an adaptive mechanism (Dix and Pearce, 1981). However, evidence to the contrary is provided by observations with carrot and Kickxia. Several carrot cell lines selected to overproduce proline (by selection for resistance to azetidine-2-carboxylic acid, a proline analog) all demonstrated increased resistance to NaCl, although the degree of resistance was not proportional to proline concentration (Riccardi et al., 1983). In Kickxia cultures, the addition of exogenous proline was shown to overcome inhibition by NaCl (Mathur et al., 1980).

Salt-resistant variants can clearly be isolated in plant cell cultures and the resistance mechanisms observed in them are similar to those known to operate in intact plants. This is encouraging evidence that salt-resistant plants can be obtained from such variant cells. The polygenic nature of salt resistance might be expected to limit the frequency with which such a genetic change could occur in cultured cells, but modifications in polygenic traits have been observed in plants regenerated from cultured cells (Larkin et al., 1984). A difficulty is presented by the physiological plasticity of cells in readily adapting to osmotic stress. Selection strategies must be employed that facilitate the identification of stable variants among adapted, but genetically unchanged, cells. It is difficult to evaluate the importance of specific ions in in vitro selection, since no field evaluation of plants regenerated from salt-resistant cells has yet been performed. However, it is reasonable to expect that the greater likelihood of eventual success in the field will be improved if the salt stress employed for selection reflects the ionic composition of the soil salts in the area for which the crop is intended.

TOLERANCE TO MINERAL STRESS

Mineral stresses include both deficiencies of essential nutrient elements and excesses of toxic elements. Such stresses are widespread and represent a major limitation to crop production worldwide (Sanchez et al., 1982; Swaminathan, 1982). The genetic control of plant response to mineral stress is well-established (Epstein, 1972) and in a number of cases genotypic differences have been attributable to a single major gene (Devine, 1982). The roles played by mineral nutrients are generally of a fundamental cellular nature (e.g. enzyme activation, components of fundamental molecules) and so are just as critical to cultured cells as to whole plants. Likewise, the primary lesions of toxic ions are also of a fundamental cellular nature (e.g. competition with essential elements for uptake, inactivation of enzymes, displacement of essential elements from functional sites) (Epstein, 1972). It is not surprising then that cultured plant cells are sensitive to artificially applied mineral stresses, both nutrient deficiencies and mineral toxicities.

The selection strategies necessary to isolate variants resistant to mineral stresses are relatively straightforward to conceive, consisting simply of subjecting cultured cells to the deficiency or toxicity of interest by modifying the culture medium. It is critical, however, that interactions with other elements, as well as such factors as pH, be considered in the design of the inorganic nutrient mixture used to produce the desired stress. Oversimplification is to be avoided. While it may lead to quick results and even new plant genotypes, these plants are likely to fail in the agricultural setting in which the mineral stress occurs unless the selection strategy employed closely simulates the conditions that exist in the soil environment.

Just as the primary lesions caused by mineral stresses are cellular, so are at least some resistance mechanisms by which plants withstand these stresses. For example, more efficient uptake of a deficient nutrient element may result from an altered membrane transport mechanism, or more efficient utilization can be achieved via higher enzyme affinity for an element (Gerloff, 1976). Ion toxicity resistance mechanisms can include exclusion at the plasmalemma, detoxification via binding to an organic molecule, or altered target sites (Foy et al., 1978).

Genetic differences in response to mineral stress have been shown to be expressed in cultured cells. Wu and Antonovics (1978) found that differences between _Agrostis_ genotypes in resistance to zinc and copper toxicity were maintained in callus cultures. Similar findings have been reported for _Anthoxanthum_ genotypes differing in resistance to zinc and lead (Qureshi et al., 1981) and soybean (_Glycine_ _max_) genotypes differing in resistance to iron deficiency (Sain and Johnson, 1983). Christianson (1979), however, did not observe the expected differences in zinc sensitivity in _Phaseolus_ genotypes.

The cellular nature of mineral nutrition suggests that mineral stresses applied to cultured cells can be expected to have an effect

similar to that in a whole plant. Likewise, the cellular nature of
many mechanisms by which plants resist mineral stresses suggests that
variants possessing resistance mechanisms that will function in whole
plants can be selected. This has yet to be proven, however. There
have been few reports of selection for resistance to mineral stresses
in cultured plant cells. Phillips and Collins (1981) have reported
obtaining red clover (_Trifolium pratense_) cells capable of growth on
low phosphorus. While plants have been regenerated, they have not
yet been fully characterized (G. C. Phillips, pers. comm.). Meredith
(1978) obtained a number of stable aluminum-resistant tomato cell
lines, but plants could not be regenerated from the cultures.
Mercury-resistant petunia cells have also been selected, but no
information regarding regenerated plants has yet been reported
(Colijn et al., 1979). More recently, cadmium-resistant _Datura_ cells
(Jackson et al., 1983) and carrot cells resistant to both aluminum
and manganese (Ojima and Ohira, 1983) have been selected. While the
nature of the regenerated _Datura_ has not yet been reported, there are
preliminary indications that the carrot plants both retain and
transmit the aluminum resistance (Ojima and Ohira, 1982).

There is ample evidence to expect that new crop genotypes with
resistance to important mineral stresses can be produced by _in vitro_
selection. The selection strategies are clear-cut, known resistance
mechanisms operate in cultured cells, and resistant cell lines have
been isolated. It is most encouraging that in many cases responses
to mineral stress are under monogenic control. Lack of success in
this area to date probably reflects the small number of investigators
with interests in this area. This in turn may be related to the only
recent general recognition of the agricultural importance of mineral
stress. The potential for the genetic manipulation of plant response
to mineral stress is not yet fully appreciated at the whole plant
level, let alone in cultured cells.

DISEASE RESISTANCE

The most obvious selection strategy by which to identify and
isolate disease resistant cells is to challenge a cell population
with the pathogen. This approach, however, may involve several
difficulties. Since plant tissue culture media are nutrient rich,
most pathogens will grow readily on them. A pathogen introduced into
a plant cell culture will often overgrow the plant cells (e.g.
Sacristan, 1982). A means of avoiding this is to introduce the
pathogen onto the plant cells only and not onto the medium, for
example, by inoculating callus pieces (Helgeson, 1976). Since a
plant cell culture is an unnatural situation, it is also possible
that a pathogen introduced into such a culture may produce unspecific
killing that is not governed by mechanisms that determine
pathogenicity in the whole plant (Shepard, 1981).

In contrast, obligate pathogens may not grow at all in plant cell
cultures. Such has been the case with several rust fungi (Puccinia
antirrhini, P. helianthi, Gymnosporangium juniperi-virginianae) on
cultured tissues of snapdragon (Antirrhinum majus), sunflower
(Helianthus annuus), and red cedar (Juniperus virginiana) (Maheshwari
et al., 1967); powdery mildew (Erysiphe graminis f. sp. hordei) on
barley (Franzone et al., 1982); and Plasmodiophora brassicae on
callus of Brassica campestris (Dekhuijzen, 1975). There is evidence
that the presence of organized plant structures may be essential for
infection with certain pathogens. Such is the case with P. brassicae
on B. napus. While callus cannot be infected, somatic embryos can
(Sacristan and Hoffman, 1979).

To avoid the difficulties inherent in introducing a pathogen into
a culture, some have looked to toxins as selective agents. There
are, however, limitations surrounding the use of toxins, not the
least of which is that there must be a toxin associated with the
disease in question. Only a very few plant diseases are known to be
associated with a toxin at present (Scheffer and Briggs, 1981). To
be effective in selecting disease resistance, the toxin must be a
significant determinant of disease--that is, it must play an important role in either the development or the severity of the disease.
While the toxin need not be the sole cause of the disease, toxin
resistance can only be expected to result in increased disease resistance to the extent that the toxin contributes to the production of
the disease. It is not necessary that a toxin be host-specific to be
effective as a selection agent (Shepard, 1981). Even if other factors
control the host specificity of the pathogen, as long as the toxin
plays a significant role in the disease, toxin-resistance can be
expected to confer a significant degree of resistance to the disease.

In those diseases where a toxin is suspected to play a
significant role but has not been isolated, crude culture filtrates
have sometimes been used as selective agents (Behnke, 1979; 1980a;
Sacristan, 1982). A danger inherent in this approach is that the
toxicity to which resistance is being selected may be due to
components of the pathogen culture medium and not to a pathogen-
produced toxin. Additionally, a pathogen may produce a toxic
principle in culture that is insignificant or absent in infected
plants (Yoder, 1981). These doubts can be erased by using purified
toxins whose role in disease has been confirmed.

The requirements for the use of a toxin to select disease resistance are met by very few plant diseases. Relatively few diseases
are associated with toxins. Of these toxins, even fewer have been
purified and conclusively shown to play a significant role in plant
disease. While many more toxins will undoubtedly be implicated in
plant disease in the future (Scheffer and Briggs, 1981), the number
is now so small that the use of toxin selection can by no means be
considered a generally applicable approach to disease resistance.

There are alternatives to the use of either the pathogen itself or its toxin to select resistant cells. When the mechanism of the toxin is known, other compounds with the same effect that might perhaps be more readily available, can be considered as selection agents. This was the case with the selection of tobacco cells for resistance to tobacco wildfire disease (Pseudomonas tabaci) using methionine sulfoximine, a compound with similar effects to those of the wildfire toxin produced by the pathogen (Carlson, 1973). In those cases in which disease resistance is associated with the biosynthesis of a specific compound by the plant (Keen, 1981), selection strategies that permit the isolation of variants that overproduce the critical compound might also result in increased disease resistance.

As with any phenotype, to be selectable in culture, disease resistance must be expressed in the cultured cells to which the selection pressure is being applied. Is it reasonable to expect such a phenotype to be expressed in cultured cells?

Resistance to disease may be associated with certain organized structures and certainly would not be expected to operate in in vitro systems where these structures may not be present. The absence of cuticle and wax in rice callus, for example, has been associated with increased susceptibility to infection by several fungal pathogens (Uchiyama and Ogasawara, 1977). Since many secondary compounds are often absent or present in reduced concentrations in cultured cells (Bohm, 1980), the expression of disease resistance that is associated with secondary compounds would also be expected to be reduced. The reduced resistance of rice callus to infection by fungal pathogens, for example, has also been attributed to the absence of lignin in this tissue (Uchiyama et al., 1983).

While there are a number of cases where resistance has not been expressed in cultured tissue derived from plants of a resistant genotype (e.g. Fett, 1983), there have been several well-documented instances in which disease resistance is clearly expressed in cultured cells. A number of these cases involve Phytophthora species. Monogenic resistance to Phytophthora infestans tomato race 0, is expressed in callus cultures of several resistant tomato cultivars (Warren and Routley, 1970). Growth of the fungus was not supported by callus from resistant cultivars to nearly the extent that it was by callus from susceptible cultivars. Similar results were obtained with callus of two near-isogenic soybean genotypes differing in monogenic resistance to race 1 of P. megasperma var. sojae (Holliday and Klarman, 1979).

Expression of resistance to P. parasitica var. nicotianae in Nicotiana callus has been investigated in several laboratories. Monogenic, race-specific resistance to race 0 is expressed in callus of N. tabacum genotypes whether the resistance is derived from N. plumbaginifolia (Helgeson et al., 1976) or N. longiflora (Maronek and

Hendrix, 1978; Deaton et al., 1982). Comparisons of plant and callus resistance over several segregating generations have clearly shown that, at least with the plumbaginifolia-derived resistance, the callus resistance and the plant resistance represent the expression of the same gene (Helgeson et al., 1976). Polygenic resistance to both races 0 and 1 (derived from yet another source) has also been shown to be expressed in tobacco callus (Deaton et al., 1982).

In maize (Zea mays), cytoplasmic resistance to the toxin of Helminthosporium maydis is also expressed in callus cultures. Callus of genotypes with N cytoplasm (resistant) was not inhibited by toxin concentrations lethal to callus of genotypes with T cytoplasm (susceptible) (Gengenbach and Green, 1975; Brettell et al., 1979).

The expression or non-expression of disease resistance should not be considered an immutable condition. Plant cell cultures are not fixed in their developmental state since morphological, physiologi- cal, and biochemical characteristics will vary in response to physical, nutritional, and hormonal factors. It stands to reason, therefore, that even in those cases where resistance does not seem to be expressed in cultured cells of resistant genotypes, it might be possible to elicit the expression of resistance by manipulating one or more of these factors. Such has been clearly demonstrated by Haberlach et al. (1978) with regard to the expression of resistance to P. parasitica var. nicotianae race 0 in tobacco callus. At low kinetin concentrations, callus from resistant cultivars exhibited resistance of a hypersensitivity type while susceptible cultivars did not. At higher kinetin concentrations, however, the expression of resistance was lost and callus from all cultivars was equally susceptible. Benzyladenine was equally effective in eliminating resistance, while two other cytokinins were not. This work should serve as a clear example that investigations concerning the expression of disease resistance in cultured cells should not be discontinued should resistance not be initially apparent, but should include the manipulation of culture variables that might permit the expression of the resistance phenotype in resistant genotypes. In selection programs to obtain resistant cells, preliminary studies with known resistant and susceptible genotypes, if available, should be undertaken so as to optimize culture conditions for the maximum expression of known resistance mechanisms.

While the availability of suitable selection strategies and the expression of resistance in cultured cells seriously limit the in vitro selection of disease resistance, there have been some successes in this area. Carlson (1973) selected tobacco cells resistant to methionine sulfoximine, which has similar effects to those of tabtoxin, a broad-spectrum toxin produced by Pseudomonas tabaci, the causal organism of tobacco wildfire disease (Yoder, 1980). Plants regenerated from the selected cells did not develop chlorotic halos, one of the disease symptoms, but did develop necrotic lesions. This partial resistance was genetically transmissable and was inherited as a monogenic semidominant trait in one case and as recessive alleles at two loci in each of two other selections.

The purified toxin of <u>Helminthosporium</u> <u>maydis</u> Race T has also been successfully employed as a selective agent (Gengenbach and Green, 1975; Brettell et al., 1979). Callus from a susceptible maize genotype was subjected to several cycles of sublethal exposure to the toxin and resistant callus sectors were isolated. Plants regenerated from resistant callus retained resistance to the toxin (but lost their original male sterility) and the resistance was inherited as a cytoplasmic trait. Resistance to the toxin was closely correlated with resistance to the pathogen (Gengenbach et al., 1977). Partially purified toxins from <u>Pseudomonas</u> <u>syringae</u> pv. <u>tabaci</u> and <u>Alternaria</u> <u>alternata</u> pathotype tobacco have been used to select resistant protoplast-derived callus of tobacco. With each toxin, some plants regenerated from selected callus were also resistant to the pathogen and resistance was sexually transmitted (Thanutong et al., 1983).

Behnke (1979; 1980a) has used crude culture filtrates of two potato (<u>Solanum</u> <u>tuberosum</u>) pathogens, <u>Phytophthora</u> <u>infestans</u> and <u>Fusarium</u> <u>oxysporum</u>, to select resistant callus. In the case of <u>P.</u> <u>infestans</u>, selected callus was resistant to all four pathotypes although the crude filtrate came from only one pathotype. The resistance was stable in the absence of the filtrate and also retained in callus derived from some regenerated plants (Behnke, 1979). Plants from resistant selections also exhibited some general resistance to a mixture of races of the pathogen itself (Behnke, 1980b). Potato callus was similarly selected for resistance to crude culture filtrate of <u>F.</u> <u>oxysporum</u>. Resistance to the filtrate was retained in some regenerated plants. Plant response to the pathogen itself was not reported (Behnke, 1980a). Callus and embryogenic cultures of <u>B.</u> <u>napus</u> have been selected for growth in the presence of a crude culture filtrate of <u>Phoma</u> <u>lingam</u>. Some regenerated plants exhibited increased resistance to the pathogen. Preliminary studies of the progeny of regenerated plants suggest a genetic basis for the resistance (Sacristan, 1982).

Limited resistance to tobacco mosaic virus has been obtained by Murakishi and Carlson (1982) by selecting green callus from leaf strips of gamma-irradiated <u>N.</u> <u>sylvestris</u> plants infected with the virus. Some of the selected callus colonies were virus-free, and of plants regenerated from those, some displayed limited resistance that was transmitted to seedling progeny. Critical factors in the success of this experiment were the careful control of experimental conditions to insure a uniform infection rather than the typical mosaic and the use of a yellow strain of the virus as a marker for infected tissue.

The preceding examples all employed a selective agent to

deliberately select a resistant phenotype. There is a rapidly
growing body of evidence, however, to indicate that disease resistant
individuals can be recovered at a high frequency in populations of
plants regenerated from cultured cells even without selection
pressure. The earliest indications of this phenomenon were observed
in sugarcane (Saccharum officinarum). Increased resistance to a
number of diseases has been noted in sugarcane plants regenerated
from cultured cells (Krishnamurthi and Tlaskal, 1974; Heinz, 1973;
1976; Liu and Chen, 1978; Larkin and Scowcroft, 1983). Individuals
with increased resistance to both early blight (Alternaria solani)
and late blight (Phytophthora infestans) have been identified in
populations of potato plants regenerated from mesophyll protoplasts
(Matern et al., 1978; Shepard et al., 1980). Resistance to Fusarium
oxysporum f. sp. apii has been observed in celery (Apium graveolens)
plants regenerated from suspension cultures (Pullman and Rappaport,
1983).

In two of the cases of deliberate selection described above,
resistance was also recovered from unselected cultures. With callus
cultures of maize that had never been challenged with T-toxin, more
than half of the regenerated plants could be shown to be toxin
resistant (Brettell et al., 1980). Similarly, some regenerants from
B. napus cultures not exposed to culture filtrate showed an increased
resistance to Phoma lingam (Sacristan, 1982).

These observations of increased frequency of disease resistance
in plants regenerated from cultured cells are most encouraging in
view of the limitations associated with in vitro selection for
disease resistance. The exploitation of such somaclonal variation
may permit the recovery of resistant variants in cases where no
selective agent is available or where resistance is not expressed in
culture. Selection is still advisable in those few cases where it is
possible, however, because it increases the recovery of resistance by
enriching the culture for resistant cells.

THE ROLE OF IN VITRO SELECTION IN CROP IMPROVEMENT

The advisability of employing in vitro selection in a crop
improvement program is dependent upon both the objective of the
program and the crop. Does the objective involve a phenotype that is
readily amenable to an in vitro approach? With regard to the crop,
how effective is the conventional breeding technology that exists for
that crop? How well-developed is the in vitro technology? Considering these questions will aid one in deciding whether selection with
cultured cells is worthy of consideration. With many agronomic
crops, the conventional breeding technology is extremely sophisticated and powerful, while cell culture manipulations are still
difficult. On the other hand, for some woody perennial crops the
power of conventional breeding technology is severely limited by such

factors as generation time and plant size, making even the current primitive state of cell culture technology associated with these crops attractive by comparison, and justifying further efforts to improve in vitro methods for these crops.

There are many limitations to the application of in vitro selection for crop improvement, the major ones being that cell culture technology is not sufficiently advanced for many crops and that most agriculturally significant phenotypes are as yet too poorly understood to permit the design of effective selection strategies. It is clear, however, that novel genotypes can be obtained with this approach and that some contributions can be made to crop improvement. Along with established plant breeding methods and other newly emerging genetic technologies, selection in plant cell cultures has both strengths and limitations. For certain crop species and certain agricultural objectives, a thorough analysis of the considerations presented here may reveal in vitro selection to be an effective genetic tool in crop improvement.

ACKNOWLEDGMENT

The valuable comments of A. J. Conner are greatly appreciated.

REFERENCES

Abel, G. H., 1969, Inheritance of the capacity for chloride inclusion and chloride exclusion by soybeans, Crop Sci., 6:697-698.

Behnke, M., 1979, Selection of potato callus for resistance to culture filtrates of Phytophthora infestans and regeneration of resistant plants, Theor. Appl. Genet., 55:69-71.

Behnke, M., 1980a, Selection of dihaploid potato callus for resistance to the culture filtrate of Fusarium oxysporum, Z. Pflanzenzuchtg., 85:254-258.

Behnke, M., 1980b, General resistance to late blight of Solanum tuberosum plants regenerated from callus resistant to culture filtrates of Phytophthora infestans, Theor. Appl. Genet., 56: 151-152.

Ben-Hayyim, G., and Kochba, J., 1983, Aspects of salt tolerance in a NaCl-selected stable cell line of Citrus sinensis, Plant Physiol., 72:685-690.

Bohm, H., 1980, The formation of secondary metabolites in plant tissue and cell cultures, in: "Perspectives in Plant Cell and Tissue Culture," I. K. Vasil, ed., International Review of Cytology Supplement 11B, Academic Press, New York, pp. 183-208.

Bressan, R. A., Hasegawa, P. M., and Handa, A. K., 1981, Resistance of cultured higher plant cells to polyethylene glycol-induced water stress, Plant Sci. Lett., 21:23-30.

Bressan, R. A., Handa, A. K., Handa, S., and Hasegawa, P. M., 1982,
     Growth and water relations of cultured tomato cells after
     adjustment to low external water potentials, Plant Physiol.,
     70:1303-1309
Brettell, R. I. S., Goddard, B. V. D., and Ingram, D. S., 1979,
     Selection of Tms-cytoplasm maize tissue cultures resistant to
     Drechslera maydis T-toxin, Maydica, 24:203-213.
Brettell, R. I. S., Thomas, E., and Ingram, D. S., 1980, Reversion of
     Texas male-sterile cytoplasm maize in culture to give fertile,
     T-toxin resistant plants, Theor. Appl. Genet., 58:55-58.
Carlson, P. S., 1970, Induction and isolation of auxotrophic mutants
     in somatic cell cultures of Nicotiana tabacum, Science, 168:
     487-489.
Carlson, P. S., 1973, Methionine sulfoximine-resistant mutants of
     tobacco, Science, 180:1366-1368.
Chaleff, R. S., 1983, Considerations of developmental biology for the
     plant cell geneticist, in: "Genetic Engineering of Plants: An
     Agricultural Perspective," T. Kosuge, C. P. Meredith, and A.
     Hollaender, eds., Plenum Press, New York, pp. 257-270.
Chaleff, R. S., and Parsons, M. F., 1978, Direct selection in vitro
     for herbicide-resistant mutants of Nicotiana tabacum, Proc.
     Natl. Acad. Sci. USA, 75:5104-5107.
Christianson, M. L., 1979, Zinc sensitivity in Phaseolus: expression
     in cell culture, Env. Exp. Bot., 19:217-221.
Colijn, C. M., Kool, A. J., and Nijkamp, H. J. J., 1979, An effective
     chemical mutagenesis procedure for Petunia hybrida cell
     suspension cultures, Theor. Appl. Genet., 55:101-106.
Conner, A. J., and Meredith, C. P., 1984, An improved polyurethane
     support system for monitoring growth in plant cell cultures,
     Plant Cell Tissue Organ Culture, 3:59-68.
Croughan, T. P., Stavarek, S. J., and Rains, D. W., 1978, Selection
     of a NaCl tolerant line of cultured alfalfa cells, Crop Sci.,
     18:959-963.
Deaton, W. R., Keyes, G. J., and Collins, G. B., 1982, Expressed
     resistance to black shank among tobacco callus cultures,
     Theor. Appl. Genet., 63:65-70.
Dekhuijzen, H. M., 1975, The enzymatic isolation of secondary
     vegetative plasmodia of Plasmodiophora brassicae from callus
     tissue of Brassica campestris, Physiol. Plant Pathol., 6:187-
     192.
Devine, T. E., 1982, Genetic fitting of crops to problem soils, in:
     "Breeding Plants for Less Favorable Environments," M. N.
     Christiansen and C. F. Lewis, eds., John Wiley and Sons, New
     York, pp. 143-173.
Dix, P. J., 1979, Cell culture manipulations as a potential breeding
     tool, in: "Low Temperature Stress in Crop Plants," J. M.
     Lyons, J. K. Raison, and P. L. Steponkus, eds., Academic
     Press, New York, pp. 463-472.

Dix, P. J., and Pearce, R. S., 1981, Proline accumulation in NaCl-resistant and sensitive cell lines of Nicotiana sylvestris, Z. Pflanzenphysiol., 102:243-248.

Epstein, E., 1972, "Mineral Nutrition of Plants: Principles and Perspectives," John Wiley and Sons, New York.

Epstein, E., Norlyn, J.D., Rush, D. W., Kingsbury, R. W., Kelley, D. B., Cunningham, G. A., and Wrona, A. F., 1980, Saline culture of crops: a genetic approach, Science, 210:399-404.

Fett, W. F., and Zacharius, R. M., 1983, Bacterial growth and phytoalexin elicitation in soybean cell suspension cultures inoculated with Pseudomonas syringae pathovars, Physiol. Plant. Pathol., 22:151-172.

Flowers, T. J., Troke, P. F., and Yeo, A. R., 1977, The mechanism of salt tolerance in halophytes, Ann. Rev. Plant Physiol., 28:89-121.

Foy, C. D., Chaney, R. L., and White, M. C., 1978, The physiology of metal toxicity in plants, Ann. Rev. Plant Physiol., 29:511-566.

Franzone, P. M., Foroughi-Wehr, B., Fischbeck, G., and Friedt, W., 1982, Reaction of microspore callus, androgenetic albino plantlets and roots of barley to Erysiphe graminis f. sp. hordei, Phytopath. Z., 105:170-174.

Gebhardt, C., Schnebli, V., and King, P. J., 1981, Isolation of biochemical mutants using haploid mesophyll protoplasts of Hyoscyamus muticus. II. Auxotrophic and temperature-sensitive clones, Planta, 153:81-89.

Gengenbach, B. G., and Green, C. E., 1975, Selection of T-cytoplasm maize callus cultures resistant to Helminthosporium maydis Race T pathotoxin, Crop Science, 15:645-649.

Gengenbach, B. G., Green, C. E., and Donovan, C. M., 1977, Inheritance of selected pathotoxin resistance in maize plants regenerated from cell cultures, Proc. Natl. Acad. Sci. USA, 74:5113-5117.

Gerloff, G. C., 1976, Plant efficiencies in the use of nitrogen, phosphorus, and potassium, in: "Plant Adaptation to Mineral Stress in Problem Soils," M. J. Wright, ed., Cornell University, Ithaca, New York, pp. 161-173.

Goldner, R., Umiel, N., and Chen, Y., 1977, The growth of carrot callus cultures at various concentrations and composition of saline water, Z. Pflanzenphysiol., 85:307-317.

Greenway, H., and Munns, R., 1980, Mechanisms of salt tolerance in nonhalophytes, Ann. Rev. Plant Physiol., 31:149-190.

Gronwald, J.W., and Leonard, R. T., 1982, Isolation and transport properties of protoplasts from cortical cells of corn roots, Plant Physiol., 70:1391-1395.

Haberlach, G. T., Budde, A. D., Sequira, L, and Helgeson, J. P., 1978, Modification of disease resistance of tobacco callus tissues by cytokinins, Plant Physiol., 62:522-525.

Handa, A. K., Bressan, R. A., Handa, S., and Hasegawa, P. M., 1983, Clonal variation for tolerance to polyethylene glycol-induced water stress in cultured tomato cells, Plant Physiol., 72:645-653.

Handa, S., Bressan, R. A., Handa, A. K., Carpita, N. C., and Hasegawa, P. M., 1983, Solutes contributing to osmotic adjustment in cultured plant cells adapted to water stress, Plant Physiol., 73:834-843.

Heinz, D. J., 1973, Sugarcane improvement through induced mutations using vegetative propagules and cell culture techniques, in: "Induced Mutations in Vegetatively Propagated Plants," International Atomic Energy Commission, Vienna, pp. 53-59.

Heinz, D. J., 1976, Tissue culture in breeding, Ann. Rep. Hawaiian Sugar Planters Assoc. Exp. Sta., pp. 9-11.

Helgeson, J. P., Haberlach, G. T., and C. D. Upper, 1976, A dominant gene conferring disease resistance to tobacco plants is expressed in tissue cultures, Phytopathology, 66:91-96.

Heyser, J. W., and Nabors, M. W., 1981, Osmotic adjustment of cultured tobacco cells (Nicotiana tabacum var. Samsun) grown on sodium chloride, Plant Physiol., 67:720-727.

Holliday, M. J., and Klarman, W. L., 1979, Expression of disease reaction types in soybean callus from resistant and susceptible plants, Phytopathology, 69:576-578.

Horsch, R. B., and Jones, G. E., 1980, A double filter paper technique for plating cultured plant cells, In Vitro, 16:103-108.

Humphreys, M. O., 1982, The genetic basis of tolerance to salt spray in populations of Festuca rubra L., New Phytol., 91:287-296.

Jackson, P. J., Roth, E. J., and McClure, P. R., 1983, Coinduction of synthesis of two metallotheionein-like, cadmium binding proteins in cadmium resistant suspension cell cultures of Datura innoxia (abstr.), ARCO Solar-UCLA Symposium on Plant Molecular Biology, April 16-22, 1983, Keystone, Colorado.

Keen, N. T., 1981, Evaluation of the role of phytoalexins, in: "Plant Disease Control: Resistance and Susceptibility," R. C. Staples and G. H. Toenniessen, eds., John Wiley and Sons, New York, pp. 155-177.

King, J., Horsch, R. B., and Savage, A. D., 1980, Partial characterization of two stable auxotrophic cell strains of Datura innoxia Mill., Planta, 149:480-484.

Kochba, J., Ben-Hayyim, G., Spiegel-Roy, P., Saad, S., and Neumann, H., 1982, Selection of stable salt-tolerant callus cell lines and embryos in Citrus sinensis and C. aurantium, Z. Pflanzenphysiol., 106:111-118.

Krishnamurthi, M., and Tlaskal, J., 1974, Fiji disease resistant Saccharum officinarum var. Pindar sub-clones from tissue cultures, Proc. Int. Soc. Sugar Cane Technol., 15:130-137.

Larkin, P. J., Ryan, S., Brettell, R. I. S., and Scowcroft, W. R., 1984, Heritable somaclonal variation in wheat, Theor. Appl. Genet., 67:443-455.

Larkin, P. J., and Scowcroft, W. R., 1983, Somaclonal variation and
     eyespot toxin tolerance in sugarcane, Plant Cell Tissue Organ
     Culture, 2:111-121.
Liu, M.-C., and Chen, W.-H., 1978, Improvement in sugarcane using
     tissue culture methods (abstr.), in: "Frontiers of Plant
     Tissue Culture 1978," T. A. Thorpe, ed., University of
     Calgary, Canada, p. 515.
Maheshwari, R., Hildebrandt, A. C., and Allen, P. J., 1967, Factors
     affecting the growth of rust fungi on host tissue cultures,
     Bot. Gaz., 128:153-159.
Maliga, P., 1980, Isolation, characterization, and utilization of
     mutant cell lines in higher plants, in: "Perspectives in Plant
     Cell and Tissue Culture," I. K. Vasil, ed., International
     Review of Cytology Supplement 11A, Academic Press, New York,
     pp. 225-250.
Maronek, D. M., and Hendrix, J. W., 1978, Resistance to race 0 of
     Phytophthora parasitica var. nicotianae in tissue cultures of
     a tobacco breeding line with black shank resistance derived
     from Nicotiana longiflora, Phytopathology, 68:233-234.
Matern, U., Strobel, G., and Shepard, J., 1978, Reaction to phyto-
     toxins in a potato population derived from mesophyll
     protoplasts, Proc. Natl. Acad. Sci. USA, 75:4935-4939.
Mathur, A. K., Ganapathy, P. S., and Johri, B. M., 1980, Isolation of
     sodium chloride-tolerant plantlets of Kickxia ramosissima
     under in vitro conditions, Z. Pflanzenphysiol., 99:287-294.
Meins, F., and Binns, A. N., 1979, Cell determination in plant
     development, BioScience, 29:221-225.
Meredith, C. P., 1978, Selection and characterization of aluminum-
     resistant variants from tomato cell cultures, Plant Sci.
     Lett., 12:25-34.
Murakishi, H. H., and Carlson, P. S., 1982, In vitro selection of
     Nicotiana sylvestris variants with limited resistance to TMV,
     Plant Cell Reports, 1:94-97.
Nabors, M. W., Gibbs, S. E., Bernstein, C. S., and Meis, M. E., 1980,
     NaCl-tolerant tobacco plants from cultured cells, Z. Pflanzen-
     physiol., 97:13-17.
Nagata, T., and Takebe, I., 1971, Plating of isolated tobacco
     mesophyll protoplasts on agar medium, Planta, 99:12-20.
Nyman, L. P., Gonzales, C. J., and Arditti, J., 1983, In vitro
     selection for salt tolerance of taro (Colocasia esculenta var.
     antiquorum), Ann. Bot., 51:229-236.
Ojima, K., and Ohira, K., 1982, Characterization and regeneration of
     an aluminum-tolerant variant from carrot cell cultures, in:
     "Plant Tissue Culture 1982," A. Fujiwara, ed., Japanese
     Association for Plant Tissue Culture, Tokyo, pp. 475-476.
Ojima, K., and Ohira, K., 1983, Characterization of aluminum and
     manganese tolerant cell lines selected from carrot cell
     cultures, Plant Cell Physiol., 24:789-797.
Orton, T. J., 1980, Comparison of salt tolerance between Hordeum
     vulgare and H. jubatum in whole plants and callus cultures, Z.
     Pflanzenphysiol., 98:105-118.

Parsons, L. R., 1982, Plant responses to water stress, In: "Breeding
    Plants for Less Favorable Environments," M. N. Christiansen
    and C. F. Lewis, eds., John Wiley and Sons, New York, pp.
    175-192.
Phillips, G. C., and Collins, G. B., 1981, Growth and selection of
    red clover (Trifolium pratense L.) cells on low levels of
    phosphate (abstr.), Agronomy Abstracts, 1981:187.
Polacco, J. C., 1979, Arsenate as a potential negative selection
    agent for deficiency variants in cultured plant cells, Planta,
    146:155-160.
Portnoy, S., and Murphy, T. M., 1983, Distribution of mutants in
    aggregates of cultured cells, Theor. Pop. Biol., 23:136-146.
Pullman, G. S., and Rappaport, L., 1983, Tissue culture-induced
    variation in celery for Fusarium yellows (abstr.), Phyto-
    pathology, 73:818.
Qureshi, J. A., Collin, H. A., Hardwick, K., and Thurman, D. A.,
    1981, Metal tolerance in tissue cultures of Anthoxanthum
    odoratum, Plant Cell Reports, 1:80-82.
Rangan, T. S., and Vasil, I. K., 1983, Sodium chloride tolerant
    embryogenic cell lines of Pennisetum americanum (L.) K. Schum,
    Ann. Bot., 52:59-64.
Riccardi, G., Cella, R., Camerino, G., and Ciferri, O., 1983,
    Resistance to azetidine-carboxylic acid and sodium chloride
    tolerance in carrot cell cultures and Spirulina platensis,
    Plant Cell Physiol., 24:1073-1078.
Sacristan, M. D., 1982, Resistance responses to Phoma lingam of
    plants regenerated from selected cell and embryogenic cultures
    of haploid Brassica napus, Theor. Appl. Genet., 61:193-200.
Sacristan, M. D., and Hoffmann, F., 1979, Direct infection of
    embryogenic tissue cultures of haploid Brassica napus with
    resting spores of Plasmodiophora brassicae, Theor. Appl.
    Genet., 54:129-132.
Sain, S. L., and Johnson, G. V., 1983, Iron utilization by iron
    efficient and inefficient soybean cultivars in cell suspension
    culture (abstr.), Plant Physiol., 72(suppl.):5.
Sanchez, P. A., Bandy, D. E., Villachica, J. H., and Nicholaides, J.
    J., 1982, Amazon Basin soils: management for continuous crop
    production, Science, 216:821-827.
Scheffer, R. P., and Briggs, S. P., 1981, Introduction: a perspective
    of toxin studies in plant pathology, in: "Toxins in Plant
    Disease," R. D. Durbin, ed., Academic Press, New York, pp. 1-20
Schieder, O., 1976, Isolation of mutants with altered pigments after
    irradiating haploid protoplasts from Datura innoxia Mill. with
    x-rays, Molec. Gen. Genet., 149:251-254.
Shepard, J. F., 1981, Protoplasts as sources of disease resistance in
    plants, Ann. Rev. Phytopathol., 19:145-166.
Shepard, J. F., Bidney, D., and Shahin, E., 1980, Potato protoplasts
    in crop improvement, Science, 208:17-24.
Smith, M. K., and McComb, J. A., 1981a, Effect of NaCl on the growth
    of whole plants and their corresponding callus cultures, Aust.
    J. Plant Physiol., 8:267-275.

Smith, M. K., and McComb, J. A., 1981b, Use of callus cultures to detect NaCl tolerance in cultivars of three species of pasture legumes, Aust. J. Plant Physiol., 8:437-442.

Swaminathan, M. S., 1982, Biotechnology research and third world agriculture, Science, 218:967-972.

Tal, M., Heikin, H., and Dehan, K., 1978, Salt tolerance in the wild relatives of the cultivated tomato: responses of callus tissue of Lycopersicon esculentum, L. peruvianum and S. pennellii to high salinity, Z. Pflanzenphysiol., 86:231-240.

Thanutong, P., Furusawa, I., and Yamamoto, M., 1983, Resistant tobacco plants from protoplast-derived calluses selected for their resistance to Pseudomonas and Alternaria toxins, Theor. Appl. Genet., 66:209-215.

Tyagi, A. K., Rashid, A., and Maheshwari, S. C., 1981, Sodium chloride resistant cell line from haploid Datura innoxia Mill.: a resistance trait carried from cell to plantlet and vice versa in vitro, Protoplasma, 105:327-332.

Uchiyama, T., and Ogasawara, N., 1977, Disappearance of the cuticle and wax in outermost layer of callus cultures and decrease of protective ability against microorganisms, Agric. Biol. Chem., 41:1401-1405.

Uchiyama, T., Sata, J., and Ogasawara, N., 1983, Lignification and qualitative changes of phenolic compounds in rice callus tissues inoculated with plant pathogenic fungi, Agric. Biol. Chem., 47:1-10.

Venables, A. V. and Wilkins, D. A., 1978, Salt tolerance in pasture grasses, New Phytol., 80:613-622.

Warren, R. S., and Gould, A. R., 1982, Salt tolerance expressed as a cellular trait in suspension cultures developed from the halophytic grass Distichlis spicata, Z. Pflanzenphysiol., 107: 347-356.

Warren, R. S., and Routley, D. G., 1970, The use of tissue culture in the study of single gene resistance of tomato to Phytophthora infestans, J. Amer. Soc. Hort. Sci., 95:266-269.

Watad, A. A., Reinhold, L., and Lerner, H. R., 1983, Comparison between a stable NaCl-selected Nicotiana cell line and the wild type, Plant Physiol., 73:624-629.

Weber, G., and Lark, K. G., 1979, An efficient plating system for rapid isolation of mutants from plant cell suspensions, Theor. Appl. Genet., 55:81-86.

Widholm, J. M., 1976, Selection and characterization of cultured carrot and tobacco cells resistant to lysine, methionine, and proline analogs, Can. J. Bot., 54:1523-1529.

Wong, C.-K., Ko, S.-W., and Woo, S.-C., 1983, Regeneration of rice plantlets on NaCl-stressed medium by anther culture, Bot. Bull. Academia Sinica, 24:59-64.

Wu, L., and Antonovics, J., 1978, Zinc and copper tolerance of Agrostis stolonifera L. in tissue culture, Amer. J. Bot., 65: 268-271.

Yano, S., Ogawa, M., and Yamada, Y., 1982, Plant formation from
        selected rice cells resistant to salts, in: "Plant Tissue
        Culture 1982," A. Fujiwara, ed., Japanese Association for
        Plant Tissue Culture, Tokyo, pp. 495-496.
Yoder, O. C., 1980, Toxins in pathogenesis, Ann. Rev. Phytopathol.,
        18:103-129.
Yoder, O. C., 1981, Assay, in: "Toxins in Plant Disease," R. D.
        Durbin, ed., Academic Press, New York, pp. 45-78.

MOLECULAR ANALYSIS OF ALIEN CHROMATIN INTRODUCED INTO WHEAT

R. Appels and Lyndall B. Moran

Division of Plant Industry, CSIRO, G.P.O. Box 1600
Canberra, A.C.T. 2601, Australia

SUMMARY

Repeated DNA sequences are examined in a number of grasses to assess the usefulness of these sequences as probes for alien chromatin after the introduction of such chromosome segments into wheat (Triticum). Various classes of sequence are shown to have evolved at a rate which is sufficiently fast to result in them being genus specific. This observation allows the development of genus specific probes.

One of the situations where genus specific DNA probes were first developed was in the detection of rye (Secale cereale) chromosomes after their introduction into wheat. The sequences which proved most useful were located in the terminal blocks of rye heterochromatin in the form of long tandem arrays. The major (350 bp repeat) heterochromatic sequence in rye has been studied in detail by isolating 10-14 kb segments of DNA containing the 350 bp repeat, using the vector λ1059. Linkage between sequence variants indicates that domains, resulting in some preferential clustering of variants, exist for these sequences. Clearly defined linkages between the 350 bp sequence and other repeated sequences were also characterized. The non-heterochromatic sequences neighboring the 350 bp sequence were more generally distributed, chromosomally, and the potential for this class of sequence for detecting alien chromatin introduced into wheat is discussed. Several analogous clones (i.e. chromosomally distributed), were isolated from Haynaldia villosa and Elytrigia elongata and a preliminary characterization of these sequences is presented.

The spacer sequences of ribosomal DNA are another example of
a sequence which is capable of providing genus specific probes.
In cereals this potential was first demonstrated by the characterization of the wheat rDNA spacer and the virtual absence of cross-
hybridization of the 130 bp spacer repeat unit with rye rDNA.
Characterization of the barley (Hordeum vulgare) and rye spacer
sequences show that the reciprocal relation is true and that the
rDNA from grasses in general can be specifically detected after
their introduction into wheat.

INTRODUCTION

     A major feature of the organization of DNA sequences in
cereals (as well as in most eukaryotes) is that many nucleotide
sequences are repeated within the genome. The average repeat unit
is 200-1000 bp long (Flavell and Smith, 1976) and can occur tandemly arranged in large blocks or dispersed among other sequences
(Appels et al., 1978; Rimpau et al., 1979). The reasons why sequence repetition is so frequent in eukaryote genomes are likely
to depend on the particular sequence under investigation. When
sequences are in tandem arrays co-operative interactions are
accentuated (Appels and Peacock, 1978). The resulting structures
are often visualized cytologically as heterochromatin and can
modulate gene expression via a nearest neighbour effect (for a
recent discussion see Hilliker and Appels, 1982). It has been suggested that such arrays of sequences may adopt characteristic
structures in a spontaneous manner (Sutton, 1972) depending on the
proteins binding the sequence. The micro-environment of a gene is
important in its expression and sequence repetition is one of the
ways in which eukaryotes can modify this micro-environment. In
this context dispersed repeated sequences have been argued to provide signals for the activation of transcription in parts of the
genome at certain stages of development (reviewed in Davidson and
Posakony, 1982).

     The analysis of the ribosomal RNA genes (rDNA) in many
eukaryotes has demonstrated the existence of tandem sequence repetition preceding the point of transcription initiation. Studies
in Xenopus in particular have demonstrated that this repetitive
sequence region is important for efficient transcription initiation in a competitive situation. The region has been interpreted
as a site for loading RNA polymerase I molecules in preparation for
transcription (Moss, 1983).

     Certain repeated sequences (e.g. those in rDNA spacers and in
heterochromatin regions) change rapidly on an evolutionary time-
scale while others are more stable within a comparable time. The
reasons underlying evolutionary stability or instability are presumably linked to the nuclear interactions in which a particular

sequence participates. An important corollary of the evolutionary
instability, is that genus specific sequences can be isolated and
utilized as probes in tracing a particular chromosome segment as
it is introduced into another genus. This type of manipulation is
now commonplace in cereal breeding programmes, as demonstrated by
several papers presented in this Symposium, and in this manuscript
we summarize recent advances in isolating DNA probes suitable for
tracing alien chromosome segments introduced into wheat.

Tandem arrays of the major heterochromatic sequence in rye identify
chromosomes

The heterochromatic sequences of rye were among the first to be
used as DNA sequence probes in the identification of alien
chromosomes (or chromosome segments) present in wheat. In this
section we characterize new isolates of the major heterochromatic
sequence as well as providing some recent applications of the
use of this sequence for identifying rye chromosomes (and segments)
in a wheat background.

i) Mapping of λ clones

Rye DNA was partially digested with the restriction enzyme
Sau 3A and DNA in the range 13-20 kb was isolated by electrophoresis
in 0.7%, low melting agarose (Langridge et al., 1980). The
extracted DNA was ligated either to purified λ1059 arms (prepared
from λ1059 DNA digested with Bam HI and Sal I) or simply Bam HI
digested λ 1059 DNA following the general cloning strategy of
Karn et al. (1980). All the clones shown in Fig. 1, with the
exception of λhet 1, were plated on the restrictive host,
Escherichia coli 359 (Karn et al., 1980), to select against
bacteriophage carrying the central fragment which is being replaced
by rye DNA in the procedure. After purification of a particular
bacteriophage it was grown in E. coli C600.

Four of the six clones carrying the 350 bp heterochromatic
sequence carry an insert which, as far as could be ascertained,
contained only the 350 bp heterochromatic sequence. The clone λSc
het1 contained a trimer flanked by Bam HI sites (Fig. 1) in addition
to the central fragment of λ1059. The 1050 bp insert was subcloned
into pBR322, and named pSc het1, to produce the map shown by a
combination of double restriction enzyme digestions and partial
digestions. The three clones λSc het3, λSc het8 and λSc het10
contained inserts 10-20 kb long and were more difficult to propagate
than λSc het1. Yields of bacteriophage were variable and multiple
bands were observed when the respective bacteriophage was centrifuged
to equilibrium in CsCl. The restriction enzymes Pst I and Hinf I
defined a 350 bp monomer within the rye DNA insert; Pst I also
defined dimers, trimers and tetramers in the proportions indicated
in Fig. 1. Although λSc het3, λSc het8 and λSc het10 were each

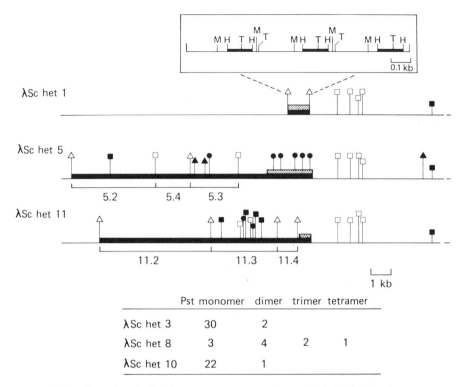

Fig. 1.   Restriction enzyme mapping of λSc het clones

	Pst monomer	dimer	trimer	tetramer
λSc het 3	30	2		
λSc het 8	3	4	2	1
λSc het 10	22	1		

λSc het1 was mapped by subcloning the Ban HI (Δ) insert
into pBR322 using standard procedures.   The insert was
recovered using Bam HI digested rye DNA (0.5 μg) ligated
to Bam HI digested λ1059 DNA (0.5 μg), packaged (Hohn,
1979) and screened after recovery on a non-selective host,
by hybridization using ^{32}P-labelled insert from pSc7235
(Appels et al., 1981), as described by Benton and Davis
(1979).   Since selection against phage carrying the central
fragment was not carried out, this fragment is also present
in addition to the 1050 bp insert.

The other λSc het phage were recovere as follows.   Rye DNA
(100 μg in 210 μl) was partially digested in 70 μl aliquots
with 1.5, 3.0 and 6.0 units of Sau 3A, after which the
aliquots were pooled and 30 μl of 5% SDS, 50 μl 0.25 M EDTA
were added followed by 2 min at 65°C.   After cooling to
room temperature the solution was vortexed with 200 μl
chloroform and the aqueous phase ethanol precipitated by

adding 420 µl ethanol. The DNA sample was electrophoresed
in a 0.7% Sea plaque)low melting) agarose gel, set in a
small transparent plastic tray, to separate the 10-20 kb
size classes of molecules. The 13-20 kb region was
excised and the DNA recovered using the extraction procedure of Langridge et al. (1981), taking care to maintain
the butanol extraction steps at 40-45°C with vigorous
vortexing (as recommended by W. Sutton). The final
aqueous phase was mixed with 2 volumes of ethanol and the
DNA recovered by centrifugation at 35,000 rpm, 2 hrs, 4°C
in a SW41 rotor. Aliquots of this DNA (0.5 µg) were
ligated to λ1059 DNA which had been digested with Ban HI
+ Sal I, denatured in 0.5 M NaOH, renatured by adding 0.25
M Tris-HCl pH 8.0 (containing enough 0.5 M HCl to neutral-
ize the NaOH) plus a 10 min incubation at 65°C, as
described in Sutton et al. (1984). The ligated DNA was
packaged as before and plated on the restrictive host
E. coli 359 (Karn et al., 1980) to select against the
central fragment of λ1059 and screened using ^{32}P labeled
pSc7235 insert.

The approximate proportions of Pst I monomers (350 bp),
dimers etc. were determined by scanning negatives from
photographing ethidium bromide stained Pst I digests of
the respective DNA's. λSc het5 and λSc het 11 were mapped
using double restriction endonuclease digestions, subcloning of fragments into pBR322, and hybridization of specific
^{32}P-labeled fragments to restriction endonuclease digests
transferred to Gene-Screen (NEN). The solid bar indicates
the rye DNA insert and the overlaying strippled bar the
portion hybridizing the heterochromatic probe (pSc7235).
The restriction endonuclease sites indicated are Eco RI
( ), Bgl II ( ), Pst I ( ), Hind III ( ), MbO II
(M), Hae III (H), Taq I (T).

The arrangement of the Pst I monomers and dimer in λSc het
5 was determined by a partial Pst I digest of Bgl II cut
λSc het5 DNA. This digest generates a series of fragments
containing the portion of the λ1059 short arm adjacent to
the inserted DNA plus varying parts of the inserted DNA.
Hybridization to such a digest with a λ1059 short arm
probe, labels the right-hand side of the inserted DNA and
thus, by length measurements allows the positioning of the
Pst I dimer within the array of Pst I monomers.

plaque purified three times before growing a culture for DNA
isolation, the final preparation showed length heterogeneity for
the inserted DNA. This is illustrated in Fig. 2 for λSc het3
where the insert was excisable by a Bgl II plus Bam HI double
digest. The multiple bands indicated in Fig. 2 hybridize to the
350 bp sequence and, from length estimates, arose by loss of
multiples of the 350 bp sequence during propagation of the
bacteriophage. This type of instability of a DNA segment composed
of a repeated sequence and cloned in bacteriophage λ, or plasmids,
has been reported by others (for example Brutlag et al., 1977;
Arnheim and Kuehn, 1979).

Two of the six λ clones analysed contained the 350 bp heterochromatic sequence linked to DNA which did not hybridize to the 350
bp sequence. The two clones, λSc het5 and λSc het11, were mapped
using a number of restriction enzymes (Fig. 1) by a combination of
double restriction enzyme digests, subcloning of DNA segments into
pBR322. Use was also made of hybridization with specific DNA
segments labelled with ^{32}P by nick-translation (Rigby et al., 1977)
after extraction from an agarose gel (Weinand et al., 1979). The
sequences linked to the 350 bp sequence were repeated in the genome
as judged from the complexity of the hybridization to restriction
enzyme digested rye DNA, (see later, Fig. 7). The Hae III digests
show that a characteristic 5.0 kb fragment present in the λSc het5
clone (indicated in Fig. 7) is present in the rye genome. Evidence
of this type argues that the rye DNA fragment as isolated is not a
cloning artefact. In this regard it should be noted that λSc het5
and λSc het11 are both stable bacteriophages as judged from the
constancy of the restriction enzyme map from a number of different
DNA preparations. This stability presumably results from the
absence of extensive sequence repetition within the inserted DNA.
In λSc het11 at most 800 bp of the heterochromatic sequence is
present. In λSc het5 the 350 bp sequence containing DNA is
approximately 2 kb long, consisting of three monomers and one
dimer as defined by Pst I, plus a part of the monomer which carried
one of the Sau 3A sites leading to the cloning of the respective
fragment.

ii) <u>Nearest neighbour relationships among members of an array of repeated 350 bp units</u>

Earlier studies on the rye heterochromatin DNA sequences
(Bedbrook et al., 1980a; Appels et al., 1981) suggested that the
tandem arrays of sequences were rather homogeneous although some
sequence variation leading to changes in restriction enzyme sites
were assayed. A particularly interesting heterogeneity was described
by Bedbrook et al., (1980b), as a result of a subgroup of "120 bp"
sequences being linked to an unrelated sequence to define a new
array of repeated sequences. Using the isolated arrays of the 350
bp sequence (Fig. 1) we determined the level of sequence variation

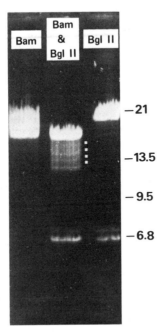

Fig. 2. Instability of repetitive DNA in λSc het bacteriophages

The rye DNA inserts of λSc het3, λSc het8 and λSc het10
were composed, as far as could be determined, only of DNA
sequences homologous to pSc7235. Although these phage
were plaque purified three times as is routinely carried
out before analysing a clone, heterogeneity of the type
illustrated here for λSc het3 persisted. Restriction
endonuclease mapping showed that the rye DNA insert was
flanked by a Bgl II and a Bam HI site but a double digest
did not reveal a single, homogeneous (in size) insert.
The ladder of bands obtained hybridized the pSc7235 probe
and, as discussed in the text, results from a loss of
multiples of the 350 bp unit associated with the presence
of sequences capable of forming cruciform structures in
λ phage (for a recent discussion see Leach and Stahl,
1983).

among 350 bp units and whether certain variants were clustered in
a particular section of a genomic array. The preliminary screen
for sequence variation was based on subcloning individual repeating
units defined by Pst I, hybridizing with a radio-active probe of
known sequence and determining the melting point (T_m) of the
hybrid. The T_m analysis allowed a rapid assessment of a large
number of units (over 100 in this study) to gauge the pattern of
sequence variation prior to carrying out an extensive DNA sequencing
study. The DNA used as a probe was obtained from the plasmid
pSc7235 (Appels et al., 1981) by digestion with Sau 3A and is 184
bp long. The data obtained is shown in Fig. 3. The striking
feature of the data is that the T_m's of the hybrids span a wide
range and some units are at the limit of detection by the hybridization assay (at least 15% base pair difference). It is thus
clear that while a sequence such as found in pSc7235 appears to
derive from a common repeating unit (T_m of the hybrid formed with
itself is the same as that with genomic DNA, Appels et al.,
1981), sampling of individual 350 bp units shows a high level of
sequence variation. Comparing the sequence variants present in
λSc het3, λSc het5, λSc het8 and λSc het10 shows a certain degree
of differentiation among the clones with regards to the variants
which are present. The differentiation between the λSc het clones
can also be visualised in restriction enzyme digests followed by
transfer to Gene-screen and hybridization with a 350 bp probe.
In λSc het8 all the 350 bp units are digested with Hae III (as is
the case with most of the 350 units in total rye DNA) while in
other λSc het clones many units have Hae III sites missing.
Digestion with Taq I reveals all the λSc het clones have 350 bp
units with a Taq I fragment which is present only as a minor band
in total rye DNA digests (indicated by an arrow in Fig 4). This
Taq I fragment is particularly prominent in λSc het3.

Although the sequence differentiation of the λSc het clones
is weak due to the overlapping distributions of sequence variants,
the data implies a degree of clustering at the genomic level.
The mechanisms generating these genomic domains of sequence
variants appears to allow considerable amounts of mixing of
variants to occur. The data is consistent with the possibility
that processes such as unequal cross-over or DNA (gene) conversion
have a large target size (> 20 kb) and thus would not result in
strong clustering at the level examined here.

The interpretation of a tandem array of a particular sequence
cloned in bacteriophage λ, in terms of the existence of domains is
complicated by the demonstrated instability of the cloned DNA. We
therefore sought evidence for the existence of domains of the 350
bp sequence in vivo. To carry out this analysis the DNA from
single rye chromosomes, or a rye chromosome arm translocated to a
wheat chromosome arm, present in wheat-rye addition or substitution
lines was hybridized with several different variants of the 350 bp

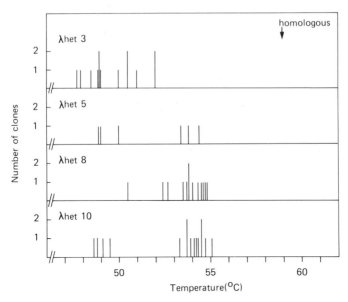

Fig. 3.   Comparison of Pst I units within and between λSc het phages.

Units defined by Pst I (see Fig. 1) were subcloned into the
Pst I site of pBR322.   In the case of λSc het3 and λSc
het10 a population of 35 plasmids from each was prepared
and screened by partial Pst I digestion to select only the
plasmids containing a single 350 bp unit.   For λSc het 5
smaller numbers of plasmids were screened since only three
350 bp units are present plus one dimer.   The latter was
also subcloned and included in the analysis.   The distribution of Pst I sites in λSc het8 is such that monomers,
dimers, trimers and a tetramer are present (see Fig. 1) and
these were all subcloned for the analysis.   The DNA (1 µg)
from each sublcone was immoblized on a nitro-cellulose
disc and hybridized with ^{32}P-labeled Sau 3A DNA, 184 bp
long (prepared from pSc7235, see Appels et al., 1981) in
3XSSC, 50% fromamide, 0.1% ficoll, 0.1% BSA, 0.1% polyvinyl pyrrolidone, 0.1% SDS, 20 µg Hae III digest pBR322.
The incubation time at 37°C, was limited to 3 hrs to eliminate hybridization to pBR322 DNA.

The melting point curves (to provide the Tm value plotted
in the bar diagrams shown) were determined as described in
Appels and Dvorak (1982a) with the bars in the figure indicating the positions of individuals Tm's.   The homologous
reaction indicated by an arrow is the Tm of the probe
hybridized to rye DNA (this is the same as the sequence
hybridized to itself, Appels et al., 1981).

sequence. The probe used to survey the individual subcloned Pst
I units of the λSc het clones did not distinguish between the
various chromosomal heterochromatin regions, except for the 1R
short arm heterochromatin, using the Tm analysis. The largest
Sau 3A fragment from pSc7235 (234 bp) allowed the heterochromatic
blocks of the 1R short arm and 6R to be reproducibly distinguished
from the other chromosomal heterochromatin regions (Fig. 5). A
similar result was obtained with the 230 bp Hae III fragment from
pSc het1. It is assumed that these differences reflect differences
in the distribution of 350 bp sequence variants in a manner
analogous to that found in the comparison of λSc het3, λSc het5,
λSc het8 and λSc het10. Although the analysis at a chromosomal
level is necessarily rendered less sensitive because of the large
numbers of repeating units on a given chromosome, the data support
the existence of domains of sequence variants in vivo. The heterochromatin blocks of the various chromosomes and chromosome arms
were also examined by restriction enzyme analyses. The
chromosomes were assayed for a minor band which is observed in
Taq I digests of rye DNA (Fig. 4) but no differences could be
defined; Appels et al., (1981) also could not distinguish the rye
chromosomes on the basis of restriction endonuclease digests.

(iii) Detecting rye chromosome segments using the 350 bp sequence

The 350 bp sequence has for several years provided a useful
probe for analysing triticales X Triticosecale (Wittmack, reviewed
in Appels, 1983) and material where specific rye chromosome segments have been introuced into wheat.

In many analyses of triticales the probe has clarified some
ambiguity which arose during a C-banding screen. It is however
easy to quantify the amount of 350 bp sequence in a triticale and
this aspect is useful in assessing the contribution that the amount
of a sequence may make to an agronomic character such as seed
shrivelling. For example a recent analysis of the triticale CN605
(L. McIntyre, R. Appels, N.L. Darvey, unpublished), which is a
smooth seed CIMMYT line, indicates no significant reduction in this
major heterochromatic sequence (relative to other triticales). This
triticale has 7 pairs of rye chromosomes which are recognisable in
other triticales (Fig. 6a). These observations run counter to
suggestions which correlate seed shrivelling with an excessive amount
of rye heterochromatin (Bennett, 1977) and are consistent with
numerous other observations indicating no correlation between these
two variables (reviewed in Appels, 1982; Varghese and Lelley, 1983).

An example where the unambiguous identification of the rye
chromosome segment of the chromosome 2RS/2BL was valuable is
given in May and Appels (1984). In a crossing programme to
determine whether the seedling lethality associated with 2RS-2BL

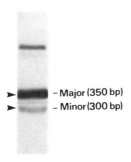

Fig. 4.   The minor and major 350 bp sequence variants visualized
          by Taq I digestion.

          Rye DNA (5 µg) was digested with the restriction endonuc-
          lease Taq I, electrophoresed in 2% agarose, transferred to
          Gene-screen and hybridized with ^{32}P-labeled pSc hetl.
          Hybridizations were carried out as described in the legend
          to Figure 3. The small amount of hybridization at a higher
          molecular weight corresponds to 700 bp (i.e. the dimer form
          of the 350 bp unit).

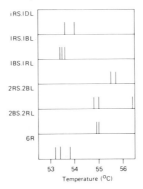

Fig. 5.   Differentiation of the heterochromatic blocks on rye
          chromosomes.

          DNA samples (10 µg) were loaded on nitro-cellulose discs
          and hybridized with a 234 bp Sau 3A DNA fragment from
          pSc7235 (Appels et al., 1981), which had been ^{32}P-labeled
          by nick-translation, under the same conditions as the
          hybridizations carried out in Figure 3. In the addition
          or substitution lines of the type analysed here there
          exists a large excess of wheat DNA and the background
          hybridization to this DNA may be an important consideration.
          However for all material analysed the background to wheat
          DNA was 1-2% of the radioactivity hybridized to the test
          DNA. The only exception was the 1RS/1BL and 1RS/1DL DNA
          where the background was 7.5% of the radioactivity hybrid-
          ized - this level is not sufficient to affect the analysis.

(May and Appels, 1980) interacted with the grass clump dwarfing genes of wheat it was found that 2RS-2BL homozygous plants (previously lethal) could be recovered as fertile plants. Plants of this type (Fig. 6b) have now provided the 2RS-2BL chromosome in a genetic background which can be studied further either from an agronomic or molecular point of view.

A final example of the utilization of the 350 bp sequence is in the investigation of the rye chromatin present in the wheat cultivar Amigo. Chromosome analyses suggest at least a portion of an arm of 1R is present (Zeller and Fuchs, 1983) and as shown in Fig. 6c it appears that the portion carrying the telomeric heterochromatin is also present as visualized by in situ hybridization. The long arm of 1R does not appear to be present, consistent with the suggestion that the 1R short arm is translocated onto a wheat chromosome 1A long (Zeller and Fuchs, 1983). The presence of the 1R short arm correlates with the presence of greenbug (Schizaphis graminum) resistance.

Dispersed repeated DNA sequences from rye, Haynaldia and Elytrigia can also be used to identify alien chromosomes

The heterochromatic sequences of rye provide useful chromosomal markers as outlined in the previous section, but are also limited in the portions of the chromosomes which they can detect. Dispersed sequences would provide probes to cover different parts of the chromosomes and in this section we present data that indicates useful probes of this type can be isolated from rye, Haynaldia and Elytrigia.

i)   Dispersed repeated DNA sequences from rye

The sequences linked to the heterochromatic 350 bp sequence in λSc het5 and λSc het11 were subcloned in pBR322. As mentioned before these sequences were repeated in the genome and it was of interest to determine whether they showed the same genus specificity as the 350 bp sequence to which they are linked. The data in Fig. 7a shows the hybridization of two regions from the λSc het clones close to the 350 bp sequence namely regions 5.3 and 5.4 from λSc het5 and 11.3 and 11.4 from λSc het11, to either rye or wheat DNA digested with Hae III. The probes in general hybridize to wheat DNA, but it is clear that for the 5.3 sequence, which was investigated in more detail, that the hybridization to wheat DNA is a cross-hybridization with a melting point lower than that with rye DNA (Fig. 7b). After washing the hybridized DNA on a filter at 52°C, the cross-hybridization is reduced (Fig. 7b). Although the addition of wheat DNA as a competitor further reduces the cross-hybridization we have not yet attained conditions which

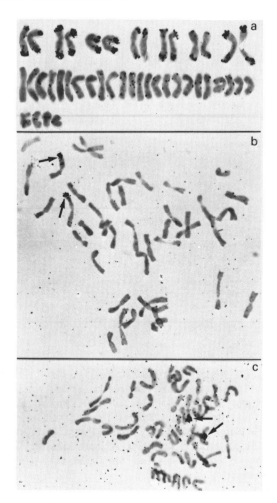

Fig. 6. Detecting rye chromosome segments using the 350 bp sequence.

The probe used in the in situ hybridizations was ^{3}H-cRNA synthesized from pSc het1.

(a) Karyotype of a smooth seeded triticale line CN605 from the CIMMYT programme (seed kindly supplied by Dr. N. Darvey).

(b) 2RS/2BL homozygous plant which proved to be viable, as discussed in the text (see also May and Appels, 1984).

(c) Rye chromosome segment in the wheat cultivar Amigo (seed kindly supplied by Dr. R. A. McIntosh).

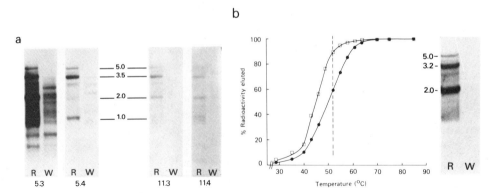

Fig. 7.   Genomic arrangement and genus specificity of sequences in
          λSc het5 and λSc het 11.

(a)   The 5.3, 5.4, 11.3 and 11.4 sequences (identified in
Fig. 1) were hybridized to rye (R) and wheat (W) DNA (10
μg) digested with the restriction endonuclease Hae III and
transferred to Gene-screen.   The numbers indicate lengths
in kb.   The X-ray films for the 5.3, 5.4 and 11.3 sequences
were exposed for the same length of time (20 h) whereas the
exposure shown for the 11.4 sequence was ten times longer
because this sequence has a much lower copy number in the
genome.

(b)   Melting curves of the 5.3 sequence hybridized to either
rye or wheat DNA (5 μg each) were determined using standard
procedures (see legend to Fig. 3).   The broken line indicates the position of 52°C.   To improve the selectivity of
the 5.3 sequence (comparing rye and wheat) the 52°C temperature determined from the melting curves was used to
carry out a post-hybridization wash.   The transfer accompanying the melting curves was obtained by hybridizing rye
and wheat DNA (10 μg each) digested with Hae III and transferred to Gene-screen as described in (a) but including a
post-hybridization wash in 3XSSC/50% formamide at 52°C
(10 min) to elute the bulk of the sequences hybridized to
wheat DNA.   The transfer shown was also prehybridized in
the hybridization mix lacking ^{32}P probe in the presence of
5 μg/ml sonicated wheat DNA (discussed in text).

provide absolute specificity of the 5.3 sequence for rye. The 5.3 sequence however is 2.3 kb long and as this is characterized at the nucleotide sequence level smaller regions of higher specificity may be derived. The Hae III bands prominent in the rye DNA digest (5.0, 3.5 and 2.0 kb) appear to describe a complex family of repeated sequences since the 5.3 and 11.3 sequences appear in these bands, although they show virtually no cross-hybridization to each other. The in situ hybridization technique shows that these linked sequences are generally distributed along chromosomes and tend not to occur in the telomeric heterochromatic blocks (Fig. 8a). The hybridization of the 5.3 sequence was sufficiently specific (relative to wheat) to allow the analysis of wheat-rye addition lines and thus confirm this more generally distributed pattern (Fig. 8b). The 5.0 kb Hae III band which was recovered in λSc het5 is present on all rye chromosomes. As shown in Figure 8c the 1R and 2R chromosomes were further analysed using the 1RS/1BL, 1BS/1RL (Shepherd, 1973; Lawrence and Shepherd, 1981) and 2RS/2BL, 2BS/2RL translocations (May and Appels, 1980) to show that the 5.0 kb band is present on both arms of these chromosomes (a very low amount is present on 1RS).

The detailed chromosomal distribution of sequences such as the 5.3 sequence has not yet been analysed. Boundaries between repeated DNA sequences and adjoining DNA are generally well defined in animal systems (Grimaldi and Singer, 1983; Maresca and Singer, 1983; McCutchan et al., 1982; Potter and Jones, 1983) and it will be of interest to examine this situation in rye by sequence analysis.

ii)   Dispersed repeated DNA sequences from Haynaldia and Elytrigia

To isolate genus specific repetitive sequences from H. villosa and E. elongata, the DNA from these species was partially digested with the enzyme Sau 3A and the 13-20 kb fraction of DNA recovered from a low melting agarose gel as described in the preceding section for rye. Initially, approximately 50,000 random DNA clones from each of the species were screened with unfractionated DNA (the same as that used to generate the clones) labelled with ^{32}P by nick-translation. The probes were hybridized with the respective λ clones to a Cot of approximately 0.1 to select repetitive sequences. Ten clones which gave a strong signal were plaque purified and the isolated DNA rendered radioactive by nick-translation. These probes were then used to hybridize Hae III digests of wheat, E. elongata, H. villosa and rye DNA to assess whether they were useful genus specific probes. Only one clone, λHav 8 containing a DNA segment which showed virtually no cross-hybridization to wheat DNA while still hybridizing significantly to the source of the DNA (in this case H. villosa), was recovered from this screen (Fig. 9).

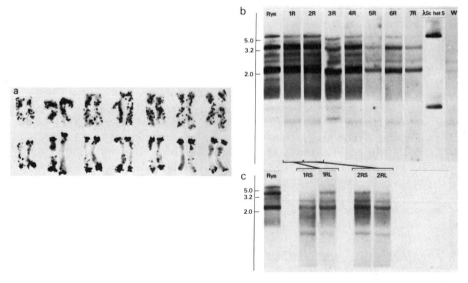

Fig. 8.   Distribution of the 5.3 sequence among rye chromosomes.

(a)   Karyotype of rye chromosomes hybridized in situ with ^{3}H-cRNA synthesized from the plasmid pSc het 5.3 (upper row of chromosomes).   Conditions of hybridization and washing have been described (Appels et al., 1981).   The rye variety used as a source of rye chromosomes is called Snoopy and its karyotype using a heterochromatic sequence probe (pSc het1) is shown for comparison (second row of chromosomes). The polymorphism seen for the chromosome tentatively assigned as 1R (left most pair) with respect to the amount of hybridization observed is typical of the type of varia- tion seen in this variety of rye (reviewed in Appels, 1982).

(b)   DNA samples (10 µg) from the rye-wheat addition lines (indicated iR etc.), as well as (10 µg, W), rye (1.5 µg,R) and λSc het5 (0.5 µg) were digested with Hae III, electro- phoresed in a 1.5% agarose gel, transferred to Gene-screen and hybridized with ^{32}P-labeled pSc het5.3 as described in the legend to Figure 6(b).   The numbers are lengths in kb.

(c)   The 1RS and 1RL designations refer to the 1RS/1BL and 1BS/1RL translocations which were substituted for 1B (see Lawrence and Shepherd, 1981; and 2RS and 2RL refer to the 2RS/2BL and 2BS/2RL translocations which were substituted for 2B (see May and Appels, 1980).   The 1RS/1BL and 1BS/ 1RL ·substitutions were obtained from Dr. K. Shepherd while all other seed was kindly supplied by Dr. C. E. May.

Fig. 9.   Distribution of λ Hav 8 sequences.

Restriction endonuclease Hae III digests of wheat (W),
Haynaldia villosa (H) and rye (R), 10 µg of each, were
electrophoresed on 1.5% agarose, transferred to Gene-screen
and hybridized with ^{32}P-labeled λHav 8. A sequence subcloned from λ Hav 8 which is responsible for most of the
hybridization shown has been successfully used to assay
Haynaldia DNA in Haynaldia-wheat addition lines produced
by Dr. E. Sears.

Enrichment for the class of sequences of interest was achieved
by replica screening the λ clones with unfractionated wheat DNA
(labeled by nick-translation) hybridized to a Cot of 0.1. This
screen was carried out on a set of approximately 50,000 clones from
E. elongata DNA and is illustrated in Fig. 10a. It can be seen from
Fig. 10a that most of the E. elongata DNA sequences hybridize both
the wheat and E. elongata probes to the smae degree. However in
this experiment five clones were clearly hybridizing the E. elongata
probe more efficiently than the wheat probe; many clones hybridized
more efficiently with the wheat probe, suggesting that numerous
repetitive DNA sequences in E. elongata are actually more prominent
in wheat. The five E. elongata clones were plaque purified and
called λEle 142, λEle 143, λEle 144, λ145 and λEle 146. The DNA
from these clones was isolated and labeled as before to allow
hybridization to the Hae III digests of wheat, E. elongata and rye
DNA for confirmation of the genus specificity. Three of the clones,
λEle 142, λEle 143 and λEle 145 were clearly genus specific while
λEle 144 and λEle 146 hybridized equally to all the test DNA's (Fig.
10b, only the λEle 146 hybridization is shown as an example of lack
of genus specificity). The E. elongata DNA's cloned in λEle 143 and
λEle 145 are different from each other as judged from the restriction
enzyme patterns obtained with Bam HI, Eco RI and Bgl II. However
the sequences hybridize to the same genomic Hae III bands (Fig. 10b).

This suggests the cloned DNA's originate from the same family of repeated sequences in a manner analogous to the non-heterochromatic rye sequences found in λSc het5 and λSc het11. We assume this is a major class of sequence since λEle 143 and λEle 145 were recovered from a group of only 5 clones analysed in detail. As discussed earlier for the λSc het5 and λSc het11 sequences, conditions for absolute specificity of the Elytrigia and Haynaldia sequences have not yet been attained. However, the probes used in this preliminary study were the entire DNA inserts of approximately 13 kb in length and further subcloning will, very likey, generate more specific probes. An analysis of several Bam HI fragments from λHav 8 has indicated that some of these fragments have a greater specificity for H. villosa DNA (compared to wheat) than others.

The conditions of hybridization for the λEle 143 and λEle 145 sequences in particular were sufficiently specific to allow an analysis of the E. elongata - wheat addition lines characterized by Dvorak (1980). The analysis shown in Fig. 11 indicates that the Hae III band which can be assigned as originating from E. elongata is found on most of the chromosomes. The presence of this Hae III band in Kite, a variety of wheat carrying the rust resistance Sr26 from E. pontica (Agropyron elongatum 2n = 70; reviewed in Knott and Dvorak, 1976), is ambiguous due to a low level of hybridization to wheat DNA in this region. Further subcloning of sequences from the isolated λEle clones and other new isolates will hopefully resolve this situation and open the way for the analysis of an agronomically important chromosome segment.

Short tandem arrays of repeated DNA sequences in ribosomal DNA spacer regions provide genus specific probes

The cloning, into the vector pACY184, of the wheat Eco Rl ribosomal DNA repeating unit by Gerlach and Bedbrook (1979) has provided the opportunity for defining the spacer region of this gene system. The dominant feature of the wheat spacer region is a tandem array of 130 bp units (Appels and Dvorak, 1982a) which change rapidly in sequence during the course of evolution (Dvorak and Appels, 1982; Appels and Dvorak 1982b). This was demonstrated by the levels of cross-hybridization between the wheat 130 bp spacer sequence and other Triticum, Secale and Elytrigia species. Melting point analyses of hybrids formed between the wheat 130 bp sequence and rye DNA indicated that essentially genus specific probes should be recoverable from the rye rDNA spacer region, to detect these regions in a wheat background. In this section we investigate the feasibility of such a detection system using cloned rDNA from rye and barley.

The rDNA clones studied are shown in Figure 12; the barley rDNA clone (pHV132) was from a collection of barley Eco Rl rDNA

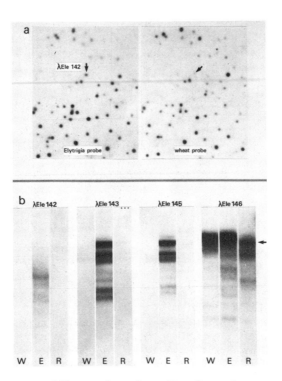

Fig. 10. Genus specific probes from E. elongata.

(a) A portion of a screen of λ phages containing cloned
segments of E. elongata DNA. The DNA from the λ phages
was transferred to nitro-cellulose filters as described
by Benton and Davis (1977) and hybridized with ^{32}P-
labeled by nick-translation and hybridized under standard
conditions to a Cot of 0.05-0.1). After an X-ray exposure
the probe was removed by two washes at room temperature
(15 min) in 50% formamide, 5 mM Tris-HCl pH 8.4, 0.5 mM
EDTA (initially this solution was at 85°C) and then soaked
in the standard prehybridization mix for 1 hr in preparation for a second hybridization. The second probe used
was ^{32}P-labeled wheat DNA, hybridized under the same conditions as the Elytrigia probe. The arrow indicates clone
λEle 142 which was seleced as containing a "genus specific"
sequences in this initial screen.

(b) Purified λEle phage DNA was ^{32}P-labeled and hybridized to 10 μg aliquots of Hae III digested wheat (W), E.
elongata (E) and rye (R) DNA. Arrow indicates a charact-
istic band detected by λEle 143 and λEle 145, as discussed
in the text.

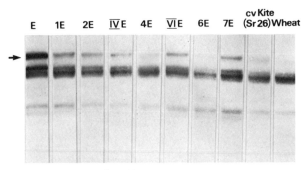

λEle 145 probe

Fig. 11.   Chromosomal distribution of λEle 145 sequences.

Elytrigia – wheat addition line DNA's (IE → 7E, 10 μg
each) were digested with Hae III for comparison with DNA
from wheat (10 μg, W), Elytrigia elongata (1.5 μg, E) and
the wheat cultivar Kite (10 μg).   The arrow indicates the
band which can be distinguished in the wheat background.
Due to a low level of hybridization in this region in the
W control, the interpretation of the Kite analysis is
ambiguous (as discussed in the text) – this particular
variety was of interest to examine because it carries a
chromosome segment for Elytrigia which confers the rust
resistance Sr26 (Dvorak and Knott, 1977).

The addition line IVE is a 4E/7E translocation and VIE is
a 2E/5E translocation (Dvorak, 1980).   The seed used was
kindly supplied by Dr. C. E. May.

units cloned into the vector pACY184 by Gerlach and Bedbrook (1979)
and the rye rDNA clone (λScR1) was one of several selected from a
library of rye DNA/Sau 3A restriction fragments cloned in the vector
λ1059.   The restriction map for pHV132 shown in Figure 12 is essentially as described by Gerlach and Bedbrook (1979) except that the
2.3 kb spacer Taq I fragment was mapped in more detail (R. Appels
and C. E. May, manuscript in preparation).   The λScR1 clone was
mapped by a combination of double restriction enzyme digests and

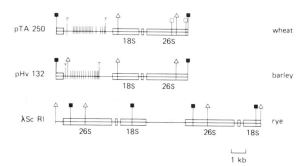

Fig. 12. Restriction endonuclease maps of ribosomal DNA clones.

The pTA250 spacer sequence shows the distribution of Hae
III sites (from Appels and Dvorak, 1982a) in addition to
the Taq I (T), Eco RI ( ), Bam HI ( ) and Bgl II sites
( ). The pHVl32 spacer sequence shows the distribution
of Hha I sites (Appels and C. E. Many manuscript in prep.)
as well as the Taq I, Bam HI and Eco RI sites. The λSc
Rl clones was one of several rDNA clones isolated from
the λ1059 library used to recover the λSc het clones
(Fig. l), as discussed in the text.

subcloning the Eco Rl and Bam HI-Eco Rl fragments into pBR322.
Assignment of these fragments to either the 18S or 26S rRNA genes
was carried out by hybridization of radioactive copies to plant
RNA transferred to nitro-cellulose after electrophoresis in a
denaturing electrophoresis system.

The 2.3kb Taq I fragment from pHVl32 is positioned centrally
in the spacer region, similar to the analogous fragment (2.7 kb)
in the wheat rDNA unit cloned in pTA250. Its internal structure
is dominated by a repeating unit of 160 bp. Hybridization of the
2.3 kb Taq I fragment to restriction enzyme digests of wheat and
rye DNA show virtually no cross-hybridization. The utilization of
the barley spacer probe to detect the barley rDNA units in a wheat
background is shown in Figure 13a. In this experiment barley DNA
(for varieties Betzes and Himalaya) as well as DNA from the available barley-wheat addition lines isolated by Islam et al. (1981)
was digested with the restriction enzyme Taq I. The Gene-screen
filter (NEN) to which the electrophoresed DNA samples were transferred was then hybridized first with a wheat rDNA spacer probe
and exposed to X-ray film in the usual way. Following this the
probe was washed off and the filter rehybridized with a barley rDNA
spacer probe. The data in Figure 13a clearly demonstrate the specificity of the probes. The low level of cross-hybridization
between, for example the barley rDNA spacer probe and wheat rDNA
spacer, also has a low Tm (47.2°C) compared with a homologous reaction (61.6°) and thus can be readily removed by a post-hybridization

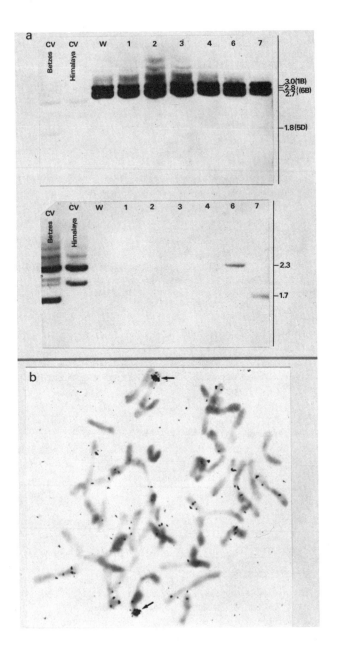

Fig. 13. The chromosomal distribution of barley ribosomal DNA
← ─────── spacer sequences.

(a) The DNA from available barley-wheat addition lines
(1,2,3,4,6,7), two barley varieites, Betzes (B) and
Himalaya (Him) and wheat (W) were digested with the
restriction endonuclease Taq I in 10 μg aliquots. The
digestion products were separated by electrophoresis in
a 1.5% agarose gel and the transferred to Gene-screen
(NEN). Fixing, prehybridization and hybridization of the
Gene-screen filter were carried out as described in Figure
3. The filter was first hybridized with ^{32}P-pTA250.4 (a
probe for the wheat spacer region, Appels and Dvorak,
1982), and exposed to X-ray film (upper panel). Following
this the filter was washed twice at room temperature for
15 min in 50% formamide, 5 mM Tris-HCl pH 8.4, 0.5 mM
EDTA (initially) this solution was at 85°C) and then
soaked in the prehybridization mix for 1 hr in preparation
for a second hybridization with ^{32}P-pHV132.2 Taq I (a
probe for the barley spacer region, see Fig. 12). The
resulting filter was again exposed to X-ray film (Lower
panel). The 1B, 6B and 5D and the associated length
markers derive from the chromosomal assignments for the
rDNA spacer lengths by Appels and Dvorak (1982a).

In the barley 7 addition line, using the barley probe, a
weakly hybridizing band is visible in a longer exposure
of the X-ray film and this corresponds to the 1.9 kb band
observed in cv. Himalaya.

(b) In situ hybridization to the barley 6 - wheat addition
line using ^{3}H-cRNA synthesized from the pHV132 spacer
region. The arrows identify the barley chromosomes.

incubation of 57°C. The data in Figure 13a confirm that the barley
chromosomes 6 and 7 carry the ribosomal genes and demonstrates that
the two loci are distinguishable by the length of the Taq I spacer
fragment. In situ hybridization allows the respective barley
chromosomes to be detected in a wheat background (Figure 13b). The
Chromosome 6 length variant is identifiable in both barley varieties
examined while the chromosome 7 length variant is visualized only
in the variety Betzes which was the source of the spacer region
and the chromosomal clustering of length variants are also found in
wheat (Appels and Dvorak, 1982a; Dvorak and Appels, 1982) and are
likely to be typical features of cereal rDNA.

The spacer region of λSc Rl (Figure 12) has been studied in
sufficient detail to show that it cross-hybridizes only at a low

level with the wheat spacer region in pTA250. This clone (and
one other which was mapped in detail) however appears to be a
rare variant from the rye genome as judged from the Eco RI site
present in the 18S gene. The sequence also fails to efficiently
assay the major 2.7 kb Taq I spacer fragment known to be present
in rye from the weak cross-hybridization reaction with the pTA250
spacer probe. A heterogeneous group of bands are assayed by the
λSc Rl sequence. The rDNA clones from the λ1059 library have not
therefore proven useful for assaying rye rDNA in a wheat background and 9 kb Eco RI rDNA units are currently being recovered
using the Charon 4A vector.

Applying DNA analyses to cereal breeding

The application of DNA analyses to a plant breeding programme
requires, at the very least, a fast method for isolating DNA from
plants in the programme without destroying the respective plants.
The ideal starting materials for isolating plant DNA are either
embryos (or young 4-10 cm seedlings) or anther tissue at a stage
when pollen mother cells are undergoing meiosis. Using embryos
or young seedlings destroys the whole plant and anther tissue is
not always convenient to obtain from cereals in amounts to do DNA
isolations. We have therefore examined the use of leaf material
as starting material and find the following procedure (based on a
protocol in Appels and Dvorak, 1982a) suitable for handling large
numbers of plants. Three leaves from the test plant are cut
into small segments and placed in a small mortar with liquid
nitrogen. Some acid washed sand (ca. 1 g) is added and the
tissue ground to a fine powder using a pestle. The powder is
transferred to another mortar (at room temperature) containing 4
mls 0.05 M Tris-HCl pH 7.5, 0.1 M NaCl, 0.1 M EDTA plus 0.5 mls
5% SDS, 0.5 mls proteinase K (Boehringer, 0.5 mg/ml in H_2O) and
immediately ground again with a pestle to ensure rapid penetration
of the solution into the tissue. The viscous slurry is incubated
at 37°C in a 15 ml or 30 ml centrifuge tube for 1 to 2 hours
after which 1 g of $NaClO_4$ is added. The $NaClO_4$ is dissolved by
vortexing, the mixture centrifuged at 8,000 rpm for 10 minutes
and the supernatant mixed slowly with 9 mls of 70% ethanol saturated
with $NaClO_4$ after transfer to a clean tube. The use of 70%
ethanol saturated with $NaClO_4$ (10 g $NaClO_4$ in a final volume of
80 ml ethanol mixed with a solution of 30 g $NaClO_4$ dissolved in
20 ml H_2O) derives from a procedure for isolating plant RNA
(Sachs et al., 1980). The DNA precipitates as a dark green knot
of material and can be scooped out with a spatula, thus avoiding
a centrifugation step. The DNA is dissolved in approximately 0.5
ml of TE (0.01 M Tris-HCl, pH 8.4, 0.001 M EDTA) and transferred
to a 1.5 ml Eppendorf tube for treatment with RNAase (40 μg/ml,
37°C, 20 min) followed by a chloroform/phenol (1:1, 200 μl of
each) extraction. The aqueous phase is almost colourless at this
stage and the DNA is recovered by adding 50 μl 2 M sodium acetate,

pH 5.5 plus 1 ml of ethanol, to this aqueous phase. The precipitated DNA is recovered by a 5 minute centrifugation, washed with 70%
ethanol and dissolved in 1 ml of TE for storage.

The yield of DNA varies from 0.25 mg to 0.6 mg depending on the
size of the leaves used. We have noted that environmentally
stressed plants (e.g. excessive heat) give poor yields of DNA.
The grinding of the leaves, which is the labor intensive step, can
be carried out at the rate of six samples an hour (per person) as
judged from over two hundred DNA preparations from embryo culture
derived triticale plants (R. Brettell, M. Jeppesen and R. Appels,
unpublished). This allows up to one hundred DNA preparations per
week to be performed on a routine basis. The analysis of DNA
from leaves has to be viewed with some caution since Hesemann and
Schröder (1982) have found the DNA content per nucleus of rye may
decrease with increasing age of the leaf. The effect of leaf age
on the content and genomic arrangement of specific DNA sequences
is unknown at present. Since the analysis of leaf DNA does not
destroy the plant, progeny testing can be done to confirm any
result obtained, should leaf age turn out to be a significant
factor affecting DNA analyses.

The restriction enzyme analysis of many DNA samples does not
present a technical problem but can be expensive if an unusual
restriction enzyme is useful for assessing the sequence of interest.
Generally the restriction enzyme which is diagnostic for a
particular sequence can be prepared conveniently by published
procedures after requiring the appropriate organism.

CONCLUSIONS

New repeated sequences from the grasses related to wheat
have been isolated either by linkage to known heterochromatic
sequences, their presence in the ribosomal DNA spacer regions and
replicate screening of λ libraries using heterogeneous probes
from wheat and the genus of interest. Chromosomal clustering and
apparent genus specificity characterize the various tandem arrays
of sequences studied. These attributes, taken together with a
simplified DNA isolation procedure from progeny in a breeding
programme, argue that contemporary DNA technology can provide an
important source of probes for specific chromosome segments
introduced into wheat from an other grass genus. The class of DNA
sequences described which are chromosomally distributed have
particular value in basic studies of the cereal genome. A
chromatin segment from E. pontica (Agropyron elongatum, 2n=70)
has for example introduced useful rust resistance (Sr26) into
many Australian wheat varieties and repeated sequence probes have
the capability of allowing this chromatin to be studied further.
It is likely that sequences will be isolated which are distributed

in this particular chromatin segment and thus allow the cloning of
other single-copy DNA by a linkage type analysis at the DNA level.
The potential therefore exists for analysing agronomically important segments of chromatin in detail.

ACKNOWLEDGEMENTS

The authors are indebted to Dr. W. Sutton for his advice on
producing λ libraries and extracting DNA from low melting agarose
gels. The authors were fortunate to have Fiona Clay participating
in the experiments reported, especially those dealing with the λ
cloning of E. elongata and H. villosa sequences and are grateful
for her dedication. This research was in part supported by a Rural
Credits grant from the Reserve Bank of Australia and a grant from
the Wheat Industry Council (to Dr. C.E. May and R.A.).

REFERENCES

Appels, R., and Peacock, W. J., 1978, Arrangement and evolution of
        highly repeated (satellite) DNA sequences with special refer-
        ence to Drosophila, Int. Rev. Cytol. Suppl., 8:69-126.
Appels, R., 1982, The molecular cytology of wheat-rye hybrids, Int.
        Rev. Cytol., 80:93-132.
Appels, R., 1983, Chromosome structure in cereals: The analysis of
        regions containing repeated sequence DNA and its application
        to the detection of alien chromosomes introduced into wheat,
        in: "Genetic Engineering of Plants", Tsune Kosuge, Carole P.
        Meredith & Alexander Hollaender, eds., Plenum Publishing
        Corporation, New York, pp. 229-255.
Appels, R., Dennis, E. S., Smyth, D. R., and Peacock, W. J., 1981,
        Two repeated DNA sequences from the heterochromatic regions
        of rye (Secale cereale) chromosomes, Chromosoma (Berl.), 84:
        265-277.
Appels, R., Driscoll, C., and Peacock, W. J., 1978, Heterochromatin
        and highly repeated DNA sequences in rye (Secale cereale),
        Chromosoma (Berl.), 70:67-89.
Appels, R., and Dvorak, J., 1982a, The wheat ribosomal DNA spacer
        region: It's structure and variation in populations and
        among species, Theor. Appl. Genet., 63:337-348.
Appels, R., and Dvorak, J. 1982b, Relative rates of divergence of
        spacer and Gene sequences within the rDNA region of species
        in the Triticeae: Implications for the maintenance of homo-
        geneity of a repeated gene family, Theor. Appl. Genet., 63:
        361-365.
Arnheim, N., and Kuehn, M., 1979, The genetic behaviour of a cloned
        mouse ribosomal DNA segments mimics mouse ribosomal gene evo-
        lution, J. Mol. Biol., 134:743-765.
Bedbrook, J., Jones, J., O'Dell, M., Thompson, R. D., and Flavell,
        R. P., 1980a, A molecular description of telomeric heterochro-
        matin in Secale species, Cell, 19:411-419.

Bedbrook, J., O'Dell, M., and Flavell, R. B., 1980b, Amplification of rearranged repeated DNA sequences in cereal plants, Nature, 288:133-137.

Bennett, M. D., 1977, Heterochromatin, Aberrant Endosperm Nuclei and grain shrivelling in wheat-rye genotypes, Heredity, 39: 411-419.

Benton, W. D., and Davis, R. W., 1977, Screening λgt recombinant clones by hybridization to single plaques in situ, Science, 196:180-182.

Brutlag, D., Carlson, M., Fry, K., and Hsieh, T. S., 1978, DNA sequence organization in Drosophila heterochromatin, Cold Spr. Symp. Quant. Biol., 42:1137-1146.

Davidson, E. H., and Posakony, J. W., 1982, Repetitive sequence transcripts in development, Nature, 297:633-635.

Dvorak, J., and Knott, D. R., 1977, Homoeologous chromatin exchange in a radiation induced gene transfer, Can. J. Genet. Cytol., 19:125-131.

Dvorak, J., 1980, Homoeology between Agropyron elongatum chromosomes and Triticum aestivum chromosomes, Can. J. Genet. Cytol., 22: 237-259.

Dvorak, J., and Appels, R., 1982, Chromsomes and nucleotide sequence differentiation in genomes of polyploid Triticum species, Theor. Appl. Genet., 63:349-360.

Flavell, R. B., and Smith, D. B., 1976, Nucleotide sequence organization in the wheat genome, Heredity, 37:231-252.

Gerlach, W. L., and Bedbrook, J. R., 1979, Clonging and characterization of ribosomal RNA genes from wheat and barley, Nucl. Acids Res., 7:1869-1885.

Grimaldi, G., and Singer, M. F., 1983, Members of the Kpn 1 family of long interspersed repeated sequences join and interrupt α-satellite in the monkey genome, Nucl. Acids Res., 11:321-338.

Hessmann, C. U., and Schroder, G., 1982, Loss of nuclear DNA in leaves of rye, Theor. Appl. Genet., 62:325-328.

Hilliker, A. J., and Appels, R., 1982, Pleiotropic effects associated with the deletion of heterochromatin surrounding rDNA on the X chromosome of Drosophila, Chromosoma (Berl.), 86:469-490.

Hohn, B., 1979, In vitro packaging of λ and Cosmid DNA, Methods of Enzymology, 68:299-309.

Islam, A. K. M. R., Shepherd, K. W., and Sparrow, D. H. B., 1981, Isolation and characterization of Eu-plasmic wheat-barley chromosome. Addition lines, Heredity, 46:161-174.

Karn, J., Brenner, S., Barnett, L., and Cesareni, G., 1980, Novel bacteriophage λ cloning vector, Proc. Natl. Acad. Sci. U.S.A., 77:5172-5176.

Knott, D. R., and Dvorak, J., 1976, Alien germ plasm as a source of resistance to disease, Ann. Rev. Phytopathol., 211-235.

Langridge, J., Langridge, P., and Bergquist, P. L., 1980, Extraction of necleic acids from agrose gels, Anal. Biochem., 103:264-271.

Lawrence, G. J., and Shepherd, K. W., 1981, Chromosomal location of
    genes controlling seed proteins in species related to wheat,
    Theor. Appl. Genet., 59:25-31.
Leach, D.R.F., and Stahl, F. W., 1983, Viability of λ phages carry-
    ing a perfect palindrome in the absence of recombination
    nucleases, Nature, 305:448-450.
Maresca, A., and Singer, M. F., 1983, Deca-satellite: a highly
    polymorphic satellite that joins α-satellite in the African
    green monkey genome, J. Mol. Biol., 164:493-511.
May, C. E., and Appels, R., 1980, Rye chromosome translocations in
    hexaploid wheat: A re-evaluation of the loss of heterochroma-
    tin from rye chromosomes, Theor. Appl. Genet., 56:17-23.
May, C. E., and Appels, R., 1984, Seedling lethality in wheat: a
    novel phenotype associated with a 2RS/2BL translocation
    chromosome, Theor. Appl. Genet., Submitted.
McCutchan, T., Hsu, H., Thayer, R. E., and Singer, M. F., 1982,
    Organization of African green monkey DNA at junctions between
    α-satellite and other DNA sequences, J. Mol. Biol., 157:195-
    211.
Moss, T., 1983, A transcriptional function for the repetitive
    ribosomal spacer in Xenopus laevis, Nature, 202:223-228.
Potter, S. S., and Jones, R. S., 1983, Unusual domains of human
    alphoid satellite DNA with contiguous non-satellite sequences:
    Sequence analysis of a junction region, Nucl. Acids Res.,
    11:3137-3151.
Rigby, P.W.J., Dieckmann, M., Rhodes, C., and Berg, P., 1977,
    Labelling deoxyribonucleic acid to high specific activity in
    vitro by Nick Translation with DNA polymerase I., J. Mol.
    Biol., 113:237-251.
Rimpau, J., Smith, D., and Flavell, R., 1978, Sequence organization
    analysis of the wheat and rye genomes by interspecies DNA/DNA
    hybridization, J. Mol. Biol., 123:327-359.
Sachs, M. M., Freeling, M., and Okimoto, R., 1980, The anaerobic
    proteins of maize, Cell, 20:761-767.
Shepherd, K. W., 1973, Homoeology of wheat and alien chromosomes
    controlling endosperm protein phenotype, Proc. 4th Int. Wheat
    Symp., pp. 745-759.
Sutton, W. D., 1972, Chromatin packing, repeated DNA sequences and
    gene control., Nature (New Biol.), 237:70-71.
Sutton, W. D., Gerlach, W. L., Schwartz, D., and Peacock, W. J.,
    1984, Molecular analysis of Ds controlling element mutations
    at the Adh1 locus of maize, Science, In press.
Varghese, J. P., and Lelley, T., 1983, Origin of nuclear aberra-
    tions and seed shrivelling in triticale: a re-evaluation of
    the role of C-heterochromatin, Theor. Appl. Genet., 66:159-
    167.
Weinand, U., Schwartz, Z., and Feix, G., 1979, Electrophoretic
    elution of nucleic acids from gels adapted for subsequent
    biological tests, Application for analysis of mRNA's from
    maize endosperm, FEBS Letters,98:319-323.

Zeller, J. F., and Fuchs, E., 1983, Cytology and disease resistance
    of a 1A/1R and some 1B/1R wheat-rye translocation cultivars,
    Z. Pflanzenzüchtg., 90:285-296.

GENETIC ASPECTS OF SYMBIOTIC NITROGEN FIXATION

Sharon R. Long

Department of Biological Sciences
Stanford University
Stanford, CA 94305

INTRODUCTION

Nitrogen fixation is the reduction of molecular dinitrogen (N_2) to ammonia. This process, along with nitrate reduction, brings nitrogen from inorganic form into the organic world, for ammonia is the form of nitrogen which can be assimilated into amino acids. Fixed nitrogen is generated chemically by the Haber process, in which gaseous nitrogen and hydrogen are combined at high temperature and pressure to yield ammonia. This industrial process provides most of the ammonium fertilizer used in agriculture, and consumes vast amounts of petrochemical fuel every year (Postgate, 1982).

A variety of prokaryotic organisms -- bacteria and cyanobacteria -- carry out biological nitrogen fixation (Sprent, 1979). The enzyme responsible for this is nitrogenase, which uses ATP and reducing equivalents to combine molecular nitrogen with protons, yielding ammonia. Nitrogenase is composed of two proteins: the iron protein (component II or nitrogenase reductase) and the molybdenum-iron protein (component I or nitrogenase). Component I has two types of subunit (the K and D gene products); component II has 1 subunit type (the K gene product). Both components are extremely sensitive to oxygen.

Many nitrogen-fixing organisms are free-living, and obtain the energy and reducing power needed for the process through photosynthesis, or through respiration or other oxidative processes (Postgate, 1982). In addition, however, there are several groups of prokaryotes which associate symbiotically with higher plants; the microbes fix nitrogen, and receive energy from plant photosynthate. One such partnership is that between the cyanobacterium <u>Anabaena</u>

azollae and the water fern Azolla (Peters and Mayne, 1974). Azolla
leaves are bi-lobed, and the dorsal lobe has an internal cavity.
Within this cavity, which is lined with specialized hair cells,
filamentous Anabeana cells proliferate and fix nitrogen. The leaf
and symbiont develop and differentiate in parallel, from young,
actively growing and dividing cells to the mature stage at which
nitrogen fixation occurs. The onset of nitrogen fixation in the
symbiotic Anabaena is similar to that in free-living cyanobacteria
in that it involves the formation of specialized heterocyst cells.

These symbiotic partners fix nitrogen effectively in wetland
agricultural areas. They have been used in parts of southeast Asia
as a means of fertilizing paddy rice. Here is a good example of
using nitrogen fixing organisms as a source of green manure for
crop growth.

Another symbiotic association which is currently being
researched is that between Gram-positive bacteria of the
Actinomycete group and various dicotyledons, often shrubs or trees,
such as Myrica, Alnus, Comptonia, and Casuarina. The bacteria
include the genus Frankia, and have only recently been cultured in
a free-living state (Callaham et al., 1978). Frankia stimulate
nodule development in their hosts, and invade the host tissue. The
bacterial cells are found in vesicles within the nodule, and will
fix nitrogen in culture if isolated together with the vesicle and
surrounding capsule material. The bacterial isolates from these
nodules generally have quite a wide host range.

The symbioses described above, together with loose associations
which occur between plants and bacteria in the rhizosphere, are
very important for the soil nutrition in their systems. By far the
most significant amount of fixed nitrogen in agriculture, though,
comes from the association of bacteria in the group Rhizobium with
host plants in the family Leguminosae. This symbiosis is of
particular importance because so many of the plants in this family
are used directly as human or animal food, in addition to being
useful as green manure or as perennials in stable ecosystems. In
the remaining part of this paper, I would like to describe this
association, focussing on how genetics may provide both the means
of analyzing how it occurs and the tools with which it may be
improved.

RHIZOBIUM-LEGUME SYMBIOSES

Root nodules are formed by Rhizobium and its legume host via a
series of interactions (Vincent, 1980). These begin with the
partners growing independently in the soil, for each is capable of
reproduction and growth in the free-living state. A particular
Rhizobium and its host recognize each other, probably by surface
interactions. It has been proposed that plant-produced lectins are
involved in recognition due to their specific interaction with
bacterial surface polysaccharides. This hypothesis is currently

the subject of extensive research (see Bauer, 1981; Dazzo and Hubbell, 1982).

Rhizobium species are grouped according to which host plant they nodulate. This classification has several problems, not the least of which is that it does not always correlate well with other taxonomic characters such as metabolic properties or DNA homology. However, it is useful to list some of the currently recognized species since they are referred to often in the literature.

TABLE 1

Bacterial species	Plant host
Rhizobium meliloti	Medicago, Melilotus
Rhizobium leguminosarum	Pisum, Vicia
Rhizobium trifolii	Trifolium
Rhizobium japonicum	Glycine
Rhizobium parasponiae	Parasponia, a non-legume
Rhizobium spp."cowpea"	Vigna, many others

The first three species are examples of the "fast-growing" Rhizobium group, and the second three are "slow-growing" -- although fast growing strains have recently been found which nodulate soybean and Parasponia. The slow growing species have been reclassified as a separate genus, "Bradyrhizobium" (Jordan, 1982).

Associated with recognition of their particular host, many bacteria cause characteristic deformations (curling) of the host's epidermal root hairs. Rhizobium invade through the root hair, progressively encased, as they invade, by a host-produced cell wall tube called the infection thread. The infection threads penetrate, occasionally branching, through several deeper layers of the host root tissue. As this invasion occurs, cells in the plant root begin dividing, and this proliferation results in the formation of the root nodule (see Dart, 1974; Newcomb, 1981).

The bacteria within the infection thread are continually dividing as the infection proceeds. While some infection threads grow all the way through host cells, or along cell walls, others penetrate partway into a host cell, stop, and release bacteria from the tip and sides of the infection thread into the plant cytoplasm. As they enter the cell, the bacteria are enveloped in a plant membrane, probably derived from the plasma membrane in a manner analogous to endocytosis. An infected plant cell may have thousands of bacteria in its cytoplasm (Newcomb, 1981).

The release of bacteria into plant cells is accompanied by the final stages of differentiation which lead to nitrogen fixation. The bacteria often enlarge and/or change shape, and in the endosymbiotic form they are referred to as "bacteroids." Not only are the bacterial genes for nitrogenase actively transcribed in bacteroids, but it is also possible that much of bacteroid metabolism is different from that of free-living cells. The plant cell also differentiates, producing new isozymes for some assimilatory metabolic functions and activating transcription of genes for an oxygen-protective protein called leghemoglobin. The two partners exhibit extensive metabolic interactions. As the symbiosis matures, the bacteria fix nitrogen, and may supply the heme moiety for leghemoglobin. The plant supplies energy for ATP and reducing power from the photosynthate translocated from the shoot system; the plant also protects the bacteroids from too high a concentration of free oxygen, yet while providing a steady supply of oxygen at low tension for bacteroidal and plant respiration (Sutton et al., 1981; Imsande, 1981).

The active, fixing nodule contains both infected and uninfected plant cells. These appear to have distinct functions in the overall process of fixing, assimilating and transporting nitrogen for the benefit of the plant. For example, infected cells are the site of nitrogen fixation by the bacteroids, and of ammonia assimilation into glutamine by plant enzymes. In some plants such as soybean, the amino acid compounds are further converted into ureides, which require fewer carbons in the skeleton per nitrogen atom transported. The enzymes which carry out this conversion are located in uninfected cells (reviewed by Newcomb, 1981; Verma and Long, 1983). Nodules have a vascular system, and often there appear to be transfer cells in the vicinity of the vascular cells.

Thus the Rhizobium-legume symbiosis is an extremely complex process, in which precise developmental and metabolic steps are carried out by two different organisms. Any attempts to alter or improve symbiotic nitrogen fixation must take into account the stages which lead to a successful association, and must find a way to improve or alternatively to circumvent them. In the following sections I will describe some of the genetic work being carried out on two aspects of this process: nitrogen fixation, and nodule formation. This work has revealed both some simplifying principles which apply to nitrogen fixation, and some further complexities.

GENETICS OF NITROGEN FIXATION IN RHIZOBIUM

Much of the fine structure work done on nitrogen fixation (nif) genetics has been carried out on free-living nitrogen fixers, in particular the enteric species Klebsiella pneumoniae (Robson et al., 1983). The genes for nitrogen fixation in Klebsiella have been mapped and cloned (Cannon et al., 1979; see also review by Robson et al., 1983). Ruvkun and Ausubel (1980) advanced nif cloning into Rhizobium when they discovered that the genes for nitrogenase were so highly conserved that they hybridized to all other nitrogen fixing organisms. This made it possible to clone and map the nitrogenase genes of Rhizobium meliloti (Ruvkun and Ausubel, 1981; Corbin et al., 1982; Banfalvi et al., 1981). This approach has also led to the successful cloning of nitrogenase genes from a wide variety of Rhizobium species, including R. trifolii (Scott et al., 1982), R. leguminosarum (Ma et al., 1982), R. japonicum (Hennecke, 1981) and others.

In any genetic analysis, each locus is defined with respect to a phenotype. This is particularly crucial when the locus in question has been identified only through physical techniques such as sequence homology. Such demonstrations have been possible for Rhizobium nitrogen fixation loci, due to the relatively recent availability of transposable elements and cloning vectors which function in Rhizobium. The technique of site-directed mutagenesis (Ruvkun and Ausubel, 1981) has blended these two approaches, and has proven useful in the study of many other genes besides the nitrogen fixation loci for which they were first used. This technique is based on the vector pRK290, derived from the incompatibility-P plasmid RK2 (see Ditta et al., 1980). This plasmid, which encodes tetracycline resistance, can be introduced into Rhizobium by conjugation, an essential attribute in this case since Rhizobium is not easily transformed. Once in Rhizobium, plasmids based on this vector are able to replicate, producing a few copies per cell.

In site-directed mutagenesis, a fragment of interest is cloned into pRK290, and is mutagenized with a transposon insertion; the transposon of choice in Rhizobium is Tn5, since its neomycin/kanamycin resistance is strongly expressed and it exhibits little transposition target preference . The position of the Tn5 can be mapped by restriction enzymes, and the clone with Tn5 therefore can be regarded as bearing a mutation (from the insertion) in a precisely mapped place. The plasmid bearing the "mutated" copy of the fragment is conjugated into a wild type recipient strain. At a low frequency, the plasmid-borne gene with its Tn5 insertion will recombine into the genome of the host Rhizobium, replacing the wild-type copy. The latter is now borne on the tetracycline resistant vector, and can be cured from the strain using plasmid incompatibility to select for bacteria which have both acquired a genomic copy of the mutated gene and lost the plasmid carrying the wild-type copy. The phenotype of this mutant can now be tested. It was by such means that Ruvkun and Ausubel (1981) showed that the R. meliloti fragments which were homologous to cloned Klebsiella nif DNA, were responsible for symbiotic nitrogen fixation in Rhizobium.

There had for some time been evidence that nitrogen fixation and nodulation genes in at least some Rhizobium strains were borne on high molecular weight plasmids. This evidence included the association of symbiotically defective phenotypes with plasmid curing, and the restoration of normal phenotype by transfer of normal plasmids into cured strains (see Beringer et al., 1980;

genetics, and especially of symbiotic phenotypes, has been the
development of techniques for transposon mutagenesis. This method
of generating mutants permits the location of the gene to be
tracked both through the transposon's drug resistance and through
detecting its presence physically, by DNA hybridization. The
pioneering use of this approach in Rhizobium resulted in the
discovery that in R. leguminosarum, genes for nodulation resided on
a plasmid (Johnston et al., 1978). Plasmid linkage of nodulation
genes and determinants of host range have since been demonstrated
in several fast growing Rhizobium strains. In the case of R.
leguminosarum, R. trifolii and R. phaseoli, this analysis has been
aided by the fact that many symbiotic plasmids are transmissible.
For example, a plasmid from R. leguminosarum can be
transferred into R. phaseoli, and this will confer upon the
recipient strain the ability to nodulate peas (Beynon et al., 1980;
for other examples, see reviews by Beringer et al., 1980; Long,
1984).

The Rhizobium meliloti system differs from the other fast
growers in several respects. The plasmid in R. meliloti -- called
the megaplasmid -- is larger even than other Rhizobium plasmids,
and is neither easily isolable nor self-transmissible (Banfalvi et
al., 1982; Rosenberg et al., 1982). Analysis of nodulation genes
in this species has proceeded by examining strains with deletions
in the plasmid, and by identification of individual genetic loci
through transposon mutagenesis.

RHIZOBIUM MELILOTI NODULATION GENES

Map location of nodulation genes. Our investigations of nodulation
loci in R. meliloti began with a single wild-type strain, Rm1021,
which was subjected to transposon mutagenesis using the technique
of Johnston et al. (1978). Individual transposon-containing
colonies were tested on plants, and analyzed for symbiotic
phenotype. Out of 6000 colonies screened, 4 mutants were found
which were defective in the ability to form nodules (Long et al.,
1981; Meade et al., 1982).

Two of these mutants were of special interest because they
failed to deform plant root hairs, the earliest visible stage of
infection. Therefore these two mutants were assumed to be blocked
at a very early stage of nodule development. By DNA-level analysis
of these strains, it was found that the mutations appeared to be
linked to the R. meliloti megaplasmid (Buikema et al., 1983), which
was indicated by the plasmid deletion studies of Rosenberg et al.
(1981) and Banfalvi et al. (1981).

To examine the nodulation gene or genes in greater detail, it
was important to obtain the gene regions as cloned fragments. This
was carried out by a method which used the plant as a selective
agent. Into one of the mutant strains, a clone bank of wild-type
R. meliloti DNA fragments, borne on a vector which could replicate

in Rhizobium, was introduced by conjugation. A certain fraction of
the mutant cells will, by chance, receive the particular clone
which bears the normal gene(s) for nodulation. This fraction of
the mutant population was therefore genetically competent to
invade the plant. The rest of the mutants could not form nodules,
and remained outside the root. By surface-sterilizing the root
nodules and squashing out the contents, it was possible to obtain
the bacteria which contain the cloned nodulation genes (Long et
al., 1982).

A clone obtained by this method, pRmSL26, was found to correct
the nodulation defect of both of the non-nodulating mutants
mentioned above. This indicated that they were either in the same
gene, or were closely linked. It was possible to analyze the
location of the fragments in pRmSL26 by comparing restriction
digests of its DNA with those of cosmid clones derived from the
megaplasmid. This revealed that the nodulation region was linked
approximately 20-30 kilobases away from the region where the nif
genes had been found (Long et al., 1982). A map of the nif-nod
region of R. meliloti strain 1021 is shown in figure 1.

The same selection-by-plant method has recently been used by
our laboratory to clone the DNA corresponding to other nodulation
loci. These loci were identified using three mutants -- WL113,
WL131, and WL188 -- which were described by Paau et al. (1981).
These mutations were induced by nitrosoguanidine mutagenesis in R.
meliloti strain 102F51. Using a clone bank of wild-type DNA, we
successfully complemented these mutants, and obtained nodules on
host alfalfa plants (S. Long, J. Tu, manuscript in preparation).
By comparing the fragments in these clones with known megaplasmid
DNA, we have found they map to both sides of the region identified
in pRmSL26 (Figure 1). Therefore, there seems to be an extended
cluster of nodulation genes on the R. meliloti megaplasmid, located
in the direction which would be downstream from nif transcription.

Nod gene function. What do these genes do? The region we know the
most about is that contained in pRmSL26. At least some of the
genes borne on this plasmid appear to be involved in "universal"
invasion functions: the cloned R. meliloti DNA in pRmSL26 is able
to restore normal phenotype to nodulation mutants in R. trifolii
(S. Long and J. Tu, manuscript in preparation). The reciprocal is
also true: a recombinant plasmid bearing the nodulation genes of R.
trifolii can restore nodulation to some R. meliloti nodulation
mutants (Djordevic, Rolfe and Shine, personal communication).
However, the host range of the corrected mutants -- whether R.
meliloti or R. trifolii -- remains the same. It therefore appears
that at least some Rhizobium genes used in invasion do not confer
host specificity. This may turn out to be quite important, since
more "universal" invasion steps could relate to the fundamentals of
how plants are invaded by microbes, and why the legumes in
particular are susceptible to Rhizobium infection.

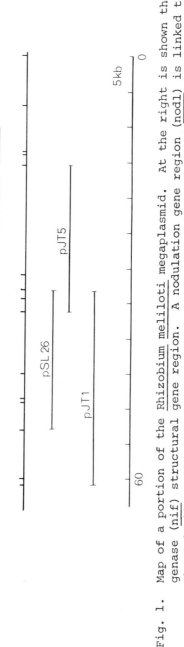

Fig. 1.   Map of a portion of the Rhizobium meliloti megaplasmid. At the right is shown the nitrogenase (nif) structural gene region. A nodulation gene region (nod1) is linked to nif,
about 20 kb to the left. This region has been subjected to fine-structure genetic analysis
as shown in figure 2. Plasmid pSL26 restores nodulation to Tn5 mutants located in the nod1
region, and to NTG mutant WL188. However, it does not restore function to NTG mutants WL133
or WL131, indicating that these are mutanted in nod genes outside the pSL26 region. These
two mutants are complemented by pJT1 and pJT5, respectively. This indicates that there may
be additional nodulation loci in regions "2" and "3".

FINE STRUCTURE OF NODULATION GENES IN R. MELILOTI

We have carried out a physical and genetic analysis of a region of the Rhizobium meliloti megaplasmid which is required for infection of alfalfa. As pointed out above, this region is likely to be only one of several which contain nodulation genes. A clone bearing a nodulation gene or genes, pRmSL26, was identified as described above. Preliminary data indicated that two nodulation mutants mapped to the 8.7 kb EcoRl fragment.

We have subcloned the fragments of the original pRmSL26 clone, and have subjected the 8.7 kb EcoRl fragment subclone to transposon mutagenesis. The site of each transposon insertions in the clones was mapped by restriction analysis, and the clones with their transposons are introduced and recombined into the Rhizobium genome by the method of Ruvkun and Ausubel. In order to test the effect of the transposon insertions, individual colonies with mutations in known positions are grown in liquid broth, and are added to sterile tubes along with aseptically grown plants. After three to four weeks, the ability of each colony to form nodules was evaluated by inoculating aseptically grown mutant plants with them and scoring for nodule formation after 3-4 weeks.

By this procedure, we obtained a series of over 90 mutants, all of which mapped to the 8.7 kb EcoRl fragment. The symbiotic assays revealed that insertions over a length of about 3.5 kb, towards the right (nif side) of.the fragment resulted in a non-nodulating (Nod⁻) phenotype. This identifies the physical extent of the nodulation genes in this particular DNA fragment. We have designated this as the nI segment, to distinguish it from the other nodulation gene regions which apparently occur in R. meliloti.

To determine the number of genes represented in nI, we have carried out complementation tests. Strains were constructed which bore a mapped Tn5 insertion in the megaplasmid nodulation region, and which contained a plasmid bearing another copy of the nodulation region with a different mutation in it. By assaying such strains for their ability to infect plants, it was possible to classify mutations in complementation groups. The preliminary data from these experiments indicates that there are at least four, and possibly six complementation groups. Two of these, nodA and nodB, located at the nif-proximal end of nI, appear to be very small, and must each represent a single peptide. It is not known at this point, however, exactly how many proteins are encoded within the 3.5 kb of the nI segment.

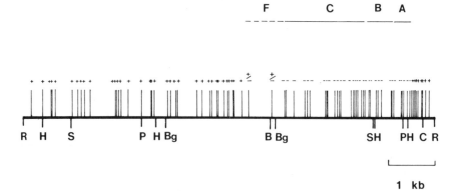

Fig. 2. A fine-structure map of the R. meliloti "nod1" region. An
        8.7 kilobase EcoRl fragment, cloned as pRmJl, was subjected
        to Tn5 mutagenesis followed by gene exchange using the
        principle of the Ruvkun-Ausubel technique (1981). The
        phenotype of the mutants was then assayed by inoculating
        alfalfa seedlings with the Tn5-containing bacterial strain.
        This is shown as a "+" or "-" immediately above the verti-
        cal line indicating position of each Tn5 insertion. The
        different mutants are grouped as follows (Jacobs and Long,
        1984): "A" and "B" are distinct regions based on comple-
        mentation data; "C" may be one or more than one comple-
        mentation group. "F" is represented by only two mutants,
        which have a leaky phenotype. The total nodulation gene
        length in this region is about 3.5 kilobases.

NODULATION AND HOST RANGE

     Clone pRmSL26 appears to have some, but not all of the genes
required for nodulation. Adam Kondorosi and co-workers, using
deletion analysis and clones from different parts of the R. meliloti
megaplasmid, have found evidence for nodulation genes between the
8.7 kb EcoRl fragment (8.5 kb in their strain) and the nif region,
and has suggested that this region may influence the host range
of the bacterium (A. Kondorosi, personal communication). This is
also where clone pRmT5, isolated by complementation of mutant WL188,
has been mapped. Further tests will be needed to

understand the organization of this region.

Host range may be very complex in Rhizobium meliloti and other Rhizobia. While between pairs of species such as R. trifolii and R. leguminosarum, host range can be changed by plasmid transfer (Beringer et al., 1980; Beynon et al., 1980; Djordevic et al., 1983; and others), transfer of such plasmids into R. meliloti does not extend its host range to include clover or pea (S. Long and J. Tu, submitted for publication). It is important to consider that the determination of host range in a bacterium may involve not only positive factors -- the ability to bind to a particular plant, for example -- but may also involve negative factors, such as bacterial properties which make an individual strain unable infect anything except its own host. Such negative factors (avirulence determinants) are hypothesized to be responsible for host specificity in pathogens; it is quite possible that they are also at work in Rhizobium host choice.

Finally, it is important to remember that host range refers to the successful completion of the entire symbiotic process, from infection through sustained nitrogen fixation. There are many levels of compatibility, and many levels at which host range is expressed (Beringer et al., 1980). The Rhizobium strains as a group form nodules on plants which, with one exception, belong to the Leguminosae. Why legumes? What distinguishes this group from other plants? Is it a property which influences an early stage of infection? Within the legume family, particular species of Rhizobium are invasive only on certain genera or species of plant. This selectivity is sometimes, but not always, reflective of taxonomic grouping of the plants. The strains which share the ability to infect one particular host do not necessarily share the ability to infect other plants. And within a group of Rhizobium strains which nodulate a particular plant, some will be found unable to fix nitrogen in those nodules. The need for compatibility extends to the molecular and attendant metabolic processes of nitrogen fixation itself.

Thus, in our efforts to broaden or control host range, we must keep in mind several points: first, host selectivity may operate partly through the genes which cause invasion, but other genes may be involved as well. Secondly, host range may be determined not only through positively acting factors, such as surface components which can bind a host plant, but through negatively acting factors, analogous to avirulence genes in pathogens, whose presence prevents a bacterium from invading most plants. Genes for host range, especially broad host range, will not act as simple dominant factors in all backgrounds. Finally, there are many levels at which compatibility must occur, including infection, formation of the nodule, proper release of the bacteria and differentiation of the partner cells into the nitrogen fixing state. "Tailoring" strains for a particular host range will require examining the genes which influence all of these steps.

FUNCTION IN THE SOIL; FUNCTION IN THE PLANT.

In considering the genes for nitrogen fixation and nodulation, we have dealt with loci which can be identified through definable phenotypes. In the case of nitrogen fixation, this extends to the biochemical level: the products of the genes are known and can be studied at the protein level. Nodulation, while biochemically elusive at this time, is measurable in macroscopic and microscopic stages. Rhizobium symbiotic properties, however, must be defined much more broadly than by these two sets of genes: successful symbiosis begins in the soil, while Rhizobium is a free-living population; and it ends, not with nitrogen fixation itself, but with the assimilation of fixed nitrogen into organic compounds and with the partition of that nitrogen nutrition into the organs of the growing plant. The genes which control these other aspects of symbiotic function are equally important to those involved in nodulation and fixation, and are likely to be much harder to define.

Primary among the concerns of Rhizobium researchers hoping to introduce engineered strains into the field is the competitiveness of particular Rhizobium strains (Bohlool et al., 1983). It has been found that strains from the laboratory, or from other areas, cannot easily be introduced into agricultural fields. Instead, the native microflora outcompete the new strains, and almost all of the nodules formed on the plants in the field are from those native bacteria. This has particularly been true of Rhizobium japonicum, the important strain which forms nodules on soybean (Glycine max), and which has many strains now indigenous to the soybean fields of the midwestern United States. Efforts to inoculate such fields with new strains have not met with great success.

Competitiveness is not an absolute measure of bacterial performance: it depends on the soil chemistry, the type of plants and other microflora which grow in the soil, the climate, and other conditions. Thus, in the soybean fields of the upper South, where Rhizobium japonicum does not survive as well over long periods of time, it has been possible successfully to inoculate the crops with cultured R. japonicum.

In addition, competitiveness is probably not simply the ability to colonize or bind to a host plant root. There appear to be very complicated interactions between plants and invading microbes, which may involve signalling between different parts of the plant (Bohlool et al., 1983).

Another factor to bear in mind is that the host plant regulates invasion by Rhizobium, in ways which are not presently understood. Pierce and Bauer have found (1983) that if bacteria invade one part of a root, then no nodules are initiated on the part of the root just below. This regulatory effect appears to occur within 12 hours of adding bacteria to the plant root, far earlier than nodule development or nitrogen fixation occur. What early events lead to this, and what the relative roles of bacterium and plant may be, are not known at present and are likely to present some of the most interesting puzzles in the study of the symbiosis.

Another aspect of regulation is the response of the symbiotic partners to nitrogen nutrition: in the presence of other sources of combined nitrogen -- such as ammonia or nitrate -- the plant does not get infected by Rhizobium. If nitrogen is added to plants after nodules are formed, then nitrogen fixation stops. This is undoubtedly a mechanism by which the plant can avoid wasting the energy which goes into building a nodule and enzymatically fixing nitrogen. This leads to an ironic problem for the grower, however: nodules do not begin to fix nitrogen and benefit the plant until a couple of weeks after inoculation; in that time period, the plants may be nitrogen deficient, and fail to grow well. However, if a shot of fertilizer is added to help along the young seedlings, then it inhibits nodulation, delaying the time at which the nodules will form and benefit the plant.

This problem of timing leads to the final consideration for improvement of nitrogen fixation: the integration of nodule function into the physiology of the plant as a whole. This is an important and very complex problem. The support for nitrogen fixation comes from photosynthate. The plant partitions energy and other resources into nodules, and it has been found that various plants will partition differently according to the bacterial strain which formed the nodule. This has a profound effect on the levels of nitrogen fixation itself, and also on the amount of bacterial activity in other processes.

An example both of the possibilities of improving the bacterium, and of the limitations of working on only the bacterium, is found in the hup, or uptake hydrogenase, system. The hup system in bacteria oxidizes molecular hydrogen to H^+, generating reducing equivalents. In nitrogen fixing cells, this can be of significance because nitrogenase produces H_2 as a by product of nitrogen reduction, consequently wasting energy and reducing power. It appears that hydrogenase activity in Rhizobium correlates with somewhat more efficient symbiotic nitrogen fixation, and one aim of improving Rhizobium is to engineer this property into highly competitive, high-fixation strains.

Here is where the effect of the plant must be considered, however. The amount of hydrogenase activity found in a nodule -- and consequently the level of energy savings in the symbiosis -- was studied by Bedmar et al. (1983). They reported that in several strain/cultivar combinations of microbe and plant, the level of hydrogenase activity in fact correlated with host cultivar as much as it did with the bacterial strain. Manipulating levels of the microbial metabolic processes will probably require tailoring of both the microbe and the host.

CONCLUSIONS

Nitrogen fixation is carried out enzymatically by a small,
diverse group of prokaryotes. Among these, the agriculturally most
important are the symbiotic nitrogen fixing bacteria in the genus
Rhizobium, which establish root nodules in association with host
plants in the legume family. These nodules are developed in a
complex multistep process, and nitrogen fixation requires
differentiation of both the bacterial endosymbionts and the host
plant. Molecular genetic analysis of Rhizobium nitrogen fixation
(nif) genes has progressed in parallel with advances in nif
genetics of other organisms. One of the most challenging research
questions is that of how Rhizobium nif genes are regulated: the
existence of non-linked nif genes in some strains, and the presence
of repeated sequences, point to intriguing differences between
symbiotic and non-symbiotic nif regulation. This question will be
of importance as we look towards improving the efficiency of
nitrogen fixation in various crops.
The bacterial genetic loci for nodule formation (nod genes)
have recently been identified in several Rhizobium species, and
are linked to the nif genes. The products of these genes,
and the mechanism of their actions on plant cells, remain unknown.
In some Rhizobium species a small DNA region appears to encode
nodulation and host range functions. However, in Rhizobium
meliloti there appear to be several regions involved in nodule
formation. Host determination for R. meliloti may involve both
positive and negative factors. Such determinants may occur in many
Rhizobium systems, and must be analyzed one locus at a time if we
are to gain the ability to manipulate host range and
competitiveness.
The ultimate success of symbiotic nitrogen fixation requires
the integrated function of both the bacterial and plant partners.
The partioning of energy and of nutrient supplies between root and
shoot, the timing of development, the response of the system to
environmental factors: an understanding of these questions, of
fundamental interest for plant biologists, will also be the
prerequisite to using molecular approaches to manipulate and
improve symbiotic nitrogen fixation.

ACKNOWLEDGEMENTS

I would like to thank the members of my research group at
Stanford - Thomas Jacobs, Robert Fisher, Mark Martin, Thomas
Egelhoff, John Mulligan, and Janice Tu - for their many
contributions to the research and ideas presented here.

REFERENCES

Banfalvi, Z., Sakanyan, V., Koncz, C., Kiss, A., Dusha, I. and
    Kondorosi, A., 1981, Location of nodulation and nitrogen
    fixation genes on a high molecular weight plasmid of R.
    meliloti, Mol. Gen. Genet., 184: 318-325.

Bauer, W.D., 1981, Infection of legumes by Rhizobia, Ann. Rev.
    Plant Physiol., 32: 407-449.

Bedmar, E.J., Edie, S.A. and Phillips, D.A., 1983, Host plant
    cultivar effects on hydrogen evolution by Rhizobium
    leguminosarum, Plant Physiol., 72(suppl.):432.

Beringer, J.E., Brewin, N.J., and Johnston, A.W.B., 1980, The
    genetic analysis of Rhizobium in relation to symbiotic nitrogen
    fixation, Heredity, 45: 161-186.

Beynon, J.L., Beringer, J.E., and Johnston, A.W.B., 1980, Plasmids
    and host range in Rhizobium leguminosarum and Rhizobium
    phaseoli, J. Gen. Microbiol., 120: 421-430.

Bohlool, B., Kosslak, R., and Woolfenden, R., 1983, The ecology of
    Rhizobium in the rhizosphere: survival, growth and competition,
    in: "Advances in Nitrogen Fixation," C. Veeger and W.E. Newton,
    eds., pp. 287-294, Nijhoff/Junk, Boston.

Buikema, W.J., Long, S.R., Brown, S.E., van den Bos, R., Earl, C.
    and Ausubel, F.M. (1983) Physical and genetic characterization
    of Rhizobium meliloti symbiotic mutants, J. Molec. Appl.
    Genet., 2: 240-260.

Callaham, D., del Tredici, P. and Torry, J.G., 1978, Isolation and
    cultivation in vitro of the actinomycete causing root
    nodulation in Comptonia., Science, 199: 899-902.

Cannon, F.C., Reidel, G.E., and Ausubel, F.M., 1979, Overlapping
    sequences of Klebsiella pneumoniae nif DNA cloned and
    characterised, Molec. Gen. Genet., 174: 59-66.

Corbin, D., Ditta, G., and Helinski, D., 1982, Clustering of nitrogen
    fixation (nif) genes in Rhizobium meliloti, J. Bacteriol. 149:
    221-228.

Dart, P., 1974, The infection process, in "Biology of nitrogen
    fixation", A. Quispel, ed., pp. 381-429, North-Holland,
    Amsterdam.

Dazzo, F.B. and Hubbell, D. 1982, Control of root hair infection. in "Nitrogen fixation, vol.2: Rhizobium," W. Broughton, ed., pp. 275-309, Oxford.

Denarie, J., Boistard, P., Casse-Delbart, F., Atherley, A.G., Berry, J.O. and Russell, P., 1981, Indigenous plasmids of Rhizobium, in "Biology of the Rhizobiaceae," K. Giles and A. Atherley, eds., pp. 225-246, Academic Press, New York.

Ditta, G., Stanfield, S., Corbin, D., and Helinski, D., 1980, Broad host range DNA cloning system for Gram negative bacteria: construction of a gene bank of Rhizobium meliloti. Proc. Natl. Acad. Sci. 77: 7347-7351.

Djordevic, M., Zurkowski, W., Shine, J., and Rolfe, B.G., 1984, Sym-plasmid transfer to various symbiotic mutants of Rhizobium trifolii, R. leguminosarum, and R. meliloti. J. Bacteriol. 156: 1035-1045.

Downie, J.A., Hombrecher, G., Ma, Q.-S., Knight, C., Wells, B., and Johnston, A.W.B., 1983, Cloning of the symbiotic region of Rhizobium leguminosarum: the nodulation genes are between the nitrogenase genes and a nifA-like gene, Embo J.,2: 947-952.

Hennecke, H., 1981, Recombinant plasmids carrying nitrogen fixation genes from Rhizobium japonicum, Nature, 291: 354-355.

Imsande, J., 1981, Exchange of metabolites and energy between legume and Rhizobium, in: "Biology of the Rhizobiaceae," K. Giles, and A. Atherly, eds., pp. 179-189, Academic Press, N.Y.

Johnston, A.W.B., Beynon, J.L., Buchanan-Wollaston, A.V., Setchell, S.M., Hirsch, P.R., and Beringer, J.E., 1978, High frequency transfer of nodulating ability between strains and species of Rhizobium, Nature, 276: 635-636.

Jordan, D.C., 1982, Transfer of Rhizobium japonicum to Bradyrhizobium, Intl. J. Syst. Bacteriol., 32: 136-139.

Long, S.R., Meade, H.M., Brown, S.E. and Ausubel, F.M., 1981, Transposon-induced symbiotic mutants of Rhizobium meliloti, in: "Genetic engineering in the plant sciences," N. Panopoulos, ed., pp. 129-143, Praeger, New York.

Long, S.R., Buikema, W.E., and Ausubel, F.M., 1982, Cloning of Rhizobium meliloti nodulation genes by direct complementation of Nod⁻ mutants, Nature, 298: 485-488.

Long, S.R., 1984, Nodulation genetics, in: "Plant-microbe interactions," E. Nester and T. Kosuge, eds., Macmillan, New York, in press.

Ma, Q.-S., Johnston, A.W.B., Hombrecher, G., and Downie, J.A., 1982, Molecular genetics of mutants of Rhizobium leguminosarum which fail to fix nitrogen, Mol. Gen. Genet., 187: 166-171.

Meade, H.M., Long, S.R., Ruvkun, G.B., Brown, S.E., and Ausubel, F.M., 1982, Physical and genetic characterization of symbiotic and auxotrophic mutants of Rhizobium meliloti induced by transposon Tn5 mutagenesis, J. Bacteriol., 149: 114-122.

Newcomb, W. 1981, Nodule morphogenesis and differentiation, in "Biology of the Rhizobiaceae," eds. Giles, K. and Atherly, A., pp.247-298, Academic Press, New York.

Paau, A., Leps, W. and Brill, W., 1981, Agglutinin from alfalfa necessary for binding and nodulation by Rhizobium meliloti, Science, 213: 1513-1515.

Peters, G.A. and Mayne, B.C., 1974, The Azolla-Anabaena azollae relationship. I: Initial characterization of the association, Plant Physiol., 53: 813-819.

Pierce, M. and Bauer, W.D., 1983, A rapid regulatory response governing nodulation in soybean, Plant Physiol, 73: 286-290.

Postgate, J., 1982, "The fundamentals of nitrogen fixation," Cambridge Press, Cambridge.

Quinto, C., de la Vega, H., Flores, M., Fernandez, L., Ballado, T., Soberon, G., and Palacios, R., 1982, Reinteration of nitrogen fixation gene sequences in Rhizobium phaseoli, Nature 299: 724-726.

Robson, R., Kennedy, C., and Postgate, J., 1983, Progress in comparative genetics of nitrogen fixation, Can. J. Microbiol., 29: 954-967.

Rosenberg, C., Boistard, P., Denarie, J., and Casse-Delbart, F., 1981, Genes controlling early and late function in symbiosis are located on a megaplasmid in Rhizobium meliloti. Mol. Gen. Genet., 194: 326-333.

Ruvkun, G.B. and Ausubel, F.M., 1980, Interspecies homology of nitrogenase genes, Proc. Natl. Acad. Sci. U.S.A.,77: 191-195.

Ruvkun, G.B. and Ausubel, F.M., 1981, A general method for
    site-directed mutagenesis in prokaryotes, Nature, 289: 75-78.

Ruvkun, G.B., Sundaresan, V. and Ausubel, F.M., 1982, Site-directed
    transposon Tn5 mutagenesis and complementation analysis of the
    Rhizobium meliloti symbiotic nitrogen fixation (nif) genes,
    Cell, 29: 551-559.

Scott, K.F., Hughes, J.E., Gresshoff, P.M., Beringer, J.E., Rolfe,
    B.G., and Shine, J., 1982, Molecular cloning of Rhizobium
    trifolii genes involved in symbiotic nitrogen fixation, J.
    Molec. Appl. Genet., 1: 315-326.

Sprent, J., 1979, The biology of nitrogen-fixing organisms.
    McGraw-Hill, London.

Sutton , W.D., Pankhurst, C.E., and Craig, A.S., 1981, The
    Rhizobium bacteroid state, in: "Biology of the Rhizobiaceae,"
    K. Giles and A. Atherly, eds., pp. 149-177, Academic Press, New
    York.

Verma, D.P.S. and Long, S.R., 1983, The molecular biology of
    Rhizobium-legume symbiosis, in: "Intracellular Symbioses," K.
    Jeon, ed., pp. 211-243, Academic Press, New York.

Vincent, J.M., 1980, Factors controlling the Rhizobium-legume
    symbiosis. in "Nitrogen fixation, vol. II: symbiotic associations
    and cyanobacteria," W.E. Newton and W.H. Orme-Johnson, eds.,
    Park Press, Baltimore.

Zimmerman, J.L., Szeto, W. and Ausubel, F.M., 1984, Molecular
    characterization of Tn5-induced symbiotic (Fix⁻) mutants of
    Rhizobium meliloti, J. Bacteriol., 156: 1025-1034.

CHLAMYDOMONAS REINHARDII, A POTENTIAL MODEL SYSTEM FOR
CHLOROPLAST GENE MANIPULATION
(chloroplast DNA/Chlamydomonas reinhardii/transfor-
mation/vectors)

J.D. Rochaix, J. Erickson, M. Goldschmidt-
Clermont, M. Schneider and J.M. Vallet

Departments of Molecular Biology and Plant
Biology
University of Geneva
30, Quai Ernest-Ansermet, 1211 Geneva 4,
Switzerland

SUMMARY

Studies on the structure, function and regulation
of genes coding for chloroplast proteins are important
for understanding the biosynthesis of the photosynthetic apparatus and the integration of chloroplasts
within plant cells. Chlamydomonas reinhardii is
particularly well suited for solving these problems
because this green unicellular alga can be manipulated
with ease both at the biochemical and genetic level.
Several genes have been identified on the physical map
of the chloroplast genome. They include genes coding
for ribosomal RNA, tRNA and several proteins including
the large subunit of ribulose 1,5 bisphosphate carboxylase (RubisCo) and several thylakoid polypeptides. The
nuclear gene for the small subunit of RubisCo has also
been cloned. Because chloroplast DNA recombination
occurs in C. reinhardii, a rare property among plants,
chloroplast genes can be analyzed by genetic means.
Numerous chloroplast photosynthetic mutations have been
isolated and several of them have been shown to be part
of a single linkage group (Gillham, 1978). We have
reached the stage where the genetic and biochemical
approaches can be coupled efficiently in C. reinhardii;
in particular, it has been possible to correlate the
physical and genetic chloroplast DNA maps at a few
sites.

A nuclear transformation system has been developed
for C. reinhardii by using a cell wall deficient
arginine auxotroph which can be complemented with a
plasmid containing the yeast ARG4 locus. Transformation
vectors have been constructed by inserting random
nuclear and chloroplast DNA fragments into a plasmid
containing the yeast ARG4 locus and by testing the
recombinant plasmids for their ability to promote
autonomous replication in yeast (ARS sites) and C.
reinhardii (ARC sites). Several plasmids have been
recovered that act as shuttle vectors between E.coli,
C. reinhardii and yeast. Four ARS sites and four ARC
sites have been mapped on the chloroplast genome of C.
reinhardii. One plasmid replicates both in C.
reinhardii and yeast. Because C. reinhardii cells
contain a single large chloroplast they offer
interesting possibilities for attempts of chloroplast
transformation by microinjection. Since appropriate
selective markers and transformation vectors are
available, this approach can now be explored.

INTRODUCTION

Considerable progress has been achieved in the
past few years in the field of plant transformation.
Special use has been made of crown gall tumours which
can be induced in dicotyledonous plants after infection
of wounded tissue by Agrobacterium tumefaciens (Van
Montagu and Schell, 1982 , for recent review). It is
now well established that oncogenic strains of these
bacteria harbor tumour inducing plasmids (Ti plasmids)
and that a defined portion of the plasmid DNA (T DNA)
is integrated into the host plant genome after
infection. Several workers have exploited this natural
transformation system and used the Ti plasmids as
shuttle vectors to introduce foreign genes into plant
cells. These genes are expressed provided that they are
placed under the control of appropriate transcriptional
regulatory sequences (Herrera-Estrella et al., 1983).
The use of this natural transformation system looks
very promising and it will undoubtedly provide new
insights into the function and regulation of plant
genomes. From a practical point of view, the
possibility of improving plants by modifying their
genome in a defined manner can now be investigated in
depth.

One important limitation of transformation studies

with higher plants is their slow growth. Weeks or
months may pass until the result of a plant
transformation experiment can be assessed. It may
therefore be appropriate to look for simpler model
systems which are easier to manipulate experimentally.
It is likely that studies with these systems may
fruitfully complement research on higher plants,
especially in the area of photosynthesis which plays a
key role in crop production. The green unicellular alga
Chlamydomonas reinhardii appears to be very useful in
this respect since its photosynthetic apparatus closely
resembles that of higher plants. For several years we
have attempted to understand some aspects of the
structure, function and regulation of genes involved in
photosynthesis in this alga.

Like higher plants, C. reinhardii contains three
distinct genetic systems located in the
nucleocytoplasm, the chloroplast and the mitochondria,
respectively.

Table I lists the major parameters of these
systems. It can be seen that the genetic contribution
of the chloroplast genome is modest, only 0.3% of the
total cellular genetic information. However, this DNA
is abundant since it constitutes 14% of the total DNA
mass. This implies that the chloroplast DNA is present
in multiple copies per cell. It is interesting to note
that the DNA of C. reinhardii found to be associated
with mitochondria (Ryan et al., 1975) is only 15 kb in
size, considerably smaller than the mitochondrial DNA
from higher plants which is over 200kb (Leaver and
Gray, 1982). Whether this 15kb DNA represents the
entire mitochondrial genome of C. reinhardii has not
yet been proven. Table I also shows that chloroplast
and cytoplasmic ribosomes differ from each other in
size, in their ribosomal RNAs and proteins and in the
spectrum of antibiotics to which they are sensitive.

C. reinhardii can be grown phototrophically
(without reduced carbon source in the light),
heterotrophically (with reduced carbon source in the
dark) or mixotrophically (with reduced carbon source in
the light). At least those chloroplast functions
related to photosynthesis are dispensable under
heterotrophic and mixotrophic growth, but not under
phototrophic growth. This property has allowed for the
isolation of numerous mutants unable to grow in the
absence of acetate (which is used as a carbon source)
because of some defect in the photosynthetic system.

Table I. Parameters of the genetic systems of chlamydomonas.

	COMPLEXITY		MASS		GENETICS
NU DNA	70 000 KB	99.7%	85%	UNIQUE	MENDEL
CT DNA	190 KB	0.3	14	50-80x	UP
MT DNA	15 KB	0.02	1	50x	BP ≠ MENDEL

		PROTEINS	rRNAs	ANTIBIOTICS
RIBOSOMES CYT	80S	60	25S, 18S, 5.8S, 5S	CHI, SPA
CT	70S	50	23S, 16S, 7S, 5S, 3S	STR, ERY, CAP
MT	?	?	?	

MRNAS	CYT	POLY A$^+$
	CT	POLY A$^-$
	MT	?

NU, nuclear; CT, chloroplast; MT, mitochondrial;
CYT, cytoplasmic; UP, uniparental; BP, biparental;
CHI, cycloheximide; SPA, sparsomycin; STR, strepto-
mycin; ERY, erythromycin; CAP, chloramphenicol.
Adapted from Rochaix (1981).

    A remarkable architectural feature of C. reinhardii
is the presence of a single chloroplast per cell which
occupies about 40% of the cell volume (Gillham, 1978).
The life cycle of this alga is shown in fig. 1.

    Vegetative cells of both mating types (mt$^+$ and
mt$^-$) can propagate by successive mitosis. Vegetative
cells differentiate into gametes once they are transferred into a medium deprived of a reduced nitrogen
source. Gametes of opposite mating type fuse to form
a zygote. Cell fusion is shortly followed by nuclear
and chloroplast fusion. After appropriate light and
dark regimes, the zygote undergoes meiosis and upon

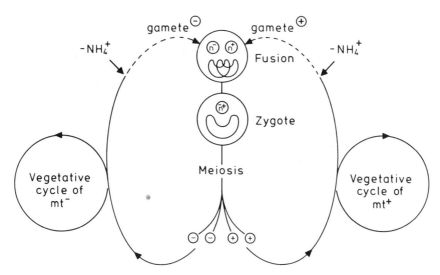

Fig. 1. Life cycle of Chlamydomonas reinhardii, mt,
        mating type, n, nucleus. See the text for
        explanations.

germination it releases four daughter cells which can
be propagated as vegetative cells. An important point
is that the chloroplast genomes of both parents mix
with each other. Chloroplast gene recombination can
therefore be studied in C. reinhardii, a feature which
is of obvious importance for genetic studies. A direct
physical proof of chloroplast DNA recombination has
been provided by the structural analysis of chloroplast
DNA from F1 hybrids between C. eugametos and C.
moewusii (Lemieux et al., 1981). These two algae are
interfertile and their chloroplast DNA restriction
patterns differ from each other (Lemieux et al., 1980;
Mets, 1980).

    Transmission of chloroplast genes does not follow
the Mendelian rules. In most cases only the chloroplast

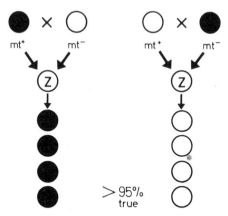

1) Transmission of ct genes is determined by the mt.

2) Exeption: Biparental Zygotes. Segregation of ct genes during the post-meiotic mitotic divisions.

Fig. 2. Uniparental inheritance in Chlamydomonas reinhardii.

DNA from the mt^+ is transmitted to the progeny as illustrated in fig. 2. In a few cases the chloroplast DNAs from both parents are transmitted to the daughter cells. The existence of these biparental zygotes has been crucial for studying chloroplast gene recombination and for constructing a chloroplast linkage map (Gillham, 1978). A second distinctive feature of uniparential genes is their segregation during the post-meiotic mitotic divisions. This uniparental inheritance of chloroplast genes was already discovered 30 years ago by Sager (1954) when she studied the transmission of streptomycin resistant mutations in C. reinhardii. Since then, numerous uniparental mutations have been isolated and characterized which fall into three major classes. The

first includes mutants resistant to antibiotics
specific for prokaryotic ribosomes (streptomycin,
erythromycin, spectinomycin, etc.), the second consists
of mutants resistant to herbicides of the s-triazine or
urea class (atrazine, diuron) and the third comprises
mutants unable to grow in the absence of a reduced
carbon source. The latter group includes a large array
of mutants unable to fix CO_2 because of a deficiency in
the photosynthetic apparatus or in the chloroplast
protein synthesizing system.

Although it appeared likely since their discovery
that the uniparental traits are encoded in the
chloroplast genome, it took many years to prove this
assumption rigorously. A definite proof of the identity
between chloroplast and uniparental genes has been
provided only recently. Grant et al. (1980) used
deletions in the chloroplast DNA as markers to
demonstrate unambiguously the uniparental inheritance
of chloroplast DNA in C. reinhardii. Analysis of the
chloroplast DNA from progeny of interspecies crosses
between C. moewusii and C. eugametos (Mets, 1980;
Lemieux et al., 1981) gave similar results. By using
deletion mutants Myers et al. (1982) could map a
uniparental locus on the chloroplast DNA of C.
reinhardii. More recently it has been possible to
correlate the uniparental linkage group and the
physical map of the chloroplast genome of C. reinhardii
at one, possibly at two sites (Dron et al., 1983;
Erickson et al., 1983; cf below).

CHLOROPLAST GENOME OF CHLAMYDOMONAS REINHARDII :
CORRELATION BETWEEN PHYSICAL AND GENETIC MAPS

The chloroplast DNA of C. reinhardii consists of
190 kb circular DNA molecules (Behn and Herrmann, 1976;
ROCHAIX, 1978). As in most higher plant chloroplast
genomes, there are two inverted repeats (Fig. 3). These
regions contain the genes of the 16S, 23S and 5S
ribosomal RNAs (Fig. 3). The 23S ribosomal RNA gene of
C. reinhardii has unique features not shared by other
chloroplast systems. Its 5' end has been split into two
small ribosomal RNAs of 7S and 3S (Rochaix and Darlix,
1982) and it contains a 888 bp intron near its 3' end
(Fig. 5) (Rochaix and Malnoe, 1978). As in other
chloroplast ribosomal units the genes of tRNA^{ala} and
tRNA^{ile} are located in the spacer between the 16S and
23S ribosomal RNA genes (Schneider, unpublished

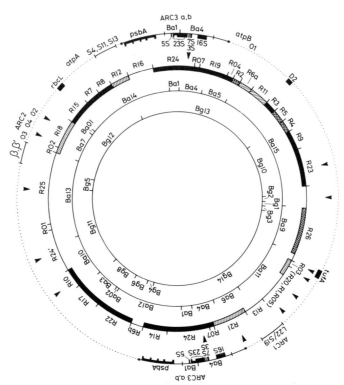

Fig. 3.   Chloroplast DNA map of <u>Chlamydomonas reinhardii</u>.
          The three inner circles from the outside to the
          inside represent the EcoRI, BamHI and BglII re-
          striction maps (Rochaix, 1978).   Dark wedges
          indicate the positions of the 4S RNA genes
          (Malnoe and Rochaix, 1978).   The two segments
          of the inverted repeat are drawn on the outside
          of the map.   They contain the rRNA genes and
          the gene of the 32 kd membrane polypeptide,
          psbA (Erickson et al., 1983).   The introns in
          the 23S rRNA gene and in psbA are drawn in
          thinner relative to the coding sequences.   D2
          is the gen for another photosystem II polypep-
          tide.   The genes for the large subunit of ribu-
          lose bisphosphate carboxylase, rbcL (Malnoe et
          al., 1978), for the α and β subunits of the
          ATP synthase, atp A and atpB, respectively
          (Gillham, personal communication; Kovacic and
          Rochaix, unpublished results) and for the elon-
          gation factor EF-Tu, tufA (Watson and Surzycki,
          1982) are also indicated.   The other gene loca-
          tions should be considered as tentative since
          they are based only on heterologous hybridiza-

tions with specific probes for the E. coli
genes of the ribosomal proteins L22 and/or S19,
for S4 and/or S11 and/or S13 and for the genes
of the β and β' subunits of E. coli RNA polymerase (Watson and Surzycki, 1983). The chloroplast DNA regions whose transcripts are present
in large   , medium   and low amounts
are shown. The four identified chloroplast
ARS sequences are indicated by 01, 02, 03 and
04 (Vallet et al., 1984). The four chloroplast
DNA sequences promoting autonomous replication
in Chlamydomonas are marked by ARC1, ARC2 and
ARC3a,b (Rochaix and van Dillewijn, unpublished
results).

results). Other regions of the chloroplast genome that
hybridize with tRNA are indicated by dark wedges in
Fig. 3 (Malnoe and Rochaix, 1978). Genes coding for
known proteins are rbcL (large subunit of ribulose 1,5
bisphosphate carboxylase RubisCo, Malnoe et al., 1979).
psbA (32 kdalton chloroplast membrane polypeptide,
Erickson et al., 1983), atpA, atpB (α and β subunits of
ATP synthase; Gillham, pers. comm.; Kovacic, Malnoe and
Rochaix, unpublished results), D2 gene (membrane polypeptide associated with photosystem II, Rochaix et al.,
1983) and tufA (elongation factor EFTu, Watson and
Surzycki, 1982). It is noticeable that, in contrast to
higher plants, the psbA gene for C. reinhardii is
located within the inverted repeat and therefore present
in two copies per chloroplast genome (Fig. 3).

Since C. reinhardii offers the opportunity of
mapping chloroplast genes by genetic means, an important task is to correlate the genetic and physical
chloroplast DNA maps. This has been achieved for the
first time recetly at the rbcL locus. Spreitzer and
Mets (1980) isolated a uniparental mutant 10-6C specifically affected in the large subunit of RubisCO. This
mutation was shown to be associated with the uniparental linkage group in C. reinhardii (Mets and Geist,
1983). A comparative sequence analysis of wild type
and mutant rbcL has revealed a single base pair substitution which results in an amino acid change near
one of the active sites of the large subunit (Fig. 4,
Dron et al., 1983). Spreitzer and Ogren (1983) have
isolated other uniparental RubisCo mutants by screening
for mutations that do not recombine with the 10-6C
mutation. Among the mutants recovered, two lack both
the large and small subunits of RubisCo. Mutants of

this sort are very valuable for studying the cooperation between the chloroplast and nucleocytoplasmic
compartments in the biosynthesis of the photosynthetic
apparatus. It is well established that RubisCo
catalyzes two competing carboxylase and oxygenase
reactions (Miziorko and Lorimer, 1983). The first leads
to CO_2 fixation while the second results ultimately in
the release of CO_2 and therefore represents a waste for
photosynthesis. Although the issue is still a matter of
debate, it may be possible to improve the
carboxylase/oxygenase ratio by subtle modifications of
the large subunit gene.

The psbA locus appears to provide a second
correlation site between the physical and genetic
chloroplast DNA maps. This locus codes for a 32 kdalton
chloroplast membrane polypeptide, associated with
photosystem II and which is believed to be the target
for herbicides of the triazine and urea class such as
atrazine and DCMU (3-(3,4-dichlorophenyl)-1,
1-dimethylurea) (Pfister et al., 1981). Several DCMU
and atrazine resistant uniparental mutants have been
isolated in C. reinhardii (Galloway and Mets, 1982;
Fellenbach et al., 1983). We have recently sequenced
the psbA locus of C. reinhardii and shown that the gene
has a mosaic structure (Fig. 5, Erickson et al., 1983).
Although the coding sequence, which is highly conserved
among different plants, contains only 1056 bp, the gene

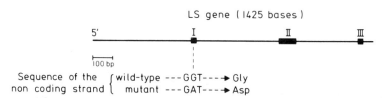

Fig. 4. Single base pair substitution in the
        uniparental mutant 106C deficient in ribulose
        1,5 bisphosphate carboxylase activity
        (Spreitzer and Mets, 1980). The gene of the
        large subunit (LS) is drawn as a line. The
        active sites I, II, III (Stringer and Hartman,
        1978; SCHLOSS et al., 1978) and the location of
        the mutation (Dron et al., 1983) are indicated.

spans 6.7 kb because of the presence of four large introns. Comparison of the psbA sequence from wild-type and from a DCMU resistant uniparental mutant, isolated by P. Bennoun, has revealed a single base change in the last exon which replaces a serine residue with alanine (Fig. 5). The same residue has also been found to be changed to glycine in an atrazine resistant biotype of Amaranthus hybridus (Hirschberg and McIntosh, 1983). The C. reinhardii mutant is resistant to DCMU and atrazine while the weed mutant is resistant mainly to atrazine. Another DCMU resistant uniparental mutant of C. reinhardii has been isolated by Galloway and Mets (1982) and shown to be associated with the uniparental linkage group (Mets and Geist, 1983). It is therefore likely that the mutant we have examined maps at the same position. Other uniparental mutants of C. reinhardii resistant to atrazine, DCMU and bromacil have been isolated and characterized (Mets, unpublished observations). A molecular analysis of these mutants is presently in progress and it should reveal more about

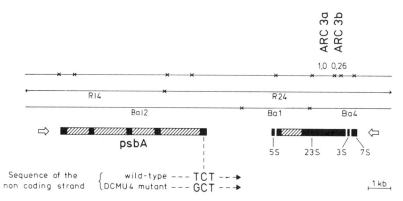

Fig. 5. Organization of the chloroplast DNA region containing the psbA and 23S ribosomal RNA genes. ▄▄ exons, ▨▨ introns. The single base substitution in the last exon of the mutant psbA gene is shown. The upper line represents the HindIII fragments. Directions of transcription are indicated by white arrows. The ARC3a and 3b sites are contained within the ribosomal HindIII fragments of 1.0 and 0.26kb, respectively.

the structure-function relationship of this important
chloroplast membrane polypeptide.

A third correlation site appears to be the ac-u-c
locus. Myers et al. (1982) examined the chloroplast DNA
restriction patterns of several photosynthetic
uniparental mutants which map at this locus and found
that several of them have deletions close to one of the
inverted repeats on BamHI fragment Ba5 and EcoRI
fragment R19 (Fig. 3). In some cases the mutation
always segregated with the deletion. Since the gene of
the β subunit of the ATP synthase has been mapped on
the same chloroplast BamHI fragment (Woessner, Masson,
Harris, Bennoun, Gillham and Boynton, pers. comm.;
Kovacic and Rochaix unpublished results) it is likely
that at least some of these mutants are ATP synthase
mutants. Table II indicates the sites of correlation
between the physical and genetic maps of the
chloroplast genome of C. reinhardii. It is obvious that
some of these sites could be used as selective markers
for attempts at chloroplast transformation.

NUCLEAR TRANSFORMATION OF CHLAMYDOMONAS REINHARDII

A few years ago when we started to work on
transformation in C. reinhardii, no unambiguous
correlation site between the chloroplast physical and
genetic maps had yet been established. It was also not
obvious how nucleic acids could be introduced into
chloroplasts through their highly selective envelope.
It therefore seemed wiser to attempt to establish first
a nuclear transformation system. One of the few nuclear
loci of C. reinhardii that have been characterized both
from a genetic and biochemical point of view is the
arg2/arg7 locus (Gillham, 1965; Loppes and Matagne,
1972). It codes for arginino succinate lyase, the last
enzyme in the arginine biosynthetic pathway which
converts argininosuccinate into arginine and fumarate.
The enzyme has been partially purified and it consists
of a tetra-or pentamer of a 38000 dalton polypeptide
(Matagne and Schlosser, 1977). Several mutations have
been isolated at this locus which is located on linkage
group I of C. reinhardii (Levine and Goodenough, 1970).
The reversion rate is on the order of 10^{-7} (Loppes and
Matagne, 1972). The first transformation attempts were
made with a yeast plasmid pYearg4 (Clarke and

Table II. Correlation sites between the physical and genetic maps of the chloroplast genome of Chlamydomonas reinhardii.

Mutation	Polypeptide affected	Site on physical map	Genetic alteration	Reference
10-6C	Large subunit of RubisCo	rbcL	Point mutation	Spreitzer & Mets, (1980) Dron et al.,(1983), Mets and Geist,(1983)
18-5B	Large subunit of RubisCo	rbcL [a]	-	Spreitzer and Ogren, (1983)
18-7G	Large subunit of RubisCo	rbcL [a]	-	Spreitzer & Ogren, (1983
DCMU4	32Kdalton membrane polypeptide	psbA [b]	Point mutation	Erickson et al.,(1983)
Dr2	32Kdalton membrane polypeptide	psbA [b]	-	Galloway & Mets, (1982) Mets and Geist, (1983)
FuD-50	βsubunit of ATPsynthase	atpB	Deletion	Myers et al.,(1982) P. Woessner, A. Masson, E. Harris, P. Bennoun, N.W. Gillham, J. Boynton, unpublished results

[a] The location of these mutations at rbcL is based on the fact that they do not recombine with the 10-6C mutation (Spreitzer and Ogren, 1983) [b] DCMU4 has not been mapped genetically and Dr2 has not been mapped physically. Since these two mutations produce similar phenotypes it is reasonable to assume that they are at the same locus. - indicates that the genetic alteration has not been determined.

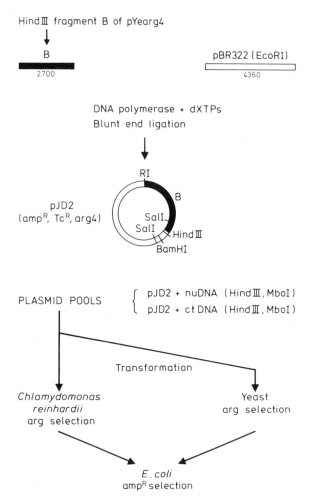

Fig. 6. Strategy for constructing plasmids that replicate autonomously in <u>Chlamydomonas reinhardii</u> and yeast. nuDNA, nuclear DNA; ct DNA, chloroplast DNA.

Carbon, 1978) containing the yeast ARG4 locus which corresponds to arg2/arg7 in C. reinhardii. The double mutant cw15 arg7 was used for these experiments. The cw15 mutant is cell wall deficient and behaves like a protoplast (Davies and Plaskitt, 1971). Treatment of these cells with poly L ornithine or polyethylene glycol in the presence of DNA allowed us to recover several arginine prototrophs with a yield of 10^{-6} to 10^{-7} transformants/treated cell. Hybridization studies revealed that most of these putative transformants had yeast DNA integrated into their nuclear genome (Rochaix and van Dillewijn, 1982).

A first important application of transformation is to use the system for isolating genes involved in specific functions by complementing known mutations. C. reinhardii is very appropriate in this respect since a large number of photosynthetic mutants have been isolated and several of them have been characterized. An approach of this sort has been extremely successful in yeast (Struhl, 1983). The availability of shuttle vectors capable of replicating autonomously both in C. reinhardii and E. coli would be of great help for this type of genetic manipulations. Since no free plasmids have been reported for C. reinhardii, except chloroplast and mitochondrial DNA, we have attempted to construct plasmids of this sort. The strategy used is outlined in fig. 6. A 2.7 kb HindIII restriction fragment containing the ARG4 locus of yeast was inserted into the EcoRI site of plasmid pBR322 (Bolivar et al., 1977), thereby producing a new plasmid, pJD2, that is resistant to ampicillin and tetracycline. Nuclear and chloroplast DNAs were digested either with MboI or HindIII and the resulting fragments were inserted into the BamHI or HindIII sites of pJD2, respectively. Pools of the recombinants were prepared and used to transform yeast or C. reinhardii by selecting for arginine prototrophy.

YEAST TRANSFORMATION

Yeast transformants could be obtained readily by using the method outlined above. The DNA of these transformants was isolated and used to transform E. coli so that the recombinant plasmids could be recovered. Hybridization of these plasmids to chloroplast DNA allowed us to map several chloroplast DNA regions which act as autonomously replicating sequences in yeast (ARS sites). These hybridizations

were verified with cloned chloroplast DNA fragments
(Vallet et al., 1984). Fig. 3 indicates the location of
these sites on the physical map of the chloroplast
genome of C. reinhardii (01, 02, 03 and 04). It can be
seen that three of these sites 02, 03 and 04 are
clustered within a region of 7 kb on the chloroplast
EcoRI fragments R02 and R18 while 01 is located on the
other side of the inverted repeat on EcoRI fragment R2.
Recently Loppes and Denis (1983) have identified three
additional ARS sites which map close to 01 on the
chloroplast EcoRI fragments R11, R5 and R4,
respectively. It is striking that those 7 chloroplast
ARS sites are not evenly distributed over the
chloroplast genome but that they are grouped into two
distinct clusters. It is likely that more ARS sites
exist on this chloroplast genome since a systematic
search has not yet been completed. Plasmids carrying
these chloroplast ARS sites are unstable in yeast when
the cells are grown under non-selective conditions. The
ARS sites 01, 02, 03 and 04 have been localized on
chloroplast DNA segments of 400, 2300, 730 and 400 bp,
respectively (Vallet et al., 1984). In order to gain
more insights into the structure of these sites, three
of them 01, 03 and 04 have been sequenced. These sites
are embedded within regions of high AT content that
contain a large number of short direct and inverted
repeats (Fig. 7, Vallet et al., 1984). Each of these
regions contains at least one 11 bp element (indicated
by C in Fig. 3) that differs by only one bp from the
yeast consensus ARS sequence 5' A/T TTT ATPu TTT A/T
which was derived from a sequence comparison of 10
yeast ARS sites (Stinchcomb et al., 1982; Broach et
al., 1982). Two other conserved sequences 5' ATT
AACAAAT and 5' PuATTTAAAT are shared between the three
chloroplast ARS sites (indicated by I and II in Fig.
7). The first element I is present once in 01 and 03
and three times in 04, of which two copies are
contained within a 26 bp inverted repeat. The possible
role of these elements in promoting autonomous
replication in yeast remains to be determined.

An important point is the relationship between
chloroplast and mitochondrial ARS elements and
authentic organellar origins of replication. Zakian
(1981) has shown that a mitochondrial DNA fragment of
Xenopus laevis which contains the origin of replication
acts as an ARS element, an observation which does not
necessarily prove that the two elements are identical.
Although mitochondrial DNA fragments from yeast (Blanc
and Dujon, 1981; Hyman et al., 1981) and from the

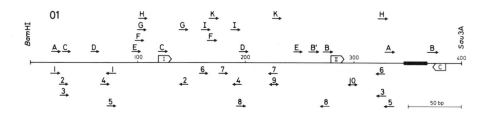

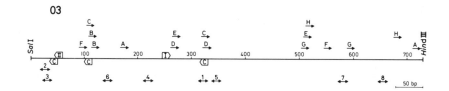

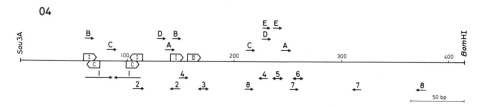

Fig. 7. Sequence organization of the chloroplast ARS region 01, 03 and 04. Short direct and inverted repeats of 8 and 11bp are indicated by arrows with letters and numbers, respectively. Element C consists of 11bp and is highly related to the yeast ARS consensus sequence (Stinchcomb et al., 1982; Broach et al., 1982). Elements I and II contain 10 and 9bp, respectively, and are shared by the three ARS regions. The thickened line in 01 represents a GC rich region (from Vallet et al., 1984 with permission).

filamentous fungus <u>Cephalosporium acremonium</u> (Tudzynski and Esser, 1983) also contain ARS sites, this property

is not shared by rat mitochondrial DNA (Zakian and
Kupfer, 1982). Because of the large size of chloroplast
DNA, it has been difficult to localize precisely their
origin(s) of replication. This site has been localized
within a small region of the chloroplast genome of
Euglena gracilis (Koller and Deluis, 1982) and it
remains to be seen whether this region contains an ARS
site. In C. reinhardii one chloroplast origin of
replication has been mapped on the EcoRI fragment R13
(Wu and Waddell, 1983) but we have not yet been able to
demonstrate that this fragment includes an ARS site. It
is interesting to note that the ability to promote
autonomous replication in yeast is a property of
eukaryotic, but not prokaryotic DNA. ARS sites could be
found readily in the DNA of Zea mays, Drosophila,
Dictyostelium, but none could be isolated from the DNA
of E. coli (Stinchcomb et al., 1979). The only
exception to this rule is the recent finding of ARS
sites in the DNA of broad host range plasmids of
Staphylococcus aureus (Goursot et al., 1983).

CHLAMYDOMONAS REINHARDII TRANSFORMATION

     Using the scheme outlined in fig. 6 recombinant
plasmids were used to transform C. reinhardii. Analysis
of the DNA from several transformants by Southern
hybridization (Southern, 1975) revealed the presence of
free plasmids. If free plasmids exist in these cells
and if they are all derivatives of the plasmid pJD2
(Fig. 6), it should be possible to recover them in E.
coli by transformation with the DNA of the
transformants, selecting either for ampicillin
resistance or arginine prototrophy. This strategy was
indeed used successfully for the isolation of several
plasmids (Fig. 8). The structure and properties of four
of these ARC plasmids (autonomous replication in C.
reinhardii) are summarized in Table III. It can be seen
that the inserts are only a few hundred bp in size. A
surprising finding is that all of the plasmids examined
hybridize to chloroplast DNA although the DNA used for
their construction was total C. reinhardii DNA. The
location of the four chloroplast ARC regions that
promote autonomous replication in C. reinhardii are
indicated in fig. 3. Of the four ARC plasmids only one,
pCA1, acts also as ARS plasmid in yeast (indicated by
04 in Fig. 3). Its ARS site is, however, weaker than
the other chloroplast ARS sites since this plasmid
transforms yeast with a lower efficiency. Comparison of
the sequences of these ARC plasmids reveals two regions

of homology which may play a role in promoting autonomous replication in C. reinhardii.

It is noteworthy that the ARC1 site is located on the chloroplast EcoRI fragment R13 on which Wu and Waddell (1983) have mapped on authentic origin of replication of chloroplast DNA by observing replication forks in the electron microscope. An interesting problem is to determine whether this origin of replication and the ARC1 site coincide with each other. The ARC3a and ARC3b sites map in the coding region of the 23S ribosomal RNA (Fig. 5). We have verified by DNA sequencing that the sequences of the 257 bp insert of plasmid pCA4 and the corresponding fragment from the cloned chloroplast ribosomal fragment Ba4 (Fig. 5) are identical. This observation is important since it demonstrates that no sequence rearrangements have occurred in this plasmid during its propagation in C. reinhardii. Furthermore the sequences of the plasmid vector that flank the inserts have been shown to be identical to the original pBR322 vector. These results are encouraging since they show that at least small DNA fragments can be propagated faithfully in C. reinhardii.

One serious problem which still limits the general use of transformation in C. reinhardii is the low transformation efficiency. Even with autonomously replicating plasmids the yield has not been significantly increased. The limiting steps could occur at different levels : delivery of DNA to the cells, maintenance of the DNA in the cell and poor expression of the yeast marker gene in C. reinhardii. An urgent task for the future is to solve this problem.

PROSPECTS FOR CHLOROPLAST TRANSFORMATION

Chloroplasts are surrounded by a double membrane which is highly selective in the uptake of chloroplast proteins synthesized on cytoplasmic ribosomes. It is well documented that these proteins are made as precursors in the cytoplasm and processed to their mature size during or shortly after their import into the chloroplasts (Chua and Schmidt, 1979). Processing involves the cleavage of a short peptide, the transit peptide, at the N terminus of the precursor. While a large number of proteins are imported into the chloroplast, there is no convincing evidence for the

Table III. Characteristics of ARC plasmids.

Plasmid	Site	Size (bp) of insert	Location			ARS activity
pCA1	ARC2	414	R18	Ba7		+
pCA2	ARC1	153	R13	Ba11		-
pCA3	ARC3$_a$	102	R24	Ba4	H1.00	-
pCA4	ARC3$_b$	257	R24	Ba4	HO.26	-

R and Ba refer to the chloroplast EcoRI and BamHI fragments shown
on fig. 3. H1.00 and HO.26 refer to two HindIII fragments displayed
in fig. 5. ARS activity is defined as the ability to promote autonomous
replication in yeast.

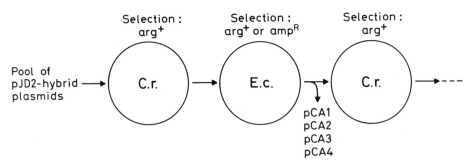

Fig. 8 Strategy for the isolation of plasmids that
replicate autonomously in <u>Chlamydomonas
reinhardii</u>.

uptake of nucleic acids into chloroplasts. Clearly, the
ability to introduce new genetic material into these
organelles would be a powerful method for modifying
photosynthetic functions in a defined way. Two

different strategies can be adopted for attempts at chloroplast transformation.

The first approach is indirect and involves two distinct steps. The gene coding for a given polypeptide is first fused with other genetic elements so that its 5' end is preceded by a strong nuclear promotor and the 5' end of the coding sequence is linked to the coding sequence of a transit peptide. It may also be appropriate to attach a polyadenylation site at the 3' end of the gene. The hope is that upon its translation in the cytoplasm, the fused polypeptide will be driven into the chloroplast through its signal peptide. This approach is certainly risky since it is by no means certain that the transit peptide is sufficient for import of polypeptides into the chloroplast. The outcome will depend to a large degree on the particular sequence configuration of the polypeptide one wishes to introduce into the organelle.

As a first step towards this goal the nuclear gene of the small subunit of RubisCo (rbsS) has recently been cloned in our laboratory (Goldschmidt-Clermont, 1983). Preliminary results indicate that there are at least two copies of this gene per genome which are located close to each other. A 600 bp fragment has been sequenced which includes the 5' upstream region of one of the rbcS genes as well as the coding regions of the signal peptide and the N-terminus of the mature polypeptide. The next step will be to insert at appropriate positions genes whose products could either complement known chloroplast mutations or confer resistance to some antibiotics. Genes of interest are those of the large subunit of ribulose bisphosphate carboxylase (rbcL), and of the β subunit of the ATP synthase (atpB) because well defined mutations exist in these genes (cf Table II). These mutants are unable to grow in the absence of a reduced carbon source. Putative transformants could therefore be readily selected for on minimal medium. Other genes of interest include those involved in herbicide resistance. Since the C. reinhardii psbA gene contains four introns it may be preferable to use the corresponding unsplit gene from an atrazine resistant biotype of Amaranthus hybridus (Hirshberg and McIntosh, 1983). Because C. reinhardii is sensitive to kanamycin, the gene of the aminoglycoside phosphotransferase from Tn5 (Beck et al., 1982) which inactivates the antibiotic is also of interest.

The second approach for chloroplast transformation is to introduce the foreign DNA through the highly selective chloroplast envelope either by permeabilizing the cell and its organelles or by microinjection. Until now we have been unable to achieve chloroplast transformation by treatment of cells with compounds that have proven efficient for nuclear transformation such as poly L ornithine, polyethyleneglycol or calciumphosphate. C. reinhardii appears to be ideally suited for microinjection because of the large size of its single chloroplast which occupies over 40% of the cell volume. The same selective markers as described above may be used. We have started to insert some of the genes whose products can be selected for into plasmids containing ARS and ARC sites. This may help in maintaining the DNA in the cells after microinjection before it can integrate into the chloroplast genome by recombination. These experiments are still at a very early stage and more work and patience is required to explore this new approach.

ACKNOWLEDGEMENTS

We thank O. Jenni for drawings and photography. This work was supported by grant No. 3.258.082 from the Swiss National Science Foundation.

LITERATURE CITED

Beck, E., Ludwing, C., Auerswald, E. A., Reiss, B., and Schaller, H., 1982, Nucleotide sequence and exact localization of the neomycin phosphotransferase gene from transposon Tn5, Gene, 19:329-356.

Behn, W., and Herrmann, R. G., 1977, Circular molecules in the β satellite DNA of Chlamydomonas reinhardii, Molec. Gen. Genet., 157:25-30.

Blanc, H., and Dujon, B., 1981, in: "Mitochondrial Genes," P. Slonimski et al., eds., Cold Spring Harbor Laboratory, Cold Spring Harbor, New York, pp. 279-294.

Bolivar, F., Rodriguez, R. L., Greene, P. J., Betlach, M. C., Heynecker, H. L., Boyer, H. W., Crosa, J. H., and Falkow, S., 1977, Construction and characterization of new cloning vehicles, Gene, 2:95-113.

Broach, J. R., Li, Y. Y., Feldman, J., Jayaram, M., Abraham, J., Nasmyth, K. A., and Hicks, J. B., 1982, Localization and sequence analysis of yeast origins of DNA replication, Cold Spring Harbor, Symp. Quant.

Biol., 47:1165-1173.

Chua, N. H., and Schmidt, G. Y., 1979, Transport of proteins into mitochondria and chloroplasts, J. Cell Biol., 81:461-483.

Clarke, L., and Carbon, J., 1978, Functional expression of cloned yeast DNA in Escherichia coli: specific complementation of Argininosuccinate Lyase (arg4) mutations, J. Mol. Biol., 120:517-532.

Davies, D. R., and Plaskitt, A., 1971, Genetical and structural analyses of cell-wall formation in Chlamydomonas reinhardii, Genet. Res., 17:33-43.

Dron, M., Rahire, M., and Rochaix, J. D., 1982, Sequence of the chloroplast DNA region of Chlamydomonas reinhardii containing the gene of the large subunit of ribulose bisphosphate carboxylase and parts of its flanking genes, J. Mol. Biol., 161:775-793.

Dron, M. Rahire, M. Rochaix, J. D., and Mets, L., 1983, First DNA sequence of a chloroplast mutation: a missense alteration in the ribulosebisphosphate carboxylase large subunit gene, Plasmid, 9:321-324.

Erickson, J. M., Shneider, M., Vallet, J. M., Dron, M., Bennoun, P., and Rochaix, J. D., 1983, Chloroplast gene function: combined genetic and molecular approach in Chlamydomonas reinhardii, in: "Proceedings of 6th Intl. Congress on Photosynthesis," Sytresma, C. ed., M. Nijhoff and W. Junk Publ., In press.

Galloway, R. E., and Mets, L.. 1982, Non-mendelian inheritance of 3-(3,4-dichlorophenyl)-1,1-dimethyl urea-resistant thylakoid membrane properites in Chlamydomonas, Plant Physiol., 70:1673-1677.

Gillham, N. W., 1978, "Organelle heredity," Raven Press, New York.

Goldschmidt-Clermont, M., 1983, Regulation of ribulose bisphosphate carboxylase gene expression in Chlamydomonas reinhardii, in: "Proceedings of 6th Intl. Congress on Photosynthesis," Sybesma, C. ed., M. Nijhoff and W. Junk Publ., In press.

Goursot, R., Goze, A., Niaudet, B., and Ehrlich, S. D., 1982, Plasmids from staphylococcus aureus replicate in yeast Saccharomyces cerevisiae, Nature, 298: 488-490.

Grant, D. M., Gillham, N. W., and Boynton, J. E., 1980, Inheritance of chloroplast DNA in Chlamydomonas reinhardii, Proc. Nat. Acad. Sci. U.S.A., 77: 6067-6071.

Hirschberg, J., and McIntosh, L., 1983, Molecular basis of herbicide resistance, Science, In press.

Herrera-Estrella, L., Depicker, A., van Montagu, M., and Schell, J., 1983, Expression of chimaeric genes transferred into plant cells using a Ti-plasmid-derived vector, Nature, 303:209-213.

Hyman, B. C., Cramer, J. H., and Rownd, R. H., 1982, Properties of a Saccharomyces cerevisiae mt DNA segment conferring high-frequency yeast transformation, Proc. Nat. Acad. Sci. U.S.A., 79: 1578-1582.

Koller, B., and Delius, H., 1982, Origin of replication in chloroplast DNA of Euglena gracilis located close to the region of variable size, EMBO J., 1: 995-998.

Leaver, C. J., and Gray, M. W., 1982, Mitochondrial genome organization and expression in higher plants, Ann. Rev. Plant Physiol., 33:373-402.

Lemieux, C., Turmel, M., and Lee, R. W., 1980, Characterization of chloroplast DNA in Chlaymdomonas eugametos and C. moewusii and its inheritance in hybrid progeny, Curr. Genet., 2:139-147.

Lemieux, C., Turmel, M., and Lee, R. W., 1981, Physical evidence for recombination of chloroplast DNA in hybrid progeny of Chlamydomonas eugametos and C. moewusii, Curr. Genet., 3:97-103.

Levine, R. P., and Goodenough, U. W., 1970, The Genetics of photosynthesis and of the chloroplast in Chlamydomonas reinhardii, Ann. Rev. Genet., 4: 397-408.

Loppes, R., and Matagne, R. F. C., 1972, Allelic complementation between arg-7 mutants in Chlamydomonas reinhardii, Genetica, 43:422-430.

Loppes, R., and Denis, C., 1983, Chloroplast and nuclear DNA fragments from Chlamydomonas promoting high frequency transformation of yeast, Current Genet., 7:473-480.

Malnoe, P. M., and Rochaix, J. D., 1978, Localization of 4S RNA genes on the chloroplast genome of Chlamydomonas reinhardii, Molec. Gen. Genet., 166: 269-275.

Malnoe, P. M., Rochaix, J. D., Chua, N. H., and Spahr, P. F., 1979, Characterization of the gene and messenger RNA of the large subunit of ribulose 1,5 biphosphate carboxylase in Chlamydomonas reinhardii, J. Mol. Biol., 133:417-434.

Matagne, R. F., and Schlosser, J. P., 1977, Purification and subunit structure of argininosuccinate lyase from Chlamydomonas reinhardii, Biochem. J., 167: 71-75.

Mets, L. J., 1980, Uniparental inheritance of chloroplast DNA sequences in interspecies hybrids of Chlamydomonas, Current Genet., 2:232-238.

Mets, L. J., and Geist, L. J., 1983, Linkage of a known

chloroplast gene mutation to the uniparental genome of Chlamydomonas reinhardii, Genetics, 105:559-579.

Miziorko, H. M., and Lorimer, G. H., 1983, Ribulose 1,5 bisphosphate carboxylase-oxygenase, Ann. Rev. Biochem., 52:507-535.

Myers, A. M., Grant, D. M., Robert, D. K., Harris, E. H., Boynton, J. E., and Gillham, N. W., 1982, Mutants of Chlamydomonas reinhardii with physical alteration in their chloroplast DNA, Plasmid, 7:131-151.

Pfister, K., Steinback, K. E., Gardner, G., and Arntzen, C. J., 1981, Photoaffinity labeling of a herbicide receptor protein in chloroplast membranes, Proc. Nat. Acad. Sci. U.S.A., 78:881-985.

Rochaix, J. D., 1978, Restriction endonuclease map of the chloroplast DNA of Chlamydomonas reinhardii, J. Mold. Biol., 126:567-617.

Rochaix, J. D., 1981, Organization, function and expression of the chloroplast DNA of Chlamydomonas reinhardii, Experientia, 37:323-332.

Rochaix, J. D., and Malnoe, P. M., 1978, Anatomy of the chloroplast rebosomal DNA of Chlamydomonas reinhardii, Cell, 15:661-670.

Rochaix, J. D., and Darlix, J. D., 1982, Composite structure of the chloroplast 23S ribosomal RNA genes of Chlamydomonas reinhardii. Evolutionary and functional implications, J. Mol. Biol., 159:383-395.

Rochaix, J. D., and van Dillewijn, 1982, Transformation of the green alga Chlamydomonas reinhardii with yeast DNA, Nature, 296:70-72.

Rochaix, J. D., Dron, M., Schneider, M., Vallet, J. M., and Erickson, J. M., 1983, Chlamydomonas reinhardii, a model system for studying the biosynthesis of the photosynthetic Apparatus, in: "15th Miami Winter Symposium, Advances in Gene Technology: Molecular Genetics of Plants and Animals," Ahmad, F., Downey, K., Schultz, S. and Voellmy, R. W., eds., Academic Press, In press.

Ryan, R., Grant, D., Chiang, K. S., and Swift, H., 1978, Isolation and characterization of mitochondrial DNA from Chlamydomonas reinhardii, Proc. Nat. Acad. Sci. U.S.A., 75:3268-3272.

Sager, R., 1954, Mendelian and non-mendelian inheritance of streptomycin resistance in Chlamydomonas reinhardii, Proc. Nat. Acad. Sci. U.S.A., 40:356-363.

Schloss, J. V., Stringer, C.D., and Hartman, F.C., 1978, Identification of essential lysyl and cysteinyl residues in spinach ribulosebisphosphate carboxylase/oxygenase modeified by the affinity label N-bromoacetylethanolamine phosphate, J. Biol. Chem.,

253:5707-5711.

Southern, E. M., 1975, Detechtion of specific sequences
    among DNA fragments separated by gel
    electrophoresis, J. Mol. Biol., 98:503-517.

Spreitzer, R. J., and Mets, L., 1980, Non-mendelian
    mutation affecting ribulose-1,5-bisphosphate
    carboxylase structure and activity, Nature, 285:
    114-115.

Spreitzer, R. J., and Ogren, W. L., 1983, Rapid recovery
    of chlorplast mutation affecting ribulose-
    bisphosphate carboxylase/oxygenase in Chlamydomonas
    reinhardii, Proc. Nat. Acad. Sci. U.S.A., 80:
    6293-6294.

Stinchcomb, J., Thomas, M., Kelly, J., Selker, E., and
    Davis, R. W., 1980, Eukaryotic DNA segments capable
    of autonomous replication in yeast, Proc. Nat.
    Acad. Sci. U.S.A., 77:4559-4563.

Stinchcomb, D. T., Mann, C., Selker, E., and Davis, R. W.,
    1981, DNA sequences that allow the replication and
    segregation of yeast chromosomes, ICN-UCLA Symp.,
    Mol. Cell. Biol., 22:473-488.

Stringer, C. D., and Hartman, F. C., 1978, Sequences of
    two active site peptides from spinach ribulose-
    bisphosphate carboxylase/oxygenase, Biochem.
    Biophys. Res. Commun., 80:1043-1048.

Struhl, K., 1983, The new yeast genetics, Nature, 305:
    391-397.

Tellenbach, M., Gerber, A., and Boschetti, A., 1983,
    Herbicide-binding to thylakoid membranes of a
    DCMU-resistant mutant of Chlamydomonas reinhardii,
    FEBS Lett., 158:147-150.

Tudszynski, P., and Esser, K., 1983, Nuclear association
    in yeast of a hybrid vector containing mitochondrial
    DNA, Current Genet., 7:165-166.

Vallet, J. M., Rahire, M., and Rochaix, J. D., 1984,
    Localization and sequence analysis of chloroplast
    DNA sequences of Chlamydomonas reinhardii that
    promote autonomous replication in yeast, EMBO J.,
    In press.

van Montagu, M., and Schell, J., 1982, The Ti plasmids
    of Agrobacterium, Current Topics in Microbiology
    and Immunology, 96:237-254.

Watson, J. C., and Surzycki, S. J., 1982, Extensive
    sequence homology in the DNA coding for elongation
    factor Tu from Escherichia coli and the
    Chlamydomonas reinhardii chloroplast, Proc. Nat.
    Acad. Sci. U.S.A., 79:2264-2267.

Wu, M., and Waddell, J. M., 1983, The replicative origins
    of chloroplast DNA in Chlamydomonas reinhardii, J.
    Cell Biochem. Suppl., 7B:286.

Zakian, V., 1981, Origin of replication from Xenopus laevis mitochondrial DNA promotes high-frequence transformation of yeast, Proc. Nat. Acad. Sci. U.S.A., 78:3128-3132.

Zakian, V., and Kupfer, D. M., 1982, Replication and segregation of an unstable plasmid in yeast, Plasmid, 8:15-28.

TOWARD AN UNDERSTANDING OF GENE EXPRESSION IN PLANTS

Roger N. Beachy

Plant Biology Program
Department of Biology
Washington University
St. Louis, MO 63130

SUMMARY

Gene expression results from a complex series of
biochemical steps that include: (1) transcription of the DNA;
(2) processing of the RNA transcript into its final form and
transport from the nucleus to the cytoplasm; (3) translation of
the RNA (in the case of the messenger RNAs) to produce a protein
product; (4) post-translational modification of the protein
product. Factors that can modify these processes include those
that control the rates of transcription and/or translation, and
that control the turnover rate of the transcript or protein.
Most research dealing with gene expression in plants includes
simply quantitating molecules that result from expression, i.e.,
RNA, proteins, catalytic reaction products, or rates or
transcription and translation. Only recently, with the advent
of gene transplantation and genetic engineering experiments, has
it been possible to site-direct the mutation of specific DNA
sequences as a means to address the fundamental processes that
control the expression of genes. The following discussion
presents examples of such studies in plant biology which,
hopefully, will lead to a better understanding of how plants
control the expression of their complex nuclear genome.

I.   INTRODUCTION

The widespread interest among scientists in understanding
how plants carry out their biological capabilities is evidenced
by the diversity of topics and research approaches that are

present in this volume. This interest is longstanding in many
disciplines of plant sciences and has been crucial to the
development of agriculture around the world. The recent advent of
the new techniques in biology and biochemistry has spurred an
additional group of biologists to study specific aspects of plant
biology that previously could not be addressed. In large part the
progress that has been made in the newly emerging plant science
fields, which has received the category title of "plant molecular
biology" is due directly to the many years of research done by
scientists in many fields of interest. If the new biology is to
be influential in future agricultural productivity the new and the
previously existing plant science must interact and be interfaced
one with another.

A common goal for many researchers is to determine how an
organism regulates its development, and how it survives changes in
its environment. Thus, a number of research laboratories are
engaged in studies of developmental processes in a number of different organisms, and under a variety of environmental conditions.
It is hoped that once the molecular basis for regulating genes is
understood, perhaps the organism can be manipulated to respond in
the fashion or manner chosen by the experimenter.

Because of technical limitations many of the studies that were
started dealt with obvious and easily defined problems. These
include studies of gene products that are present in high amounts,
such as seed storage proteins and fraction one protein (1,5-
ribulose bisphosphate carboxylase/oxygenase, RuBisCo) in leaves.
Other, more difficult studies include identifying gene products
that accumulate as a response to heat or anaerobiotic stress, in
response to mechanical or pathogen induced wounding, in response
to a variety of types of light, or in response to added hormones.
Most of the studies dealt solely with measurements of molecules
because that is what current techniques allow. In this paper I
will briefly describe examples of the types of studies being done
with genes encoded by the nuclear genome, dealing in detail with
the work that we are doing with the soybean seed storage proteins,
and listing questions that need to be addressed prior to reducing
our science in the laboratory to practice in field agriculture.

II. Genes that are regulated by light

Some of the earliest molecular studies carried out in higher
plants were designed to identify the products of genes expressed
following the exposure of etiolated leaf tissues to full light.
During the greening period that follows, the major new gene
products expressed by the nuclear genome are two proteins used in
photosynthesis in the chloroplast, the small subunit (ssu) of
RuBisCo and the chlorophyll a/b binding protein (for review see
Ellis, 1981). Both proteins are translated in the cytoplasm

from messenger RNA (mRNAs) transcribed in the nucleus, and are
proteolytically processed during transport across the chloroplast
membrane. Once inside the chloroplast eight ssu molecules
interact with eight large subunits (encoded by the chloroplast
genome) in the stroma to form active RuBisCo. By contrast, the
major chlorophyll a/b binding proteins are transported to the
internal membrane system of the chloroplast where they are
inserted into the membrane, providing the site on which
chlorophylls a and b assemble by a non-covalent mechanism. While
research done in the 1970's was concerned primarily with
determining how polypeptides were transported from the cytosol
into the chloroplast, there has been increasing interest in
characterizing the gene(s) that encode the proteins, reasoning
that by comparing genes from widely diverse plant species, one
could distinguish features of the proteins that were conserved
during evolution, and which represent regions that are important
to the function of the protein. During these studies it became
apparent that each plant species contains multiple copies of genes
that encode the major chlorophyll a/b protein and the ssu of
RuBisCo. Dunsmuir et al. (1983) determined that there are at
least ten different genes that encode the ssu proteins in Mitchell
petunias, a diploidized haploid cultivar. The amino acid
sequences of the proteins themselves as deduced from DNA sequence
analysis of DNA copies of mRNAs (cDNAs) are nearly homologous.
However, each ssu gene expressed differed from others in the 3'
untranslated portion of its mRNA. Similar results were obtained
with genes encoding the ssu in Lemma gebba (Wimpee et al., 1983),
soybeans (Glycine max, Berry-Lowe et al., 1982), pea (Pisum
sativum, Coruzzi et al., 1983a), and in wheat (Triticum aestivum,
Broglie et al., 1983). Each of the mRNAs is encoded by a unique
single gene. Comparison of the amino acid sequence of the ssu
proteins of peas and wheat revealed that although there are long
regions of amino acid identity (16 amino acids in one section),
the amino acid sequence of the entire proteins differ by about
30%. It remains to be seen whether each of these genes is
expressed under all conditions during the life of the plant, or
whether different conditions of light, temperature or other
environmental or nutritional conditions, cause the preferential
expression of a specific gene or set of genes.

In a similar manner, the genes encoding the chlorophyll a/b
binding proteins (Cab) are encoded by a family of genes in
Petunia, estimated to be at least least 16 different genes
(Dunnsmuir et al., 1983). The nucleotide sequences within the
protein coding regions of at least some of these genes are similar
to each other (about 10% differences in nucleotides), while
sequences in the 3' untranslated regions are significantly
diverged from each other (between 50 and 70% differences in their
nucleotides, Dunsmuir et al., 1983). The "family" of Cab genes
can be subdivided further into five smaller gene families of two

to five members based on a high degree of sequence homology
between individuals. Several of the two-gene families were
examined in detail, and it was found that, in some instances, the
two genes are physically linked to each other, and are separated
by less than 2,000 nucleotides. In one case the genes are
inverted with respect to each other, and in the other case they
are in a direct repeat (Dunsmuir et al., 1983). Expression of
individual genes is currently being measured. Such studies are
made possible because nucleotide sequences in regions of the genes
that are not translated are different on each gene family. To
measure the expression of individual genes, a segment of cloned
cDNA representing the 3' untranslated portion of the mRNA will be
used as a specific probe in molecular hybridization reactions,
i.e., using cDNA:mRNA hybridizations to quantitate the amount of
mRNA present in a given tissue at any time chosen by the
experimenter.

There is a great deal of interest in understanding how the
chlorophyll a/b and small subunit of RuBisCo proteins are
regulated in amount such that the functions of the 50 to 100
chloroplasts in each cell, with each chloroplast containing
multiple copies of the chloroplast genome, can be maximally
efficient. In at least one system, Lemna gibba, the amounts of
mRNAs of both proteins are regulated by the action(s) of
phytochrome (Tobin, 1981). With the development of experimental
techniques to dissect genes, to produce chimeric genes, to
transfer to, and express chimeric genes in, "foreign plants" via
modified tumor inducing plasmids of Agrobacterium tumefaciens (see
general reviews by Nester and Kosuge, 1981; Hoekema et al., 1983;
see Fraley et al., 1983 and Murai et al., 1983 for specific
examples of plant transformations) and to regenerate plants from
such transformed plants (Zambryski, 1983; Horsch et la., 1984) it
is now possible to identify the nucleotide sequences that regulate
the amount of mRNA transcribed from a specific transcriptional
promoter. Such approaches have been used to define
transcriptional promoters from viral and cellular genes in a
number of lower and higher eucaryotic systems. N.H. Chua and
colleagues at the Rockefeller University, in collaboration with
researchers at Monsanto Company in St. Louis, MO, including R.T.
Fraley and S. Rogers, recently demonstrated that the
transcriptional promoter sequences of a gene encoding the ssu of
pea function in petunia cells (Broglie et al., 1984). In these
experiments a fragment of DNA from pea that contained a ssu
structural gene as well as several hundred nucleotides 5' of the
gene was placed into the Ti-plasmid, and transferred to petunia
cells by co-cultivation with A. tumefaciens carrying the modified
Ti-plasmid. Transformed callus cells produced the pea ssu protein
only if grown in the light. Furthermore the pea protein was
transported to the chloroplast where it apparently assembled with
the large subunit of RuBisCo of petunia. Thus, these workers have

isolated a promoter that responds to an added stimulus in a
foreign plant cell as it does in its native cell. It will be
important to further dissect the nucleotide sequences that control
transcriptional activity of these light inducible genes as a
stepping stone to understanding precisely how such regulation is
effected.

In the co-ordinated expression of genes in the nucleus with
those of the chloroplast it is crucial that proteins produced in
the cytoplasm be efficiently transported to the chloroplast. The
requirements for transport in in vitro reactions have been
described by a number of workers (Chua and Schmidt, 1978; Ellis,
1981; Grossman et al., 1982). Since the uptake of proteins by
chloroplasts is selective, i.e., only proteins needed by the
chloroplast are taken up, it is expected that there are common
features that characterize the "transit peptide" of these
proteins. Coruzzi et al. (1983) compared the transit sequences of
the ssu protein from wheat and peas, which diverged from each
other as long as 150 million years ago. Both subunits can be
taken up by pea chloroplasts in in vitro reactions. The transit
sequence is 47 amino acids long in the wheat protein and 33 amino
acids long in pea; both transit peptides are basic in net charge.
There has been strong conservation of amino acids near the cys↓
met site that is recognized by the enzyme which cleaves the
transit sequence from the remainder of the protein. These types
of studies indicate that such interspecific comparisons may be
very useful in intimating important features of these proteins.

Among the responses regulated by light those known to be
affected by the level of phytochrome have been examined by plant
scientists for a number of years (for general reference, see Song,
1984). Using the tools of molecular biology and biochemistry the
presumed active form of the phytochrome molecule has recently been
elucidated (Vierstra and Quail, 1982). This work also
demonstrated that phytochrome purified by previous researchers
represented only part of the active molecule. Concurrent studies
on the expression of the genes encoding the phytochrome mRNA
revealed that the level of mRNA is autoregulable, a unique
phenomenon in light regulated genes. To study the regulation more
precisely, P. Quail and colleagues generated cloned cDNAs for the
phytochrome mRNA to use as probes for molecular hybridization to
quantitate gene expression (P. Quail, personal communication).
Nucleotide sequencing of the cloned cDNAs will reveal the amino
acid sequence of this interesting protein and will be helpful in
elucidating its role in controlling plant growth and development.

III.   Genes that are expressed during plant development.

For years studies of plant development have included
descriptions of pattern formation, fate maps, gametogenesis,

fertilization, embryology, germination, and growth and development
through the life of the plant. These early studies laid the
groundwork for many studies of molecular aspects of plant development. The most thoroughly studied phase of the plant life cycle
is probably embryogeny. It is in these stages that embryo-
specific gene products, including the seed storage proteins, are
produced in abundance. Extensive chemical characterization of
many of the seed proteins by food scientists and nutritionists
provided a good background for molecular biologists who chose to
study the accumulation of specific seed storage proteins and their
respective mRNAs. Because of the abundance of both mRNAs and
proteins, and the technical advantages of using embryogenic
tissues for experimentation, a number of researchers have chosen
to study storage protein accumulation. These have included
studies on the seed proteins in a number of legume seeds
(Derbyshire et al., 1976), including P. sativum (Gatehouse et al.,
1981; Spencer and Higgins, 1980), Phaseolus vulgaris (Sun et al.,
1974; Slightom, 1983), G. max (Thanh and Shibasaki, 1975),
Gossypium hirsutum (Galau et al., 1983; Dure, 1983), and Conavales
eusiformis (Smith et al., 1982), as well as in monocots, such as
Triticum, Zea mays, Avena, Hordeum, Oryza (for an extensive review
of the seed proteins, see Gottschalk and Müller, 1983), and,
undoubtedly, others. The stated goals of many researchers is to
study the "regulation" of the genes encoding the storage reserves,
and to learn how to manipulate their expression, and, perhaps, to
improve the nutritional value of specific proteins by altering
their amino acid content. However, because gene regulation
results from a complex and as yet unidentified series of determinants, we find ourselves waiting for new insights to help
clarify how genes are activated to produce the storage proteins.
Experiments are underway to modify the amino acids in storage
proteins by changing the nucleotide sequences of the genes, but
until the modified genes can be re-inserted into their original
plant species (i.e., soybean genes returned to soybean) and
directed to be expressed only in embryos, all of these experiments
remain preliminary in nature. As an example of the types of
experiments that have been and are currently being done on
regulation of seed proteins I will describe the ongoing work on
the globulin storage proteins of soybean in my laboratory, making
reference to work in other laboratories as necessary to make
specific points. It will become apparent that there is a great
deal of work to be done before a real understanding of gene
regulation can be achieved.

A.   THE SEED STORAGE PROTEINS.

Seed storage proteins accumulate in large amounts in a number
of types of seeds, including cereal and legume grains, to be used
during seedling growth. Since the storage proteins of most of the
agronomically important seeds have been chemically characterized

(Gottschalk and Müller, 1983), and are produced in large amounts, seed proteins have been extensively studied by developmental and molecular biologists as gene products that characterize the stage of development of the embryo. Seed proteins are usually categorized on the basis of their solubility. Prolamines are soluble in high concentrations of alcohol and are typified by the zeins found in the endosperm of Z. mays. Zeins, and other prolamines, are synthesized on membrane bound polyribosomes and sequestered in protein bodies derived from the endoplasmic reticulum (Larkins and Hurkman, 1978; Burr and Burr, 1976). Genetic studies mapping chromosomal location of some of the many genes encoding the zeins (there may be as many as 100 genes, Viotti et al., 1982), and nucleotide sequencing of a number of cloned cDNAs and genomic clones encoding the zeins demonstrated that different genes encode protein subunits of 22,000, 19,000 and 15,000 molecular weight (Pedersen et al., 1982; Burr et al., 1982). The genes can, on the basis of nucleotide sequence relatedness, be associated into subfamilies of genes encoding each of the groups of subunits. The structure of the zeins has been predicted by computer analyses using amino acid sequences derived from DNA sequences (Argos et al., 1982), and partially explains the insolubility of the proteins. Both intermolecular and intramolecular interactions were predicted in this study.

In contrast to genetic studies of the zeins, genetic studies with the globulin seed proteins of G. max are in very early stages. Currently, less than ten linkage groups of G. max have been described, and studies on inheritance of the seed proteins are just beginning (N. Nielsen, work in progress). Therefore, studies at the level of gene organization and expression in G. max are done in a partial vacuum of information.

Globulin storage proteins, those soluble in 0.4 M NaCl, are most abundant in legume seeds, and are often subdivided into the vicilins (β-conglycinin, 7S proteins) and the legumins (glycinins, 11S proteins) as described by Derbyshire et al. (1976) and others. Since a number of reviews describing the physical structure of the globulins have been written, they will not be discussed here. Instead, I will go through the rationale for addressing questions about the regulated expression and accumulation of the soybean globulins. The discussion will revolve around our studies of the β-conglycinins. The β-conglycinins are comprised of three major subunits, α' and α (with apparent molecular weights of about 76 Kd and 72 Kd, respectively), and the β-subunit (about 50 Kd). In addition, other minor subunits of between 52 Kd and 70 Kd co-purify with the major subunits when β-conglycinin is purified by standard methods (Bryant, Ladin and Beachy, unpublished). Each of these subunits is produced on membrane bound polyribosomes where core glycosylation (Sengupta et al., 1981) occurs (Figure 1). Proteins are then transported to Golgi bodies for further

modification, and finally are deposited in the vacuole which
accumulates these proteins (Beachy et al., 1979; Yao and
Chrispeels, 1980), budding off smaller membrane-bound structures
that become protein bodies. Although exo-, and endo-peptidases
are present in cotyledons during embryogenesis (for example, see
Bond and Bowles, 1983) they apparently are inactivated during seed
development.

Different subunits of β-conglycinin accumulate at different
times during embryogenesis of soybeans, and in varying amounts in
embryonic axes and cotyledons (Meinke et al., 1981; Ladin et al.,
in preparation). This suggests that β-conglycinins are comprised
of varying combinations of subunits at different times in
development. There is also evidence to support the hypothesis
that nutritional status of soybean affects the types of subunits
produced (Holowach et al., 1984). Similar observations were made
with seed proteins of pea (Chandler et al., 1983). This may be a
significant adaptive advantage for at least some plants.

The β-conglycinin is a trimeric protein of about 7S under
high salt conditions (0.4 $\underline{M}$ Na$^+$) and a double-trimer under low
salt conditions. Electron microscopy of purified preparations of
these proteins verify this structure (Bryant, Hauser and Beachy,
unpublished).

The complete amino acid sequence of the α' subunit (as well
as portions of the α-subunit, and of other β-conglycinin subunits)
has been determined (Schuler and Beachy, unpublished), and is presented in Figure 2 by the single letter designation. The mature
α'-subunit as found in dried seeds begins at the Val residue at
position number 63 from the amino terminus. On the basis of in
vivo and in vitro experiments and computer modeling studies, it is
thought that signal peptide cleavage occurs between amino acid
residues number 22 and 25. Thus, at least two cleavage reactions
occur during biosynthesis of the α' (and α) subunit. Signal peptide cleavage occurs concurrent with translation on membrane bound
polyribosomes (Hosangadi and Beachy, unpublished), but we are un-
sure where the second cleavage reaction occurs, although in vivo
pulse-chase experiments indicated that it occurs between 12 and 24
hrs post-synthesis (Bryant and Beachy, unpublished), suggesting
that the cleavage reaction occurs in protein bodies. An intriguing observation is that there are four cysteine residues between
the 24th and the 64th residues, while the remainder of the protein
contains a single cys residue. This raises questions about whether
the 40 amino acid long peptide generated by the second cleavage
has some biological function, or whether its removal is required
in order for the remainder of the polypeptide to assemble into
β-conglycinin holoproteins. There are a number of examples of
cleavage of storage protein subunits during their biosyntheis
(Spencer and Higgins, 1980; Barton et al., 1982). The pea vicilin

undergo the most extensive cleavages studied to date (Gatehouse et al., 1981). Storage proteins of most seeds do not undergo extensive degradation until seed germination and seedling growth.

While assembly of subunits to form the β-conglycinins may be semi-random in nature, degradation of the β-conglycinins is highly precise, occurring during the first seven days of seedling growth. Site specific endoproteases cleave the subunits into progressively smaller polypeptides while the protein is in the trimeric, 7S configuration. This degradation is observed by six hours of imbibition in the embryonic axes, but not until 3 or 4 days later in cotyledons (Fig. 1). This pattern of degradation is probably reflective of the difference in structure of protein bodies in axes and cotyledons (protein bodies in axes contain much less densely packed protein than those in cotyledons, Ladin et al., manuscript in preparation), and the need for a rapid release of amino acids in axes during early growth of the radicle.

The accumulation of the soybean storage proteins is largely regulated at the level of transcription (Meinke et al., 1981; Goldberg et al., 1981). Once transcribed, the α' and α-mRNAs may be stable for long periods of time, perhaps for as long as six days (J. Madison, unpublished data), representing a type of post-transcriptional regulation. There is also some evidence of translational regulation of some mRNAs, and of post-translational regulation, in the form of changes in stability of the β-conglycinins during embryogenesis (Shattuck-Eidens and Beachy, unpublished). Although suggestive evidence is indicative that several types of regulation are operative during embryogenesis rigorous proof of such regulation remains to be completed.

As discussed earlier in the case of transcription of the ssu of RuBisCo, light regulated transcription requires the presence of specific DNA sequences in front of the gene. We expect that transcription of the β-conglycinin genes is likewise controlled by DNA sequences in front of the structural gene; however, it remains to be proven that there are unique sequences that are responsible for the very high levels of transcription of those genes in embryos, but which silence the gene in other tissues. To test the hypothesis that there are such specific sequences in front of these genes, we have carried out the following experiments, which, at this date, remain in progress. A fully sequenced gene encoding the α'-subunit, pGmg 17.1, is flanked to the 5' side by about 8.5 Kb of DNA (Schuler and Beachy, unpublished). Using specific restriction endonucleases, and the exonuclease Bal 31 a series of deletion mutants were produced (Schuler, unpublished) which left different amounts of DNA sequence in front of the structural gene as shown in Fig. 3. The promoter sequence "TATAAA" (the sequence involved in positioning RNA polymerase II) is positioned at about -30, and the "CCAAAT" promoter sequence is at about -70.

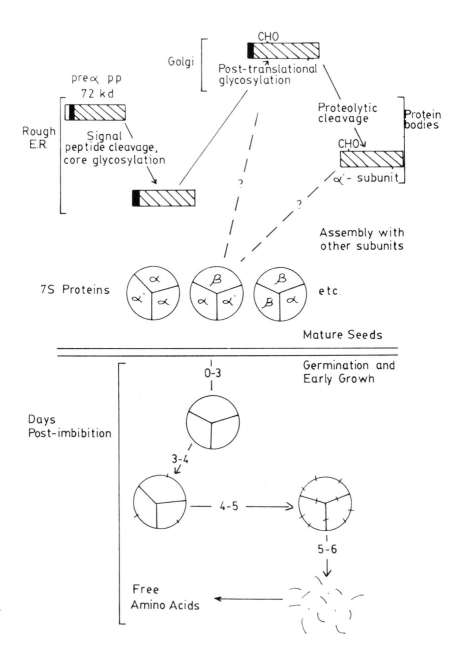

Also indicated on Fig. 3 is the position of the sequence
GTGG$_{TTT}^{AAA}$G, a proposed consensus sequence that may be involved in
in "enhancing" the expression of genes in a tissue specific manner
(Khowry and Gruss, 1983; Gillies et al., 1983). Although a number
of enhancer like elements have been described in animal viruses
and other cellular genes, no such element has been shown to play a
role in causing the high level tissue-specific expression of genes
in plants. Since several of the gene mutations shown in Fig. 3
lack the enhancer consensus sequence, we hope to determine whether
these or other sequences control the level of expression and/or
tissue specific expression of the α'-subunit gene. An
experimental test of each of the mutants is made possible by the
current technologies in genetic engineering discussed above. The
level of expression of each of the mutant genes in somatic tissues
and in new embryos will be determined, thus assessing the
functional role of specific DNA sequences in regulating
transcription of the α-subunit gene. This type of experimental
approach assumes that a promoter from a leguminous plant will
function accurately in a solonaceous plant. Since in a recent
paper by Broglie et al. (1984), it was demonstrated that a
promoter from the ssu of RuBisCo from ssu functions as expected in
petunia, there is some reason to be optimistic about the
expression of the soybean α'-subunit gene in petunias. By these
types of experiments it is hoped that we will learn more about the
regulation of genes that are normally expressed in a tissue
specific manner.

These studies are also prerequisite to any attempts to change
the amino acid composition of storage proteins. Since protein
accumulation requires that the structural gene be transcribed,

<hr>

Fig. 1. Summary of the steps involved in the biosynthesis,
         assembly, and degradation of the β-conglycinins of
         soybeans. Synthesis of α-subunit is specifically
         described, but a similar pathway exists for the
         α'-subunit: synthesis of the β-subunit may or may not
         include a second proteolytic cleavage in the protein
         bodies as in the case of the α' and α subunits. The
         sites of cleavage of the signal peptide and the secondary
         cleavage that occurs in protein bodies are indicated in
         Figure 2. It is not known whether these subunits
         assemble into 7S, β-conglycinin proteins in Golgi bodies,
         in transport vesicles, or in protein bodies. Degradation
         of the 7S proteins during early seedling growth is
         indicated as days post-imbibition in moist soil. This
         summary is derived from published data by Sengupta et al.
         (1981) and Beachy et al. (1981), and from unpublished
         work by Bryant, Hosangadi, Schuler and Beachy.

```
         +                                              ↓
    NH₃ M M R A R F P L L L L G V V F L A S V S V S F G I A Y W E
K Q N P S H N K C L R S C N S E K D S Y R N Q A C H A R C N L L K ↑
V E E E E C E E G Q I P R P R P Q H P E R E R Q Q H G E K E E D
E G E Q P R P F P F P R P R Q P H Q E E E H E Q K E E H E W H R K
E E K H G G K G S E E E Q D E R E H P R P H Q P H Q K E E E K H E
W Q H K Q E K H Q G K E S E E E E E D Q D E D E E Q D K E S Q E S
E G S E S Q R E P R R H K N K N P F H F N S K R F Q T L F K N Q Y
G H V R V L Q R F N K R S Q Q L Q N L R D Y R I L E F N S K P N T
L L L P H H A D A D Y L I V I L N G T A I L T L V N N D D R D S Y
N L Q S G D A L R V P A G T T F Y V V N P D N D E N L R M I A G T
T F Y V V N P D N D E N L R M I T L A I P V N K P G R F E S F F L
S S T Q A Q Q S Y L Q G F S K N I L E A S Y D T K F E E I N K V L
F G R E E G Q Q Q G E E R L Q E S V I V E I S K K Q I R E L S K H
A K S S S R K T I S S E D K P F N L G S R D P I Y S N K L G K L F
E I T Q R N P Q L R D L D V F L S V V D M N E G A L F L P H F N S
K A I V V L V I N E G E A N I E L V G I K E Q Q Q R Q Q Q E E Q P
L E V R K Y R A E L S E Q D I F V I P A G Y P V M V N A T S D L N
F F A F G I N A E N N Q R N F L A G S K D N V I S Q I P S Q V Q E
L A F P R S A K D I E N L I K S Q S E S Y F V D A Q P Q Q K E E G
N K G R K G P L S S I L R A F Y-COO⁻
```

Fig. 2. Amino acid sequence of the α'-subunit of β-Conglycinin as
derived from DNA sequence of the gene GmG 17.1 (see Fig.
3) (Schuler and Beachy, manuscript in preparation).
↓ indicates the proposed position of signal peptide
cleavage; ↑ indicates the position of the second
proteolytic cleavage. The amino terminus of α'-subunits
isolated from mature seeds, as determined by Moreiria and
Nielsen (unpublished) begins at the Val residue (no. 63).

that the mRNA be transported to the cytoplasm where it is translated, and where protein synthesis and post-translational
processing, and packaging occurs, any changes in the structural
gene must not affect any step in this series of events. The
greatest potential problem probably lies in choosing the site at
which mutations, for example to increase the levels of sulfur-
containing amino acids, are to be made such that assembly of the
subunits into molecules of β-conglycinin, and the packing of
multiple molecules of β-conglycinin are not affected. Finally,
such mutations should not affect the degradation of the
β-conglycinins during seedling growth.

      B. STUDIES OF OTHER PLANT GENES.

     As evidenced by the discussion above there is a great deal of
research effort directed toward understanding how genes are
regulated during seed development and as the plant establishes
autotrophy. Significant advances have also been made in

Fig. 3. Diagram of the gene encoding the α'-subunit of
β-conglycinin with pertinent features. ☐ =
untranslated regions of the messenger RNA (mRNA); ■ =
coding sequences of the mRNA; ⊡ = intervening
sequences; nucleotide sequences at -30 and -80
(nucleotides 5', in front of, the first nucleotide of
the mRNA) are suspected to be involved in recognition,
binding and transcription of RNA polymerase II (see the
review McKnight and Kingsbury, 1982 for more
information). Sequences at -560 are indicated because
of the marked similarity to cellular enhancer sequences
(see minireview by Khoury and Gruss, 1983).

understanding how plants respond to changes in their environment.
The following discussion is to serve only as an introduction for
the reader to the types of research that are in progress on this
topic, and to stimulate the interest of others who can contribute
to this important area of plant biology.

A. The plants' response to abiotic stresses. Of the studies in
gene expression that result as a response to abiotic stresses, the
most advanced relate to how the plant responds to high temperature
(heat shock), flooding (anaerobiosis) and ultraviolet light.

(1) The heat shock (HS) response in plants has been examined
in several plant species including G. max (Barnett et al., 1980;
Key et al., 1983), and Z. mays (Cooper and Ho, 1983). Each of
these plants respond to high temperatures (usually above 35°C, but
differing from plant to plant) by producing a number of new
proteins, or increased amounts of specific proteins, within one to
three hours of the shift to high temperatures. Plants that have

been held at high temperature producing HS proteins, can withstand
normally lethal temperatures, of as much as 45°C (as was demonstrated for soybean seedlings by Key et al., 1983). After
returning the plant to normal temperatures, the synthesis of
normal proteins returns within 30 min, suggesting that these mRNAs
are not degraded during high temperatures, but are somehow
protected until the organism returns to permissive temperatures.

Exposure to elevated temperatures increases the synthesis of
several high molecular weight polypeptides, between 70 and 90 kd,
and a complex group of polypeptides, generally between 12 and 20
kd. The high molecular weight HS polypeptides in plants are
antigenically related to those of a number of other eucaryotic
organisms. The low molecular weight proteins may be unique to a
given organism or groups of related organisms. The role of HS
proteins in protecting a cell or organism, including plants, to
high temperatures is unknown, but a number of hypotheses have been
presented (for a comprehensive review see the proceeding of a
conference held at Cold Spring Harbor, 1982, eds. Schlesinger,
Ashburner, and Tissieres).

The current status of studies of HS in plants are at the
level of cataloging the appearance of new proteins during HS,
isolating and identifying cloned cDNAs representing mRNAs that
encode these proteins, and isolating and characterizing their
respective genes. It is now known that products of some of the
heat shock genes in a given organism are related to each other
(Schoffl and Key, 1983), but as yet no function is ascribed either
to the individual gene products or to the group of gene products.

(2) When plants become anaerobic, such as occurs during
flooding, they produce a number of unique gene products (Sachs et
al., 1980). Concurrently, synthesis of normal proteins is
dramatically decreased. During the anaerobic response of maize
the major newly synthesized gene product is alcohol dehydrogenase,
representing as much as 10% of new protein synthesis. The
genetics of the Adh 1 and 2 genes in maize has been well described
by D. Schwartz and his colleagues (1971) who developed a pollen
sensitivity assay to screen plants containing Adh$^+$ and
Adh$^-$ phenotypes. This easy screening assay was then used by
others to identify plants containing genes into which a
transposable element has inserted into the Adh 1 and 2 genes.
Standard cDNA cloning techniques were used to isolate Adh gene
sequences, as well as the genomic clone encoding Adh 1 (Dennis et
al). It is hoped that study and characterization of and with this
gene will lead to an understanding of how the expression of
"anaerobic response" genes are controlled. The other less well
described genes that are preferentially expressed during
anaerobiosis are also being studied.

(3) Many plant species respond to wounding, either mechanical damage, or that caused by a bacterial or fungal pathogen, by accumulating new gene products that wall off the invading pathogen, or to heal the damaged area. Part of the wound repair mechanism in carrots (<u>Daucus carota</u>) for example, includes the production of large amounts of a hydroxyproline-rich glycoprotein (as well as other glyco-proteins) that is secreted into the extracellular space (reviewed by Lamport and Catt, 1981). The hydroxyproline rich cell wall protein, a rigid rod composed of a polypeptide ensheathed in carbohydrate, is then cross-linked with other cell wall substituents (Varner and Cooper, 1983). The amino acid sequence of the carrot cell wall protein has been inferred from DNA sequence analysis of a cloned cDNA and appears to be comprised of repeated sequences (Chen and Varner, manuscript in preparation). The regulation of the genes encoding these proteins is extremely interesting, since one expects that these genes, or other related genes, will be expressed during plant growth, as well as a response to wounding. Since wound responses are often correlated with ethylene production, an understanding of how these genes are controlled may also shed light on the role of this plant hormone in regulating gene expression.

In addition to physical wounding, interactions between a host and pathogen can elicit a series of biochemical events leading to the accumulation of metabolites that have the potential to slow or halt the invasion of the pathogen, in some instances. Known collectively as phytoalexins, these metabolites are, in the leguminoseae, products of the shikimate-polymalonic acid biosynthetic pathway, and are pterocarpans. Among the enzymes whose activities controlling the synthesis of phytoalexins are phenylalanine ammonia lyase (PAL), chalcone synthase (CHS), and 4-coumarate:CoA ligase (4CL). A number of biological and non-biological factors can cause the induction of genes encoding these enzymes, including UV-light treatment (Hahlbrock et al., 1982) and purified "elicitors" derived from cell walls of phytopathogenic fungi. Recently Kreuzaler et al. (1983) and C. Lamb and co-workers (personal communiction) were able to detect increased activity of some of the enzymes involved in phytoalexin biosynthesis within 1 hr of the induction treatment. This increase in activity was due, at least in part, to expression of genes encoding these proteins. Cloned cDNAs for some these genes have recently been isolated and characterized (Reinold, et al, 1983). Using these cloned cDNAs as hybridization probes, nuclear genes will be isolated for use in studies to determine precisely how the elicitors cause the increase in gene expression that leads ultimately to phytoalexin accumulation. Although probably not the sole determinant, or even a major determinant, in disease resistance or other insults, phytoalexins may play a significant role in the plants response to its environment.

CONCLUDING REMARKS

The goal of this discussion was to introduce the reader to
examples of the types of resarch that is currently ongoing in a
number of laboratories to begin to understand how plants control
the expression of their nuclear genomes. Many areas of research
have not been discussed which are of keen interest to large
numbers of plant molecular biologists, such as those concerning
the expression of plant and bacterial genes during Rhizobium-
induced nodulation and symbiotic nitrogen fixation, the expression
of plant and bacterial genes during transformation of plants by
Agrobacterium tumefaciens, the expression of genes in chloroplasts
and mitochondria, and the regulation of the nuclear genomes by the
organelles, to name a few. It is clear from the studies done to
date that regulation is complex and includes regulation of
transcription, translation, and post-translational events.
Because of this complexity reducing the expectations of the
molecular biologist into practice that integrates the studies of
other plant scientists will require large amounts of communication
and cooperation between molecular biologists, physiologists, and
breeders and agronomists. Without those factors, the role of
molecular and biology and the "new genetics" in crop productivity
will be severely limited.

REFERENCES

ARGOS, P., PEDERSON, K., MARKS, M. D., and LARKINS, B. A., 1982, A
    structural model for maize zein proteins, J. Biol. Chem., 257:
    9984.
BARNETT, T., ALTSCHULER, M., MCDANIEL, C. N., and MASCARENHAS, J.
    P., 1980, Heat shock induced proteins in plant cells.
    Dev. Gen., 1:311.
BARTON, K. A., THOMPSON, J. F., MADISON, J. T., ROSENTHAL, R.,
    JARVIS, N. P., and BEACHY, R. N., 1982, The biosynthesis and
    processing of high molecular weight precursor of soybean
    glycinin subunits, J. Biol. Chem., 257:6089.
BEACHY, R. N., THOMPSON, J. F., and MADISON, J. T., 1979, Isolation
    and characterization of messenger RNAs that code for the
    subunits of soybean seed proteins, in: "The Plant Seed,
    Development, Preservation and Germination," I. Rubenstein, R.
    L. Phillips, C. E. Green, and B. G. Gingenbach, eds.,
    Academic Press, N.Y. pp. 67-84.
BEACHY, R. N., JARVIS, N. P., and BARTON, K. A., 1981, Biosynthesis
    of subunits of the soybean 7S storage protein, J. Mol.
    Appl. Gen., 1:19.
BERRY-LOWE, S. L., MCKNIGHT, T. D., SHAW, D. M., and MEAGHER, R.
    B., 1982, The nucleotide sequence, expression and evolution
    of one member of a multigene family encoding the small subunit
    of ribulose-1,5-bisphosphate carboxylase in soybean, J. Mol.
    Appl. Gen., 1:483.
BOND, H. M., and BOWLES, D. J., 1983, Characterization of soybean
    endopep-tidase activity using exogenous and endogenous
    substrates, Plant Physiol., 72:345.
BROGLIE, R., CORUZZI, G., LAMPPA, G., KEITH, B., and CHUA, N.-H.,
    1983, Structural analysis of nuclear genes coding for the
    precursor to the small subunit of wheat ribulose-1,
    5-bis-phosphate carboxylase, Bio/technology, 1:55.
BROGLIE, R., CORUZZI, G., FRALEY, R. T., ROGERS, S. G., HORSCH, R.
    B., NIEDERMEYER, J. G., FINK, E. C. F., FLICK, J. S., and
    CHUA, N.-H., 1984, Light-regualted expression of a pea
    ribulose biophosphate carboxylase small subunit gene in
    transformed plant cells, Science, in press.
BURR, B., BURR, F. A., ST. JOHN, T. P., THINAS, M., and DAVIS, R.
    W., 1982, Zein storage protein gene family of maize. An
    assessment of heterogeneity with cloned messenger RNA
    sequences, J. Mol. Biol., 154:33.
BURR, B., and BURR, F. A., 1976, Zein synthesis in maize endosperm
    by polyribosomes attached to protein bodies, Proc. Natl.
    Acad. Sci. USA, 73:515.
CHANDLER, P. M., HIGGINS, T. J. V., RANDALL, P. J., and SPENCER,
    D., 1983, Regulation of legumin levels in developing pea seeds
    under conditions of sulfur deficiency, Plant Physiol., 71:47.

CHUA, N.-H., and SCHMIDT, W., 1978, Post-translational transport into intact chloroplasts of a precursor to the small subunit of ribulose-1, 5-biphosphate carboxylase, Proc. Natl. Acad. Sci. USA, 75:6110.

COOPER, P., and HO, D., 1983, Heat shock proteins in maize, Plant Physiol., 71:215.

CORUZZI, G., BROGLIE, R., CASHMORE, A., and CHUA, N.-H., 1983a, Nucleotide sequences of two pea cDNA clones encoding the small subunit of ribulose 1,5-biphosphate carboxylase and the major chlorophyll a/b-binding thylakoid polypeptide, J. Biol. Chem., 258:1399.

CORUZZI, G., BROGLIE, R., LAMPPA, G., and CHUA, H.-H., 1983b, Expression of Nuclear genes encoding the small subunit of ribo-lase-1, 5-biphosphate carboxylase, in: "Structure and Function of Plant Genomes," O. Ciferri and L. Dure III, eds., Plenum Press, N.Y. pp. 47-59.

DENNIS, E. S., GERLACK, W. L., PRYOR, A. L. BENNETZEN, J. L., INGILIS, A., LEWELLYN, D., SACHS, M. M., FERL, R. J., and PEACOCK, W. J., The Adh 1 gene of maize, submitted for publication.

DERBYSHIRE, E., WRIGHT, D. J., BOULTER, D. 1976, Legumin and vicilin, storage proteins of legume seeds, Phytochemistry 15: 3.

DUNSMUIR, P., SMITH, S. M., and BEDBROOK, J., 1983, The major chlorophyll a/b binding protein of petunia is composed of several polypeptides encoded by a number of distinct nuclear genes. J. Mol. Applied Gen., 2:285.

DUNSMUIR, P., and BEDBROOK, J., 1983, Chlorophyll a/b binding proteins and the small subunit of ribulose bisphosphate carboxylase are encoded by multiple genes in petunia, in: "Structure and Function of Plant Genome," O. Ciferi and L. Dure III, eds., Plenum Press, N.Y. pp. 221.

DURE, L., PYLE, J. B., CHLAN, C. A., BAKER, J. C., and GALAU, G. A., 1983, Developmental biochemistry of cottonseed embryogenesis and germination XVII. Developmental expression of genes for the principal storage proteins, Plant Molec. Biol., 2:199.

ELLIS, R.J., 1981, Chloroplast proteins: synthesis transport and assembly, Ann. Rev. of Plant Physiol., 32:111.

FRALEY, R. T., ROGERS, S. G., HORSCH, R. B., SSADERS, P. R., FLICK, J. S., ADAMS, S. P., BITTNER, M. L., BRAND, L. A., FINK, C. L., FRY, J. S., GALLUPI, G. R., GOLDBERG, S. B., HOFFMANN, N. L., and WOO, S. C., 1983, Expression of bacterial genes in plant cells, Proc. Natl. Acad. Sci. USA, 80:4803.

GALAU, G. A., CHLAN, C. A., and DURE, L., 1983, Developmental bio-chemistry of cottonseed embryogenesis and germination XVI. Analysis of the principal cotton storage protein gene family with cloned cDNA probes, Plant Molec. Biol., 2:189.

GATEHOUSE, J. A., CROY, R. R. D., MORTON, H., TAYLER M., and
BOULTER, D., 1981, Characterization and subunit structures of
the vicilin storage proteins of pea (Pisum sativum L.) Eur. J.
Biochem., 118:627.

GILLIES, S. D., MORRISON, S. L., OI, V. T., and TONEGAWA, S., 1983,
A tissue-specific transcription enhancer element is located in
the major intron of a rearranged immunoglobulin heavy chain
gene, Cell, 33:717.

GOLDBERG, R. B., HOSCHEK, G., DITTA, G. S., and BRIEDENBACH, R. W.,
1981, Developmental regulation of cloned superabundant embryo
mRNAs in soybean, Dev. Biol., 83:218.

GOTTSCHALK, W., and MULLER, H. P., 1983, Seed Proteins Biochemistry, Genetics, Nutritive Value, Martinses Nijhoff/Dr. W.
Junk, Boston, MA.

GROSSMAN, A. R., BARTLETT, S. G., SCHMIDT, G. H., MULLET, J. E.,
and CHUA, N.-H., 1982, Optimal conditions for
post-translational uptake of proteins by isolated
chloroplasts. In vitro synthesis and transport of
plastocyanin, ferridoxin-NaDp$^+$ oxidoreductase, and fructose-1,
6-biphosphatase, J. Biol. Chem., 257:1558.

HAHLBROCK, K., BOUDET, A. M., CHAPPELL, J., KREUZALER, F., KUHN,
D. N., and RAGG, H., 1982, Differential induction of mRNAs by
light and elicitor in cultured plant cells, in: "Structure and
Function of Plant Genomes," O. Ciferri and L. Dure II, eds.,
Plenum Press, N.Y.   p. 15.

HOEKEMA, A., VON HAAREN, M. J. J., HILLE, J., HOGE, J. H. C.,
HOOYKAAS, R. J. J., KRENS, F. A., WULLEMS, G. J., and
SCHILPEROORT, R. A., 1983, Agrobacterium tumefaciens and its
Ti-plasmid as tools in transformation of plant cells, in:
"Plant Molecular Biology," R. B. Goldberg, ed., Alan Liss,
Inc., N.Y.   p. 3.

HOLOWACH, L. P., THOMPSON, J. F., and MADISON, J. T., 1984, Effects
of exogenous methionine on storage protein composition of
soybean cotyledons cultured in vivo, Plant Physiol., 74:576.

HORSCH, R. B., FRALEY, R. T., ROGERS, S. G., SANDERS, P. R., LLOYD,
A., and HOFFMANN, N., 1984, Inheritance of functional foreign
genes in plants, Science, 223:496.

KEY, J.L., LIN, C.Y., CEGLARZ, E., and SCHOFFL, F., 1983, The heat
shock response in soybean seedlings, in: "Structure and
Function of Plant Genomes," O. Ciferri and L. Dure III,
eds., Plenum Press, N.Y.   p. 25.

KHOURY, G., and GRUSS, P., 1983, Enhancer Elements--A Minreview,
Cell, 33:313.

KREUZALER, F., RAGG, H., FAUTZ, E., KUHN, D. N., and HAHLBROCK, K.,
1983, UV-induction of chalcone synthase mRNA in cell
suspension cultures of Petroselinum hartense, Proc. Natl.
Acad. Sci. USA, 80:2591.

LAMPORT, D. T. A., and CATT, J. W., 1981, Glycoproteins and enzymes of the cell wall. in: "Plant Carbohydrates II, Encyclopedia of Plant Physiology," New Series 13B, W. A. Tanner and F. A. Loewis, eds., Springer-Verlag, N.Y. pp. 133-165.

LARKINS, B. A., and HURKIMAN, W. J., 1978, Synthesis and deposition of zein in protein bodies of maize endosperm, Plant Physiol., 62:256.

MCKNIGHT, S.-L., and KINGSBURY, R., 1982, Transcriptional control signals of a eucaryotic protein-coding gene, Science, 217:316.

MEINKE, D. W., CHEN, J., and BEACHY, R. N., 1981, Expression of storage protein genes during soybean seed development, Planta, 153:130.

MURAI, N., SUTTON, D. W., MURRAY, M. G., SLIGHTOM, J. L., MERLO, D. J., REICHERT, N. A., SENGUPTA-GOPALAN, C., STCOCK C. A., BARKER, R. F., KEMP, J., and HALL, T. C., 1983, Phaseolin gene from bean is expressed after transfer to sunflower via tumor-inducing plasmid vectors, Science, 222:476.

NESTER, E. W., and KOSAGE, T., 1981, Plasmids specifying plant hyperplasias, Ann. Rev. Microbiol., 35:531.

PEDERSON, K., DEVEREUX, J., WILSON, D. R., SHELSON, E., and LARKINS, B. A., 1982, cloning and sequence analysis reveal structural variation among related zein genes in maize, Cell, 29:1015.

REINOLD, U., KROGER, M., KREUZALER, F., and HAHLBROCK, K., 1983, Coding and 3' non-coding nucleotide sequence of chalcone synthase mRNA and assignment of amino acid sequence of the enzyme, The EMBO Journal, 2:1801.

SACHS, M. M., FREELING, M., and OKIMOTO, R., 1980, The anaerobic proteins of maize, Cell, 20:761.

SCHLESINGER, M. J., ASHBURNER, M., and TISSIERES, A., eds., 1983, Heat shock--from Bacteria to Man, Cold Spring Harbor Laboratory, N.Y.

SCHOFFL, F., and KEY, J. L., 1983, Identification of a multigene family for small heat shock proteins in soybean and physical characterization of one individual gene coding region, Plant Molec. Biol., 2:269.

SCHWARTZ, D., 1971, Genetic control of alcohol dehydrogenase--A competitive model for regulation of gene action, Genetics, 64:411.

SENGUPTA, C., DELUCA, V., BAILEY, D. S., and VERMA, D. A. S., 1981, Post-translational processing of 7S and 11S components of soybean storage proteins, Plant Mol. Biol., 1:19.

SLIGHTOM, J. L., SUN, S. M., and HALL, T. C., 1983, Complete nucleotide sequence of a French bean storage protein gene: phaseolin, Proc. Natl. Acad. Sci. USA, 80:1897.

SMITH, C. S., JOHNSON, S. ANDREWS, J., and MCPHERSON, A., 1982, Biochemical characterization of canavvlin, the major storage protein of jack bean, Plant Physiol., 70:1199.

SONG, P.-S., 1984, Phytochrome, in: "Advanced Plant Physiology," M. B. Wilkins, ed., Pitman Publ. Inc., Marshfield, MA. pp. 354-379.

SPENCER, D. and HIGGINS, T. J. V., 1980, The biosynthesis of legumin in maturing pea seeds, Biochem. Internat., 1:502.

SUN, S. M., MCLESTER, R. C., BLISS, F. A. and HALL, T. C., 1974, Reversible and irreversible dissociation of globulins from Phaseolin vulgaris seed, J. Biol. Chem., 249:2118.

THANH, V. H., OKUBO, K., and SHIBISAKI, K., 1975, Isolation and characterization of the multiple 7S globulins of soybean proteins, Plant Physiol., 56:19.

TOBIN, E. M., 1981, Phytochrome mediated regulation of mRNAs for the small subunit of ribulose-1, 5-biphosphatase carobxylase and the light-harvesting chlorophyll a/b-protein in Lemna gibb., Plant Mol. Biol., 1:35.

VARNER, J. E., and COOPER J. B., 1983, Hydroxyproline-rich glycoproteins extracted from the cell walls of aerated carrot slices. in: "Structure and Function of Plant Genomes," O. Ciferri and L. Dure III, eds., Plenum Press, N.Y. p. 463.

VIERSTRA, R. D., and QUAIL, P. H., 1982, Natural phytochrome, Inhibition of proteolysis yields a homogeneous monomer of 124 kilodaltons from Avena, Proc. Natl. Acad. Sci. USA, 79: 5272.

VIOTTI, A., ABILDSTEN, D., POGNA, N., SALA, E., and PIROTTA, V., 1982, Multiplicity and diversity of cloned zein cDNA sequences and their chromosomal localization, The EMO Journal, 1:53.

WIMPEE, C. F., STIEKMA, W. J., and TOBIN, E. M., 1983, Sequence hetero-geneity in the RuBP carboxylase small subunit gene family of Lemma gibba. in: "Plant Molecular Biology," R. B. Goldberg, ed., Alan R. Liss, Inc., N.Y. p.p. 391- .

YAO, B. Y., and CHRISPEELS, M. J., 1980, The origin of protein bodies in developing soybean cotyledons, Protoplasma, 103: 201.

ZAMBRYSKI, P., JOOS, H., GENETELLO, C., LEEMANS, J., VAN MONTAGU, M., and SCHELL, J., Ti plasmid vector for the introduction of DNA into plant cells without alteration of their normal regenreation capacity, The EMBO Journal, 2:2143.

PERSPECTIVES ON GENETIC MANIPULATION IN PLANTS

John R. Bedbrook
Advanced Genetic Sciences, Inc.
6701 San Pablo Avenue
Oakland, California 94608

SUMMARY

Progress in efforts directed toward the genetic engineering of plants is documented. Likely progress in the near future is summarized and the prospects for the involvement of the technology of genetic engineering in plant breeding is briefly discussed.

INTRODUCTION

Evaluating the importance of genetic engineering in plant improvement is too premature a topic for a scientific paper. The developments in the field in the past year have been substantial and significant, and certainly wet the lips for excited speculation. The purpose of this paper is to summarize the state of field and not to venture too far into the unknown. Tangible prospects for genetic engineering experiments of the immediate future are summarized and technological developments which will enable accelerated progress are placed in perspective.

DNA TRANSFER SYSTEMS

Dicotyledonous Plants and *Agrobacterium tumefaciens*

DNA transfer to dicotyledonous plants can be mediated by the plant pathogen *Agrobacterium tumefaciens* (Bevan and Chilton, 1982; Ream and Gordon, 1982) which causes crown gall disease. *A. tumefaciens* induces a neoplastic transformation of plant tissue. Virulence of *A. tumefaciens* encoded by Ti plasmids (Van Larebeken

627

et al., 1979; Zaenen et al., 1974; Watson et al., 1975) is the consequence of the stable, covalent transfer of a portion of this
plasmid (the T region) to the plant chromosome (Chilton et al.,
1977; Thomashow et al., 1980). Insertion of foreign DNA sequences
within the T region of Ti plasmids can result in cotransfer of the
foreign DNA into the plant genome (Hernalsteens et al., 1980). The
transferred T region is expressed in plant cells (Drummond et al.,
1977) and encodes for several transcripts (Gelvin et al., 1981;
Willmitzer et al., 1983; Bevan et al., 1982). These transcripts
appear to encode products which cause the tumorous growth in infected plants as well as the production of unusual compounds called
opines.

Importantly, none of these transcripts are essential for the
stable integration of T DNA into the host chromosome. (Garfinkel
et al., 1981; Leemans et al., 1982; Joos et al., 1983). Further,
it has been shown (Zambryski et al., 1983) that DNA transfer via
A. tumefaciens can be achieved without altering the phenotype of
the plant. In this work, the entire region of T DNA encloding products inducing the tumorous phenotype were replaced with the common
cloning vector, pBR322. The only portion of T DNA retained in the
Ti plasmid were the border regions and the gene for nopaline
synthase.

The information above, taken together, allowed the development
of useful, if not cumbersome, vectors for the introduction of
foreign DNA into plant chromosomes. These vectors are known as
intermediate or shuttle vectors (Leemans, 1981). The vectors consist of typical Escherichia coli cloning plasmids, such as pBR322,
containing sequences homologous to T DNA itself, or sequences introduced between the T DNA borders. Foreign sequences are cloned
into these vectors in the normal fashion. The vector plus cova-
lently linked foreign sequence are introduced into A. tumefaciens
by in vivo homologous recombination between the Ti plasmid and the
vector resulting in the transfer of foreign sequence between the
borders of T DNA. Infection by this modified A. tumefaciens results in cotransfer of the foreign DNA sequence to the plant.

In addition to the T region, another portion of the Ti plasmid
called the vir region is essential for tumor-induction (Ooms et
al., 1980; Garfinkel and Nester, 1980; De Greve et al., 1981). The
vir region appears to contain functions required for stable transfer of T DNA but in itself is not stably transferred to the plant
chromosome. Hoekema et al., (1983) have constructed an A. tumefaciens in which the vir functions and the T DNA are on separate
plasmids. They found that such a construct maintains the normal
tumor-inducing capacity. This observation has led to the potential
for the development of more sophisticated vectors known as binary
vectors. Binary vectors consist of two plasmids - one resident in
A. tumefaciens which contains all the vir functions and the other

a modified permissive R factor which can replicate in both *E. coli* and *A. tumefaciens*. The modified R factor, e.g., RK290, contains the T DNA border sequences. Foreign DNA sequences are inserted between the borders. Such constructs are then mobilized into the *A. tumefaciens* containing the "vir" plasmid and the exconjugants used to inoculate plants or plant cells. The trans-acting vir function and the T DNA borders mediate the transfer of the foreign DNA sequences to the plant cell. The binary DNA vector provides considerably more flexibility and considerably less potential comlications than the recombinative intermediate vector system mentioned above. In summary, modified forms of the Ti plasmid allow for stable transfer of foreign DNA to plants. Such transfer can be achieved in the absence of neoplastic transformation of the plant cell.

MONOCOTYLEDONS

At this time, successful stable DNA transfer to monocotyledons has not been reported. The host range of *A. tumefaciens* does not include monocotyledonous plants though Zambryski et al. (1983) speculate that "with increased knowledge of the process of crown gall tumor formation, monocots can be expected to become susceptible to genetic exchanges similar to those induced by *Agrobacterium*". Other approaches for DNA transfer to monocots which are being attempted include the use of transposable elements such as the controlling elements in maize (*Zea mays*), (Federoff, 1983). Such strategies are derivative of that described by Rubin and Spradling (1982), for *Drosophila*. In these experiments, the gene for the wild type rosy phenotype was transferred to rosy mutant flies by inserting the appropriate chromosomal DNA fragment into the P transposable element. The rosy transposon (ryl) DNA, injected into mutant embryos, transformed germ line cells in 20 to 50 percent of the injected rosy mutant embryos. For maize, various schemes involving controlling element-mediated DNA transfer have been proposed. As yet, to the author's knowledge, no one has demonstrated that any of the various transposons of maize are capable of integration from an extrachromosomal state.

Selectable Markers for DNA Transfer (Mediating Foreign Gene Expression in Plants)

In the following section, the discussion will be limited to dicotyledonous plants. In the previous section, technological developments in the use of *A. tumefaciens* for DNA transfer to plants were discussed. It is clear that modified Ti plasmids retaining DNA transfer capacity, which are incapable of inducing tumorous development, are the ideal DNA transfer system. The initial drawback with such modified Ti plasmids was the fact that they no longer conferred a selectable phenotype on the recipient

cell or tissue. Recently, considerable effort has been devoted to
constructing genes conferring a trans-dominant selection for DNA
transfer. Herrera-Estrella et al. (1983a), reported the expression of the bacterial gene for chloramphenicol acetyltransferase
in plants. Herrera-Estrella et al. (1983b), reported the expression of bacterial genes for aminoglycoside phosphotransferase from
Tn5 and the expression of bacterial dihydrofolate reductase from
the R67 plasmids. These latter two genes served as dominant selectable markers in plants. (Bevan et al. (1983), and Fraley et al.
(1983), similarly reported the expression of the bacterial gene
for aminoglycoside phosphotransferase in plants and its use in
selection for DNA transfer to plant cells. In all this work the
bacterial gene expression was mediated using the transcription
control regions of the gene for nopaline synthase from Ti plasmid
which had been previously characterized (Depicker et al., 1982)
and shown to be typical of a eukaryotic gene. The advantage of
this gene is that it is expressed constitutively in host plant
cells (Wullems et al., 1981). These plant-bacterial chimaeric
genes confer resistance to to antibiotics kanamycin, neomycin and
methotrexate. These dominant selectable markers are of enormous
significance in the genetic engineering of plant cells. They enable, by cotransfer, the selection of any DNA sequence in plant
cells using vector systems which, in themselves, do not confer an
altered genotype. The experimental approach described above, used
to develop dominant selectable markers in plants, can be used to
mediate or alter the level of expression of any gene in plants.
At this time, many groups are using the control regions of characterized plant genes to mediate the expression of bacterial and
other genes in plants. In this work, due consideration must be
given to the fact that we still lack any detailed knowledge of the
true mode of the control of expression of genes in plants. Factors
such as chromosomal location of the introduced gene, stability of
the chimaeric mRNA, translational control of the chimaeric mRNA,
and stability of the foreign protein, all must affect the level
of the final expression of the introduced foreign DNA.

GENETICALLY ENGINEERED PLANTS

Central to the usefulness of genetic engineering is the stability of the introduced DNA and its retention through sexual generations. There are now several reports confirming that DNA sequences
introduced via A. tumefaciens are retained through sexual generations. Barton et al. (1983) reported the passage of the yeast
alcohol dehydrogenase gene and the bacterial neomycin phosphotransferase gene through to the progeny of the self-pollination of the
transformed plants. In this experiment the genes were inserted
in the "rooty" locus of T DNA allowing regeneration of fertile
tobacco (Nicotiana tobacum) plants. Zambryski et al. (1984) studied
the transmission of introduced DNA in tobacco. The DNA was intro-

duced using a non-oncogenic vector. In two plants derived from
in planta wound callus they checked 550 seedlings for the presence
of nopaline and found 100% transmission of the nopaline character.
They offer several explanations for this unexpected result. Firstly, the possibility that there are T DNA copies on several chromosomes, secondly that the plant becomes homozygous for the T DNA,
or thirdly that the regenerated plants have an abnormal chromosome
complement as the consequence of the loss of the T DNA homologue.

These workers also describe sexual transmission of plants derived from cocultivation with *A. tumefaciens* containing a non-oncogenic vector conferring resistance to methotrexate. The first
plant studied exhibits normal mendelian segregation of the introduced traits. Although these studies on the sexual transmission
of introduced traits are in their infancy, the results thus far
are of substantial and obvious significance to the prospects of
this technology for genetic engineering in plants.

Single Genes for Crop Improvement

In the preceding discussion, evidence is provided that it is
possible to transfer and express foreign genes in otherwise normal
plants and that the introduced traits are passed stably to subsequent generations. There are, at this time, several obvious foreign
genes to express in plants which may have commercial impact. These
include genes for the detoxification of herbicides, bacterial genes
for protein toxin of insects, enzymes such as chitinases confer
specific antifungal properties. All of these are genes which have
been isolated and characterized. The present challenge is to obtain expression of these genes in plants and determine their
efficacy *in planta*.

Methods for Identifying and Isolating Important Plant Genes

What will rapidly become a rate limiting factor to meaningful
genetic engineering in plants is the isolation of genes which confer agronomically important properties on crop plants. A major
obstacle in this endeavor is our lack of knowledge concerning the
biochemical basis of important properties such as disease resistance, yield components, or physiological aspects of growth habit,
fertility, etc. Most gene identification techniques depend on a
biochemical knowledge of the gene product. Therefore, traditional
gene isolation and identification schemes are not necessarily useful in searching for agronomically important genes. Various groups
are attempting to adopt a "molecular genetic" approach to the isolation of such genes using various strategies. One such approach
is to use transposon-induced mutations affecting important traits.
If the transposon has been isolated and characterized, it can be
used as a probe for sequences surrounding its point of insertion.
This way one can obtain any transposon mutated gene and then use

the mutated gene sequence to probe for the wildtype sequence.

   Antoher method for plant gene isolation and identification is
to be able to select directly for its transfer to a plant cell
system.   This possibility has been made closer to a reality by the
development of efficient systems for the infection of single cells
by *A. tumefaciens* (Fraley and Horsh, 1983).   Such a technology will
enable the direct selection of plant genes which confer resistance
to natural toxins and toxic chemicals.

PROSPECTS

   The prospects for genetic engineering in crop improvement
has been the underlying theme for many symposia and often the
basis for considerable vacuous debate.   There is a natural resistance among some breeders to be generous and forward-thinking
toward the technology and a virtually ubiquitous lack of patience
and realism among molecular biologists concerning the application
of their technology to plant breeding.

   The above discussion documents the developments in the technology of plant genetic engineering over the past few years.   It
shows that it is now possible to transfer foreign genes to plants,
have them propagated stably through several generations and mediate their expression in some tissues.   The burning question is:
will this technology (exciting and fun as it is) ever be used suc-
cesssfully in the improvement of crop varieties?   One thing is
clear, in order for genetic engineering to gain general acceptance
as a breeding tool, it must present to the breeder useful applications.   As Duvick (1983) points out, the traits rated with high
priorities by plant breeders for crop improvement are traits predominantly covered by many different genes of unknown location and
function, rather than single genes encoding well-defined properties.   Further, he points out that detailed genetic physiological
investigations have not been useful to the practical plant breeder
because of the lack of knowledge of the basis of plant variety
performance.   He concludes that most of the proposals made for the
use of genetic engineering are either so far-fetched as to be
impossible, or already within the scope of more conventional breeding techniques.   It is undeniable that our present understanding of
the detailed basis of crop performance is very limited and that
most traits manipulated by plant breeders are genetically complex.
However, in my view these facts in themselves do not warrant dis-
carding genetic engineering as a useful breeding tool.   To some
extent plant breeding, as with any technology, is self-serving
and, in a sense, creates its own goals and its own limitations.
Therefore, it is expected that manipulation of single genes or,
for example, the expression of a bacterial gene in a plant to
create varieties with different capabilities, are likely to have

been been approaches not seriously entertained by breeders over the years. There is a dearth of accurate market research assessing the market interest and potential profitability of the successful introduction and expression of single genes in crop plants. Our limited research indicates that the initial impact of genetic engineering will probably be in crops in which there is limited product differentiation, ie.e., in crops where traditional breeding has not been sufficiently intense, or of sufficient duration, to give rise to varieties which clearly distinguish themselves in performance at a particular location. Under these circumstances, singular changes, e.g., herbicide resistance, affording the farmer certain savings without loss in yield, may provide the basis for particular seed companies to gain increased market share and, therefore, increased profitability.

It is my view that genetic engineering techniques can be used profitably in the plant breeding program. Genetic engineering clearly provides novel approaches to solving age-old problems such as resistance to virus diseases and increased resistance to fungal and bacterial diseases. We are now at the stage in the development of the technology where specific proposals can be tested out in the laboratory on existing important varieties in species.

Finally, I would propose that the next symposium on the role of genetic engineering in plant breeding be held after the publication of a paper demonstrating conclusively the genetic engineering of an altered plant with a commercial value.

REFERENCES

Barton, K. A., Binns, A. N., Matzke, A.J.M., and Chilton, M-D., 1983, Regeneration of intact tobacco plants containing full length copies of genetically engineered T-DNA, and transmission of T-DNA to R1 progeny, Cell, 32:1033-1043.
Bevan, M. W., and Chilton, M-D., 1982, Multiple Transcripts of T-DNA detected in nopaline crown gall tumors, J. Mol. Appl. Genet., 1:539-546.
Bevan, M. W., and Chilton, M-D., 1982, T-DNA of the Agrobacterium Ti and Ri plasmids, Ann. Rev. Genet., 16:357-384.
Bevan, M. W., Flavell, R. B., and Chilton, M-D., 1983, A chimaeric antibiotic resistant gene as a selectable marker for plant cell transformation, Nature, 304:184-187.
Chilton, M-D., Drummond, M. H., Merlo, D. J., Sciaky, D., Montoya, A. L., Gordon, M. P., and Nester, E. W., 1977, Stable incorporation of plasmid DNA into hgiher plant cells: the molecular basis of crown gall tumorigenesis, Cell, 11:263-271.
De Greve, H., Decraemer, H., Seurinck, J., Van Montagu, M., and Schell, J., 1981, The functional organization of the octopine Agrobacterium tumefaciens plasmid pTi B6S3, Plasmid,

6:235-248.

Depicker, A., Stachel, S., Dhaese, P., Zambryski, P., and Goodman, H. M., 1982, Nopaline synthase: transcript mapping and DNA sequencing, J. Mol. Appl. Genet., 1:561-574.

Drummond, M. H., Gordon, M. P., Nester, E. W., and Chilton, M-D., 1977, Foreign DNA of bacterial plasmid origin is transcribed in crown gall tumors, Nature, 269:535-536.

Duvick, D. N., Qualset, C. O., Hollaender, A., Cutter, M., Fox, J. E., Garcia, R. L., Jaworski, E. G., and Lawrence, R. H., 1983, Genetic engineering in plants, an agricultural perspective - roundtable discussion on research priorities, Pp. 467-485, in: "Genetic Engineering of Plants," Kosuge T., Meredith, C. P., Hollaender, A., eds., Plenum Press, New York, London.

Federoff, N. V., 1983, Controlling elements in maize, Pp. 1-63, in: "Mobile Genetic Elements," Shapiro, J. A., ed., Academic Press, New York.

Fraley, R. T., and Horsch, R. B., 1983, In vitro plant transformation systems using liposomes and bacterial cocultivation, Pp. 177-194, in: "Genetic Engineering of Plants," Kosuge, T., Meredith, C. P., Hollaender, A., eds., Plenum Press, New York, London.

Fraley, R. T., Rogers, S. G., Horsch, R. B., Sanders, P. R., Flick, J. S., Adams, S. P., Bittner, M. L., Brand, L. A., Hoffman, N. L., and Woo, S. C., 1983, Expression of bacterial genes in plant cells, Proc. Nat. Acad. Sci. U.S.A., 80:4803-4807.

Garfinkel, D. J., and Nester, E. W., 1980, Agrobacterium tumefaciens mutants affected in crown gall tumorigenesis and octopine catabolism, J. Bacteriol., 144:732-743.

Garfinkel, D. J., Simpson, R. B., Ream, L. W., White, F. F., Gordon, M. P., and Nester, E. W., 1981, Genetic analysis of crown gall: fine structure map of the T-DNA by site-directed mutagenesis, Cell, 27:143-153.

Gelvin, S. B., Gordon, M. P., Nester, E. W., and Aronson, A. A., 1981, Transcription of the Agrobacterium Ti plasmid in the bacterium and in crown gall tumors, Plasmid, 6:17-29.

Hernalsteens, J. P., Van Vliet, F., De Beuckeleer, M., Depicker, A., Engler, G., Lemmers, M., Holsters, M., Van Montagu, M., and Schell, J., 1980, The Agrobacterium tumefaciens Ti plasmid as a host vector system for introducing foreign DNA in plant cells, Nature, 287:654-656.

Herrera-Estrella, L., De Block, M., Messens, E., Hernalsteens, J. P., Van Montagu, M., and Schell, J., 1983b, Chimeric genes as dominant selectable markers in plant cells, The EMBO Journal, 2:987-995.

Herrera-Estrella, L., Depicker, A., Van Montagu, M., and Schell, J., 1983a, Expression of chimaeric genes transferred into plant cells using a Ti-plasmid-derived vector, Nature, 303: 209-213.

Hoekema, A., Hirsch, P. R., Hooykaas, P.J.J., and Schilperoort, R. A., 1983, A binary plant vector strategy based on separation of *vir* and T-region of the *Agrobacterium tumefaciens* Ti-plasmid, Nature, 303:179-180.

Joos, H., Inze, D., Caplan, A., Sormann, M., Van Montagu, M., and Schell, J., 1983, Genetic analysis of T-DNA transcripts in nopaline crown gall, Cell, 32:1057-1067.

Leemans, J., Deblaere, R., Willmitzer, L., De Greve, H., Hernalsteens, J. P., Van Montagu, M., and Schell, J., 1982, Genetic identification of functions of T_L-DNA transcripts in octopine crown galls, The EMBO Journal, 1:147-152.

Ooms, G., Klapwijk, P. M., Poulis, J. A., and Schilperoort, R. A., 1980, Characterization of Tn904 insertion in octopine Ti plasmid mutants of *Agrobacterium tumefaciens*, J. Bacteriol., 144:82-91.

Ream, L. W., and Gordon, M. P., 1982, Crown gall disease and prospsects for genetic manipulation of plants, Science, 218:854-859.

Rubin, G. M., and Spradling, A. C., 1982, Genetic transformation of *Drosophila* with transposable element vectors, Science, 218:348-353.

Thomashow, M. F., Nutter, R., Montoya, A. L., Gordon, M. P., and Nester, E. W, 1980, Integration and organization of Ti plasmid sequences in crown gall tumors, Cell, 19:729-739.

Van Larebeke, N., Engler, G., Holsters, M., Van Den Elsacker, S., Zaenen, I., Schilperoort, R. A., and Schell, J., 1974, Large plasmid in *Agrobacterium tumefaciens* essential for crown gall inducing ability, Nature, 252:169-170.

Watson, B., Currier, T. C., Gordon, M. P., Chilton, M-D., and Nester, E. W., 1975, Plasmid required for virulence of *Agrobacterium tumefaciens*, J. Bacteriol., 123:255-264.

Willimitzer, L., Dhaese, P., Schreier, P. H., Schmalenbach, W., Van Montagu, M., and Schell, J., 1983, Size, location and polarity of T-DNA-encoded transcripts in nopaline crown gall tumors: evidence for common transcripts present in both octopine and nopaline tumors, Cell, 32:1045-1056.

Wullems, G. J., Molendijk, L., Ooms, G., and Schilperoort, R. A., 1981, Retention of tumor markers in F1 progeny plants from *in vitro* induced octopine and nopaline tumor tissues, Cell, 24:719-727.

Zaenen, I., Van Larebeke, N., Teuchy, H., Van Montagu, M., and Schell, J., 1974, Supercoiled circular DNA in crown gall-inducing *Agrobacterium* strains, J. Mol. Biol., 86:109-127.

Zambryski, P., Goodman, H. M., Van Montagu, M., and Schell, J., 1983, *Agrobacteria* tumor induction, Pp. 505-535, in: "Mobile Genetic Elements," Shapiro, J. A., ed., Academic Press, New York.

Zambryski, P., Herrera-Estrella, L., De Block, M., Van Montagu, M., and Schell, J., 1984, The use of the Ti plasmid of *Agrobacterium* to study the transfer of foreign DNA in plant cells: new vectors and methods, In press.

Zambryski, P., Joos, H., Genetello, C., Leemans, J., Van Montagu,

M., and Schell, J., 1984, Ti plasmid vector for the introduction of DNA into plant cells without alteration of their
normal regeneration capacity, In press.

GENE MANIPULATION AND PLANT BREEDING

N. W. Simmonds

Edinburgh School of Agriculture
West Mains Road, Edinburgh
EH9 3JG, Scotland

INTRODUCTION

Some 23 specialist papers have been delivered to this Symposium and I have been honoured by the organisers by their invitation to try to summarise, in a sort of 'wrap-up' talk, the general trend of opinions as to the place and nature of gene manipulation in plant improvement. I have a difficult task. It is relatively easy to discourse at length about one's own chosen specialisms and the audience has perhaps cause to be grateful that I was not invited to talk about bananas or potatoes or sugarcane, subjects on which I am virtually unstoppable! Instead, I am charged to talk very generally, even somewhat philosophically, about the subject at large, steering an uneasy, and probably erratic, course between the Scylla of detail and the Charybdis of bland generalisation. In choosing to introduce the subject historically and to go on to consider the sciences to which plant breeding technology appeals, I hope I shall have done no worse than strike either Scylla or Charybdis a mere glancing blow.

I had the privilege of reading most of the papers before I had to write my piece and, for simplicity, refer to them here simply by the author's name in capitals (thus, Burton, meaning Burton's chapter in this Symposium).

THE NATURE OF PLANT BREEDING

Plant breeding was, until quite recently, a practical art that owed much to empirical experience, to a 'feel' for plants (and love of them) and to agricultural understanding, but little or nothing

637

to science. The essential feel for and love of plants is still
essential and it comes out very well in Burton. The science,
indeed, was quite inadequate until genetics and cytology emerged,
a mere 60-70 years ago. But plant breeding goes back at least
9000 years in the hands of myriads of farmers who kept seed or
planting material of this plant because they liked it and threw
away that one because they did not. Plant breeding became profe-
ssionalised in the 18th century, spurred on by the agricultural
revolution which was itself provoked by the move, in Europe and
North America, away from agrarian societies towards industrial
ones; the urban masses and the armies had to be fed somehow.
Frederick the Great promoted the potato, William Cobbett raged
against it and breeders bred it; Napoleon promoted sugar beet to
mitigate the effects of the British naval blockade and thereby
encouraged the professionals of the day to develop a new crop.

In the industrial countries, professionalisation was pretty
complete by the middle of the 19th century. Collectively, the
great men such as Knight, Laxton, Shirreff, le Couteur, the
de Vilmorins, Luther Burbank (and a host of other, less famous,
figures) knew very well what they were doing. So did the animal
breeders, with a very similar body of experience. They knew that
both parents contributed to progeny, that inbreeding tended to
'fix the type' but that too much inbreeding could be deleterious,
that F_1 uniformity was succeeded by F_2 segregation, that 'sports'
occurred, that genetic potential was not identical with phenotypic
performance and that systematic seed production systems, often with
continued selection, were necessary if good seed was to be delivered
to farmers. There were even hints of formal economic elements in
breeding programmes (for example, reference to the idea of feed-
conversion efficiency in stock in the mid-19th century).

The word 'selection' was in standard use by practical breeders
back into the 18th century and Darwin drew it from this source in
inventing the term 'natural selection' as the driving force of
evolution. But Darwin's theory, though convincing within its
limits, lacked any real genetics and he retreated fatally into his
woolly 'gemmules' and 'pangenesis'. In retrospect, he and his contemporaries had in their hands, from the experience of the breeders,
all the essentials save one: quantitative (ie mathematical) understanding. When Mendel in 1866 used probabilistic arguments to prove
the particulate nature of (some) inheritance, Darwin simply did not
understand him and neither did anyone else at the time. Mendel's
achievement arose out of statistical imagination applied to the
practical experience of the 'plant hybridists', the breeders, but
the emergence of genetics into general understanding had to await
still wider probabilistic thinking. This was initiated by Francis
Galton who, no mathematician himself, carried the physicists' ideas

of 'errors of observations' over into biology, invented regression and correlation (1877-88), fired Karl Pearson with biometrical enthusiasm and thus started biometry. The ensuing unseemly (and, in retrospect, unnecessary) squabble between the 'Mendelians' and the 'Biometricians' in the early years of this century was but a hiccup in a profoundly important process: the assimilation of statistical thinking into the general framework of biology. By the early 1920s, East and Fisher had reconciled the particulate with the continuous in genetics, the chromosomal basis of nuclear inheritance was established, the existence of non-nuclear maternal systems was known, there were hints of biochemical genetics, there was a satisfactory genetical basis for evolution and Fisher was embarked on the 'biometricisation' (to coin a word) of biological experimentation. All this constituted, I believe, an intellectual revolution no less important that that which had been wrought by Darwin and his contemporaries (though a less public one).

This revolution, the emergence of genetics as the central science of biology, had many scientific elements but one can argue in retrospect, I believe, that the practical experience of the breeders and the intrusion of statistical thinking were the crucial elements. If breeding now appeals primarily to genetics as its 'core' science, genetics itself emerged from breeding experience informed by statistical theory.

The title of this Symposium includes the phrase 'gene manipulation'. The foregoing historical sketch shows that the breeders of domestic plants and animals were manipulating genes, systematically and successfully and in a highly professional way, long before genetics emerged to help them to do so. Somewhere in the 19th century (no precise point can be defined) breeding made the transition from being a professional art to being a professional technology, a technology with a secure body of empirical experience to guide it, even though its practice far outran the capacity of the science of the time really to understand how it worked. Since the 1920s, breeding has become a science-based technology but it is one which still outruns the capacity of science fully to interpret. The early hybridists were gene-manipulators; breeders are still gene manipulators today and the diverse sciences which underlie plant breeding are still trying to catch up with the practice.

Lest I be accused of an extreme 'technology-pull' view of the history of science, I had better state that I think that this view is as silly as the one which imputes all technology to 'science-push'. Plainly, the relation is reciprocal. A working, empirical technology evokes good scientific questions and, subsequently, answers to those questions enhance the technology. Breeding

experience lays at the root of genetics but cytogenetics and biometry are at the heart of modern breeding. As a semi-political
aside, I add here that I believe that the (not common) contemporary view, that technology should answer its own scientific questions,
is just silly. Technology needs independent scientific imagination
to ask penetrating questions as much as science needs technology to
post the questions. The productive feedback only works if both
elements are there.

SOCIO-AGRICULTURAL BACKGROUND

    Plant breeding is a technology that works in a socio-
agricultural ambience. It works well if it fits that ambience but
not if it doesn't. The breeder can hardly know too much about the
agriculture he seeks to serve, especially in framing objectives
within the constraints of what he judges to be biologically feasible.
Generally, because time-scales are long, he has to guess 10-20 years
ahead, but always in the knowledge that he is not merely trying to
meet the needs of agriculture but often helping to form agricultural
practice; varieties are as much a determinant of husbandry as husbandry is of varieites; they are mutually co-adapted. Thus chemical
weed control, precision planting, high plant population and generous
fertilizing grew up with hybrid maize but which determined what?
There is plenty of room for the plant breeder to imagine possible
objectives before his customers know that they will want them.

    That plant breeding is a successful technology is not in doubt
but the effects have been surprisingly poorly analyzed in economic
terms. Crop yields in the rich countries have mostly doubled (or
thereabouts) in the past 40 years and it is generally agreed that
breeding (genotype, G), husbandry (environment, E) and interactions
(GE) have all been significant elements in the change. But we have,
I think, no critical analyses of the components, though it is generally thought that the G plus a share of the GE bit are of the order
of 50 per cent of the whole. A very similar interpretation would
apply to the tropical wheat (Rajaram *et al.*) and rice (Oryza) (Khush)
areas that participated in the Green Revolution. But there are limits
to enhanced husbandry and we are probably now entering upon a phase,
in our technology-based agricultures, of diminishing returns based
essentially upon plant breeding along, or nearly so (Duvick).

    Herein lies a formidable challenge; to continue genetic improve-
ment *per se* without, or with little, recourse to the exploitation of
positive GE interactions. Indeed, the breeders of the future may
have to face the prospect of maintaining yields in the face of
declining inputs. Their task may become 'protective', even in the
agricultures of rich countries. In serving low-input, small-farmer

agricultures in poor countries, this is already their situation:
though forbidden the attainment of really high yields by environmental constraints, there is much that breeders can do to fit
varieties to the socio-economic circumstances of small farmers
through adjustments of maturity and quality features, adaptation
to local husbandry and tolerance of or resistance to unavoidable
environmental stresses due to drought, wetness, weeds, bad soils
and diseases (Khush; Rajaram *et al.*).

Economically, plant breeding is immensely attractive because
the prime cost is nearly all and it is effectively free of adverse
side effects; the same cannot be said of enhanced husbandry which
nearly always costs something to apply and often has adverse 'exter-
nalities'. The benefits of plant breeding do not, of course, long
stay with farmers. In a market-economy, at least, the benefits of
lower unit production costs are quickly passed on to consumers in
the form of lower food prices. I calculated recently that UK wheat,
barley and potato yields had all approximately doubled in the years
1950-80, that cereal prices had halved and that potato prices had
fallen by two-thirds. Probably, rather more than one half of the
effect was due to plant breeding. The poorer a society, the more
important is this effect; a 10 per cent fall in the price of rice
in Asia would be socially far more weighty than a similar fall in
the price of European barley, wheat or potatoes.

Plainly, plant-breeding is socially important, though societies
and governments do not seem very conscious of the fact. Maybe plant
breeders and economist colleagues should do more to advertise the
wares?

THE UNDERLYING SCIENCES

Plant breeding is a technology that, in the beginning, often
generated the science as appealed to it; now, over a century later,
it appeals to and is supported by diverse sciences and it continues
to generate interesting, important and often difficult questions.
Genetics, of course, is central but at least eight other disciplines
(as more or less conventionally defined) contribute (Figure 1).
This diagram should really be multi-dimensional to portray all the
possible (and existing) intersections of interest but it shows well
enough what a diverse business plant breeding really is. It would
be possible to give examples of all the intersections in the diagram
but tedious to do so: in the following paragraphs I simply try to
sketch what I take to be the main features.

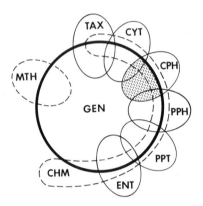

Figure 1.   The sciences that underlie plant breeding, a simiplified
            two-dimensional view of their relationships.   GNE, gene-
            tics; MTH, mathematics; TAX, taxonomy; CYT, cytology;
            CPH, physiology at the cell/tissue level; PPH, physiology
            at the whole-plant level; PPT, plant pathology, ENT,
            entomology; CHM, chemistry.   The shaded area is the area
            of acute current interest in 'biotechnology' and 'genetic
            engineering', where genetics, chemistry, cytology, cell
            physiology and whole-plant physiology intersect.

Mathematics

Mathematics, in the form of statistical thinking and biometry, is, I believe, nearly as important as genetics. Certainly, it underlies all selection schedules and trials systems and should certainly yet go further, in some form of operational research, in optimising breeding plans. Biometrical genetics (Baker), as such, seems to me to have helped to interpret what has already been done and to point questions, especially about the all-important matter of response to selection, but to have had little impact on the actual practice of plant breeding: hybrid maize *works* but we are still trying to understand how and we cannot even be quite sure that the best open-pollinated population would not be equally good. It is worth noting, too, that the purely biometrical device of 'combining abilities', making no appeal to formal biometrical genetics, is a widely used and perfectly workable basis for much plant breeding.

Taxonomy

Taxonomy appears in plant breeding whenever a crop is diverse and/or has wild relatives: we need classifications and names, after all. Within a crop, one or more species, with Latin names attached, can often be recognised on the basis of good cytogenetic information: little or no ambiguity attaches to *Triticum aestivum*, *Hordeum vulgare*, *Lycopersicum esculentum* or *Gossypium barbadense* and *G. hirsutum*. But in some complex hybrid and/or polyploid groups such as bananas (*Musa*), sugarcanes (*Saccharum*) and potatoes (*Solanum*) conventional taxonomy simply breaks down and the practice of using informal group/race nomenclatures based on cytogenetic/evolutionary criteria is becoming universal. Similarly, within one variable crop species, we now tend to recognise informal groupings and discard the multitude of ill-applied Latin names that were once all too common (for example in *Sorghum* and *Zea*, where informal 'races' now prevail). Taxonomically the best known plant groups are the relatives of crops, for the good reason that plant breeders have needed to understand them as potential donors of economic characters. Thus, thanks to the economic impulse, we have an incomparably better knowledge of the wild wheats (Dewey) than would ever have been achieved if they had merely been another group of weedy grasses. As ever, the plant breeding has generated the science, here in the form of cytogenetically based taxonomy, often enough with chemical overtones (Figure 1).

Cytology

The practical importance of cytology in plant breeding, at least in crops in which polyploidy, interspecific hybridity and wide crossing obtrude, needs no emphasis. Many interspecific crosses need knowledge (and often manipulation) of ploidy levels, sometimes with the object of interspecific genetic transfer, sometimes with new allopolyploid combinations in view (eg *Triticum* x *Secale*, *Brassica* x *Raphnus*). At a more refined level, the manipulation of individual chromosomes and segments is well established, especially in the wheats, *Triticum* (Kimber; Riley and Law; Sears). Again, the plant breeding called forth the science because it needed it. At a yet more refined level, Bennett's evidence that chromosomes occupy specific sites in relation to each other suggests a spatial differentiation of metabolic activity within the cell which looks like linking chromosome behaviour with molecular aspects of differentiation. Thus there is here a striking and potentially very fruitful link between cytology (CYT) and cell physiology (CPH) (Figure 1).

Plant physiology

Until about 40 years ago, plant physiology was essentially whole-plant physiology (PPH in Figure 1). Since then, biochemistry, at the cell/tissue/organ level has generated an essentially new kind of plant physiology (CPH in Figure 1). This is the meeting ground of plant physiology, biochemistry, genetics and cytology, the area reviewed in this Symposium by Beachy who shows, I believe, that we have the beginnings of understanding of normal development and of response to environmental stresses; it is one thing to be able to relate a DNA sequence to a protein, another, and perhaps a more difficult, task to relate the protein to a phenotype. This area is roughly coextensive with the area of acute current interest in 'biotechnology' and 'genetic engineering', though with somewhat different objectives. That some components of the former are already well established and developing strongly is clear but that the field of plant breeding as a whole is vastly wider than this one area of research is also clear.

Whole plant physiology (PPH) has a substantial area of interaction with plant breeding (which goes back to ideas on yield components in the 1920s) but has only developed really strongly in the past 30 years or so, mostly centered around the ideas of 'ideotypes' (Rasmusson), parition and photosynthetic efficiency (Evans). Early hopes that breeders would somehow be provided with easily observed and highly heritable proxies for yield potential have not, alas, been realized; nor is there yet any evidence that net photosynthesis, and thereby biomass, can be enhanced by selection. What has emerged, I think, is that the 'ideotype' remains a useful general concept (if a less sharply defined one than used to be thought), that favorable partition (generally leading towards dwarfer plants) is a major component of yield advance and that the idea of simple additive yield components collapses under the ever-present negative correlations. In genral one has to conclude, I think, that whole-plant

physiology has tended to interpret what plant breeders have already
done rather than guide them to new methods. The intersection in
Figure 1 between plant breeding, genetics and whole-plant physiology
(PPH) is nicely illustrated by Evan's insistence upon the importance
of isogenic lines for physiological analysis and Rasmusson's exploitation of them in barley.

Pathology and entomology

I shall comment upon plant pathology and entomology jointly
because, in the plant breeding context, they have much in common
and are treated jointly by Hooker in this Symposium. Coping with
diseases is a substantial part of most plant breeding programmes,
not infrequently a dominant part (cf. Khush; Rajaram et $al.$).
Nowadays, of course, resistance is the preferred method of pathogen
control because, if successful, it is cheap (there are no 'implementation costs') and environmentally attractive. In general, I should
suppose that fungal, bacterial and viral pathogens have taken a far
larger share of the effort than the animal pathogens, the insects
and eelworms. The relation between plant pathology, then, and
plant breeding is particularly close and has been since the early
years of this century when it was first appreciated that systematic
disease resistance breeding was feasible. As ever, the relation
is reciprocal. Breeding programmes need pathological/entomological
expertise to develop and operate testing schedules and the products
of breeding and genetic studies pose new pathological questions.
The investigation of pathotype specificty and its chemical nature,
of pathotoxins and host responses, of phytoalexins, of population
aspects of disease response and other topics is only possible on the
basis of construction and genetic knowledge of specific genotypes.

Chemistry

Finally, I refer to chemistry which intersects with virtually
all the other sciences that related to plant breeding (Figure 1).
At a fairly crude level, the plant breeder nearly always needs
quick, cheap assays of quality characters or of proxies for them:
sugar contents of beet ($Beta$) and cane ($Saccharum$); protein contents
of cereal grains and legume seeds; amino-acid spectra; digestibili-
ties of forages; active contents of drugs and narcotics; undesirable
toxins in diverse crops such as $Brassica$ and potatoes ($Solanum$) and
others come to mind. Some good quick assays have been available for
many years (refractrometric Brix) but very many more have been
developed in the past two decades and progress continues. Thanks
to a combination of automation and instrumentation (I an thinking
of micro-colorimetric and infra-red reflectance equipment), the
plant breeder new has access to a volume and accuracy of analysis
far beyond anything that his predecessors could deploy. (Valuable

as these techniques are in enhancing the precision of selection,
though, we should perhaps recall that chemical success has not been
universal: smells and flavors have always (so far as I know) proved
to be too subtle for the gas chromatograph). At all events, the
principle is plain enough: quick, cheap assays, even roughish ones,
are extremely valuable and ever more widely used and useful.

Beyond the comparatively crude uses of chemistry just outlined,
there has developed over the past 30 odd years a very fruitful bio-
chemical-genetic interaction with plant breeding. This has taken
diverse forms, for example: a multiplicity of studies of isozymes
(with relevance for evolutionary and taxonomic understanding and
genotype identification); specification of biosynthetic pathways of
economically important metabolites (eg *Brassica* fatty acids); seed
protein characteristics; phytotoxins and phytoalexins (intersecting
with PPT and ENT, of course); and many more. The provocations
offered to biological research by plant breeding problems are many
and obvious.

LEVELS OF GENE MANIPULATION

Three levels of gene manipulation can usefully be distinguished,
I think; the macro-, the micro- and the molecular.

The macro-level

As a working technology, plant breeding gets along very well
most of the time at the macro-level of the whole-plant phenotype
treated as though it were a resultant of genetic variance (G), due
to segregation, compounded with environmental variance components
(E), with or without GE interactions. Nearly all plant breeders
spend nearly all their time coping with polygenically inherited
characters; true, they quite often see Mendelian segregation in
their progenies and occasional major genes are of economic importance (conspicuously, disease resistances but also some morphological, physiological and biochemical characters, such as plant stature,
day-length response and content of specific metabolites). Nevertheless, most plant breeding can and does proceed very successfully on
the basis of practical empiricism and biometrical awareness, with
little or no reference to formal biometrical genetics or Mendelian
analysis. At one extreme, sugarcane (*Saccharum*), one of the outstanding success stories of plant breeding (*contra* Orton), is a
high polyploid of complex interspecific hybrid constitution in
which formal genetic analysis of any kind is simply impossible.
At the other extreme, we have maize (*Zea*), barley (*Hordeum*) and
tomato (*Lycopersicum*), all diploids with excellent genetic maps,

and the wheats (*Triticum*), mostly polyploids, with not-such-good
maps but with very beautiful cytology (Dewey; Kimber, Riley and
Law; Sears). Micro-level genetic analysis has contributed something
to the breeding of all these crops but nevertheless their improvement broadly remains of a macro- nature. No formal genetic research
is required to isolate, test and exploit a new maize inbred (even
though this now fairly routine process poses a crowd of very difficult genetic questions). I conclude that the great bulk of plant
breeding rests on the macro-level manipulation of genes *en masse*.

The micro-level

At the micro-level, the manipulation of specific genes does
occur, as I remarked above. Disease resistances are surely the
most numerous class and here, perhaps paradoxically, there are
signs of a retreat (if that is the right word) from micro to macro.
At least, major-gene resistances against the obligate air-borne
pathogens are falling into some disfavor as it is realized that
more stable ('durable') resistances may often be provided by polygenic systems. Apart form disease resistances, a total list of
major genes used in plant breeding would be a long one, even though
they underlie but a small fraction of the effort. The list would
include nuclear genes that govern morphological, physiological and
chemical characters of direct economic importance, and others that
govern useful marker characters. It would also include a few
cytoplasmic genes, notable mitochondrial male sterility (as in
maize). In time, probably, the chloroplast genetic system will be
practically exploited too and we have the beginnings of understanding of it (with *Chlamydomonas* as a sort of model system - Rochaix
et al.)

At the micro-level of gene manipulation we should also include
the elegant control of genome, chromosome and segment behaviour,
transfer and substitution made possible in the wheats (*Triticum*)
by cytological methods; and in a few other allo-polyploids too,
such as tobacco (*Nicotiana*) and cotton (*Gossypium*), though with less
precision and elegance than in the wheats.

As an aside, I note here that a few major genes have recently
been creeping into animal breeding, an activity that is dominated
by macro-level genetics, though with, perforce, greater appeal to
biometrical refinement than plant breeding makes. And, as another
aside: animal breeders, it could be argued, have it easy because
they only have to deal with rather few species, with fertile,
diploid outbreeders, can often ignore GE interactions, have rather
few breeding options, and do not have to worry (yet) about non-
nuclear genetic systems.

The molecular level

The molecular level of gene manipulation has yet to touch the
practice of plant breeding, though it is widely expected that, in
the form of 'genetic engineering', it will do so in time. By
'genetic engineering' I mean the identification, isolation and
multiplication of specific DNA sequences, followed by their transfer
(by whatever means) into an alien genetic environment, followed, in
turn, by the integration of that DNA into the host genome and its
replication, transcription, translation and phenotypic expression
in the new genetic environment. Such molecular manipulations are,
of course, well established in micro-organisms and bid fair to
revolutionize a good deal of the pharmaceutical industry. It is
now almost commonplace to persuade *Escherichia* to incorporate and
express genes from diverse animal sources; and the molecular manipulation of *Rhizobium* proceeds space (Long). A few such operations
have been achieved in higher plants (for example the well-known
nopaline gene from *Agrobacterium* in *Nicotiana*) but there are, as
yet no economic achievement (Bedbrook). The prospects are there
but have yet to be realized in practice; I return later to what
those prospects might be. Meanwhile, one notes that the principal
impact of molecular studies so far has been to interpret, *post hoc*,
the micro-level understanding already achieved by the breeders:
cytoplasmic male sterility (CMS) in maize and sorghum was a well
understood technology at this level decades before the mitochondrial-
molecular basis was known.

BIOTECHNOLOGY AND GENETIC ENGINEERING

The delineation of 'genetic engineering' given above is a
limited one. It excludes what can broadly be called 'biotechnology'.
Under this word I would include the various *in vitro* cell/tissue/
organ manipulations which have grown to prominence in the past
three decades but which, in the form of embryo-culture go back
some 60 years as a more or less routine adjunct to breeding.

Biotechnology

What seem to me to be the currently most interesting aspects
of biotechnology are listed in Figure 2 and brief comments upon
them follow here.

   1. Embryo culture (Collins *et al.*) is an old technique, of
   great value for making 'difficulty' hybrids that suffer from
   endosperm deficiency; no doubt restraints on age of embryo
   will be pushed back as techniques improve. Ovule culture is
   a useful variation (Collins *et al.*)

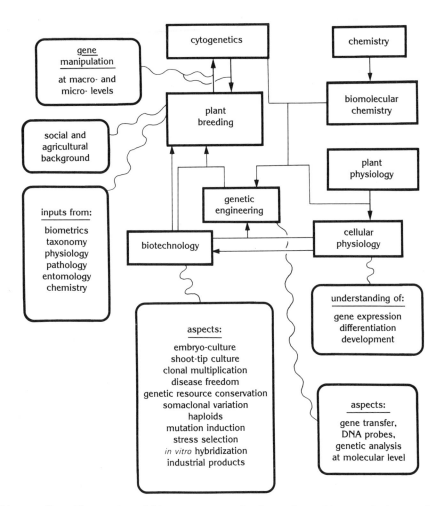

Figure 2.   The scientific context of plant breeding and its relationships with 'biotechnology' and 'genetic engineering'.

2.   <u>Shoot-tip (meristem) culture, clonal multiplication and
disease freedom</u> are all aspects of the same technology, of
well proven and ever increasing value; already, scores of
economic species are being multiplied thus (by shoot-tips)
and scores more will follow. The value of the technique for
surmounting the obstacles of international plant quarantine
can hardly be over-emphasized. Problems remain of course:
techniques for routinely inducing shoots and/or embryoids
in vitro are far from secure; many woody plants present
difficulties of rooting (related to the yet ill-understood
matter of 'phase-change'); the degree to which disorderly
cell multiplication can be allowed to precede differentiation
without the supervention of undesirable somaclonal variation
in vitro is uncertain; present indications, I think, are that
disorderly divisions can be, at most, few (Orton). At present,
anyway, a distinction has to be drawn between shoot-tip
cultures in which orderly meristems are maintained (and which
present no somaclonal problems) and regeneration from disorderly cultures (which do); cell/callus cultures with
abundant viable embryoids but genetic stability would be very
attractive indeed.

3.   <u>Genetic resource conservation</u> goes back to work many years
ago, with grapes (*Vitis*); several important collections of
herbaceous clonal plants potatoes (*Solanum*), sweet potatoes
(*Impomoea*), cassava (*Manihot*) are already stored and transmitted *in vitro*. Woody plants [see (2)] present problems
but will no doubt prove amenable in due course. The technique
requires only fairly elementary facilities and is certainly
cheaper and safer than field or glasshouse maintenance.
Cryogenics may make it cheaper still but yet with many residual
doubts as to what plants can safely be frozen.

4.   <u>Somaclonal variation</u> (Orton) occurs generally in regenerates from cycles of disorderly cell division *in vitro* and is
certainly genetically diverse in character. The genetic
mechanisms responsible are yet unknown but there is an enticing possibility that we are looking, in part at least, at the
consequences of changes in controlling elements rather than
of base-pair changes in structural genes. If so, somaclonal
variation would offer a different spectrum of variability from
conventional mutation-induction (which has not, in practice,
been notably useful). The usefulness of somaclonal variation
in generating, *per se*, new varieties of clonal crops has yet
to be properly tested; high hopes expressed for sugarcane
(*Saccharum*) and potatoes (*Solanum*) have not, to my knowledge,
actually yet eventuated in successful commercial varieties.
One looks forward with interest to learning the outcome of

the big effort on potatoes now being put in at the Plant
Breeding Institute, Cambridge. Maybe the main use will
ultimately lie in generating new variability for conventional
plant breeding to exploit? Perhaps, again, if controlling
systems are at the root of it, somaclonal variation will mimic
the spectrum of variation available in plant collections and
thus generate some rethinking of received ideas on genetic
resource conservation? At all events, it is plain, I believe,
that this phenomenon, lying as it does, at the intersection of
plant-breeding, cytogenetics, cell physiology and molecular
biology is of central biological interest, whatever the practical outcomes.

5. Haploids, in this context (Baenziger), follow from the
preceding. As a means of expediting the production of homozygous lines (in appropriate species), they have undoubted
attractions. But doubled haploids from pollen-grain culture
('anther culture' is surely a misnomer) have been dogged by
somaclonal variation which renders most of the products inviable
or useless. Unless or until this problem is cracked, it looks
as though useful haploids will continue to be the products of
long-established cytogenetic tricks (as in *Zea* and *Hordeum*).
Nevertheless, homozygous double haploids from pollen grains
in vitro remain an attractive possibility.

6. Mutation induction and stress selection (Meredith), in
this context, have much in common and imply the production
and selection, *in vitro*, with or without the use of mutagens,
of cell lines resistant to or tolerant of specific chemical
hazards such as fungal phytotoxins, herbicides or osmotic
stress. If and when produced, there remain the problem of
regeneration of whole plants from the cell lines and the
question as to whether or not the character expressed in the
cell is also expressed in the whole organism.

7. *In vitro* hybridization (Cocking; Collins *et al.*) has been
done many times between somatic cells of very diverse plants
and a few allopolyploids have been regenerated from such
cultures (needless to say, based upon nearly-related species
combinations). It seems possible, but perhaps unlikely, that
allopolyploids that are really wanted but cannot be got by
conventional means will be constructed thus. It seems far more
probable that the long term value of *in vitro* hybridization
(or, perhaps, more generally, protoplast manipulation - Cocking)
will lie in the selective transfer of genetic material (perhaps
nuclear, perhaps cytoplasmic) and will thus lie nearer the
area of 'genetic engineering' as I define the term.

8. Industrial products (eg costly secondary metabolites such
as scarce alkaloids) from cell cultures remain a possibility
though perhaps a slim one; higher plant cells are genetically
unstable and slowly growing. Engineering of tractable bacteria
to do the job seems far more likely.

There is no doubt that biotechnology has 'arrived', that it is
already having a very substantial scientific and technological
impact and that its uses and usefulness are growing rapidly. These
developments are inherently independent of genetic engineering (as
I define the term). They will continue whatever the impact of the
latter, whether it initiates useful genetic change in crop plants
or merely interprets, *post hoc*, the changes which have already been
wrought. Indeed, genetic engineering itself depends upon biotechnology and its pace is arguably limited by the restraints imposed
by the many uncertainties of growth, genetic constancy and regeneration of higher plant cells and protoplasts *in vitro*. In short, a
vigorous biotechnology is a precondition for the exploitation of
genetic engineering (Figure 2).

Animal breeding in this context offers a fairly exact parallel
to plant breeding. There, too, biotechnology has arrived in the
form of artificial insemination (AI) (practiced for decades), with
the newer, and certainly important, techniques of multiple ovula-
tion, embryo transfer and cloning all coming on apace and all with
very attractive prospects for the enhancement of animal breeding.
The prospects for genetic engineering are as uncertain as in plant
breeding and for the same reasons: most economic characters are
polygenic and the identification, delivery and expression of useful
foreign DNA sequences remain uncertain. Again, it seems clear that
a vigorous biotechnology will be a precondition for successful
genetic engineering.

Genetic engineering

If 'genetic engineering' is to affect the practice, as distinct
from the understanding, of plant breeding, it will have to move
specific bits of DNA across wide genetic boundaries and get them
expressed in new genetic backgrounds. The genes may be nuclear or
cytoplasmic (herbicide resistances in chloroplasts, maybe?) but they
will have to be single genes rather than macro-level polygenic
systems, the stuff of ordinary plant breeding. Perhaps the most
enticing possibility is the identification and transfer of pathogen-
(*not* pathotype-) specific major genes for resistance, if such can
be identified. Most plants do not get most diseases: is this due
to pathotype-non-specific genes or to diffuse genetic background
effects? If to the latter, it is hard to see how 'genetic engineer-

ing' could be useful but, if single or very few genes were responsible (and could be identified, transferred and expressed), then the
prospects would indeed be exciting.

The preceding paragraph outlines what might be called the
central practical problem for genetic engineering in plant breeding.
The problem seems to lie more in diciding what genes to transfer
rather than in how to do it; at least, it is clear that both the
necessary biotechnology and the development of diverse plasmid
vector systems (Bedbrook) are making spectacular progress and we
cannot be too far away from a generalized transfer technology.
There will no doubt also be peripheral uses, of which at least one
is already apparent, namely molecular (DNA) probes. They are finding uses for identifying viruses in clonal stocks with a sensitivity
and precision far beyond what is possible by conventional methods.
The power that comes from being able to identify the DNA rather
than the product is evident in the use by Appels *et al.* of molecular
probes to detect alien chromatin in wheat, a sort of ultra-refined
cytotaxonomy one might say.

CONCLUSIONS

Plant breeding is a well established technology of great social
power. It appeals primarily to genetics as its underlying science
and, historically, was a major source of genetics, the central
science of contemporary biology. It appeals also to diverse other
scientific disciplines, notably to biometrics, taxonomy, cytology,
plant physiology, pathology, entomology and chemistry; in every case
the relationship has been and is reciprocal, the technology posing
scientific problems and itself using the solutions, while enhancing
the sciences. Most plant breeding operates at the macro-level of
gene manipulation, of whole phenotypes dominated by polygenic
systems; the micro-level of single genes and chromosome segments
is also significant but less pervasive; the molecular level, that
of genetic engineering, holds promise but has yet to achieve
practical impact.

Biotechnology has emerged in recent decades from the meeting
ground of plant breeding with cytogenetics, cell physiology and
molecular biochemistry. It includes diverse manipulations of cells,
tissues and organs *in vitro* and has already enhanced plant breeding
technology, especially by way of embryo culture and *in vitro* clonal
multiplication and storage. Progress is rapid and potential for
further enhancement is good. If genetic engineering, in the sense
of specific DNA transfer from alien sources into cultivated species,
is to be successful, it will depend upon biotechnological techniques
in some form to achieve the delivery. The most attractive objective

for genetic engineering that is now apparent would be the transfer
of alien pathotype–non-specific disease resistance genes, if such
could be identified.

    The practical possibilities of biotechnology and the potential
attractions of genetic engineering noweithstanding, the mainstream
of plant breeding is likely to continue much as it is (and has been
for decades): a prolonged and fruitful dialogue between the working
technology and the realted sciences, each informing and enhancing
the other. Plant breeding (the technology of gene-manipulation) is
a complex business that yet poses many difficult scientific problems:
it is not about to be transformed by any single area of scientific
endeavour. 'More of the same only bigger, better and deeper' might
be the motto for its future.

Arulsekar, S., Parfitt, D. E., and McGranahan, G.   Isozymes in
     Juglans Species.
Bechtel, D. B., and Barnett, B. D., U.S. Grain Marketing Research
     Laboratory, ARS-USDA, Manhattan, Kansas 66502. Hard Red
     Winter Wheat Endosperm Development.
Bietz, J. A., and Burnouf, T., Northern Regional Research Center,
     USDA, ARS, Peoria, Illinois 61604. Chromosomal Coding of
     Gliadin Proteins in the Wheat Cultivar 'Chinese Spring':
     Analysis by Reversed-Phase HPLC.
Briggs, K. G., Plant Science Department, University of Alberta,
     Edmonton, Alberta, Canada T6G2H1. Deployment of Genes for
     Resistance to Stem Rust in Kenyan Wheats.
Burnouf, T., and Bietz, J. A., Northern Regional Research Center,
     USDA, ARS, Peoria, Illinois 61604. Chromosomal Coding of
     Glutenin Subunit Proteins in the Wheat Variety 'Chinese
     Spring': Analysis by Reversed-Phase HPLC.
Carson, M. L., and Wicks, Z. W., III, Plant Science Department,
     South Dakota State University, Brookings, South Dakota
     57007. Selection for Maize Grain Yield in Disease Stress
     and Non-Stress Environments.
Chen, L. F. O., and Palmer, R. G., Department of Agronomy, Iowa
     State University, Ames, Iowa 50011. Frequency of Polyembryony
     and Polyploidy from Male-Sterile (MS_1) Soybean Gene = Cytology.
Conner, Anthony J., and Meredith, Carole P., Department of Viticul-
     ture and Endogy, University of California, Davis, California
     95616. Strategies for the Selection and Characterization of
     Aluminum Resistant Variants from Cell Culture.
Cooper, D. B., Sears, R. G., Guenzi, A. C., Lapitan, N. L. V.,
     Jones, B. L., and Lookhart, G. L., Kansas State University,
     Department of Agronomy, Manhattan, Kansas 66502. Somaclonal
     Variation in Gliadin Proteins of Wheat Plants Derived from
     Embryo Callus Culture.
Crossway, Anne, 1910 Fifth Street, Davis, California 95616. DNA
     Microinjection of Tobacco Protoplasts by Holding Pipette
     Technique and Hanging Drop Microculture.
Fedak, George, Ottawa Research Station, Agriculture Canada, Ottawa,
     Ontario K1A0C6 Canada. Intergenetic Hybrids Between Hordeum
     vulgare and Three Agropyron Species.

Florence, L. Zack, Department of Forest Science, University of
    Alberta, Edmonton, Alberta T6G 2EO Canada. Analysis of the
    Chloroplasts Genome in Pines: Preliminary Results of DNA-DNA
    Hybridizations.

Gill, B. S., Raupp, W. J., Rayburn, A. L., Duffens, K. L., Snyder,
    E. B., Kam-Morgan, L., and Stoddard, S. L., Department of
    Plant Pathology, Kansas State University, Throckmorton Hall,
    Manhattan, Kansas 66506. An Overview of Wheat Germ Plasm
    Enhancement Research at KSU.

Gleddie, S., Keller, W. A., and Setterfield, G., Carleton University,
    Department of Biology, Ottawa Research Station, Ottawa,
    Ontario K1A 0C6, Canada. Somatic Hybridization between
    Solanum melongona L. and Solanum sisymbrifolium Lam.

Groose, R. W., and Bingham, E. T., Department of Agronomy, University
    of Madison, Madison, Wisconsin 53706. An Unstable Anthocya-
    nin Mutation Recovered from Tissue Culture of Alfalfa.

Guenzi, A. C., Sears, R. G., and Jones, B. L., Kansas State Univer-
    sity, Department of Agronomy and USDA, USGMRL, Manhattan,
    Kansas 66506. Increased Lysine Accumulation in Wheat Var-
    iants Recovered from AEC-Resistant Cell Lines.

Haleikav, Ed, Rhoads, Marsha, and Bidney, Dennis, Advanced Genetic
    Science, P. O. Box 1373, Manhattan, Kansas 66502. Regenera-
    tion of Plants from Merophyll Protoplasts of Rapeseed.

Hasenkampf, Clare A., Florida State University, Tallahassee, Florida
    32306. Examination of Meiosis, Using Whole Mount Spreading
    of Synaptonemal Complexes in Tradescantia.

Helentjaris, Tim, King, Gretchen, Siedenstrang, Chris, Wegman,
    Sharon, NPI, 417 Wakara Way, Salt Lake City, Utah 84108.
    Development of DNA Molecular Markers in Crop Plants and
    Their Potential Application to Breeding Programs.

Hood, Elizabeth E., Department of Biology, Washington University,
    St. Louis, Missouri 63130. Restriction Enzyme Map of pTi
    Bo542 from Agrobacterium tumefaciens Strain A281.

Jacobsen, Haus-Jörg, Kysely, Wilfried, Salha, Ahmed Abou, and Wou,
    Joug-Lak, Institut für teuetik, Universität Bonn, Kirschallee
    1, D-53 BONN, West Germany. Somatic Embryogenesis in Seed
    Legumes - A Genetical and Hormone-Physiological Approach.

Jones, B. L., Lookhart, G. L., and Cooper, D. B., USDA, ARS, U.S.
    Grain Marketing Research Laboratory and Department of Agron-
    omy, Kansas State University, Manhattan, Kansas 66502.
    Protein (Purothionin) Amino Acid Sequences and Wheat Evolution.

Jones, Davy, Jones, George A., and Jackson*, D. Michael, Department
    of Entomology, University of Kentucky, Lexington, Kentucky
    40506, and *USDA-ARS, Tobacco Research Laboratory, Oxford,
    North Carolina 27565. Insects and Tobacco: A model system
    for genetic engineering of plants for insect resistance.

Kennell, John C., and Haner, Harry T., Department of Botany, Mole-
    cular, Cellular and Developmental Biology, Iowa State Univer-
    sity, Ames, Iowa 50011. Influence of the Soybean Male
    Sterile Gene (ms_1) on Female Gametophyte: Microscopy.

Kenny, J. R., Dancik, B. P., and Florence, L. Z., Department of
    Forest Science, University of Alberta, Edmonton, Alberta,
    Canada T2H 2EI. Hybridization Between Lodgepole Pine Actin
    and Soybean Actin.

Kothari, S., and Widholm, J., Department of Agronomy, University
    of Illinois, Urbana, Illinois 61801. Selection of Intra-
    specific Carrot Hybrids Based on Amino Acid Analog and
    Herbicide Resistance Complementation.

Lapitan, N. L. V., Sears, R. S., and Gill, B. S., Departments of
    Agronomy and Plant Pathology, Kansas State University,
    Manhattan, Kansas 66506. Translocation and Other Karyotypic
    Structural Changes in Wheat X Rye Hybrid Plants Regenerated
    from Tissue Culture.

Lazar, M. D., Baenziger, P. S., and Schaeffer, G. W., Biotechnology
    Department, Alberta Research Council, 11315 87th Avenue,
    Edmonton, Alberta T6G 2C2, Canada. Combining Abilities and
    Heritability of Callus Formation and Plantlet Regeneration
    in Wheat (<u>Triticum</u> <u>aestivum</u> L.) Anther Cultures.

Lelley, Tamas, Institute of Agronomy and Plant Breeding, University
    of Göttingen, Von Siebold Str 8, D-3400 Göttingen, Federal
    Republic of Germany. Triticale Breeding, A New Approach.

Metz, S. G., Varnum, J. M., and Schubert, K. R., Monsanto, 800 N.
    Lindbergh, Missouri 63167. Factors Affecting Wheat
    Androgenesis.

Perez de la Vega, M., Vences, J., Vaquero, F., and Garcia, P.,
    Departamento De Genetica, Facultad De Biologia, Universidad
    De Leon, Leon, Madrid, Spain. Estimation of Rye Population
    Parameters and Phylogenetic Relationships.

Quemada, Hector, and Lark, Karl G., University of Utah, Salt Lake
    City, Utah 84112. Soybean 5S Genes.

Rayburn, A. L., and Gill, B. S., Department of Plant Pathology,
    Kansas State University, Throckmorton Hall, Manhattan, Kansas
    66506. Use of Biotin-Labeled DNA Probes to Map Specific DNA
    Sequences on Cereal Chromosomes.

Salerno, Juan C., Department of Genetics, INTA, Castelar, Buenos
    Aires, Republica Argentina. Relationship Between Genetic
    Load and Plant Improvement in Corn Populations.

Schank, S. C., and Smith, Rex L., Agronomy Department, University
    of Florida, Gainesville, Florida 32611. Associative N$_2$-
    Fixation and Contribution of Plant Roots.

Schnabelraugh, Linda, and Hirschberg, Joseph, Department of Soil
    Sciences, Michigan State University, East Lansing, Michigan
    48824. Detection of mtDNA in Protoplast-Derived Calli and
    Fusion Products of Cytoplasmic Male Sterile and Fertile
    Maintainer Lines and Sugarbeet (<u>Beta</u> <u>vulgaris</u> L.).

Seguin-Swartz, G., and Dean, C., Agriculture Canada Research Station,
    Saskatoon, Saskatchewan, Canada. Anther Culture Studies in
    <u>Brassica</u> <u>campeslies</u> L. Line 7B3.

Shoemaker, R. C., Palmer, R. G., and Atherly, A. G., Department of
    Genetics, Iowa State University, Ames, Iowa 50011. Plastome
    Variation Among Soybeans and Related Species.

Sigurbjörnsson, S., Micke, A., Maluszynski, M., and Donini, B.,
    Joint FAO/JAEA Division, 1400-Vienna, P. O. Box 100, Austria.
    Results of Plant Mutation Breeding - 1984.

Sitch, L. A., and Snape, J. W., Plant Breeding Institute, Cambridge,
    England. Intrachromosomal Mapping of Crossability Genes in
    Wheat.

Skogen-Hagenson, M. J., Robertson, D. S., and Morris, D. W., Depart-
    ment of Genetics, Iowa State University, Ames, Iowa 50011.
    A Transposable Element in Maize Chloroplast DNA.

Smith, R. H., Bhaskaran, S., Newton, R., and Miller, F., Department
    of Soil & Crop Sciences, Texas A & M University, College
    Station, Texas 77843. Screening of Sorghum Genotypes on
    PEG and Physiological Studies.

Staszewski, Zygmunt, Plant Breeding and Acclimatization Institute
    at Radzikov, P. O. Box 1019, 0-059 Warsaw, Poland. Male
    Sterile Mutations of Gramineae and Papilionaceae for Hybrid
    Breeding.

Stelly, David M., and Peloquin, Stanley J., Department of Soil &
    Crop Sciences, Texas A & M University, College Station,
    Texas 77843. Cytology of 2n Female Gametophytes in Potato.

Stommel, J. R., and Simon, P. W., USDA - ARS, Department of Hort-
    iculture, University of Wisconsin, Madison, Wisconsin 53706.
    Genotype, Carbohydrate and 2-deoxy-d-glucose Interaction in
    Carrot All Suspension Cultures.

Vogel, K. P., Anderson, B. E., Ward, J. K., Britton, R., Haskins,
    F. A., and Gorz, H. J., USDA-ARS, 332 Keim Hall, East Campus,
    University of Nebraska, Lincoln, Nebraska 68583. Improving
    Animal Performance by Breeding for In vitro Dry Matter Digest-
    ibility in Switchgrass.

Wang, Richard R. C., Dewey, D. R., and Hsiao, C., USDA-ARS, Crops
    Research Laboratory, USU, UMC63, Logan, Utah 84322. Inter-
    generic Hybrids of Agropyron and Pseudoroegneria at Diploid
    and Tetraploid Levels.

Weber, David F., and Plewa, Michael, Illinois State University,
    Normal, Illinois 61761. Effect of B Chromosomes on Mutation
    of the Yg2 Locus in Maize.

Zimmer, E. A., and Schaal, B. A., Department of Biology, Washington
    University, St. Louis, Missouri 63130. Ribosomal Gene
    Variation and Inheritance: Patterns in Interspecific Hybrids
    and in Natural Populations.

DATE DUE

MAR 2 0 1995		
MAR 1 3 2000		
MAR 0 8 2001		
APR 1 4 2001		